University of Chicago School Mathematics Project

DEVELOPMENTS IN SCHOOL MATHEMATICS EDUCATION AROUND THE WORLD

APPLICATIONS-ORIENTED CURRICULA AND TECHNOLOGY-SUPPORTED LEARNING FOR ALL STUDENTS

Proceedings of the UCSMP
International Conference on Mathematics Education

The University of Chicago
28–30 March 1985

Edited by

Izaak Wirszup and Robert Streit

National Council of Teachers of Mathematics
1906 Association Drive, Reston, VA 22091

Organization of the UCSMP International Conference on Mathematics Education and publication of these *Proceedings* were made possible by the generous support of the Amoco Foundation, Inc.

Printed in the United States of America

Contents

Preface v

Welcoming Remarks and Presentation of UCSMP: Organization, Aims, Programs 1

Part 1: School Mathematics Education Worldwide

The Second International Mathematics Study: An International Perspective 39
  Kenneth J. Travers

The Second International Mathematics Study: A Look at U.S. Classrooms 62
  F. Joe Crosswhite and John A. Dossey

On the Mathematical Curriculum for Grades K–3 in Poland 83
  Zbigniew Semadeni

The Bulgarian Academy of Sciences Research Group on Education Project 98
  Boyan Penkov and Blagovest Sendov

Computers in Elementary School: Some Experiments 109
  Renée de Graeve

The Introduction of Computers in Elementary and Secondary School in France:
Their Effect on Mathematics Teaching 120
  Anne-Marie Pastel

Mathematics in Swedish Schools 130
  Hans Brolin

An Example of a Long-Life Curriculum Project and Its Underlying Philosophy 156
  Maxim Bruckheimer and Rina Hershkowitz

Mathematics in Junior and Senior High School in Japan: Present State and Prospects 172
  Tatsuro Miwa

The Present State and Current Problems of Mathematics Education at the Senior
Secondary Level in Japan 191
  Hiroshi Fujita

SMP 11–16: The Most Recent Work in Curriculum Development by the
School Mathematics Project, and Its Relation to Current Issues in Mathematical
Education in England 225
  John Ling

Post-"New Math" Since 1963: Its Implementation as a Historical Process 237
  Tamás Varga

Action and Language—Two Modes of Representing Mathematical Concepts 250
  Elmar Cohors-Fresenborg

iii

Part 2: General Principles, Curriculum Design, and Instructional Strategies

Mathematics Starting and Staying in Reality 279
  Hans Freudenthal

Reflections on Mathematics Education 296
  Felix E. Browder

The Role of Applications in the Pre-College Curriculum 307
  Peter J. Hilton

Curricula for Active Mathematics 321
  Hugh Burkhardt

Understanding and Teaching the Nature of Mathematical Thinking 362
  Alan H. Schoenfeld

Statistics and Probability in the School Mathematics Curriculum:
A Review of the ASA/NCTM/NSF Quantitative Literacy Program 380
  Richard L. Scheaffer

Cognitively Guided Instruction: The Application of Cognitive and Instructional Science
to Mathematics Curriculum Development 397
  Thomas P. Carpenter, Elizabeth Fennema, and Penelope L. Peterson

Lessons Learned from the First Eighteen Months of the Secondary
Component of UCSMP 418
  Zalman Usiskin

Teaching Addition, Subtraction, and Place-Value Concepts 430
  Karen C. Fuson

Mathematical Preparation for College 460
  Jeremy Kilpatrick

On the Value of Examinations 476
  Heini Halberstam and Anthony Peressini

Organizing for the Classroom Ideas Drawn from the History of Mathematics 494
  Roland J. K. Stowasser

Coded Graphical Representations: A Valuable But Neglected Means of
Communicating Spatial Information in Geometry 514
  Claude Gaulin and Ewa Puchalska

Instructional Processes in the Teaching of Mathematics 540
  Alphonse Buccino and James W. Wilson

Part 3: Technology-Supported Learning

The Impact of Computer Science on Pre-College Mathematics: A Research Program 551
  Anthony Ralston

The WICAT Computer-Based Elementary Mathematics Project:
A Curriculum in Development 562
  Warren D. Crown

Learning Mathematics Through Modern Media and the THE System,
A Prototype Videodisc Learning System 576
    Fumiyuki Terada

Using the Computer to Teach Geometry 587
    Mary Grace Kantowski

Microcomputer-Based Courses for School Geometry 604
    Max S. Bell

THE GEOMETRIC SUPPOSER:
Using Microcomputers to Restore Invention to the Learning of Mathematics 623
    Judah L. Schwartz and Michal Yerushalmy

Technology-Supported Learning 637
    Alan Hoffer

Dienes Revisited: Multiple Embodiments in Computer Environments 647
    Richard Lesh, Thomas R. Post, and Merlyn Behr

Computers and Applications in Secondary School Mathematics 681
    James T. Fey and Richard A. Good

About the Authors 691

Participants in the UCSMP International Conference 719

v

Preface

In response to the growing need for a mathematically literate public in a technological age, the University of Chicago School Mathematics Project (UCSMP), sponsored by the Amoco Foundation, has been established to help upgrade mathematics education for American students, including all children in grades K-6 and the great majority of students in grades 7-12. International comparative studies have identified several shortcomings in traditional American school mathematics. At least two of them are pervasive and serious: 1) most of our students spend nine years on arithmetic, while in other industrialized countries children cover arithmetic—combined with intuitive geometry—in six years; 2) fewer than half of American students ever take any plane geometry, because the one-year "packaged" geometry course generally offered is inaccessible to the majority of students, as shown by recent educational psychology.

To meet these and other challenges, in 1983 UCSMP began to develop a comprehensive school mathematics program aimed at preparing the vast majority of students for the 1990's and beyond, a program that could serve as a national model. UCSMP curriculum goals are: 1) to design an applications-oriented curriculum spanning grades K-12; 2) to integrate the use of calculators and microcomputers throughout the curriculum; 3) to cover arithmetic together with introductory geometry in grades K-6; 4) to facilitate the teaching of algebra and geometry; 5) to give increased emphasis to "newer" mathematics, especially statistics and probability; and 6) to structure effective programs for teacher development at the elementary level so that the planned curriculum changes can be implemented.

America's mathematics education problems cannot be resolved without serious commitment from the mathematics research community and the continuing leadership and inspiration provided by the teaching profession. From its inception, UCSMP has been staffed by professors from the Department of

Mathematics and the Department of Education and experienced classroom teachers. A detailed description of UCSMP, its organization, aims, and programs is given in the Opening Session Summary Transcript.

The aim of the UCSMP Resource Development Component is to provide previously unexploited ideas and materials to the project's Elementary and Secondary Components. Since awareness of successes and innovative approaches in other countries is essential for American mathematics education, the Resource Component surveys foreign mathematics textbooks and teacher support literature, journals, and research monographs for the most significant developments around the world. Its translations of outstanding foreign school mathematics publications, including texts and workbooks, are not published, but used exclusively in UCSMP-related research and experimentation. In this way, the Resource Component offers a first-hand look at curricula, achievement standards, approaches, and methodologies that are different from those in the United States. This input has challenged UCSMP to raise its expectations of children's mathematical abilities and constitutes a vital new resource to help it implement its goals.

To advance its programs and activities, the Resource Component promotes international professional cooperation in mathematics education. In order to provide UCSMP staff and American mathematics educators and curriculum developers in general with direct scholarly contacts and a forum for the discussion of ideas and programs from various educationally advanced countries, the Resource Development Component organized the UCSMP International Conference on Mathematics Education held at the University of Chicago March 28-30, 1985.

The Conference was devoted primarily to two major topics closely connected with UCSMP: 1) development of an applications-oriented mathematics curriculum for all students; and 2) innovative instructional strategies applying new technologies and research findings.

Thirty-six foreign and American mathematicians and educators presented papers and participated in panel discussions. There were delegations from Bulgaria, Canada, France, West Germany, Great Britain, Hungary, Israel, Japan, Mexico, the Netherlands, Poland, and Sweden.

In order to achieve a more productive exchange of information and stimulate discussion, attendance at the Conference was by invitation only, and the total Conference audience was limited to 250. At the Conference every measure was taken to build a working partnership, like the one at UCSMP, among mathematicians, educators, and classroom teachers.

By all accounts, the Conference was a major event in mathematics education. The world's most prominent mathematics educator, Professor Hans Freudenthal of the Netherlands, gave a lecture at a special session in his honor. As a group, the American and foreign speakers presented some of the best new thinking in the field. Rarely can one find the work of such an array of distinguished mathematics educators in a single volume.

Illuminating talks by twelve foreign scholars described the mathematics programs in their respective countries, providing American educators and government officials with clear evidence of achievements in other nations and the indispensability of overall curricular changes in the U.S. The International Conference brought a wealth of ideas not only to the leaders and staff of UCSMP, but also to the American and international mathematics education communities.

The response to UCSMP's International Conference was enthusiastic throughout its audience, which represented a broad range of groups concerned about mathematics education. A university mathematics professor wrote later that the event was "the most exciting conference I have ever attended" and "a turning point in my career—from passive participation to active involvement in the mathematical education community." A high school teacher stated, "I found it particularly encouraging to see university mathematicians joined with elementary and secondary school teachers in a common effort to improve the quality of math education. For me the conference was a source of inspiration and concrete teaching suggestions." One of the leading mathematics educators in the U.S. termed the Conference "a resounding success" and "a milestone in America's slowly evolving recognition of the contributions to be made to American mathematics education by our colleagues in other countries."

UCSMP is very pleased that the <u>Proceedings</u> are being published by the National Council of Teachers of Mathematics. We are most appreciative for the support and encouragement of the NCTM Educational Materials Committee and Board of Directors. I would like to extend my special thanks to Mr. Charles R. Hucka, Director of NCTM Publication Services, for his expert and generous assistance with this project.

I would like to thank my co-editor, Mr. Robert Streit, and also Dr. George Fowler, Ms. Angela Harris, and Ms. Birute Tamulynas for all their help in preparing the manuscripts.

Finally, our gratitude is due to the Amoco Foundation, whose generous and continued support has been the very cornerstone of the success of UCSMP and its International Conference on Mathematics Education.

Izaak Wirszup

Principal Investigator, University of Chicago
School Mathematics Project

Director, Resource Development Component

Welcoming Remarks and
Presentation of UCSMP: Organization, Aims, Programs

Opening Session Summary Transcript

Izaak Wirszup:

It is a distinct honor to welcome you to the International Conference on Mathematics Education organized by the Amoco-sponsored University of Chicago School Mathematics Project. Thirty-six speakers representing 12 foreign countries and the United States will present papers related to two major topics of concern to UCSMP. The first topic is the development of an applications-oriented mathematics curriculum for all children in grades K through 6 and the great majority of students in grades 7 through 12. The second major topic is the creation and implementation of innovative instructional strategies.

As you know, participation in the conference is by invitation only. We have tried to limit the total conference audience to 200 in order to promote a more effective exchange of ideas. The response has been enthusiastic and we at UCSMP would like to thank you all very much for coming. We are extremely pleased to see that despite its relatively small size, this gathering constitutes a virtual cross-section of American professionals and institutions concerned about mathematics education. We hope that the conference will be as enjoyable for you as it is instructive for all of us. You will now be hearing some welcoming words on behalf of the University of Chicago and UCSMP.

It is a pleasure to present Stuart A. Rice, the Frank P. Hixon Distinguished Service Professor of Chemistry and Dean of the Division of the Physical Sciences. Dean Rice, a member of the National Academy of Sciences and the National Science Board, lent his enthusiastic support to our efforts during the organization of UCSMP and continues to do so. We are very grateful to him.

1

Stuart Rice:

Let me bid you all good morning. Welcome to the University of Chicago. I am really not the person to tell you about the state of precollegiate education, precollegiate mathematics education in particular, and how seriously it has to be treated, especially in the United States. As a consumer, that is, as a chemist, I am well aware of a large number of problems. I think for the first time now in many years we have the attention of a number of individuals in the executive branch who are in a position to plan and implement programs which can lead to changes. Certainly it is the case that the National Science Board in the last two years has expressed a more public and a more vigorous interest in precollegiate mathematics education, indeed all precollegiate science education, than it has at any time within my memory. Now that we have the attention of at least some of these figures, it is the time to act, prudently, yet swiftly. Of course, the problem is that we really need a great diversity of ideas, because of the heterogeneity of the school population. There is not going to be any simple way that one program or even one class of programs would be suitable for everybody, and part of the task of this and other conferences is going to be to invent and innovate on a scale which permits the highest quality precollegiate mathematics education for the greatest diversity of the school population. But you are much more expert at that than I, and all I can do now is to wish you a successful conference.

Izaak Wirszup:

Next, I would like to introduce Felix E. Browder, the Max Mason Distinguished Service Professor and Chairman of the Department of Mathematics. Professor Browder is a member of the National Academy of Sciences and serves as a member of the UCSMP Advisory Board.

Felix Browder:

Let me welcome the International Conference on Mathematics Education sponsored by the University of Chicago School Mathematics Project and the Amoco Foundation in my role as Chairman of the Mathematics Department of the University of Chicago. I shall begin by commenting on its location on the physical premises of our Department. To some of you this may seem somewhat unusual. After all, the Chicago Mathematics Department is one of the leading research departments of mathematics in the United States. We are not accustomed in recent times to see major meetings on primary and secondary mathematics education in such surroundings.

Of course, the reason for its presence here is that it arises from the efforts of some of our colleagues and especially the efforts of Izaak Wirszup. This fact and this location are not as anomalous as conventional stereotypes might suggest. These stereotypes are based upon the sharp separation that has tended to develop over the past half-century in the United States between the mathematical research establishment and those who work in and think about precollegiate mathematics education. Even in those special circumstances when efforts have been made to bring the separate tribes together (as in the "New Mathematics" movement), the consciousness of separateness of the tribes has remained quite acute.

However, it is a historical fact that it was not always thus. In the Mathematics Common Room on the second floor of this building, a room which serves as one of the principal foci of this meeting, you will find on the east wall the portrait of the founder of the Chicago Mathematics Department, Professor Eliakim Hastings Moore, who became Acting Chairman of the Department in 1892 and remained its Chairman till 1929. E.H. Moore was one of the most distinguished American research mathematicians of his time, and can be regarded as the most important figure in the creation of the American mathematical research community. Together with his student L.E. Dickson and others, he founded the strong mathematical research tradition of the University of Chicago. Others of his students and disciples helped establish the other major traditions of American mathematical research, Veblen at Princeton and the Institute for Advanced Study, Birkhoff at Harvard, R.L. Moore at Texas, and a number of others. In the first decade of the twentieth century, when American Men of Science listed the practitioners of each major field in order of distinction according to the vote of their colleagues, E.H. Moore's name was the first in mathematics.

You might well expect a figure of such stature to have taken a prominent part in the creation of the major mathematical professional societies, as Moore in fact did in founding the Chicago Mathematical Society which, in 1893, merged with the New York Mathematical Society to become the American Mathematical Society. In the field of collegiate teaching of mathematics, Moore collaborated with his colleague Slaught in 1919 to found the Mathematical Association of America. What many find surprising is that E.H. Moore was one of the founders of the National Council of Teachers of Mathematics, the principal organization in the field of precollegiate teaching of mathematics, and that the NCTM gave him its first major award for distinguished service to mathematics teaching.

E.H. Moore had a sharply defined philosophy of mathematics education, which he expounded and practiced with great vigor at all levels of mathematics education. That philosophy assigned a central role to the process of arousing and stimulating the independent creative faculties of the student and to focusing them in simple concrete forms the student could assimilate and master. He was a forceful advocate of the mathematics laboratory, understood in very imaginative terms as involving very concrete forms of activity intended to mold the student's grasp of mathematical intuition and process. E.H. Moore's philosophy clearly had an influence on that of his namesake R.L. Moore, who gave it vivid expression and practice in the Texas school of topology.

Moore clearly belonged to the current of the Chicago pragmatist tendency of the turn of the century, represented in its more publicized forms by men like John Dewey and George Herbert Mead. Moore's pragmatism took a sharply different tack in education from Dewey's, however. Whatever Dewey may have been ultimately aiming at (and it is not so easy to say in view of the inimitable obscurity of his writing), the principal thrust of the progressive education movement which he founded lay in learning by doing, represented by children's visits to dairies and settlement houses. His disciples saw no great virtue in the cultivation of intellectual skills as such. Moore placed his emphasis on the mathematical laboratory, where students could develop such skills through insightful projects of their own. Let me suggest that these two conceptions corresponded to two contrasting concepts of the citizenry of the society: Dewey's to a society of consumers and tourists, Moore's to a society of craftsmen, technicians, and scientists.

It has become increasingly clear in the past decade that it is Moore's craftsmen, technicians, and scientists that we need in America today, not Dewey's tourists. It is this growing realization that makes Moore's concept increasingly relevant to our present problems, and furnishes the theme and thrust of the symposium which we are opening today.

Izaak Wirszup:

I would now like to introduce Professor John Craig, Director of the Comparative Educational Center and Acting Chairman of the Department of Education.

John Craig:

What he did not mention is that I am an historian by discipline and that as such feel almost totally out of my element in this company ... although perhaps not totally. My interests in comparative education and my training in European history have long alerted me to the differences across countries and to the interesting problems that arise through comparisons, across time and societies, in particular in terms of my

5

current research interest in patterns of diffusion of innovations. Clearly this is a conference devoted to at least diffusing innovations or information about innovations. What will happen in practice when those innovations are tested in different kinds of systems (centralized, decentralized, and others) remains to be seen. I suspect this is not the appropriate place for me to share my own ideas on the subject with you. Perhaps some of us will have an occasion to do that later.

I think my real responsibility at this time is just to welcome you on behalf of the Comparative Education Center, but particularly on behalf of the Department of Education, the home base for UCSMP along with the Department of Mathematics, and to wish fervently that you have an exceedingly enjoyable and successful, productive conference.

Izaak Wirszup:

Allow me now to introduce Dr. Keith W. McHenry, Jr., Chief Scientist and Vice-President for Research and Development at the Amoco Corporation. Dr. McHenry is the ideologue and architect of the project. Without his efforts and advice, there simply would be no UCSMP.

Keith McHenry:

It is really an honor and a pleasure for me to bring you greetings from the management of the Standard Oil and Amoco companies, and our Chairman Richard Morrow joins his colleagues and me in wishing you great success in this truly unique international conference.

It was Izaak Wirszup's dream of holding such a conference, and of the advances in mathematics education which might flow out of it, that attracted our company's interest in early 1982. With the idea that the Amoco Foundation might provide support for a well-conceived program, Izaak worked tirelessly to come up with such a program during most of 1982 and the early part of 1983. And there were some false starts, but with strong support from the Mathematics and Education Departments

6

and the University, initial phases of the University of Chicago School Mathematics Project were in place and ready to go in the beginning of the 1983-84 school year.

Because I have spent my professional life doing and directing research and development, I was personally intrigued with the concept of taking the best research on methods of teaching mathematics from around the world and adapting them to the educational culture of our country. A research director's nightmare is that his team will reinvent something that has been invented before. This Conference is obvious proof that the UCSMP team is not as much concerned with authorship of ideas as it is with the concrete results that those ideas could bring to the classroom. And you, the conference participants, are demonstrating by showing the results of your research that you are similarly concerned.

With respect to the Project itself, I can claim no expertise as an educator to evaluate it. However, my background in research does allow me to recognize a well-designed experimental program. The emphasis that the Project is placing on classroom trials, feedback, testing of results, and revision of the teaching materials is in the best traditions of science. There is, however, another element present which is found in only the best of research teams - enthusiasm. I will leave it to those of you who represent the social sciences to explain why that element is a necessary condition for successful research. Having met with both the Mathematics and Education Departments' representatives, I can only assure you that enthusiasm in its best form is thriving on the UCSMP team.

And finally, what do we at Amoco see as a desired result of this conference, and with its help, the School Mathematics Project? We see ultimately a population which is better equipped to deal empirically and quantitatively with an increasingly technological world, and with those skills, a population which can leave superstition and hearsay behind. We believe that healthy international cooperation and its sister, healthy international competition, can lead to healthy international relations. That population which can leave superstition and hearsay behind should ultimately be the world's population. I would be less than candid, however, if I did not admit that one of our motivations is to improve the competitive position of the United States. I am sure that all of you have already arrived at that conclusion.

As educators, you may feel uncomfortable with that motivation, but with my background in industry I do not, because I have learned that if my competition has equal skills, then both of us will be better off for having applied those skills to the task at hand. When that task is the education of young people, then it's hard to see how there can be any losers.

So let me thank you again for your unselfish participation in the conference, not only on behalf of my company, but also on behalf of the ultimate winners, all of our children.

Izaak Wirszup:

Now I am pleased to present Mr. Donald G. Schroeter, Executive Director of the Amoco Foundation. We are extremely grateful to Mr. Schroeter for his continuing advice and encouragement. He has been instrumental in bringing this project to fruition.

Donald Schroeter:

On behalf of Amoco Foundation, I am most pleased to be here with you as you present and discuss significant research and experiences with elementary and secondary school mathematics. I am glad to represent Amoco Foundation.

We were founded in 1952 and have developed one of the ten largest corporate giving programs in the United States. The Foundation has historically placed major emphasis on education. We are convinced that the University of Chicago School Mathematics Project is the most important educational program we have ever funded. This conference is clearly of basic importance to the success of the School Mathematics Project. The Foundation is indeed proud to be playing a part in it.

Izaak Wirszup:

I was happy to hear that the President of the American Federation of Teachers, Dr. Albert Shanker would make a special effort to come from New York to address our conference. It goes without saying that America's educational problems cannot be resolved without the continuing dedicated efforts of our teachers. The success of this partnership between the University of Chicago and a major industrial corporation depends on the leadership and inspiration provided by the teaching profession. I am delighted to introduce to you Albert Shanker.

Albert Shanker:

I just want to say that I am very excited about the prospects of this conference. With all of the reports that we had last year — about thirty of them, with about another ten reports to come — there is somehow a belief out there in the world of politics that education can be improved by adopting lengthy pieces of legislation of 150 or 200 pages, and that changing the political structure of schools essentially places all sorts of mandates on school systems. In almost none of those reports was there any discussion of the fact that improvement in education requires knowledge of what people are doing in their fields. That is what is missing. There is no other field in which there is not a good deal of time made available for experts to discuss things with each other, to reach agreement as to what the critical problems are, and to share information about how to solve some of these problems. That is very much missing in education. The notion that we will solve our problem if a state legislature throws out one set of textbooks and brings in another set of textbooks (with a few harder or easier words) and mandates it for the next ten years is ridiculous. The only way we will make real improvements is if there is an opportunity for those in the field to meet and to discuss particular disciplines and find better ways of doing things. So when the history of all this is written, what you are doing here today, this conference and a few others like it in other fields will go down as being much more important events in terms of the improvement of education during this period than all of those reports. Good luck.

Izaak Wirszup:

Finally, it is a great pleasure to introduce Paul J. Sally, Jr., Professor in the Department of Mathematics and overall Director of the University of Chicago School Mathematics Project.

Paul Sally:

I would like to extend my welcome to you. I think we should immediately change our tune here. These statements about how great the project is and how great we all are I'd like to change a little bit to: "Let's get down to work." I'm delighted to see you all here. One of the major reasons I'm delighted to see you all here is that I believe you can give us some help in running this project. It's very nice to come to a conference and exchange ideas and to talk to one another, but I do hope that as the day and the conference go on, you will extend to us your ideas about what we are doing. No matter how critical they are, we would like to hear them. One does not like to operate in a vacuum. It is a little too early to break out the champagne about our project, I might say. We have been underway for a year and a half, and we are beginning to get the feel of the ground beneath us. So your help would be greatly appreciated.

It is my pleasant duty to give you a very brief summary of what goes on in UCSMP and then to introduce the people who are really doing the work. But I thought you should hear their names at least once before they step up here to the podium and tell you about their components.

The University of Chicago School Mathematics Project has four components. It actually has five components, the fifth one being the Administrative Component, which is my component, but we try to keep a low profile, because the point of the project is education, and the education and the effects on education are taking place in these four components.

The Elementary Component is developing into a three-pronged component, which will be explained to you by Max Bell. The leaders of the three parts are Max Bell, Professor of Education, University of Chicago; Alan Hoffer, Professor of Mathematics at the University of Oregon; and Sheila Sconiers, who has spent quite a few years in the Chicago school system and has joined the project as a research associate. We can only regard it as our excellent good fortune that she is with us.

The Secondary Component is led by Zalman Usiskin, Professor of Education, and he will tell you, as will the others, about their particular phase of UCSMP.

The Evaluation Component is led by Professor Larry Hedges of the Department of Education; and the Resource Component (or "international component," I might call it) is led by Professor Izaak Wirszup of the Department of Mathematics, whom you will see regularly throughout the conference.

I would like to take a moment to express my gratitude to Don Schroeter and Keith McHenry and the Amoco Foundation in general, especially Richard Morrow, for their strong support of this project. They have been unstinting in their support and I hope they will continue to be so as the years pass. Without them, there is no project. It's that simple.

Our focus is the average student. Many of the programs that have come up over the past twenty years have been addressed to the very talented student or to the student who needs remedial help. It might not be an exaggeration to state that education for the average student has been somewhat of a wasteland in the schools over the past ten or fifteen years. This program, at least at the secondary level, is designed to address the problems in mathematics of the average student, the middle-range student.

As I mentioned, the Elementary Component has three parts. All of these descriptions are, of necessity, quite general, and therefore somewhat imprecise. Alan Hoffer is leading the efforts to develop a K-6 curriculum, at least partially based on the adaptation of foreign materials as provided by Izaak Wirszup and the Resource Component (the best of materials from foreign sources). Sheila Sconiers is leading teacher development in the classrooms, especially in the city of Chicago. She is developing many programs for teacher training and retraining, and Max will have something to say about those. Max Bell is leading the effort to develop teacher and classroom resources. There is so much more needed than simply writing a curriculum or training and retraining teachers, and Max is making the effort to address those problems.

In the Secondary Component led by Zalman Usiskin, the major effort is the development of an up-to-date grade 7-12 curriculum, which he will outline for you.

In the Evaluation Component we have Larry Hedges and the traditional and non-traditional methods of evaluation. It would be presumptuous of me to try to explain either to you. I'll leave that to Larry.

In the Resource Component, as I have already pointed out, Izaak Wirszup is providing to the project truly fine foreign materials which we have found to be remarkable. These will be used in the development of the curriculum for K-6 especially, and for other purposes. I would like to take this opportunity to thank Izaak for organizing this conference; it is a magnificent job, and I would like to give him a hand.

I am not quite through. The emphasis in UCSMP, if there is one, is on teachers and what they do in the classroom. Through curriculum writing, we are providing materials to teachers. But in every facet of our effort, one can discern that, whatever we're doing here at the University, ultimately we want to work with the teachers in the schools. From day one of the project we have been in the schools with the teachers. As a matter of fact, Zalman Usiskin for the last academic year was in a school every day writing the first book in his series, a volume entitled Transition Mathematics. So it is very important to realize both for us and for those who view us, that our goal is to get into the schools and to stay in the schools.

This project is designed to last until 1989, and perhaps beyond, but I have to be careful of what I say with Amoco people in the room, because we have a commitment from them through 1989, and that will conclude six years of the project. Where can one get in mathematics education in six years? Well, you're a very experienced crowd; many of you have been in education for many years. At the end of six years you can begin to see what the problems are. You may find your feet firmly on the ground at the end of six years. How much wide-scale success will you have had in the schools? Not much. You may know what you want to do. It always comes back to me, Winston Churchill's surge of optimism in the middle of World War II, when he said, "This is not the end, nor is it the beginning of the end, but it is the end of the beginning." And that's where I hope we'll be at the end of six years.

How long will this effort take to do something serious about mathematics education in the elementary and secondary schools of this country? Easily fifteen years — by the turn of the century, if we keep our enthusiasm, if we work very hard, despite the fact that it is obvious that the current wave of enthusiasm in the general public will wane considerably. Perhaps some TV program or something else will catch the public's attention. We have to sustain our efforts for fifteen years, and we must obtain the financial support to do that. And a number, just a round figure showing what it will cost to reform and revitalize mathematics education in this country at the precollege level, is 100 billion dollars. Now that's rather cheap if you think about it. If you think about the number of teachers that need serious attention; if you think about the fact that teachers — the emphasis is always on teachers — need time to reflect and to learn, that's a very small amount of money. It's now practically one third of the annual defense budget, and I'm just asking for it over fifteen years. So I feel that I'm being very parsimonious in my demands.

Again, I welcome you here, and I urge you to assist us in our efforts to do something about mathematics education in Cook County. You see, everyone might think we have national designs. Well, that would be all well and good. But if by the turn of the century mathematics education in Cook County is in good shape, we can be really proud of ourselves, I'll tell you. Now somebody else can take Suffolk County, or other counties, and we'll join you in your efforts, okay? Thank you for coming, and please give us whatever suggestions you have.

Now it is my turn to introduce people. Our first presenter for UCSMP will be Professor Max Bell of the Department of Education, who is the Director of the Elementary Component.

Max Bell:

Roughly speaking, the Elementary Component of the University of Chicago School Mathematics Project is characterized by three periods of development — two of them nearly finished, the third and most difficult still ahead of us. The first period was the decade before beginning the UCSMP Elementary project. During that time each of us worked with children 5 to 11 years of age, schools, teachers, and prospective teachers, and solidified the convictions that have guided the project in its formative stages.

The second period, now drawing to a close, comprised the first two years of the six-year project that has been assured by the Amoco Foundation. This phase has been pretty much a matter of learning new roles, establishing urban school laboratories, working closely with teachers, building a knowledge base, and adapting or inventing essential curriculum materials. As a result of that work, we have a substantial knowledge and resource base from which to work.

In the third period, now beginning, we will work from our convictions and from the knowledge and resource base accumulated in the first two periods to specification and production of management procedures, instruction procedures, and curriculum resources for teachers and students in grades K-6. We will be concerned with building a mathematics experience for children that is considerably richer than is common now, yet teachable by most of today's teachers. We will also develop and model programs for retraining of relatively large numbers of teachers.

The First Phase: Solidifying Basic Convictions

In the years before actually beginning with UCSMP we formulated some basic convictions that form a sort of axiomatic basis for what we are now doing. First, we generally conceive of curriculum building, whether for teacher development or for children, as working out the interrelationships of the classic Tyler curriculum diagram.

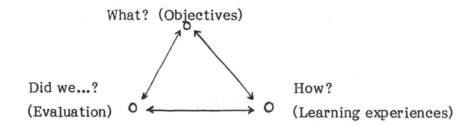

As to objectives — what exactly we want to accomplish — our work is guided by the general imperative of giving children a rich and playful early mathematics experience, with content pretty much as outlined in some of my earlier publications.[1] Specification of learning experiences and evaluation procedures to accomplish and validate those objectives and to make them practicable in real classrooms is the present task of this component.

A second part of our foundation from the pre-UCSMP years was neatly summarized 17 years ago by the guest of honor at this conference, Hans Freudenthal:

> We can no longer keep silent about teaching mathematics so as to be useful... mathematics often figures as the paragon of a disinterested science... but we can no longer afford to stress this point if it keeps our attention off the widespread use of mathematics and the fact that mathematics is needed not by a few people but virtually by everybody.[2]

As a third consideration from past experience, a variety of lines of evidence indicate that no more than 15 to 20 percent of American children survive the school mathematics experience with understandings that make mathematics truly useful to them. There is considerable evidence that our failure with the other 80 percent or more has its roots in generally pessimistic and often wrong-minded approaches to children's early school mathematics experience.

Fourth, past experience teaches us that the problems of improving early mathematics schooling are complex, highly interrelated, and very resistant to solution. Here are some of the most troublesome of these

---

[1]Bell, M.S. "Teaching Mathematics as a Tool for Problem Solving." Prospects, Vol. IX, No. 3, 1979, 311-320.

Bell, M.S. "What Does 'Everyman' Really Need from School Mathematics?" The Mathematics Teacher, 67 (March 1974), 196-202.

[2]Freudenthal, H. "Why to Teach Mathematics So As to Be Useful." Educational Studies in Mathematics, Vol. 1, No. 1 (May 1968) 3-8.

persistent problems:

- New content and methods, however good they might be, do not become standard practice.

- Practical realities of classroom life invalidate many reforms. If we ignore management difficulties and limits on teachers' time, energy, and knowledge, failure is certain.

- Pessimistic and narrow objectives dominate K-3 instruction. Most school textbooks reflect and perpetuate this pessimism.

- Children's intuitive and informal abilities are often ignored, and this is especially true in early schooling.

- Time allocations to mathematics learning are terribly inadequate, in official guidelines and in actual school practice.

- Problems of scale defeat most teacher retraining efforts. That is, many of us know how to work effectively with 20 to 50 teachers at a time, but only effectiveness with thousands, tens of thousands or hundreds of thousands of teachers will support genuine reform. Mathematics education has never faced up to that reality.

- Standard school practice is very resistant to new technologies. Hence, the potentially revolutionary effects of video, calculators, and computers may go the way of previous technological revolutions — pervasive in everyday life but ignored in schooling.

In sum, these convictions — axioms — from the first phase of this project in the years before its actual beginning both support our subsequent efforts and give us some idea of the magnitude of the task on which we have embarked.

Second Phase: Building Bases of Knowledge, Experience, and Resources

In the past two years we have made excellent progress in relatively small scale attacks on many of the problems just outlined. We have established school laboratories and close links to a group of teacher-collaborators. We have gained quite a lot of insight into why innovations seldom work beyond experimental trials. Based on that, we are formulating management technologies to help teachers so that excellent rather than mediocre methods and materials are practical for use in the real life of classrooms. We have developed and piloted teacher development methodologies and, lately, teacher support networks.

Turning from schools and teachers to children, we now have quite a lot of information about the capabilities of K-4 children via analyses of data from interviews with about 500 children, in a wide range of school settings. Karen Fuson has developed some efficient links between children's verbal counting capabilities and doing symbolic arithmetic. Jim Stigler has helped us understand differences in arithmetic achievement of U.S. and Asian children.

With respect to development of teaching technologies, we are giving teachers the practical means to at least double the amount of time spent on mathematics each day. We have developed and pilot-tested a number of calculator exercises. We have developed some computer based problem collections. We are exploring uses of computers in two quite different ways: one or two per classroom and a computer instruction laboratory to which teachers bring all their children.

Also in this initial two years, each of us has been learning new skills and gaining experience in new roles.

Third Phase: New Initiatives in Elementary School Curriculum and Teacher Training

We in the elementary component of UCSMP will now move from establishing our resource and experience base to creation and implementation of immediately practical management and curriculum resources. At the same time, we will move from intensive work with just a few teachers to development of model training programs that can be effective with large numbers of teachers. Future-oriented research and development will continue.

For those purposes, we are making rather substantial changes in what we do and how we do it. For 1985-86, there will be three subgroups of the Elementary component. A subgroup directed by Alan Hoffer, collaborating with Izaak Wirszup, will attempt to draw on some of the best materials already proven effective in foreign countries and recast them for use by U.S. children.

The Teacher Development subgroup, directed by Sheila Sconiers, will work on model teacher retraining packages that reflect the actual things we want teachers to teach. One target will be K-3 teachers, normally teaching all school subjects in self-contained classrooms. A second target is conversion of willing but inexpert grade 4-6 teachers into credible "specialists" to teach all the mathematics in those grades.

The Teacher and Classroom Resources subgroup, directed by me, will engage in relatively speculative proof of concept development. We will aim for materials and methods that are "practical" immediately on a small scale but future-oriented with respect to widespread implementation. For example, Everyday Mathematics will be a series of easy-to-use, day-by-day teacher guides to considerably enriched mathematics instruction for primary school children, usable with minimal in-service training. Experiments in the computer laboratory of a Chicago public elementary school will involve essentially all the children in grades K-8 in a full-year trial of exemplary and fairly comprehensive microcomputer-based curriculum materials. A few special microcomputer programs for support of Everyday Mathematics will be programmed. Jim Stigler and Karen Fuson will continue their research and teaching experiments.

Summary

In summary, we in the Elementary Component of UCSMP believe that the present elementary school mathematics experience is generally very inadequate and that it is characterized by a complex, interacting, collection of barriers to improvement. We propose to attack those barriers both by development of fundamentally new procedures and by finding ways to bring into common practice existing but neglected resources.

Paul Sally:

Our next presenter will be Zalman Usiskin, Professor of Education, who is the Director of the Secondary Component.

Zalman Usiskin:

The Secondary Component of UCSMP is a curriculum development component of the project. Its goal is the design and the production of implementable and effective materials for a complete six-year mathematics curriculum for students in grades 7 through 12.

We describe the target population in various ways. Sometimes we describe it as consisting of all students who are neither gifted nor learning disabled. Sometimes we describe it as consisting of the middle x% of the population, where x is some number between 50 and 150. And at still other times we describe it as consisting of those students who will graduate high school, but do not plan to major in the mathematical sciences. We have different descriptions, not because we're trying to be elusive, but because we're talking about a target population. We are trying to convey that target. We don't want to make our decisions as if our students are gifted or learning disabled or future mathematics majors. Yet it still may be that our curriculum will be best for those students. In fact, we would hope it would serve as the starting point for curriculum that would be adapted for those students, but our target is the middle.

Curriculum development involves the selection, content, sequence, timing, and approaches and the translation of these into materials. Our global guidance comes from recommendations of various national committees over the past ten years. Many of these committees have gone to international sources for their ideas. The point here is we are not trying to make new recommendations. Some of these committees you are familiar with: the NACOME Report of 1975; the National Council of Supervisors of Mathematics "Position Paper on Basic Skills"; the National Council of Teachers of Mathematics and Mathematical Association of America position paper on the high school mathematics that students need to know for college; NCTM's Agenda for Action, and, more recently, "A Nation At Risk"; the Conference Board of Mathematical Sciences' "What Is Still Fundamental and What Is Not"; the College Board's Academic Preparation for College; and the report of Jim Fey's conference on Computing and Mathematics.

The consensus recommendations in these reports are of two types. There is a desire to "upgrade" students by increasing graduation requirements and by improving problem solving performance. And there is a desire to update the curriculum by increasing the amount of attention given to applications, to statistics and probability, by utilizing calculators at all times, and by changing content and approach due to computers. Two recommendations in the consensus are contradictory. On the one hand, there is a strong movement to improve the algebra skills of college-bound students. At the same time, we are hearing recommendations to decrease attention to these manipulations, because of the existence of symbol manipulator computer software. We are trying to satisfy both sides.

We have been guided in our deliberations by many individuals and groups. Obviously, there are the UCSMP faculty and staff. But there are others. The Secondary Component has its own Advisory Board of college and university professors, until now consisting of seven people: Art Coxford, Dave Duncan, Jim Fey, Glenda Lappan, Tony Ralston, Jim Schultz, and Sharon Senk. This board has met as a group and has also split to discuss with other consultants specific aspects of the curriculum. Last year we invited 27 school personnel ranging from state supervisors to classroom teachers to discuss with us issues related to revising the curriculum in grades 7 and 8. This year a similar number has been invited to discuss with us issues related to changing standards in school algebra. We've held meetings with a half-dozen publishers whose advice we also seek. We are initiating a dialogue with directors of employment and business and industry to obtain their views, particularly as regards appropriate mathematics for those who do not attend college. We welcome advice from all quarters, and one of the obvious reasons for this meeting is to get advice from you.

Let me turn to the decisions we have already made. A first conclusion made even before the project began is that to implement the recommendations requires expansion of the standard four-year senior high school mathematics curriculum to five years. Over a long period of time and many meetings we considered the possibility of integrating our curriculum. That is, teaching geometry along with other subjects because it turns out geometry is the key to integration. Everything else in the United States is integrated.

Three major factors influenced us to stick with the curriculum whose course is parallel to existing ones. First, we saw great difficulty in getting test sites for an integrated curriculum. And if it were difficult to test the curriculum, we felt it would be even more difficult to implement it.

Second, we thought that even within the parameters of today's course titles, we could develop a curriculum that was rather integrated by keeping geometry alive through algebra and vice versa.

Third, we wished to maximize the number of entry points into our curriculum. In theory, we would like a school to be able to use any of our materials without having used the previous years' materials.

It is obvious that we have a difficult task of balancing many priorities, but if the task were easy it would have been done before, and it would not be worthy of the effort that so many people have put and are putting into it.

Here's the broadest possible outline of the curriculum. In year 1, which we call <u>Transition Mathematics</u>, two-fifths of the year is applied arithmetic, a half of the year is pre-algebra, and two-fifths of the year is pre-geometry. There is obvious overlap. For example, the formula for the area of a rectangle falls into all three categories.

The first draft of this course was written last year, and a substantial revision was begun last summer and is just now being finished. The table of contents is on a handout. <u>Transition Mathematics</u> is currently being used in nineteen classes in twelve schools in the city and suburbs of Chicago. All of these schools are part of the pilot-testing program. Next year the materials will be available for any school that wishes to use them. We have just made the decision to do a larger-scale testing of the materials next year, and this requires that second revisions be done within the next couple of months. Revisions will be made as a result of the opinions of teachers from this year. We are looking for test sites and users, and if you are interested, a form is included with one of the handouts.

<u>Transition Mathematics</u> is a course to precede algebra. Its prerequisite is that students have facility with whole number arithmetic and have a passing acquaintance with fractions and decimals. They should know some things about that. Though we would hope that <u>Transition Mathematics</u> would be an appropriate course for all students in grade 7 (<u>in grade 7</u>), the plain fact is that only about half of the students today are ready for those materials then. So we are testing these materials at three different grade levels: the 50th to 90th percentile in grade 7, the 30th to 70th in grade 8, and the 15th to the 40th (or 50th) in grade 9 or above. The last of these includes students who do not get into algebra in grade 9 but will take algebra the next year. So in this curriculum algebra occurs one year earlier than usual. This multi-level approach is not as radical as it might seem. Currently, about 11% of eighth grade courses

are algebra. We would propose they might be seventh grade. And about a third of all first year algebra students in high school are tenth graders or above, so algebra is not as much a ninth grade course as it is often classified. So we are asking students to take algebra a year earlier.

It can't be the same algebra course that currently destroys many students. The UCSMP algebra course will have a lot of statistics and applications and quite a bit less with complicated manipulations and contrived word problems. It is planned to be much in the spirit of my Algebra Through Applications currently distributed by NCTM. But we anticipate many major changes. I am rather convinced it will not look a bit like that other stuff.

The algebra will be written this summer by a team of writers selected in a nationwide competition. The method we used for the selection was unprecedented. We brought in finalists and asked them to plan and write under pressure and then picked only some of them. On a handout is the list of the writers who will be joining us for eight weeks, and you will see that we are still looking for one writer. The algebra course will be piloted next year by its authors or, when an author is not a secondary school teacher, by a teacher monitored closely by that author. We will then invite some or all of the algebra authors back for the following summer and revise for a larger testing program two years from now.

A second writing team this summer will be working on advanced algebra. We still have two openings for that team. We can give application forms to anybody who is here, but we may not accept all applicants, even from as distinguished a group as this.

We plan to begin the writing for the courses for years 3 and 5 next year, that is, a year from this summer and the year 6 course the following year. The geometry as presently envisioned will flip-flop the usual order of topics so that area and volume formulas, coordinates and transformations are done first. You might say that would be reviewing stuff that had been done earlier, and the proof work will be done mostly during the second semester. Years 5 and 6 are called pre-college mathematics rather than pre-calculus to emphasize that they are designed to prepare students for all of the mathematics they may encounter later. In particular, year 5 has a strong statistics component and year 6 a strong component in discrete mathematics. The precise detailing of the statistics and discrete mathematics components for these years is one of our tasks for next year. Notice that though manipulations are de-emphasized in years 2 and 4, they are re-emphasized in year 6. This is our way of dealing with those conflicting recommendations that I mentioned earlier.

There are certain features that run throughout this curriculum. We require scientific calculators; we do not want to create technology for the curriculum, but we want to take advantage of the existing technology. We feel it exceedingly important that students learn at all levels applications outside mathematics of almost everything they study. And we feel it exceedingly important that they learn to read mathematics. In every lesson we give students materials worth reading and ask them questions on it.

Finally, we have two features designed to increase performance. At the end of each chapter is a sequence of activities in mastery learning fashion, a formative test followed by feedback and review, because we're asking for some mastery not just of the typical skills, but of mathematical underpinnings of applications, of graphing, and other representations. We feel we are going to increase performance particularly in those areas.

A second feature is that in each lesson about a quarter of the questions are reviewed. Review is from the previous lessons or from previous chapters, and in later years, from previous years. We have some evidence that this feature will greatly improve performance, particularly on those topics that are usually studied in a chapter and then dropped. It is planned that these features will run throughout the six years. More detailed rendering of the curriculum is on a handout. I welcome the opportunity to discuss these ideas with you over the next few days. I would be happy to try to answer any questions that you might have.

Paul Sally:

I might warn you that Zal has indicated that there are some positions open for writers in the summer. Now you might wish to apply, but he is a stern taskmaster, so that one has to be very careful about that.

The other thing I should say is that Zal mentioned some of these national reports. We have set as one of our goals to see that no more reports are written decrying the state of mathematics and science education in the United States. We probably have had too many and we do not need more. What we need is some action, and that is what we are hoping for here.

Now in that vein, let me present Professor Larry Hedges, from the Department of Education, who will talk on the Evaluation Component.

Larry Hedges:

I would like to start with an apology. I am not a mathematics educator. I am primarily a statistician, but I have had the privilege of working with a distinguished group of mathematics educators in the last two years. It has been a pleasure, because I have learned a great deal about mathematics education and about the problems of mathematics teaching in the schools.

The evaluation of any project that promises profound educational change forces us to confront fundamental issues of what should be taught, how it should be taught, and how we will know if it is learned. It also forces us to confront seriously the forces of organizational inertia that thwart most efforts at fundamental changes. Lest anyone underestimate the importance of inertia, I remind you that the vast majority of educational innovations in this country have failed. However well conceived, they quietly disappear, leaving little if any permanent change. I think in that vein that it is helpful to consider innovations in two stages.

The first stage is a laboratory stage, and the second stage is what we might call a mass-dissemination stage. Now in the laboratory stage most innovations work. In this stage innovators work closely with school personnel to make sure that the first experiences with the new curriculum materials are as positive as possible. Expert consultants are available at a moment's notice to provide information, guidance, creative problem solving, financial resources, and above all enthusiasm for the intervention. School personnel involved have the feeling of real meaningful participation in the project. The best projects listen to school teachers and administrators, and their comments make a tangible difference. And most of the time, of course, innovations work at this stage.

In the mass-dissemination stage things are usually different. Innovators are remote from the school personnel, and school personnel may have only the vaguest understanding of the details of the new curriculum. There is no input from outsiders anxious to make sure that the initial experiences are positive. Our consultation is often unavailable and, in my experience, what consultation there is may actually be providing incorrect information to the school people. There are no extra financial resources, and creative problem solving actually imposes an extra burden on already overburdened personnel. Enthusiasm is dimmed, and there is really no ready source of outside stimulation or encouragement. The reality is that innovation is harder than continuing to do the same thing. Thus, it is not really surprising that so many innovations fail at the mass-dissemination stage.

My role and that of my staff is to help others in this project examine both the laboratory and the mass-dissemination stages of their work and to try and insure that the project actually works at both stages. In the laboratory stage, new curriculum content calls for new ways to measure educational achievement. One of the greatest dangers to a new curriculum is that it will be evaluated on objectives that it did not attempt to teach. Developing tests that match explicit curriculum objectives is essential.

It is also essential to develop assessment procedures that match the implicit curriculum objectives. When we develop curricula that stress learning mathematics from reading, for example, we are finding concrete representations for mathematical abstractions. We must not cheat ourselves by failing to assess these important but non-traditional outcomes.

Perhaps more important is the idea of matching assessment tools to the philosophy of the curriculum. The question is often how can we assess whether the curriculum is affecting the way students think. If we are trying to give students a clear grasp of mathematical thinking, then we need to find assessment tools, assessment devices that require the sort of thinking that we are after. Of course, the more global a goal, the more difficult it is to find such assessment procedures. Many traditional tests do a poor job because they can be foiled by rote application of memorized algorithms. For example, we struggle with the problem of ways to assess the articulate use of concepts such as models and representations of mathematical ideas.

How can we assess such abstract notions? Some creative colleagues have suggested several possibilities in the context of the Transition Mathematics materials. For example, we can ask the students to describe the meaning of mathematical concepts. Alternatively, we might ask the student to tell us some different concrete situations that could lead to a particular problem, a particular abstract representation of a problem; or we can ask the students to show us why an incorrect solution is wrong conceptually; or we might ask the students for a model of a multi-step problem, and then we could ask them to show us how the model could be manipulated to lead to different solutions. No doubt there are many other possibilities. What is interesting is that you will find none of these on traditional standardized achievement tests. The best questions may be hard to administer and equally hard to score, but our feeling is that even a little bit of data on the right question is preferable to a lot of data on the wrong question.

By working to develop means of assessment that match the philosophy of the curriculum we hope to attain better ways of showing the impact of the curriculum on the way students think. This is obviously important for fairly evaluating the merit of the curriculum, but it may be more important in raising the consciousness of mathematics teachers and educators, who should probably think less about repetitious performance on low level skills and more about the way students reason. Selecting appropriate curricula and assessment procedures may bring marvelous success in the laboratory stage of an innovation. But that alone cannot ensure success at the mass-dissemination stage. The sad truth is that just being good enough is not enough to ensure success at the mass-dissemination stage.

Another focus of this project is in teacher development and the dissemination of new philosophies, technologies and ideas throughout the schools. We have been assisting our colleagues in trying to understand the barriers to innovations in schools. As all of you know, the number and variety of potential barriers has been distressingly high, ranging from the trivial to the very deep, I think. The model of evaluation research that we are applying in this effort requires intensive analysis of diverse sources of information in the form essentially of case studies. A taxonomy of stumbling blocks is needed along with ways of regaining lost balance. Problems of assessment in the most global sense are crucial here. How do the proponents of an innovation determine what the problems are likely to be and set in motion a course of events that is likely to correct those problems?

Which key actor's support is most crucial in various administrative arrangements that we find in schools? How do individuals become committed to the innovation in the first place? On a more mundane level, we are addressing some of these questions in our evaluation of the teacher development activities. We are watching teachers trained by the project as they act to spread innovation in their schools. We are trying to find minimally obtrusive ways to gather data on what happens if they try to encourage curricular innovation. We are also trying to find ways to assess how rapidly their efforts are blocked and how they deal with these blocks. We also hope to find feasible ways of making the same sorts of assessments when our teacher networks are enlarged by one to three orders of magnitude, as Max mentioned.

Now there is yet another component to the activities of the evaluation of this project. We are trying to systematically examine the research literature on mathematics education to see what can be learned from it by using yet another set of assessment tools. The literature on mathematics education, as you all know, has increased dramatically in the last 25 years. For example, a survey of empirical research by Ed Begle revealed literally thousands of articles that contained empirical research results. He argued that although there is a great deal of information, the information did not add up to much usable knowledge because it was too widely scattered and disorganized. Begle argued that in order for the information to become useful knowledge it had to be pulled together, organized, and made available in some convenient form. No doubt much of this research will not prove useful. Some of the studies may be uninteresting or of such poor quality as to be virtually worthless. On the other hand, there is reason to believe that there will be at least some important findings of such an exercise in research reviewing. Even if a small fraction of the research can provide reliable evidence, then the exercise of synthesis would probably be worthwhile. The technology for integrative research reviews has advanced rapidly in the last ten years, and I would argue that the principal breakthrough is the notion that reviews should maintain the same standards of methodological and, I should add, conceptual rigor as do original research studies. I also argue that that is, in fact, a relatively new idea.

Thus, rigorous reviews must devote considerable attention to several crucial issues. One is problem formulation. Just what is the review about? How do we operationalize concepts, in fact, theoretical constructs like treatments and outcomes? The second issue is data collection, including the idea of sampling of studies to be reviewed. The third issue is data evaluation, which is the problem of how we determine which studies provide valid and therefore reliable research evidence. Finally, there is the question of data analysis and interpretation, which is how do we synthesize the evidence provided in the reasonably valid studies? Quantitative methods are helpful in this latter endeavor, but solid qualitative thinking is always indispensable.

The purpose of reviewing research, I might add, is not to obviate the need for new research, but to set it, or more practical efforts, in a firmer context. On the other hand, new research does not obviate the need for research synthesis. The myth of the "superstudy" that will settle all questions dies hard, I think, but I hope it will be laid to rest soon.

Real educational situations are usually just too complicated for a single study to capture all the complexities. The variation among research studies and their outcomes is one of the richest sources of information on educational processes, and modern methods of research synthesis can tap this source of information. We have to carry out a comprehensive synthesis of research in mathematics education to enhance our efforts in research development and evaluation in this project. It may be a step toward fulfilling Begle's optimistic belief that there exists a solid body of information about mathematics education that, when dug out and organized, can suggest new directions for mathematics education. At the very least, we hope it will suggest which directions we should not try to go in the future.

Paul Sally:

I'd like to say that we take the efforts of Larry and his staff very seriously as a component that is equal to all of the other activities. We call him "The Enforcer," in fact, because one of the major words in his vocabulary is "nonsense." We'd like to keep it that way.

So, finally, we have Professor Izaak Wirszup, who is one of the originators of this project and the Director of the Resource Development Component.

Izaak Wirszup:

Despite American achievements in science and technology that are known around the world, international comparative studies in mathematics and science education have clearly shown that the United States is in the midst of an educational crisis. Research has pointed to numerous shortcomings in our educational system, particularly in school arithmetic, geometry, physics, and chemistry. In the past two years a number of collective and individual reports have addressed various aspects of this crisis and contributed, as you know, to constructive analysis of the situation. However, none of these reports has focused on overcoming the most critical impasses. Solutions require curricular changes, perhaps the most extensive in the history of American mathematics education.

Meanwhile, the calculator and microcomputer present major challenges and opportunities. The potential of these new technologies is not yet fully explored. Over the past forty years the United States has consistently produced the most advanced computer hardware in the world. Scientists and engineers have created extraordinary instruments for programmed instruction. But as we know, lack of success has tarnished even the phrase "programmed instruction" and is already squeezing out of usage its successor, "computer-assisted instruction." To succeed with new technology, as with new curricula, we need a finer understanding of children's mental processes.

We must improve the teaching of arithmetic from kindergarten on and combine it with intuitive geometry so that all children can complete it by the sixth grade. As you are aware, this is a major goal of UCSMP. We must put an end to the present situation in which the majority of our school population studies arithmetic for nine years and is not adequately prepared for secondary mathematics and hard sciences. Students often satisfy high school requirements by taking so-called consumer mathematics, general mathematics or business mathematics, which are merely other forms of arithmetic. For real improvement to be possible we must introduce new content, new text materials, new methods of instruction, and new instructional technologies based on research and sustained experimentation in the psychology of learning and teaching mathematics. These are precisely the goals of the University of Chicago School Mathematics Project, an applications-oriented program with a broad base in research, experimentation, and evaluation.

Because his name has been mentioned here already, I would add that Ed Begle dreamed of developing a literature on methods of teaching mathematics based on psychology research, a literature which is sadly lacking in the United States. However, UCSMP is a short-term, six-year program which does not have the time, the cumulative expertise, and the manpower to discover or rediscover what is needed in these areas. The aim of the Resource Development Component is to provide the project with samples of the best programs, materials and teaching methods the world mathematics community has to offer.

The main purpose of this International Conference is to give us an opportunity to learn from you and to extend and strengthen our international and national cooperation in educational matters of common concern. In this regard we do have considerable experience. It was work done under the auspices of the National Science Foundation Survey of Recent East European Mathematical Literature that served as a catalyst for the University of Chicago School Mathematics Project. It is here in

Eckhart Hall, home of the Department of Mathematics, that the Survey prepared 43 adapted translations of Soviet mathematics literature for students and teachers, including the Popular Lectures in Mathematics, which served as a model for the School Mathematics Study Group's New Mathematical Library.

While the United States is a world leader in several branches of psychology, educational psychology, unfortunately, has not been given a high priority here. The fourteen volumes of Soviet Studies in the Psychology of Learning and Teaching Mathematics published jointly by S.M.S.G. and the University of Chicago NSF Survey and V.A. Krutetskii's Psychology of Mathematical Abilities in Schoolchildren have exercised a major influence here and abroad and given new directions to research.

It was in this building that we prepared the adaptation of Kutuzov's Geometry, Volume IV of S.M.S.G's Studies in Mathematics, which first inspired the School Mathematics Project in Great Britain to introduce geometric transformations as a basis for geometry courses. Only years later was this approach accepted in our country.

Again, it was research at the University of Chicago that enabled us to call attention to the remarkable work of Hans Freudenthal and his students, the Van Hieles, on the five development levels in geometry learning. Hans Freudenthal's work is world-renowned. One paper "La pensée de l'enfant et la géométrie", published by Pieter Van Hiele in 1959 in a French journal, marked a significant breakthrough in the psychology of learning geometry and an ingenious refinement of the theories of Jean Piaget. But it went practically unnoticed, in the United States, and even in Europe. We were fortunate to be able to bring it to the attention of the American and European public. Now several projects in this area have sprung up around the United States with funding from the National Science Foundation and the National Institute of Education. The directors of four of these programs are here today. The Freudenthal-Van

Hiele theory has also served as a basis for developing promising software for elementary and secondary geometry courses.

By surveying and translating outstanding school mathematics publications from around the world for both students and teachers, the Resource Development Component offers UCSMP a first-hand look at curricula, achievement standards, approaches, and methodologies that are different from those in the United States. These inputs challenge us to raise our expectations of children's mathematical abilities and constitute a valuable resource for developing materials to implement UCSMP's goals.

To advance its programs and activities, the Resource Component maintains professional relations with scholars and institutions in other countries. This conference, organized by the Resource Component, is an expression of our strong commitment to international cooperation. We believe that the Resource Development Component gives UCSMP a perspective and a dimension that are unique among current American mathematics education research and curriculum projects. The Component is approaching its new tasks with anticipation and optimism, despite the magnitude of the undertaking.

We are deeply grateful to the leaders of Amoco and Standard Oil (Indiana) for their acute understanding and the uncommon courage it took to sponsor this bold program. To help tackle our educational crisis, to launch pilot programs in urban schools, shows truly extraordinary commitment on the part of the Amoco Foundation. We hope that the UCSMP partnership between the University of Chicago, Amoco and Chicago-area schools will produce a national model whose usefulness will extend beyond the 1990's and beyond the field of mathematics education.

Paul Sally:

I will close with a couple of brief statements. First of all, as you realize, what we have embarked on sounds like a very heady project, but in fact, we are into it a year and a half and it is just plain old simple hard work. Some days we come to the job with "dungarees and a lunchbox." And one should not be at all reluctant to state that there is nothing but hard work ahead. Part of what we are doing, of course, depends not on us but on the political structure. For example, when the lawyers at the Baker and McKenzie law firm are patrolling the lunchroom, and the doctors at Cook County Hospital are patrolling the parking lot, then they will be in the same professional situation as the teachers in the schools...and then, perhaps, society will do something about it.

Finally, you might ask, what is the role of the Administrative Component in all of this, because clearly the other four components are carrying forth with all the activities that we need. Well, we do a little managerial work, but mostly my job is to stand up and say over and over again: "Don't promise what you can't deliver."

Part 1: School Mathematics Education Worldwide

The Second International Mathematics Study:
An International Perspective

Kenneth J. Travers

University of Illinois at Urbana-Champaign

The Second International Mathematics Study has involved some two dozen countries in a comprehensive study of school mathematics.[1] The goal was to provide detailed information from each country about the content of the mathematics curriculum, how mathematics is taught, and what mathematics students really learn. This information is intended to help policy analysts and mathematics educators in each country analyze their school programs, identify areas of strength and weakness, and provide data of use to national officials as they plan for future directions in their school mathematics programs.

The Study was conducted by members of the International Association for the Evaluation of Educational Achievement, or IEA, an international network of educational research institutions.[2] In each country, a national committee of specialists was appointed to monitor the conduct of the study. An international committee of specialists in mathematics education and measurement had the overall responsibility for this study. Key persons involved in the study are listed in Appendix B.

The data were collected between 1980 and 1982, and are now being analyzed. Japan, Canada (British Columbia), and the United States have released their national reports. Other countries are doing so at this time. Until that process has been completed, or until the international reports are published, national data are not publicly available.

Components of the Second International Mathematics Study

The conceptual design of the Second International Mathematics Study has three components, as represented in the figure below:

THREE ASPECTS OF A CURRICULUM

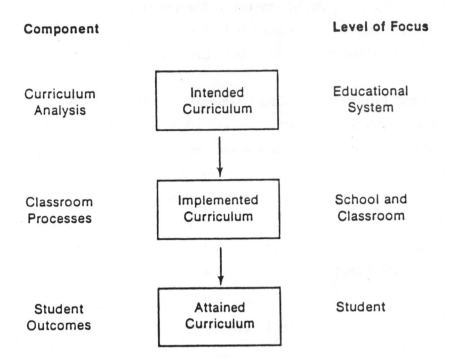

Each of these components is the subject of an international report, scheduled for publication by Pergamon Press, London, in 1986.

The <u>intended</u> curriculum is reflected in curriculum guides, course outlines, syllabuses, and textbooks adopted at the educational system level. In most countries, nationally defined curricula are issued by a ministry of education or similar national body. In the U.S., such statements of intended goals or specifications of curricular content are developed in state Departments of Education or at the local school district level.

The <u>implemented</u> curriculum focuses on the classroom level, the level at which the intended curriculum is taught by the teacher. Teachers may exercise their own judgments in translating curriculum guides or textbooks into an actual program. Thus, their selection of topics or patterns of emphasis may not be consistent with those intended. To identify the implemented curriculum, a number of questionnaires were developed for completion by the classroom teacher. For example, teachers were asked whether they had provided instruction on each of the items in the achievement tests to the target class. They were also asked to provide detailed information on the number of class periods devoted to specific subtopics and specify the interpretations of selected concepts and processes actually utilized. Such highly specific information on curriculum coverage, coupled with similarly detailed information on instructional strategies, yields a rather comprehensive characterization of what mathematics was taught to the target populations and how it was taught.

The third component of the study addressed the attained curriculum — what students had learned as measured by tests and questionnaires. Extensive achievement tests were given to measure student knowledge and skill in areas of mathematics designated as important and appropriate for each population. The "fit" of these tests to the curriculum in individual countries varied substantially and is a limiting factor in international comparisons. Quite obviously, the tests contain items less appropriate in some countries than in others and may not always contain an adequate range of items to fully represent all curricula. Because of this factor, any cross-national comparisons that might be drawn are of necessity quite limited, and similarly the study can hardly be interpreted as an international contest. Student outcome measures also included a number of opinionnaires and attitude scales intended to elicit student views as to the nature, importance, ease, and appeal of mathematics and selected mathematical processes.

This design was implemented in a subset of eight countries (including the U.S.) and includes the following sources of data:

1. Questionnaires completed by school officials concerning school, teacher, and mathematics program characteristics; organizational factors; school and departmental policies affecting mathematics instruction.

2. Questionnaires completed by teachers to provide background information on experience, training, qualifications, beliefs, and attitudes. Additional questionnaires to provide general information on instructional patterns (allocated time, ability of class, classroom organization and activities related to individualization of instruction, resources used; goals and factors affecting instructional decisions) and beliefs about effective teaching. Additional questionnaires related to instruction on selected specific topics.

3. Ratings by teachers (opportunity to learn) as to whether the content needed to respond to each item of the achievement tests had been taught that year, in prior years, or not at all, to their students.

4. Questionnaires completed by students providing background information (e.g., parents' education and occupation), time spent on homework, and attitudes and beliefs related to mathematics.

5. Achievement tests completed by students at the beginning and end of the year.

Target Populations

The study included a survey of the mathematics curriculum provided for two groups of students, identified as target Populations A and B.

> <u>Population A:</u> All students in the grade in which the modal number of students have attained the age of 13.0–13.11 years by the middle of the school year.

In the United States, this corresponded to the eighth grade. In Japan, the seventh grade was chosen. A major reason for Japan's decision was the suitability of the content of the international test for the younger age group.

> <u>Population B:</u> All students who are in the normally accepted terminal grade of the secondary education system and who are studying mathematics as a substantial part (approximately five hours per week) of their academic program.

The Intended Curriculum

A pool of items was developed for the study according to a content-by-behavior grid (one for each target population; see Appendices C and D). The importance ratings, which represent a consensus of ratings between the countries participating in the Study, served as weights for determining how many items to select for the content classifications and behavioral levels in the grid. For Population A, a pool of 199 items was arrived at, and for Population B, 136 items. Each national center was then asked to rate each of these items as

- Not appropriate (0);
- Acceptable (1); or
- Highly appropriate (2).

From these data indices of intended coverage for each of the content areas (for example, arithmetic, algebra, geometry, statistics, and measurement) were first produced, and then an "index of intended coverage" for each country and each content area, expressed as the proportion of items for a given content area judged to be at least appropriate for mathematics students in the given target population. These indices were then used to generate an importance profile of the various content areas across countries. Figure 1 presents these data for Population A.

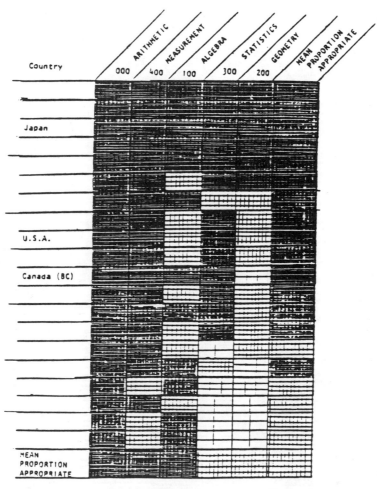

Figure 1. Intended Coverage for Content Areas: Population A
(Note: Darkness is directly proportional to size of coverage index.)

43

The overall impression with respect to appropriateness is that of considerable variation between countries in the mathematical content of the curriculum. The mean index of appropriateness (intended coverage) for the countries is .80, but the range is from .60 to .96. Indices for British Columbia, U.S., and Japan are .81, .82, and .93, respectively. The topics of arithmetic and measurement are most appropriate to the curriculum of the various countries, while geometry is rated least appropriate.

It must always be kept in mind that these ratings, and the data in general, refer to the content of the IEA grids and the IEA tests which were based upon those grids.

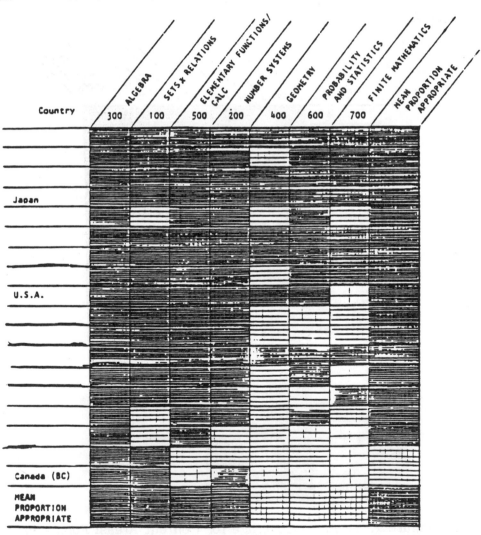

Figure 2. Intended Coverage for Content Areas: Population B

The appropriateness data for Population B are presented in Figure 2. Here the mean index of appropriateness is a little higher — .86 — but the range is also a little greater, from .57 (for British Columbia) to .96. The index for Japan is .93 and for the United States .91. The content areas of algebra, sets and relations, elementary functions/calculus, and number systems share a rather high commonality across the countries, with mean appropriateness indices ranging from .89 to .95 for these areas, respectively.

The Implemented Curriculum

How much mathematics is actually taught was measured by teacher questionnaires. For each of the items on the international test taken by their class, teachers were requested to respond to the following questions:

1. During this school year, did you teach or review the mathematics needed to answer the item correctly?

    a) No   b) Yes

2. If, in this school year, you did <u>not</u> teach or review the mathematics needed to answer this item correctly, was it because:

    a. It had been taught prior to this school year

    b. It will be taught later (this year or later)

    c. It is not in the school curriculum at all

    d. For other reasons

Corresponding to the index of intended coverage there is an index of implemented coverage, which expresses, for a given content area (such as arithmetic), the proportion of items taught prior to or during the year in which the data are collected. Such opportunity-to-learn (OTL) data can be portrayed at the between-country level, as indicated in Figure 3.

The curriculum as taught exhibits great variation between countries — greater than that for the intended curriculum. The OTL data have an overall mean of .62, with indices ranging from a low of .30 to a high of .77 for Japan. The OTL indices for British Columbia and the United Sates are .67 and .68, respectively. The content areas most taught across the countries are arithmetic, measurement, and algebra, which correspond to the intended emphases for these subject matter areas.

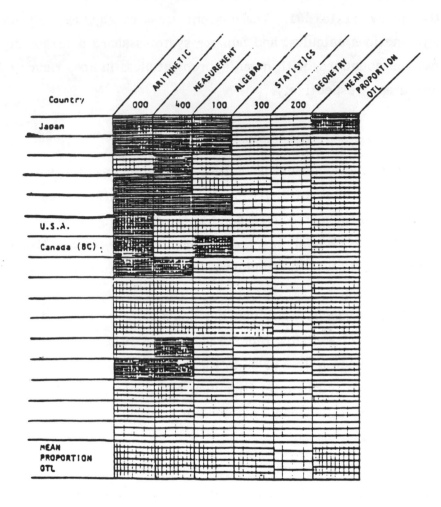

Figure 3. Implemented Coverage (OTL) for Content Areas: Population A

At the Population B level (see Figure 4), the mean OTL index is .73 (somewhat higher than the figure of .62 for Population A). This higher proportion of coverage may reflect the "mathematics specialist" nature of Population B as compared with the broad range of students found in Population A. Mean OTL varied from a low of .51 for British Columbia to a high of .91 for Japan. The content areas of algebra, elementary functions/calculus, and number systems have the highest OTL indices, with values from .84 and .75, respectively.

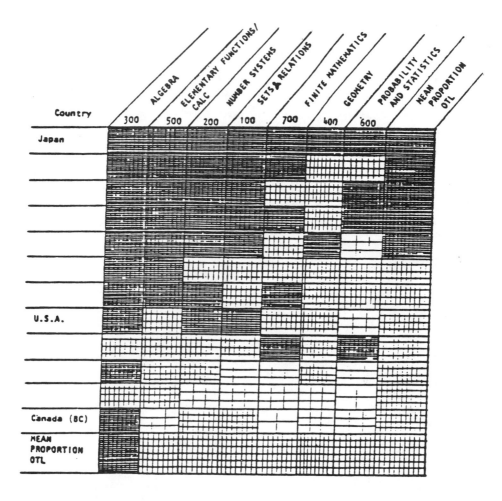

Figure 4. Implemented Coverage for Content Areas: Population B

Relationship between Intended and Implemented Curriculum

The extent to which teacher coverage of various content areas is related to intended coverage may be portrayed by scatter plots. Figures 5 and 6 show plots of intended coverage-versus-taught coverage (OTL) for Population A geometry and algebra, respectively. A low positive relationship is indicated for algebra and a stronger positive relationship for geometry.

For Population B, scatter plots of intended coverage versus opportunity-to-learn are given in Figures 7 and 8 for algebra and elementary functions/calculus,

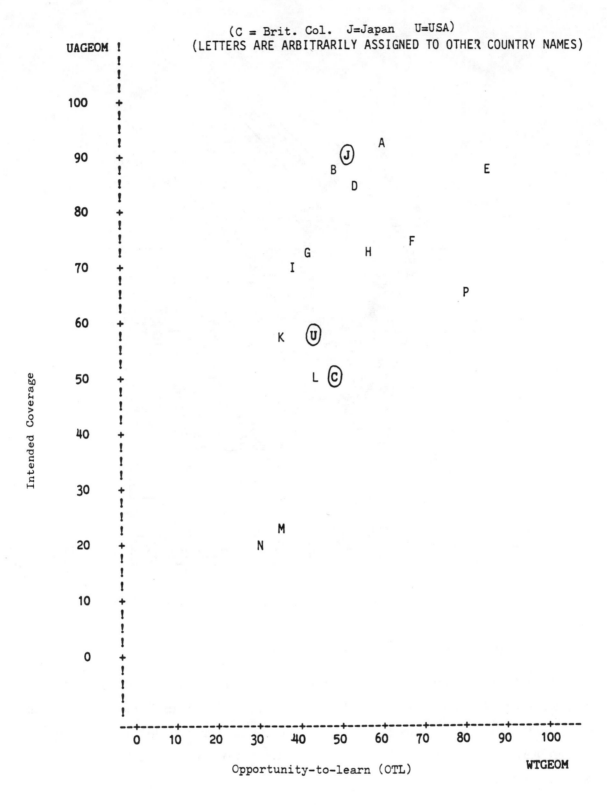

Figure 5. Intended Coverage vs. OTL for Geometry: Population A

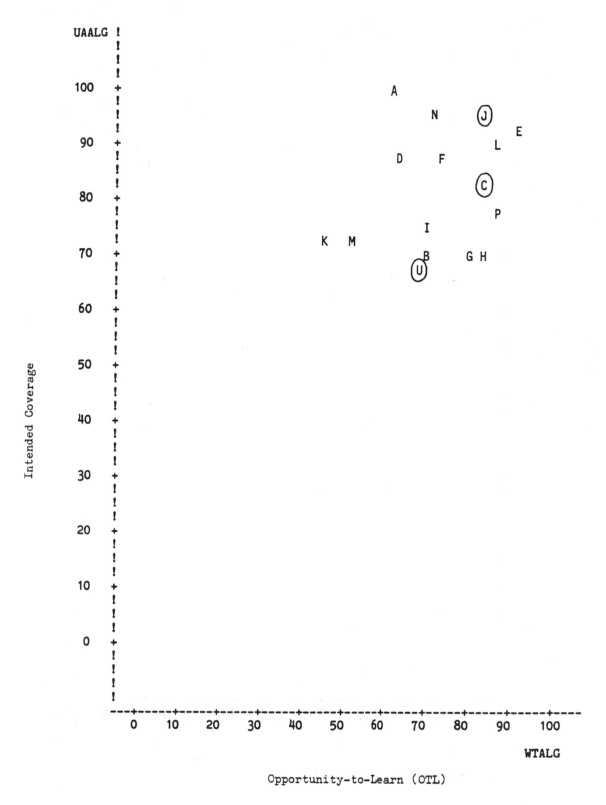

Figure 6. Intended Coverage vs. OTL for Algebra: Population A

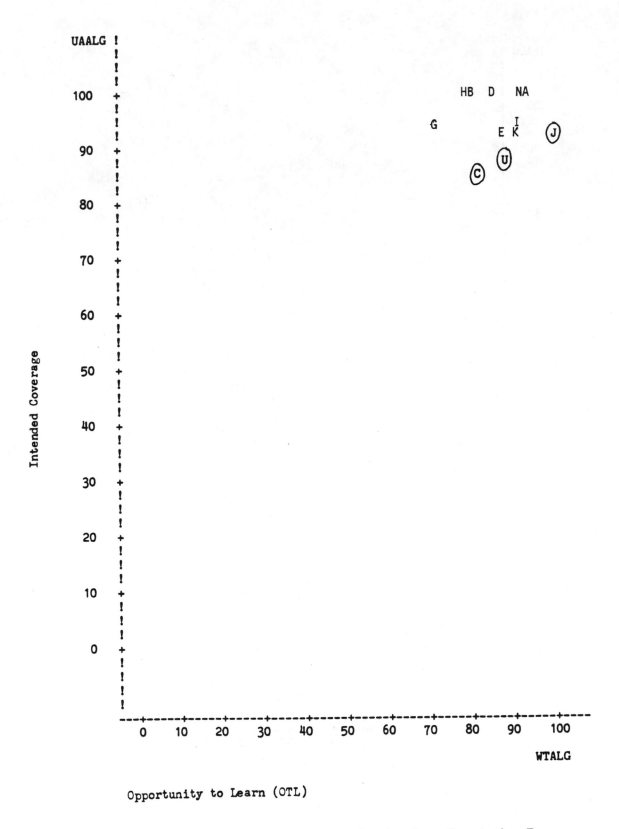

Opportunity to Learn (OTL)

Figure 7. Intended Coverage vs. OTL for Algebra: Population B

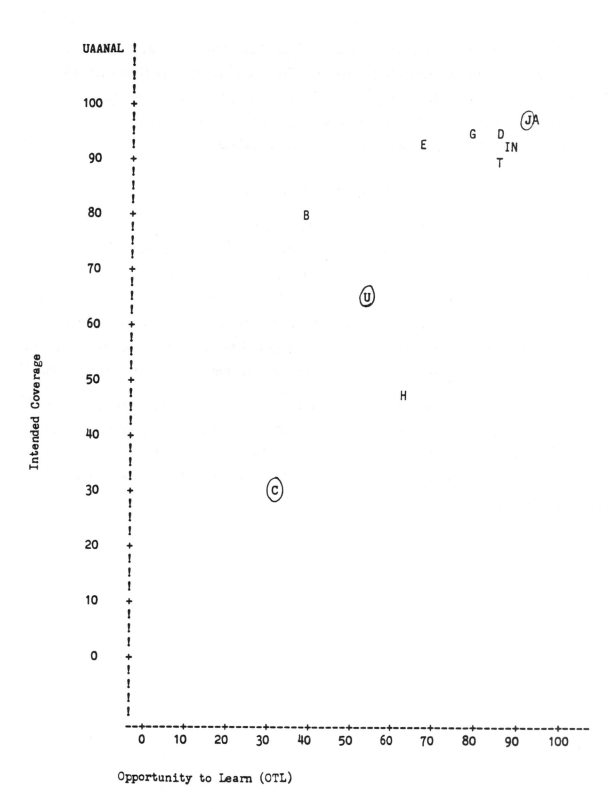

Figure 8. Intended Coverage vs. OTL for Elementary
Functions/Calculus: Population B

respectively. As can be noted in Figure 7, both intended coverage and OTL for algebra are high for almost all countries. For elementary functions/calculus, on the other hand, a rather striking, positive relationship can be seen between intended coverage and OTL, with a small cluster of countries, including Japan and excluding the U.S., having high coverage on both indices.

Concluding Comments

Patterns of variation in the content of the intended and the implemented (taught) curricula exist for both target populations in the Second International Mathematics Study. At the Population A level, there is a significant common core of content for the countries in arithmetic and measurement. In algebra, however, there is considerable between-country variation both in what is intended to be taught at the system level and in what is reported by the teacher as having been taught. A rather high relationship exists between intention and implementation.

Similar patterns were found in Population B. For some topics (e.g., algebra), there is a common core of content at both the intended and implemented levels. For elementary functions and calculus, on the other hand, countries differ considerably in emphasis. There is again, however, a rather strong association between intention and implementation.

Appendix A

The International Association for the Evaluation of
Educational Achievement

The International Association for the Evaluation of Educational Achievement (IEA) is an international, nonprofit scientific association incorporated in Belgium for the principal purposes of: (a) undertaking educational research on an international scale; (b) promoting research aimed at examining educational problems in order to gather information that can help in the ultimate improvement of educational systems; and (c) providing the means whereby research centers in the various member countries of IEA can undertake cooperative projects. The current chairman of the IEA Council is T. Neville Postlethwaite of the University of Hamburg, Federal Republic of Germany.

The Mathematics Project Council, which has been responsible for the Second Mathematics Study, is chaired by Roy W. Phillipps of the New Zealand Department of Education. Robert Garden, also of the New Zealand Department of Education, is International Project Co-ordinator for the Study. Kenneth J. Travers is Chairman of the International Mathematics Committee (IMC) which designed the Second International Mathematics Study and developed the international instruments. Other members of the IMC are: Sven Hilding, Sweden; Edward Kifer, United States; Gerard Pollock, Scotland; Tamas Varga, Hungary; and James Wilson, United States. A.I. Weinzweig, United States, is consulting mathematician and Richard Wolfe, Canada, is consulting psychometrician to the IMC.

Appendix B
Participating Countries

Country	National Research Coordinator	Council Member	Center
Australia*	Malcolm Rosier	John Keeves	Australian Council for Educational Research, Hawthorn, Victoria
Belgium (Flemish)	Christiana Brusselmans-Dehairs	A. De Block (1976-81) J. Heene	Seminaire en Laboratorium voor Didactiek, Gent
Belgium* (French)	George Henry	Gilbert De Landsheere	Universite de Liege au Sart Tilman
Canada (B.C.)	David Robitaille	Joyce Matheson	Ministry of Education, Victoria
Canada (Ontario)	Leslie McLean	Bernard Shapiro	Ontario Institute for Studies in Education, Toronto
Chile	Maritza Jury Sellan	Marino Pizarro	University of Chile, Santiago
England*/ Wales	Michael Cresswell (1978-83) Derek Foxman	Clare Burstall	National Foundation for Educational Research, Berkshire
Finland*	Erkki Kangasniemi	Kimmo Leimu	Institute for Educational Research, University of Jyvaskyla, Jyvasklyla
France*	Daniel Robin	Daniel Robin	Institut National de Recherche Pedagogique, Paris
Hong Kong	Patrick Griffin (1976-83)	M.A. Brimer	University of Hong Kong, Hong Kong
Hungary	Julia Szendrei	Zoltan Bathory	Orszagos Pedagogiai Intezet, Budapest
Ireland	Elizabeth Oldham	Elizabeth Oldham	Trinity College, Dublin

Israel*	Arieh Lewy (1976-83) David Nevo	Arieh Lewy (1976-83) David Nevo	Tel-Aviv University, Tel-Aviv
Ivory Coast	Sango Djibril	Ignace Koffi	Service D'Evaluation, Abidjan
Japan*	Toshio Sawada	Hiroshi Kida	National Institute for Educational Research, Tokyo
Luxembourg	Robert Dieschbourg	Robert Dieschbourg	Institut Pedagogique, Walferdange
Netherlands*	Hans Pelgrum	Egbert Warries	Twente University of Technology, Enschede
New Zealand	Athol Binns	Roy W. Phillipps	Department of Education, Wellington
Nigeria	Wole Falayajo	E.A. Yoloye	University of Ibadan, Ibadan
Scotland*	Gerard Pollock	W. Bryan Dockrell	Scottish Council for Research in Education, Edinburgh
Swaziland	Mats Eklund (1976-81) P. Simelane	Mats Eklund (1976-81) P. Simelane	William Pitcher College, Manzini
Sweden*	Robert Liljefors	Torsten Husen (1976-83) Inger Marklund	University of Stockholm, Stockholm
Thailand	Samrerng Boonruangrutana	Pote Sapianchai	Institute for the Promotion of Teaching Science and Technology, Bangkok
United States*	Curtis McKnight	Richard Wolf	Teachers College, Columbia University New York

*Also participated in First International Mathematics Study. The Federal Republic of Germany took part in the First Study only.

Appendix C
Population A: Importance of Content Topics and
Behavioral Categories for Instrument Construction

Content Topics	Behavioral Categories* Computation	Comprehension	Application	Analysis

000 Arithmetic

		Computation	Comprehension	Application	Analysis
001	Natural numbers and whole numbers	V	V	V	I
002	Common fractions	V	V	I	I
003	Decimal fractions	V	V	V	I
004	Ratio, proportion, percentage	V	V	I	I
005	Number theory	I	I	-	-
006	Powers and exponents	I	I	-	-
007	Other numeration systems	-	-	-	-
008	Square roots	I	I	-	-
009	Dimensional analysis	I	I	-	-

100 Algebra

		Computation	Comprehension	Application	Analysis
101	Integers	V	V	I	I
102	Rationals	I	I	I	I
103	Integer exponents	Is	-	-	-
104	Formulas and algebraic expressions	I	I	I	I
105	Polynomials and rational expressions	I	Is	-	-
106	Equations and inequalities (linear only)	V	I	I	Is
107	Relations and functions	I	I	I	-
108	Systems of linear equations	-	-	-	-
109	Finite systems	-	-	-	-
110	Finite sets	I	I	I	-
111	Flowcharts and programming	-	-	-	-
112	Real numbers	-	-	-	-

*The following rating scale has been used: V = very important; I = important; Is = important for some countries. A dash (–) indicates that the topic was not considered important enough to warrant locating or constructing trial items.

Appendix C (cont.)

Content Topics	Behavioral Categories* Computation	Comprehension	Application	Analysis
200 Geometry				
201 Classification of plane figures	I	V	I	Is
202 Properties of plane figures	I	V	I	I
203 Congruence of plane figures	I	I	I	Is
204 Similarity of plane figures	I	I	I	Is
205 Geometric constructions	Is	Is	Is	–
206 Pythagorean triangles	Is	Is	Is	–
207 Coordinates	I	I	I	Is
208 Simple deductions	Is	I	I	I
209 Informal transformations in geometry	I	I	I	–
210 Relationships between lines and planes in space	–	–	–	–
211 Solids (symmetry properties)	Is	Is	Is	–
212 Spatial visualization and representation	–	Is	Is	–
213 Orientation (spatial)	–	Is	–	–
214 Decomposition of figures	–	–	–	–
215 Transformational geometry	Is	Is	Is	–
300 Statistics				
301 Data collection	Is	I	I	–
302 Organization of data	I	I	I	Is
303 Representation of data	I	I	I	Is
304 Interpretation of data (mean, median, mode)	I	I	I	–
305 Combinatorics	–	–	–	–
306 Outcomes, sample spaces and events	Is	–	–	–
307 Counting of sets, (P(A B), P(A B), independent events	–	–	–	–
308 Mutually exclusive events	–	–	–	–
309 Complementary events	–	–	–	–
400 Measurement				
401 Standard units of measure	V	V	V	–
402 Estimation	I	I	I	–
403 Approximation	I	I	I	–
404 Determination of measures: areas, volumes, etc.	V	V	I	I

Appendix D
Population B: Importance of Content Topics and
Behavioral Categories for Instrument Construction

Content Topics	Behavioral Categories* Computation Comprehension Application Analysis			
100 Sets and relations				
101 Set notation	I	I	–	–
102 Set operations (e.g., union, inclusion)	I	I	–	–
103 Relations	–	–	–	–
104 Functions	V	V	V	I
105 Infinite sets, cardinality and cardinal algebra (rationals and reals)	–	–	–	–
200 Number systems				
201 Common laws for number systems	I	I	I	–
202 Natural numbers	I	I	I	I
203 Decimals	I	I	I	I
204 Real numbers	I	I	I	–
205 Complex numbers	V	I	I	I
300 Algebra				
301 Polynomials (over R)	V	V	V	I
302 Quotients of polynomials	I	I	I	–
303 Roots and radicals	V	V	I	–
304 Equations and inequalities	V	V	V	I
305 System of equations and inequalities	V	V	V	I
306 Matrices	Is	Is	Is	Is
307 Groups, rings and fields	–	–	–	–

Content Topics	Behavioral Categories* Computation	Comprehension	Application	Analysis
400 Geometry				
401 Euclidean (synthetic) geometry	I	I	–	–
402 Affine and projective geometry in the plane	–	–	–	–
403 Analytic (coordinate) geometry in the plane	I	I	V	I
404 Three–dimensional coordinate geometry	–	–	–	–
405 Vector methods	I	I	I	I
406 Trigonometry	V	V	V	I
407 Finite geometries	–	–	–	–
408 Elements of topology	–	–	–	–
409 Transformational geometry	Is	Is	Is	Is
500 Elementary functions and calculus				
501 Elementary functions	V	V	V	V
502 Properties of functions	V	V	V	I
503 Limits and continuity	I	I	I	–
504 Differentiation	V	V	I	I
505 Applications of the derivative	V	V	V	I
506 Integration	V	V	V	I
507 Techniques of integration	V	V	I	I
508 Applications of integration	V	V	V	I
509 Differential equations	Is	Is	Is	Is
510 Sequences and series of functions	–	–	–	–
600 Probability and statistics				
601 Probability	V	V	I	–
602 Statistics	I	I	I	–
603 Distributions	I	I	I	–
604 Statistical inference	Is	Is	–	–
605 Bivariate statistics	–	–	–	–
700 Finite mathematics				
701 Combinatorics	I	I	I	–
800 Computer science	Is	Is	I	–
900 Logic	–	–	–	–

59

Notes

1. For a list of countries involved, see Appendix B. The First International Mathematics Study, involving twelve countries, took place in 1964.

2. For information on the IEA, see Appendix A.

References

McKnight, Curtis C., Kenneth J. Travers, F. Joe Crosswhite, and Jane O. Swafford. 1985 (April). "Eighth-Grade Mathematics in U.S. Schools: A Report From the Second International Mathematics Study." Arithmetic Teacher 32(8): 20-26.

McKnight, Curtis C., Kenneth J. Travers, and John A. Dossey. 1985 (April). "Twelfth-Grade Mathematics in U.S. High Schools: A Report From the Second International Mathematics Study." Mathematics Teacher 78(4): 292-300 (cont. 270).

Murray, Asa and Robert Liljefors. 1983. Matematik i Svensk Skola. Skoloverstyrelsen.

Pelgrum, H., T. Eggen, and T. Plomp. 1984. Results of the Second International Mathematics Study in the Netherlands. Enschede.

Robitaille, David F., J. O'Shea Thomas, and Michael Dirks. 1982. The Teaching and Learning of Mathematics in British Columbia. Victoria: Ministry of Education, Learning Assessment Branch.

Sawada, Toshio, ed. 1981-82. Mathematics Achievement of Secondary School Students. 2 vols. Tokyo: National Institute for Educational Research.

Sawada, Toshio, ed. 1983. Mathematics Achievement and Teaching Practice in Lower Secondary Schools (Grade 7), vol. 3. Tokyo: National Institute for Educational Research.

Travers, Kenneth J. and Curtis C. McKnight. 1985 (February). "Mathematics Achievement in U.S. Schools: Preliminary Findings From the Second IEA Mathematics Study." Phi Delta Kappan: 407-413.

Travers, Kenneth J., Curtis C. McKnight, F. Joe Crosswhite, and Jane O. Swafford. 1985 (September). "Eighth-Grade Math: An International Study." Principal 65(1): 37-40.

Travers, Kenneth J., Curtis C. McKnight, and John A. Dossey. 1985 (November). "Mathematics Achievement in U.S. High Schools From an International Perspective." Bulletin: 55-63. National Association of Secondary School Principals.

United States. 1985. United States Summary Report: Second International Mathematics Study. Champaign, IL: Stipes Publishing Company.

The Second International Mathematics Study:
A Look At U.S. Classrooms

F. Joe Crosswhite

Ohio State University

John A. Dossey

Illinois State University

In 1981-82, students and teachers in about 500 mathematics classrooms across the United States joined their counterparts in some two dozen countries around the world in a comprehensive study of school mathematics. This study, the Second International Mathematics Study, was designed to provide detailed information from each country about the content of the mathematics curriculum, how mathematics is taught, and how much mathematics students learn. It was hoped that this information would help mathematics educators in the individual nations analyze their school programs, identify areas of strength and weakness, and plan for the future.

The Study (hereafter referred to as SIMS) was conducted by members of the International Association for the Evaluation of Educational Achievement, a network of leading education research organizations in the various countries. The study was the second to take place in mathematics in a twenty-year period. In 1964, twelve countries, including the United States, took part in a similar study, the First International Mathematics Study (hereafter referred to as FIMS).

Components of the Study

The conceptual design for SIMS has three components.

The intended curriculum is reflected in curriculum guides, course outlines, and syllabuses and textbooks adopted or endorsed at the national, or system-wide level. In the U.S., such statements of intended goals are difficult to describe, as they most often occur at the state or local district level.

The implemented curriculum focuses on the classroom level, at the point where the teacher is responsible for realizing the intended curriculum. It is here that teachers exercise their own judgment about the selection of topics, their appropriateness, and patterns of emphasis.

The third component of the study addresses the <u>attained curriculum</u>, the study of what students learned, as measured by tests and questionnaires. Extensive achievement tests were designed to measure student knowledge and skill in areas of mathematics designated as important and appropriate for each population. The "fit" of these tests to the curriculum in individual countries varied substantially. This variation places limitations on the kind of cross-national comparisons that can be drawn and also argues against interpreting the results of SIMS as international competition. Student outcome measures also included a number of opinionnaires and attitude scales. These were intended to elicit student views as to the nature, importance, ease, and appeal of mathematics and selected mathematical processes.

Target Populations

The study included a survey of the mathematics curriculum provided for two groups of students, identified as target Populations A and B. The students in Population A were enrolled in normal mathematics classes in the grade with modal age of thirteen; in the United States, this was the eighth grade. Population B consisted of students enrolled in advanced mathematics classes in the terminal year of secondary school, which in the United States consisted of students enrolled in classes intended for twelfth graders in the areas of pre-calculus, trigonometry, analytic geometry, and advanced placement calculus.

United States Sample

The United States sample for the Study was composed of students and teachers from approximately 500 classrooms in about 250 public and private schools across the United States. These classrooms were selected through a stratified random sampling process similar to that used in National Assessment Studies.

One matter of concern in an international study of schooling is whether the other countries involved have comparable proportions of their students in the target populations defined. At Population A, virtually all students are still in

63

school in all the countries in the study (see Appendix for participating countries). At the Population B level, there is more variation. The information in Table 1 presents for several of the countries in the Study the percentage of the age cohort in Population B, the percentage of the grade cohort in Population B, and the percentage of the age cohort still in school.

Table 1
Percentage of Population B Students in Different Age Groups
and Grade for Each Country: 1981

Country	Age group in Population B	Grade Group in Population B	Age Group in School
Belgium	9–10	25–30	65
Canada (British Columbia)	30	38	82
Canada (Ontario)	19	55	33
England	6	35	17
Finland	15	38	59
Hungary	50	100	50
Israel	6	10	60
Japan	12	13	92
New Zealand	11	67	17
Scotland	18	42	43
Sweden	12	50	24
U.S.A.	10–12	12–15	85

While the U.S. is among the leaders in the percentage of the age group still in school, its Population B is seen to be among the more selective. This is especially true if one considers only the 2 to 3 percent group in the advanced placement calculus classes.

Eighth Grade Findings

Class Types in the Eighth Grade Sample

Teachers were asked to characterize the main subject matter taught in the sampled classes as remedial, typical, or enriched. On the basis of course and textbook titles supplied by teachers, a fourth category of classes, first-year algebra, was identified, largely from those classes identified as enriched.

Of the 236 classes in the final eighth-grade sample, 155 (66%) were classified as typical, 24 (10%) as remedial, 26 (11%) as enriched, and 31 (13%) as algebra classes. While these proportions may be representative, the actual numbers of all class types other than the typical classes are so small that generalizations for these class types must be considered tentative.

Class Size

Population A mathematics classes had a median size of 26 students, with 70% of the classes having between 20 and 30 students enrolled. The median size of the remedial classes was 21 students.

Class Periods and Hours of Instruction

Mathematics was taught 5 days per week with a median class length of 50 minutes. This resulted in a median number of 145 clock hours of instruction per year. Ninety percent of the classes had between 115 and 180 hours of instruction during the school year.

Teaching Assignments

The typical weekly teaching assignment for the eighth-grade teacher included 25 to 30 class periods. Approximately 25% of the teachers had an additional teaching assignment outside of mathematics, but few of these were in the area of a science.

Teacher Characteristics

The sample for the Study was intended to be representative of U.S. classrooms at each of the two grade levels studied. Thus, the following teacher characteristics and teaching modes should represent those present in a typical U.S. classroom at the two different grade levels.

Gender

The ratio of female to male teachers for the national sample of eighth-grade teachers was almost 1 to 1. However, a slightly higher percentage of teachers of algebra classes were women (55%) and a markedly greater percentage of teachers of enriched classes were men (63%).

Age/Experience/Training

According to median data, the typical eighth-grade mathematics class had a teacher who was experienced and welltrained. This teacher was about 37 years old with 13 years of experience in teaching, eight of which involved teaching classes similar to the target class. This teacher's collegiate record included nine or ten semester courses in mathematics, two courses in the teaching of mathematics, and four courses in more general methods and pedagogical issues.

That the teaching corps was experienced and well trained is demonstrated by the age and experience of this sample: The national sample included relatively few younger teachers (only 15% below the age of 30), few inexperienced teachers (only 15% with less than seven years of experience), and few with limited training in mathematics (less than 20% with fewer than five semester courses and only 4% with fewer than three semester courses).

Beliefs About Effective Teaching

The teachers were asked to rate a series of 41 items concerning effective teaching. The teachers emphasized factors such as establishing and enforcing clear-cut rules for acceptable student behavior (rated very important by 67% of the teachers); making encouraging remarks to individuals as they worked (62%); getting materials, equipment, and space ready before class (55%); and reviewing tests with students shortly after they have been graded (55%). They tended to rate much lower specific techniques of instruction such as outlining and summarizing lessons, planning transitions, calling on non-volunteers, discussing feelings directly, etc.

Goals in Teaching Mathematics

Teachers were asked to rate the relative emphasis that should be given to each of nine objectives in mathematics instruction. The highest rated goals were developing a systematic approach to solving problems and developing an awareness of the importance of mathematics in everyday life (rated as relatively more important by 63% and 61% of the teachers responding). These were followed closely by performing computations with speed and accuracy (58%) and knowing mathematical facts, principles, and algorithms (55%).

The results differed for some of the class types separately. Teachers of remedial classes reported that their most important goal was developing an awareness of the importance of mathematics to everyday life (84%), followed by the goals of knowing facts, principles, and algorithms (64%); or performing computations with speed and accuracy (60%) and becoming interested in mathematics (60%). Problem-solving was in fifth place (48%), rather than in first.

Teachers of other eighth-grade classes reported that their most important goal was problem-solving (69%) followed by the goals of developing an attitude of inquiry (48%), understanding the logical structure of mathematics (48%), becoming interested in mathematics (41%), and performing computations with speed and accuracy (41%).

How Eighth-Grade Teachers and Students Spent Their Time

There was considerable variation in total time commitment and the way time was used by the teachers participating. Making a profile of the typical teacher, one might describe the typical week as having one to two hours spent outside of class preparing for class and another one to two hours spent on grading papers. Routine administrative duties such as taking attendance, making announcements, and setting up equipment required less than one-half hour per week. The typical teacher spent from one to two hours of class time each week explaining content new to the students and about half as much time reviewing old material. This target class was, in general, only one of five classes taught by the teacher. Hence, these time estimates must be multiplied by some factor of the total number of classes (preparations) to profile the entire time load on the teacher per week.

Teachers also estimated the average time spent by their students in various classroom activities. These reports showed that the majority of student time was spent in seatwork, blackboard work, or listening to lectures or explanations. A median of 40 minutes, about one class period per week, was spent taking tests and quizzes. Very little time was spent in individual or small group work. This pattern differed very little from one type of class to another.

Homework

The teachers estimated that students in their classes typically spent 2.3 hours per week outside of class on assigned homework. Students from 75% of the classes were estimated as spending three or less hours per week on homework.

Use of Calculators

Use of calculators in mathematics classes was low. Only 4% of the teachers reported calculator use for two or more periods per week. About two-thirds of the teachers reported calculators were never used or were not allowed in mathematics classes. When used, the most common applications of the calculator in the eighth grade were in recreational settings, checking exercises, or conducting projects. Six percent of the classes reported the use of calculators during test taking.

What Mathematics Was Taught

Teachers reported the approximate number of class periods they expected to have spent by the end of the year on each of a selected list of topics. The list shows a great variation in the content of eighth-grade mathematics taught, as shown in table 2. Generally speaking, topics in arithmetic predominated, with a median of 30 class periods devoted to common fractions and decimals. Relatively little time was given to probability and statistics. More detailed studies show that this coverage was fragmented into a number of smaller pieces. The international classroom process data indicates less fragmentation in some other countries. A frequent U.S. pattern of instruction was that of a large proportion of teachers devoting only a single lesson to a topic. The result was a "low intensity" coverage of many topics in the eighth-grade curriculum.

In the algebra classes, the teacher reports indicated that very limited attention was given to topics outside the content of algebra. When it was, it tended to focus on quasi-algebraic aspects of the topic area, i.e., coordinates in geometry, algebraic forms of formulas, and use of equations in solving arithmetic problems.

The distinctions between the items taught in typical and enriched classes were not as great. The enriched classes seemed to spend less time on arithmetic topics, devoting more attention to formulas and equations.

Table 2

Median Anticipated Number of Periods for Selected Topics
in Eighth-Grade Classes by Class Type: 1981-82

Median Number of Periods

Topic	Total Sample	Remedial	Typical	Enriched	Algebra
Common Fractions	15	20	20	10	2
Decimal Fractions	15	20	15	10	1
Ratio and Prop.	10	10	10	6	4
Percentages	15	15	15	10	2
Measurement	10	10	12	10	1
Geometry	15	10	15	15	1
Formulas/Equations	20	2	20	26	50
Integers	10	10	15	10	8
Prob. and Stat.	4	0	5	5	0
Classes per Type	236	24	155	26	31
Periods Noted*	114	97	127	102	69

* Note that the topics identified in the list account for much less time than is found in the total year — especially in the algebra classes.

The curriculum for the remedial classes was clearly characterized by a strong emphasis on common and decimal fractions. Fifty percent of these classes saw no treatment of algebraic topics, while 25% saw little if any work with geometry, measurement, or descriptive statistics.

What Mathematics Is Learned

Students were tested in the following five content areas: arithmetic, algebra, geometry, measurement, and statistics. The same items were taken by the students at the beginning and end of the 1981–82 school year, thus providing a measure of how much growth occurred over the span of the year. Table 3 contains information showing teachers' indications of their students' opportunity to learn the material reflected in the test items (which had been taught to them in the eighth grade or earlier), their pre- and post-test scores, gain scores, and the international post-test scores.

Table 3
Eighth–Grade Mean Opportunity to Learn and Achievement by Subtopic:
U.S. and International, 1981–82

| Topic (# of Items) | U.S. | | | | |
	Opportunity to Learn[1,2]	Mean Pretest Score	Mean Post-test Score	Mean Post-test Score for 19 Countries[2]	Mean OTL Score for 19 Countries[3]
Arithmetic (62)	84	42	51	50	73
Algebra (32)	68	32	43	43	67
Geometry (42)	44	31	38	41	45
Measurement (26)	74	35	42	51	73
Statistics (18)	70	53	57	55	51

(1) Opportunity-to-learn by the end of eighth grade — that is, up to and including eighth grade.
(2) The international means are based on a restricted set of 157 items in common between the international test and the U.S. National version of the test. The number of items by topic on the 157-item international test were: Arithmetic, 46; Algebra, 30; Geometry, 39; Measurement, 24; and Statistics, 18. In all cases the U.S. results differ less than 1% from those in the table above when restricted to the set of 157 items. The countries included, in addition to the United States, were: Belgium (Flemish); Belgium (French); Canada (British Columbia); Canada (Ontario); England and Wales; Finland; France; Hong Kong; Hungary; Israel; Japan; Luxembourg; Netherlands; New Zealand; Nigeria; Scotland; Sweden; Thailand.
(3) The international mean OTL was 62, the mean U.S. OTL was 68.

Table 3 (cont.)

Eighth–Grade Mean Achievement of Each Content Area by Class Type:
U.S., 1981–82

Mean Percentages by Class Type, Pre- and Post-test

Content Area (# of items)	Remedial				Typical		
	Pre	Post	Difference		Pre	Post	Difference
Arithmetic (62)	25	32	+ 7		38	47	+ 9
Algebra (32)	20	24	+ 4		28	38	+ 10
Geometry (42)	22	26	+ 4		28	35	+ 7
Measurement (26)	23	27	+ 4		32	39	+ 7
Statistics (18)	34	39	+ 5		50	55	+ 5
All items (180)	24	29	+ 5		34	42	+ 8

Content Area (# of items)	Enriched				Algebra		
	Pre	Post	Difference		Pre	Post	Difference
Arithmetic (62)	54	64	+ 10		67	72	+ 5
Algebra (32)	40	57	+ 17		56	70	+ 14
Geometry (42)	39	50	+ 11		47	53	+ 6
Measurement (26)	43	52	+ 9		53	57	+ 4
Statistics (18)	65	68	+ 3		74	75	+ 1
All items (180)	48	58	+ 10		59	65	+ 6

Averaging across the entire set of 180 items, only 70% of the teachers reported that the relevant mathematics had been taught by the end of the eighth grade. This limited coverage was most notable in the area of geometry, where less than one-half of the items had been taught by this time.

The average achievement of students in the U.S. in arithmetic, algebra, and probability and statistics was at the average level of performance for all the countries. Achievement was slightly above the international mean for computational arithmetic (ability to calculate), but below the international mean for arithmetic items involving comprehension and the ability to solve problems. Achievement in geometry for the U.S. was low, exceeded by three-fourths of the other countries. The overall OTL was about the same as that for other countries. Within geometry, however, knowledge of transformational geometry, which is not a part of the curriculum in many U.S. schools, was at the international average. Achievement in measurement was very low compared to other countries. While the international test used metric units for all items involving units of measure, this was not sufficient to completely explain the low level of performance, since many items involved measurement concepts in a way that did not require knowledge of metric system measures.

Twelfth-Grade Findings

Class Types in the Twelfth-Grade Sample

To simplify the diversity of curricular patterns noted at the twelfth-grade level, the 237 classrooms in the U.S. Population B sample were classified as either precalculus or calculus classes. The precalculus classes, 191 in number (81%), were those that focused on topics drawn from "college" algebra, analytic geometry, and elementary functions. The calculus classes, 46 in number (19%), were those that completed a full-year course in calculus, following the outline provided by the College Entrance Examination Board Advanced Placement syllabi for the topic. This was determined by an examination of the textbooks used, the teachers' responses to classroom processes questionnaires, and coverage of topics in the calculus.

Class Size

The student enrollment in the sample classes had a median size of 20. The male-female breakdown of the overall Population B student sample showed 57% of the students to be male and 43% female.

Class Periods and Hours of Instruction

Mathematics was taught five days per week with a median class length of 54 minutes. There was a slight difference in the total yearly median clock hours of instruction received by the precalculus and calculus students. The precalculus classes had a median of 155 clock hours of instruction, and the calculus classes, 150 clock hours.

Teaching Assignments

The teaching loads in the twelfth grade varied more than at the eighth-grade level. The most typical teaching load was five classes per day. However, the precalculus teachers ranged from two to seven periods of teaching per day and the calculus teachers from one to six periods per day. Very few teachers in either category taught anything outside of mathematics. When they did, it was invariably in one of the sciences. A large proportion of the calculus teachers (44%) also served as department chairs for their schools.

Teacher Characteristics

As at the eighth-grade level, the sample of classes was selected in such a way as to be representative of the typical twelfth-grade mathematics class nationwide. Thus, the following teacher characteristics and teaching modes should represent those in the typical twelfth-grade classroom.

Gender

The teachers participating in the Population B phase of the Study were almost evenly divided between male (52%) and female (48%) teachers. However, some differences were noted in their assignments in the two types of curricula. In the precalculus classes, the ratio of male to female teachers was nearly 1 to 1. In the calculus classes the ratio of male to female teachers was nearly 2 to 1.

Age/Experience/Training

According to median data, the typical twelfth-grade mathematics class had a teacher who was experienced and well trained. This teacher was about 41 years of age with 18 years of teaching experience. Ten of these years had involved teaching mathematics at the Population B level. The teacher's academic background showed a median of 16 semester courses in mathematics, two semester courses in the teaching of mathematics, and four semester courses in general methods and pedagogical issues. Unlike the eighth-grade teachers, about 8% of the teachers showed little or no preparation in the areas of teaching mathematics or general pedagogy. However, all of the teachers were reported to certified in mathematics.

Beliefs About Effective Teaching

Like the eighth-grade teachers, the twelfth-grade teachers were asked to respond to a set of items dealing with possible components of effective teaching. Here there were differences in the level of endorsement given statements by teachers in the two class types. Seven of the maxims were selected as among the highest in importance by at least 50% of the precalculus teachers, while only one was selected by at least 50% of the calculus teachers. Both groups endorsed the maxim that one should review tests with students shortly after they have been graded and this was the only maxim both groups saw as important. Sixty-five percent of the precalculus teachers and 63% of the calculus instructors were of this opinion.

Other statements selected as important by precalculus teachers dealt with making encouraging remarks to students (53%), establishing and enforcing clear-cut rules of classroom behavior (54%), clearing up problems from a previous lesson (60%), and avoiding criticism of a student in front of the class (51%). These responses were also among the highest rated by calculus teachers, but their response levels indicate a more conservative approach to endorsing general maxims for effective teaching.

Goals in Teaching Mathematics

Teachers were asked to rate the relative emphasis given to a set of nine goals characterizing objectives for the teaching of mathematics. The precalculus teachers chose to give relatively more emphasis in their target classes to the development of systematic approaches to problem-solving; knowing mathematical facts, principles, and algorithms; and understanding the logical structure of mathematics.

Like the eighth-grade teachers, both of the groups gave their lowest emphasis rating to the goal of understanding the nature of proof. Only 12% of the precalculus teachers and 15% of the calculus teachers gave it relatively more emphasis than other objectives, while 38% and 41% of the same groups of teachers, respectively, noted that it received relatively less emphasis than the other goals with their target classes.

How Twelfth-Grade Teachers and Students Spent Their Time

The teachers were asked to estimate the number of minutes per week devoted to generic teaching tasks related to the teaching of their target classes. The median precalculus teacher spent about two hours in preparing instruction, 1.5 hours in grading papers, two hours in explaining new content to the class, one hour reviewing content taught previously, and about a half-hour on classroom administrative and management details. The median calculus teacher followed much the same pattern, except the time allocations show an additional half-hour of preparation, only 50 minutes for reviewing previously taught material, and 15 minutes on classroom administrative and management tasks. As these responses deal only with a teacher's sampled class, they need to be multiplied by a factor related to the total number of classes (preparations) to get the total load borne by the teacher.

Overall, about 40% of the total allocated in-class instructional time was devoted to developing new material, 20% to reviewing previously taught material, 10% percent to administrative/management tasks, and 30% to supervising student work in the classroom.

Like the eighth-grade students, the students at the twelfth-grade level spent the major portion of their class time listening to teacher presentations (130 minutes a week on the average). Doing seatwork (60 minutes per week) and taking tests and quizzes (45 minutes per week) accounted for other large blocks of time. While on average little time was spent on small-group work, there was a great range in time devoted to this activity across the sampled classrooms. Some classes spent as much as 80 minutes per week on small group activities while others spent virtually none.

Homework

Homework expectations differed markedly between the precalculus and the calculus classes. In precalculus classes, teachers reported that the student in a typical class was expected to do about four hours of homework per week, with the middle 50% of the students expected to do between three and five hours. In calculus classes, teachers reported that the students were expected to complete five hours of homework per week, with the middle 50% of all classes having from four to eight hours of homework per week.

Use of Calculators

The use of calculators was more prevalent at the senior high-school level than at the eighth-grade level. About 33 percent of the classes used them two or more periods per week. Another 28 percent of the classes used them one period per week or less. Teachers of 11% of the classes reported that their students did not use calculators, 9% indicated that calculators were not allowed in their classes, and 20% failed to provide information.

In the twelfth-grade, the most commonly reported instructional uses of calculators in the classroom were checking work, doing homework, and solving mathematics problems in class. In contrast to eighth graders, about 50% of the twelfth graders were allowed to use calculators on examinations.

Applications in the Twelfth Grade Curriculum

Teachers' responses to extensive classroom processes questionnaires reflected their sources of, feelings about, and extent of use of applications in teaching twelfth-grade mathematics.

The 161 precalculus teachers responding to the classroom processes questionnaire in the area of trigonometry felt that applications are an important objective in trigonometry, 54% of them reporting that the study of trigonometric applications are a primary goal for their courses. When asked to describe these applications, the teachers reported that they were basically selected from surveying (61%), vector-related situations (43%), and periodic motion (41%).

Of the 148 precalculus teachers responding to the classroom processes questionnaire in the area of college algebra, 73% agreed that applications should play a greater role in college algebra than development of proof skills. However, their responses to other items dealing with applications of logarithmic functions and complex numbers indicated that they actually taught very few applications of college algebra topics.

The 46 calculus teachers reported a greater emphasis on applications, but were somewhat split on a question that asked about the relative merits of applications versus theory in the initial calculus course. Sixty-two percent favored more emphasis on applications, while 32% supported more emphasis on theory in the initial calculus courses. When asked about the breadth of applications included in their courses, it appeared that about two-thirds of the applications were still being drawn from the natural and physical sciences. However, some social science and economics applications were being taught in most classes. When asked to report the sources of their applications, the teachers most frequently cited problems in their textbooks and in supplementary textbooks. Few teachers noted sources such as the UMAP Modules or articles from professional journals. Hence, it appears that much remains to be done to bring applications of mathematics into precalculus and calculus classrooms at the twelfth-grade level. While it seems that teachers are somewhat receptive, the actual teaching of applications has not become an integral part of the twelfth-grade curriculum at the present time.

What Mathematics Was Taught

Teachers were asked to report whether the content needed to respond to the items on the international test had been taught to their classes. The responses about the opportunity to learn are reported in Table 4. The results show that the content taught to the precalculus and the calculus groups differed in distinctive ways. The greatest difference between the 191 precalculus classes and the 46 calculus classes was, as expected, in elementary functions and calculus, which included differential and integral calculus. In number systems and algebra, for example, the precalculus classes covered more than the calculus classes. However, other results showed that the calculus students had been taught much of this material in preceding years. For elementary functions and calculus, relatively little of the material was reported to have been taught to the calculus classes prior to the twelfth grade, though content for 83% of the items did get taught during the senior year.

Table 4

Percentage of Cognitive Items Taught to

Twelfth-Grade Students 1981-82

(Average Percentage Across Items)

Content Area (# of Items)	Precalculus (n = 191)			Calculus (n = 46)		
	Taught Before	Taught This Year	Never Taught	Taught Before	Taught This Year	Never Taught
Sets & Rel. (7)	31	50	19	50	40	10
Number Sys. (17)	39	42	19	75	14	11
Algebra (26)	34	52	14	53	41	6
Geometry (26)	21	40	39	41	26	33
Elem. Fun. & Calculus (46)	8	37	55	9	83	8
Probability - Statistics (7)	29	14	57	50	6	44
Finite Math (4)	29	21	50	62	8	30

What Mathematics Is Learned

Students were tested in seven content areas: sets and relations, number systems, algebra, geometry, elementary functions and calculus, probability and statistics, and finite mathematics. Table 5 contains results for the entire Population B sample and by the individual class types for both the pre- and post-tests, as well as a comparison with the international mean performance for each of the test areas.

Table 5
Twelfth Grade Mean Opportunity to Learn and Achievement by Subtopic:
U.S. and International, 1981-82
(Mean Percent Correct)

Content Area	United States All Classes (Number of classes = 237)				International	
	OTL	Pre	Post	Difference	Post-test*	OTL**
Sets & Relations	83	52	56	+ 4	62	70
Number Systems	83	35	40	+ 5	50	75
Algebra	88	38	43	+ 5	57	84
Geometry	62	26	31	+ 5	42	62
Elementary Functions/ Calculus	54	19	29	+10	44	75
Probability & Statistics	45	39	40	+ 1	50	59
Finite Mathematics	53	27	31	+ 4	44	63

* The countries included, in addition to the U.S., are: Belgium (Flemish); Belgium (French); Canada (British Columbia); Canada (Ontario); England and Wales; Finland; Hong Kong; Hungary; Israel; Japan; New Zealand; Scotland; Sweden; Thailand.

** The international mean OTL was 73, the mean U.S. OTL was 65.

Table 5 (cont.)

United States

Content Area	Precalculus (Number of classes = 191)				Calculus (Number of classes = 46)				
	OTL	Pre	Post	Diff.	OTL	Pre	Post	Diff.	Int'l Post.*
Sets & Relations	81	48	54	+ 6	90	66	64	− 2	62
Number Systems	81	33	38	+ 5	89	43	48	+ 5	50
Algebra	87	35	40	+ 5	94	53	57	+ 4	57
Geometry	60	24	30	+ 6	67	35	38	+ 3	42
Elementary Functions/ Calculus	45	18	25	+ 7	92	26	49	+23	44
Probability & Statistics	43	36	39	+ 3	57	48	48	0	50
Finite Mathematics	50	24	29	+ 5	71	36	38	+ 2	44

Averaging across the entire set of 136 items, teachers reported that only about 65% of the items dealt with content the students had been exposed to in their studies of mathematics. This again was below the international exposure mean of 73%. This limited coverage was most notable in the areas of geometry, probability and statistics, and finite mathematics. Students in the precalculus classes were also at a disadvantage in calculus-related areas of the elementary functions and calculus test.

The average achievement of students in the U.S. fell below the international average for each subtest. In the case of the precalculus classes, the performance levels were near the 25th percentile in most areas. The calculus classes, on the other hand, performed at or near the median level in all areas except geometry and finite mathematics.

Summary

The overall results sketched for Population A are disturbingly mixed. By comparison to the other participating countries, the U.S. eighth-grade sample never achieved better than the median national performance, and sometimes were only at the 25th percentile or below. The U.S. performance over the period between the First and Second International Studies showed a modest decline, especially in student comprehension and application tasks.

Compared to the curricula of other countries, the U.S. eighth-grade curriculum was dominated by arithmetic (except for the small portion of first-year algebra classes). Geometry did not appear to be a significant part of the mainstream of the eighth-grade curriculum in mathematics. This lack perhaps explains some of the low overall scores of U.S. students in geometry, but, in spite of the emphasis, U.S. scores in arithmetic are not outstanding. An analysis of international OTL scores shows also that the U.S. students were not significantly disadvantaged on an international level in their exposure to geometry. Finally, the performance of the U.S. sample on the measurement items was very disturbing.

When the twelfth-grade results are examined in an international context, the picture is no better. The content coverage of the curriculum is still lacking and performance is even lower. The calculus classes, where some of our best students may be found, were able to achieve only international median level performances, even though the comparison here is between the top 2-3 percent of U.S. students and much larger percentages of the age cohort in other countries. However, the comparison of scores with the First International Study indicates that the twelfth-grade students have maintained about the same level of performance with some minimal gains in comprehension tasks.

Overall, the pattern is much like what is seen at the eighth-grade level, suggesting a curriculum lacking clear direction and a common core of topics. A pattern of focused instruction was more typical of other countries participating in the Second International Mathematics Study.

Only a sketch of the results from the U.S participation in the Second International Mathematics Study can be presented here. Yet even these brief glimpses raise a number of questions as to the direction, strength, and viability of the present mathematics curricula and learning environments. Some of the most important are "What patterns of instruction lead to significant increases in student learning?" "What might be done to redirect the present eighth- and twelfth-grade mathematics curricula?" "What particular goals are we trying to achieve through our mathematics education programs?"

Appendix

Participating Countries

Australia	Israel
Belgium, Flemish	Ivory Coast
Belgium, French	Japan
Canada, British Columbia	Luxembourg
Canada, Ontario	Netherlands
Chile	New Zealand
England	Nigeria
Finland	Scotland
France	Swaziland
Hong Kong	Sweden
Hungary	Thailand
Ireland	United States

On the Mathematical Curriculum for Grades K-3 in Poland

Zbigniew Semadeni

Institute of Mathematics, Polish Academy of Sciences

Warsaw, Poland

Introduction

The purpose of this paper is to describe our curriculum briefly and to highlight some of the issues concerning primary mathematics teaching in Poland. This is not an easy task. The issues are complex and various factors are involved. Poland is a country with a centralized system of education; however, what actually happens in an individual classroom may run counter to the guidelines of the official curriculum. Moreover, Polish educational philosophy differs considerably from that in America (e.g., by not taking a behavioristic approach).

Since the late 1960's I have been following various discussions on primary mathematics teaching (including those of working groups at ICME 2 through ICME 5). Originally I was surprised by the noticeable difference in attitudes of two groups of educators: those from North America and Britain on the one hand, and those from Continental Europe (France, Germany, Poland, and so on) on the other. The prevailing attitude of Anglo-Saxon writers is what I would label as "didactic pessimism", expressed by stressing how hard mathematics is for children and how poor the fruits of learning are, whereas continental publications appear much more optimistic, with focus on how bright children are if taught properly. This didactic pessimism (towards children) is often combined with apparent confidence in the teaching methods, whereas didactic optimism is associated with blaming teachers.

For all these years I have been asking myself what the main reason for this difference was. One possible explanation was that British and American educators were more occupied with the evaluation of actual school achievements and could better see reality, while educators of the other group were more likely to be centered around their new teaching methods and were looking for evidence of success. Another possible explanation was that the results of mathematics teaching in Continental Europe were indeed better (in some sense) than those in North America (in spite of the outstanding research of Thomas Carpenter, Leslie

Steffe and many others, whose influence on educational practice is still limited). I do not know of any data to support or refute either explanation, but gradually, after many years, I am inclined towards the second one. Conceivably didactic pessimism may contribute to poor results through setting objectives which are too narrow.

At any rate, on average, Polish publications on primary mathematics teaching are more optimistic than those in America, and a likely reason is that rote learning in my country appears less overwhelming. Some specific hints will be discussed later.

In 1975 a new mathematics curriculum for grades 1-3 officially went into effect in Poland. It brought a radical change in content (with some features of new math) and, deliberately, a substantive change in recommended teaching methods. The latter turned out to be much more difficult than had previously been anticipated. In 1981 and 1983 the content was revised, and many topics were deleted, reduced or toned down. The present syllabuses will be described in what follows.

It is very difficult to give a balanced description of the achievements and failures of the 1975 curriculum change. First, there are no hard data to assess the actual effects. Second, other powerful factors greatly affected the schools in the same period: an abandoned attempt by the authorities to change the whole structure of the Polish educational system, the economic crisis and the state of upheaval in the country, a dramatic decrease in the number of elementary school teachers (retirements, leaving the profession). Too much was definitely attempted in 1975. The curriculum designers were influenced by a desire to improve teaching methods, but they failed to develop their methods in consultation with the teachers involved.

In 1984 a leading Polish weekly, "Polityka," organized a public forum on issues in mathematics education. After four months of interesting and heated discussions, the main question of a global evaluation of the new math component of the curriculum still seems to be unanswered. In the present paper I present my personal point of view.

An Outline of the Present Polish Educational System

Polish children start compulsory schooling in grade 1 at the age of 7 (some children are allowed to start at 6), and attend so-called "basic school," which comprises grades 1-8. After grade 8 they may be admitted to a university-directed 4-year secondary school, or to another type of school (e.g., vocational). In each of grades 1-3 in basic school all (or almost all) of seven subjects (Polish, math, environment, gym, music, arts, crafts) are taught by a single teacher-generalist. From grade 4 on, in principle, mathematics is taught by a teacher who specializes in this subject and who presumably has a college degree or some post-secondary training.

Children from 3 to 6 may attend so-called "preschool" (the words "kindergarten" and "nursery school" may be misleading, so I will not use them); however, the number of existing preschools is not sufficient for the needs, particularly in rural areas. Authorities make efforts to at least provide preschool activities for all 6-year olds, either in a regular preschool or in so-called "grade zero" in a nearby basic school. Grade zero has the same program in reading and mathematics as the last year of preschool, but it offers only three hours of activities a day and lacks the playtime, meals, etc. typical of an 8-hour preschool.

The curriculum for all children attending regular basic schools is centrally worked out for the whole country and issued as a document of the Ministry of Education. Possible variation may concern the style of teaching and details of the order of topics. Providing the same basic, comprehensive education for all children is one of our educational principles. Yet, in spite of the uniformity of the curriculum, there are differences between what students from different schools actually learn, depending on the qualification of the teachers, family background, and so on.

Uniformity of the curriculum is combined with uniformity of the textbook. Students in the same grade use the same mathematics textbook throughout the country. Early efforts to change this were abandoned for a variety of reasons, primarily economic.

There are no regular nationwide assessments of student achievement in Polish schools; the only nationwide examination is the secondary school exit exam (there may also be an entrance examination for secondary school). The marks recorded on the student's certificate are at the discretion of the teachers and are based on the student's performance during the school year. There are, however, systematic surveys of student achievement in various grades, based on random samples, which are organized by school authorities (and sometimes by universities). In addition, each school is the subject of a thorough inspection every few years and then each student's achievements are tested.

Around 1981 the Polish educational administration abandoned the controversial policy of promoting almost all students in grades 1-3 even if their records showed insufficient progress in reading, writing and mathematics. Previously a student was able to repeat a grade only under special circumstances, like prolonged illness. That policy was established several years ago on the advice of experts; they argued that separation of failing students from their group was psychologically catastrophic and made it even more difficult to overcome emotional blockages. However, a significant portion of society objected to that promotion policy, and it was unpopular among teachers themselves, mainly because it undermined their traditional authority. It is now believed that this so-called "automatic promotion" resulted in a significant decline in student achievement. Thus, the present position of our authorities is that students in grades 1-3 whose performance in Polish or mathematics is judged definitely unsatisfactory should repeat the grade.

During the school year, slow learners and students with difficulties have extra activities in small groups organized by the school; in severe cases they are sent to a local remedial center for help.

The Mathematics Curriculum for Preschools

Mathematics is explicitly mentioned for each age level in the official guidelines for activities in Polish preschools. At age 3, it consists of simple activities, e.g., sorting toys, playing with shapes, and using the numerals "one" and "two". At age 6 children learn to read simply-spelled words (but are not taught writing). Their mathematical activities include comparing objects; classifying and arranging them according to such attributes as shape, size or color; dealing with

86

positional relationships such as inside, outside, before, after, between, left, right, above, below, first, and last; one-to-one matching of small sets of objects; and counting. The 6-year-olds also learn to recognize the symbols for numerals up to ten and to perform addition and subtraction by joining or separating and by counting. The preschool activities have three main goals:

1. To stimulate the cognitive development of the child
2. To discover possible specific conceptual difficulties of the child and to try to overcome them
3. To help children reach a state of readiness for learning in basic school.

In principle, the preschool program is informal and exploratory; children should use physical objects found in their environment or materials especially designed to help them develop certain mathematical ideas. In practice, I am afraid, most of the learning is not of the type officially recommended; teachers are more likely to adopt the traditional school teaching style (some of them appear proud of this): inappropriate, scholarly language is often heard, and children are found "doing sums."

A General Description of the Mathematics Curriculum for Grades 1-3

The curriculum consists of three parts: (1) goals (2 pages); (2) content (9 pages), and (3) comments concerning the interpretation of the syllabus and recommended teaching methods (22 pages). The number of pages (jointly for grades 1-3) gives a hint of how detailed the descriptions are in each part.

The goals for grade 1-3 may be divided into groups:
* general aims (contribution to the child's development, especially cognitive development, and to the development of the child's personality, critical thinking, and so on.
* forming concepts (the concept of a whole number, with its various aspects, the four arithmetical operations and some other preliminary, intuitive concepts)
* preliminary development of desirable attitudes, appreciation and competence for applying the acquired knowledge to practical problems, mathematizing, verbalizing, using graphical schemata, handling symbols,

reading simple mathematical texts, spatial exploration

- operational goals: a list from "writes and reads numerals to 1,000,000" to "recognizes perpendicular line segments" to "calculates simple intervals of time."

The goals listed in the curriculum are only informational in nature; they are guidelines or statements of intent, rather than objectives to be achieved, and are not detailed enough to be performance tasks. Their influence on actual teaching is difficult to determine. A typical teacher "covers the topics," that is, follows the scope and sequence of the syllabuses or, more likely, just follows the textbook.

The syllabuses consist of a rather detailed list of topics to be covered grade by grade (given below in condensed form). The first paragraph of the syllabuses explains their significant new feature: some entries (distinguished in print by boldface type) concern the <u>fundamental material</u> to be mastered by students in the given grade, with the acquisition of certain skills. Other entries play different roles; they may be

- <u>auxiliary topics</u> (to help the student learn the fundamental material; e.g., in grade 1 equations are given to help the child understand the operations performed during problem solving, and should not be regarded as an aim in themselves)
- <u>preparatory topics</u> (activities aimed at intuitive preparation for the learning to be introduced in the following grades)
- <u>stimulating topics</u> e.g., certain activities with shapes, sorting and classification of objects (using the term "set").

A topic of one of these three types may become a fundamental topic later, in another grade. The distinction between fundamental topics and the others was established for the best of reasons, but it was not achieving the desired outcome. It was not understood properly by teachers after it was introduced in 1975, and a possible reason for this was that the fundamental topics were not marked clearly enough (boldface was used for the first time in the syllabuses in 1983; before that the teacher had to search through "Comments" to find pertinent information). A certain perplexity was felt by teachers and supervisors faced with the problem of assessing the students' performance in auxiliary or preparatory topics. The prevailing attitude was to treat all the entries of the syllabuses with the same weight. This causes undesired effects, e.g., it did not leave enough time for developing basic computational skills. Whether the new explicit way of distinguishing the fundamental topics from the others will help remains to be seen.

Before listing the contents of the curriculum, let us recall that first-grade students in Poland are 7 years old, that is, they are one year older than American first graders.

Outline of the Syllabuses

Grade 1 (5 lessons of 45 minutes per week)

1. Spatial relations (activities related to concepts such as "before", "behind", "by", "between", "to the left", "near", "outside", and so on).

2. Comparing and sorting given objects according to size, length, and so on (activities concerning concepts such as "longer", "the longest", "narrower", "less heavy", and so on).

3. Simple geometric figures (activities with shapes: recognizing, constructing, drawing).

4. Sets (classifying objects according to a feature, e.g., size or color; activities with two features considered at the same time, leading, e.g., to comprehending inclusions and distinguishing the intersection of two sets in easy concrete cases; joining sets).

5. Comparing and sorting given sets according to their cardinalities (one-to-one matching of elements; constructing a sequence of sets which increase in number of elements; activities related to transitivity).

All activities belonging to the above five groups are assumed to be preliminary in nature; it is not expected that any skills or specific knowledge in these areas will be the result in grade 1 (except for the concepts less, more, as many).

The following groups concern the core of mathematics in grade 1; fundamental topics are underlined.

6. Numbers 0–10 (<u>counting</u> in various ways; <u>using ordinals</u>, e.g., distinguishing between "three apples" and "the third apple"; <u>writing and reading numbers</u>; the number zero; <u>comparing numbers</u> and <u>using the symbols</u> >, <, =; ordering given numbers from the smallest to the largest or vice versa).

7. Addition and subtraction up to 10 (<u>finding sums</u> by counting and joining given elements; <u>writing sums with the symbol +</u>, decomposing numbers into addends, e.g., 5 = 3 + 2; <u>subtraction as an operation inverse to addition, taking away and finding unknown subtrahends, writing differences with the symbol −</u>; practical use of commutativity; adding and subtracting zero; the use of various types of tables and graphs with arrows).

8. Simple word problems (distinguishing what numbers are known and what is being looked for; <u>solving word problems</u>; <u>formulating word problems</u> for a given situation, picture, formula, and so on; changing given word problems to new ones).

9. Numbers as a result of measuring (comparing, adding and subtracting lengths, volumes, etc.; <u>measuring segments with a ruler, comparing lengths</u>). Finding the numbers 0–10 on the number line.

10. Equations (writing equations for given word problems; solving equations of the types $x + 5 = 8$, $5 + x = 8$, $x - 2 = 4$ by manipulating objects; formulating word problems for a given equation).

11. Numbers up to 20 (<u>writing and reading numbers</u>; their places on the number line; comparing).

12. <u>Addition and subtraction up to 20 with or without carrying</u> (using grouping; writing parentheses).

13. Multiplication and division up to 20.

14. Numbers up to 100 (<u>counting; writing, reading and comparing numbers; understanding the place value</u>; addition of the types 20 + 60, 30 + 4, and corresponding subtraction).

15. Practical skills (telling time; <u>the days of the week</u>; counting money; measurement).

The above list gives the scope, but not necessarily the sequence (e.g., topics of groups 6, 7, 8 and 9 are somehow to be intertwined). Equations are regarded as an entirely auxiliary topic (no skills required), and are supposed only to serve better understanding of addition and subtraction. In grade 1 multiplication is a preparatory topic only.

Grade 2 (5 lessons per week).

1. Sets (activities concerning the intersection, union and difference of given sets; the empty set).

2. Addition and subtraction up to 20 (including practical use of associativity). Word problems on comparing (with "differs by" and so on).

3. Addition and subtraction up to 100 (computing sums and differences using various methods).

4. Multiplication and division up to 30 (finding products by different methods, including practical use of distributivity and commutativity; division as an operation inverse to multiplication; word problems; equations of the type 6·x=12).

5. Geometry: recognizing perpendicular and parallel lines; activities with geoboard or quadrille paper; the number of unit squares which fill a given rectangle (when each side is equal to an integral number of units).

6. Multiplication and division up to 100 (including word problems with two operations; activities preparatory to division with a remainder and divisibility).

7. A half and a quarter (e.g., 1/2 meter = 50 cm, 1/4 hour = 15 min).

8. Numbers up to 1000.

9. Practical skills (measurement, money, clock, calendar).

Grade 3 (5 lessons per week)

1. Addition and subtraction up to 100 (including basic properties of the operations).

2. Multiplication and division up to 100 (including memorization of the multiplication table, using parentheses); word problems (including "how many times", "so many times", and so on). Division with a remainder.

3. Addition and subtraction up to 1000.

4. Geometry (distinguishing between curves and straight lines; various

activities; <u>computing the perimeter of a rectangle</u>).

5. Multiplication and division up to 1000 (simple cases, including <u>multiplication and division by one-digit numbers</u>).

6. Simple fractions.

7. Expressions of the type "2m 75 cm" and analogous expressions for money, etc. (including the notation with a decimal point, e.g., 2.75m).

8. Numbers up to one million.

9. Practical skills.

Pocket calculators are not mentioned in the Polish curriculum for grades K-3. (In the textbook for grade 4 there are calculator problems). Microcomputers are still very rare.

The curriculum encourages teachers now and then to give children intentionally ill-formulated verbal problems from grade 1 on, i.e., problems with superfluous or irrelevant data, with not enough data, or even with contradictory data. In the textbook for grade 1 by Ewa Puchalska and Marek Ryger (the first edition was published in 1972 as a pilot textbook for the new curriculum) a boy (nicknamed "Gapcio", connoting an absent-minded child who is easily taken in) formulated problems for his sister to solve; the students' task was to say what was wrong with the problem and to suggest how the problem could be fixed up. Here is a sample of Gapcio's problems: "There were two women in a shop. The first woman bought 3 eggs. How many eggs did the other woman buy?". At first, the children are bewildered; but after some explanation they grasp the point and seem to like such "funny problems."[1]

One of the most dramatic differences between the previous curriculum and the present one is that several topics are now started in earlier grades than under the traditional approach. In so doing, the topics are carried along throughout several grades and are adequately formed over several years, as opposed to the previous approach of delaying topics, possibly past their germination periods, and then trying to cover them too rapidly. These two sentences, paraphrased from the "K-6 Curriculum Development" section of the 1984 description of the University of Chicago School Mathematics Project, apply perfectly well to the rationale of the 1975 Polish curriculum changes. Classroom practice has been rather disappointing, though; instead of introducing exploratory, less formal methods and extending the formation of concepts and skills over several years, teachers were more likely to imitate the long-standing practice of their colleagues from the upper grades and again cover the topics too rapidly, this time with younger children. All potential adopters of the above-described splendid idea should be warned that it is difficult to implement and is liable to be distorted and cursed. Yet, it is certainly worth further study.

A Comparison between the Polish and the American Approaches to Teaching Arithmetic

There are conspicuous differences between the content of the Polish syllabuses for grades 1-3 and the curriculum of a typical U.S. school. But more significant, in my opinion, may be the differences in attitudes towards teaching methods. American educators stress the importance of learning "number facts," which are the basis for performing arithmetic operations. In Poland, educators are aware of the warnings of psychologists (especially those of A. Szeminska, a Polish collaborator of Piaget) that multiple repetition of the same mental action leads to its mechanical performance, and the earlier the action becomes mechanical the more difficult it is for a child to be aware of what s/he is doing. In short, early memorizing kills understanding. Learning in "number facts" is dangerously close to the condemned rote learning.

Consequently, the Polish curriculum recommends that memorization of number facts be the last stage of a lengthy process of learning the given arithmetic operation (after a prolonged period of exploration, computing results making use of concrete materials, measurement, solving problems, learning the properties of the operation, learning how to derive number facts, and so on). Thus, first the child should keep on getting the results of operations using various methods, and incidentally memorizing certain number facts. After a year or two, the remaining number facts, if any, can be acquired by rote, after the child has grasped the idea of how to handle the operation.

In practice, however, Polish teachers often prefer to have a quick way for the child to remember the sums — and later the products — of the one-digit numbers. They want to maximize the current performance of their students. This is what the parents expect; they pressure teachers, or teach the children themselves at home, not being aware that early rote learning is harmful for their children in the long term.

Let me give some examples. In grade 1, the concept of a product is based on concrete examples only (the expression of, e.g., 3 · 4 as the sum 4 + 4 + 4 comes a posteriori). The introduction of the symbol for the product (like 3 · 4) should follow a period of preliminary activities: counting in groups (e.g., 3,6,9,12,...), making enactive and iconic representations of multiples (e.g., children are given a picture of 4 shirts and are told to draw 3 buttons on each shirt and answer the question "How many buttons do we need for 4 shirts?"), and so on. In grade 2, students compute products using various methods, learn the properties of multiplication and solve problems. The multiplication table is to be mastered (i.e., the students are to memorize all the products of one-digit numbers) during the first semester of grade 3 at the latest, before going on to the products of multi-digit numbers.

Division starts in grade 1 with concrete problems, e.g., "15 candies are to be shared equally by 5 children. How many candies will one child get?" Students deal counters, find the answer and fill in a multiplicative formula of the type 5 · ... = 15. In my opinion, the symbol for division should be introduced a few days later, after a series of such preparatory exercises.

The next example shows the present Polish approach to addition with carrying. In the second semester of grade 1 children encounter sums like 8 + 5. The prevailing method is grouping: 8 + 5 = 8 + 2 + 3 = ... (based on concrete activities with sticks, beads, Cuisenaire rods, etc.). In grade 2 children first review sums up to 20 and then do arbitrary sums up to 100 in increasing order of difficulty. It is recommended that the crucial computation of sums of the type 16 + 27 be done by regrouping, with the help of concretization. Previous generations of educators advocated the use of only one (always the same) method of regrouping, e.g.,

16 + 27 = 16 + 20 + 7 = 36 + 7 + ...,   28 + 34 = 28 + 30 + 4 = 58 + 4 = ...

They were afraid that a change in the method might confuse the children.

The present curriculum points out that the teacher should not impose a particular method of computation and that it is advisable to try various methods, e.g.,

$$16 + 27 = 16 + 7 + 20 = 23 + 20 = \ldots$$

and $16 + 27 + 10 + 6 + 20 + 7 = 30 + 6 + 7 = \ldots$ as well. Students should be encouraged to suggest other methods. These computations are written horizontally, with the equality sign. The vertical notation

$$\begin{array}{r} 16 \\ +27 \\ \hline \end{array} \; ,$$

with its prescribed order of operations should come much later, after the children have firmly grasped the principle of adding two-digit numbers by regrouping. The Polish curriculum stresses the algorithmic, mechanical nature of adding two-digit numbers with vertical notation, which therefore is not suitable as the only early method of addition.

These examples provide an insight into the change in approach between the former (prior to 1975) and the present Polish curricula for the lower grades. Previously, when an arithmetic operation was taught, the teacher (or, more likely the textbook author) chose a method of handling it that would be optimal for the adult, and devised a hierarchical sequence of performance tasks which would presumably lead to mastery of the method. We are now in favor of a different approach: first the teacher should introduce — in simple cases — a method which is optimal for the child (natural, easy to comprehend) and then — step by step — should suggest modification of the method so as to include all possible cases and make it maximally efficient; the modified method should be introduced when the previous one either does not work or could be made more efficient, so that it will be appreciated by the child as an improvement.

Vertical addition is an example of a method optimal for the adult, but not necessarily for the child. To be accepted by the latter, it should stem from activities with concrete materials and regrouping.

I would conjecture that Polish students utilize their fingers for sums such as $7 + 8$ to lesser extent than American children learning the same material. I have not heard of any comparative studies of this phenomenon, but a possible explanation is that for American children counting on their fingers is an escape from prematurely introduced computation drills and a premature leap from pictures and other concrete material to number facts.

When operations with multi-digit numbers are taught, money problems serve as a wonderful motivation. For instance, each step of the algorithm for division of the type 741:3, $3\overline{)741}$ in American notation, is natural and easy when children act it out with play money (7 hundreds, 4 tens and 1 to be shared by 3 people). The children perform the operations themselves (changing a one-hundred bill into 10 tens, etc.) and find the result; then their actions are expressed in the standard notation, e.g.,

$$741 = 600 + 120 + 21, \ 600 : 3 = 200, \ 120 : 3 = 40, \ 21 : 3 = 7$$

(this is how the money was grouped and distributed); finally, the vertical method of writing is introduced as a shorthand notation for what the children have already done.

The above examples provide an illustration of some tendencies in primary mathematics education in Poland. A detailed description for teachers' use is published in Primary Mathematics Teaching (three volumes, in Polish), comprising 309, 285 and 400 pages, with colorful pictures. In this book, which has several authors, one can find psychological and pedagogical discussions of problems of mathematics teaching and learning relating to concrete operational children (in Piaget's sense).

In conclusion, may I suggest that the core of the problem of unsatisfactory results in mathematics learning consists in the choice between two basic philosophies of education: behavioristic (with scientifically sound theories of learning and memorizing given information, standardized tests, and so on) and humanistic (where children are regarded as subjects rather than objects, treated with love before objectivity, and with more trust in their own capacities, in their own responsibility, as Hassler Whitney would put it). The Polish mathematical curriculum for grades 1-3 was designed at least partially with the latter philosophy in mind. Nevertheless, the practice in schools is often behavioristic.

Notes

1. See E. Puchalska and Z. Semadeni, "Verbal Arithmetical Problems with Missing, Surplus or Contradictory Data," submitted to <u>Educational Studies in Mathematics</u>.

The Bulgarian Academy of Sciences Research Group on Education Project

Boyan Penkov, Blagovest Sendov

Research Group on Education, Sofia, Bulgaria

Abstract

This paper gives a brief description of the experimental work done in Bulgaria by the Research Group on Education (RGE), affiliated with the Bulgarian Academy of Sciences and the Department of National Education. BARGEP is a project covering school education in general, and the paper deals with issues outside the two topics of the conference. We present here only the main guiding ideas. Nor are names of the numerous contributors and collaborators given; these can be found in the various textbooks.

Historical Background

A few words on the educational situation in Bulgaria will be needed to begin with. Compulsory and free primary education were introduced in the 1890's, and at the same time the first university of Sofia was founded. New problems, not only quantitative but also qualitative, arose under the new socialist system established in the 1940's. In the 1950's, the number of secondary schools grew enormously, and there was a great need for teachers. After several reforms, the Government reconsidered the education problem as a whole in its Education Act of July 1979. This document provides the current guidelines and will be in effect for more than a decade. Having resolved once and for all such fundamental issues as the elimination of illiteracy (which still existed in the forties), a high percentage of dropouts, the problem of appropriate and modern school equipment, and the problem of teacher training, Bulgarian society has now to prepare, in a humanistic way, its educational system for a computerized future.

The latest reform — the most fundamental in our recent history — is described as "radical" in the wording of the Education Act itself, and has among its goals the achievement of a 12-year education for the entire population, a school age of 6 years (instead of 7, as was the case up to 1979), full-day schooling for the majority of students (especially the younger ones), and a broad spectrum of professional opportunities offered to everyone in tandem with a general education.

The RGE — Foundation of Development

The RGE was created within the framework of the July 1979 Education Act. The idea was to start a small-scale experiment coordinated with the larger reform movement but implementing more radical changes and being freer than traditionally accepted models. Starting in the fall of 1979, three pilot classes of six year olds were at our disposal, one in the capital Sofia, one in a small nearby village, and one in a town of medium size. There was no special consideration in choosing children or teachers. This principle of impartiality has continued since.

The initial core of the RGE consisted of a handful of university professors (two linguists, two mathematicians, and a psychologist), but did not include professional educators. This was done, almost, if not entirely, on purpose. The experiment had to be a fresh one.

The first and foremost task was to produce and publish a new textbook (or several) for the six-year olds. The initiators had to decide this issue within weeks. The decision was taken on the eve of the 1979-80 school year. We may say that the fundamental principles (to be discussed at greater length below) were established in these early, enthusiastic days. The result was a primer for the first semester.

Since that time more than five years have passed. The children of our pilot classes are now eleven year olds. The RGE conducts its experiment with about 2% of the Bulgarian school children, in 27 schools throughout the country. Our Research Group works under a contract between the Academy of Sciences and the Department of Education, which provides financial support to the Project. It has no permanent research staff, but includes about 20 administrative personnel. The RGE is unusual in acting as its own editor, without the help of a publishing house, and in preparing its printed teaching materials for color offset printing. This procedure has advantages but, of course, consumes both time and energy. At present (the beginning of 1985) we have published about 100 textbooks and various supplementary manuals.

Basic Educational Policy

Each of the founders of the RGE has had his own experience in reforming his own school discipline. This applies mainly to the linguists and to the mathematicians. The unanimous conclusion drawn by everyone from previous experience was the following statement, which became one of the cornerstones of our philosophy: <u>You cannot reform an isolated discipline; either you reform everything or you are doomed to failure.</u> As a consequence, this paper deals not only with teaching mathematics.

The reforms had to be radical by definition, so our next fundamental move was to accept an <u>objectives-oriented approach.</u> We are convinced that one should not confine oneself to, say, curricular issues, but review the objectives of the general school education. We should put all the knowledge and skills needed to address these objectives into a "melting pot" and let the various elements (the concepts, propositions, theories, disciplines, the whole curriculum) crystallize in an intrinsic and natural way, as if "cooling." Unfortunately, this noble aim is easy to formulate, but almost impossible to be put into practice in a satisfying way.

<u>Integration</u> is our second fundamental principle. It was applied almost totally in our primary cycle. The argument for integration goes like this: the child is an integrated individual and perceives the world as a whole; the child also has only one teacher in this period. Any differentiation of learning into disciplines will be felt as unnatural and will not be understood by children aged 6-10. We therefore could not permit ourselves the luxury of having a whole period (35 minutes) devoted to a single discipline. Moreover, children of this age cannot concentrate for very long periods of time. They need not only to be taught, but to be entertained as well. Thus, various activities are mixed and follow each other in a mosaic pattern that keeps the child interested.

The integrated approach leads to an interesting "paradox": on the one hand the interaction of disciplines saves time and seems to be a way out of the notorious "overburdening" of modern schoolchildren. On the other hand, the curriculum can be intensified and the textbooks can become quite a bit richer than the usual ones.

Textbooks should also be redundant for other reasons. In our view, when

100

formulating educational objectives, one should keep in mind two levels of achievement. The first — call it the "sine qua non" level, contains what every student must achieve, and the evaluative result is "yes" or "no." For normal children, things must be designed so as to get only "yes" answers. The second level is an encyclopedic one. It is not compulsory and may be left to the individual interests of the children. Therefore the textbook has to contain more than is needed by the "sine qua non" level. Finally, there is one more argument for redundancy of considerable weight. Today's children are overloaded with unordered information coming through many channels; the school remains the only institution trying to impart systematic information. We cannot permit this kind of information to remain poor.

We live today in a non-static world, and the average period of change is shorter than a human life span. No one can design an educational system that would continue to work over a very long time. This recognition leads to the well known idea of continuing education. The previous notion of the school preparing a citizen once for all of his or her future has to be abandoned. The new task of the school should be not only to teach, but to teach one how to learn by oneself. The student should be shown that knowledge is infinite, everchanging, and that no one can possess it totally (including the teacher). Schooling is only an initiation that will stimulate individual interest and inclinations. One aspect of all this is the principle of nonexplicitness. Things should not be presented to the youngsters in a ready, apodictic and final form. Situations have to be created that will push the children to find out, to ask and to satisfy their natural curiosity. Learning has to be an active process, and action should come before understanding. Thus we arrive at two similarly well known phenomena: the so-called "Monsieur Jourdain syndrome," and the "A-ha effect."

What was said above has important consequences for the teacher. He has to create a new atmosphere in the class, to act more as a sports trainer, where he is not always expected to perform better than the students. This analogy should not be pursued too far, of course — schoolchildren are not trainees. The teaching staff that the RGE began collaborating with in 1979 was to some extent reluctant to adopt a philosophy that gives a greater degree of freedom to teachers and students, but increases the teachers' and students' responsibility. Although the

RGE publishes teachers' manuals for every textbook, the teachers are in no way restricted to certain didactic patterns, and need to be inventive themselves if they want their pupils to be. After two years of work without the usual methodological restrictions and prescriptions that they were used to, our teachers have shown quite a positive development. They now feel more self confident and are prepared for new initiatives and increased responsibility. To sum up: do not prescribe everything for the teacher, give him more freedom to act on his own to achieve the stated objectives.

The Primary Cycle

According to the views accepted by BARGEP, the entire period of schooling should be divided into the three classic cycles: 1) primary (4 years, age 6-9); 2) secondary junior (3 years, age 10-12); secondary senior (4 years, age 13-16); and terminal class (one year, age 17-18). It is the primary cycle for which the RGE has developed a more or less final version of its teaching materials and greatest experience in implementation and evaluation.

The system of textbooks starts with a primer to be used during the first semester of the first year. There is no other textbook during that time (a single textbook is characteristic for the entire primary cycle), and the primer is used only in school rather than taken home. The main objectives of this first book are elementary reading, writing with block letters and digits, and adding and subtracting integers up to 20. Although there is only one textbook, we differentiate among 8 activities, according to the following curriculum (lessons are 30 long in the morning and 40 minutes in the afternoon).

Activity	Lessons per week	Lessons per year
Reading, writing, calculating	10	302
Singing and instrument playing, concert visit	4	122
Drawing and molding	3	90
Design and construction	5	130
Sports (physical culture)	8	238
Reading books	4	122
Studying one's homeland	1	32
Excursions	16 days per year	

The first item represents the "hard" studies and the other items the "soft" ones. The daily schedule is roughly as follows:

8:00–11:30	Classroom activity
11:45–12:15	Lunch
12:15–13:00	Free time
13:00–13:30	Sleep (6-year-olds only)
13:30–16:30	Classroom "soft" study activities

In grades 2-4 the schedule remains essentially the same. The classroom activities are 35 minutes long in the morning and 40 minutes long in the afternoon.

An important new factor in the third grade is the first foreign language, Russian. Russian starts with an intensive course of several weeks using a special primer, after which the regular integrated textbook becomes bilingual. The third year is also marked by introducing handwriting in tandem with block letters.

The tables above show an emphasis on the arts and on physical culture and a decrease in time for "hard" studies. This is possible because of the intensification brought about by the integrated approach.

A recent evaluation of BARGEP by an independent government educational laboratory has shown that the students of the RGE perform no worse than the average student, but have broader interests, are not afraid of school, are not intimidated and are more self-confident.

To sum up: the primary cycle has as its objective providing initial, non-systematic knowledge about the child's environment, and achieving a basic literacy in Bulgarian and Russian. An attempt is made to reach this goal through an integrated approach and a relaxed atmoshpere, stimulating the child's activity. Neither marks nor homework are given. All the teaching materials (books, slides, tapes, etc.) remain in the school and are not brought home. Time at home, after a whole "working day" at school, belongs to the family. Every class is assigned two teachers: one senior teacher, who typically takes over the "hard" lessons, and a junior teacher, who is responsible for the rest. There is usually a professional sports teacher available at most of the the schools.

The Secondary Junior Cycle

In this cycle the first differentiation of disciplines emerges, although the system still remains integrated horizontally among disciplines and vertically among grades. The next important feature is the advent of a second foreign language, English.

The main disciplines (the "hard" ones) are "Language and Mathematics," "Nature," and "Society."

"Language and Mathematics" is a most unusual discipline and is intended as an alloy of general and comparative linguistics and mathematics. Linguistics (not just the native language) through phonetics, morphology and syntax logically comes first, and an attempt is made to use simple mathematical tools to describe linguistic phenomena. This was the most difficult textbook to conceive, and it has not yet reached its final form. Incidentally, the same applies to a lesser degree to all our teaching materials, which were planned to be continually revised. A great difficulty to overcome was that the language and literature teachers on the one hand and math teachers on the other usually form disjoint sets and it is not at all easy to find and convince the same person to teach according to such a hybrid textbook. In practice, two teachers enter the class and team-teach, hopefully in a concerted way. Linguistics should be comparative, in view of the synchronized study of Bulgarian, Russian, English and some programming languages.

The next discipline, "Nature," deals with the sciences. It includes topics from physics, chemistry, biology, ecology, medicine, and related areas, in a more or less unified, systematic approach, according to the increasing complexity of natural phenomena.

Broadly speaking, the third discipline, "Society," represents the humanities, but it is in fact a history of civilization. The backbone of this discipline is history, but not in the narrow sense of a political history of war, conquests, and empires. The main emphasis is on the development of human society, its productive forces, the history of social formations and relations, the history of ideas, and especially the history of arts. So "belles lettres" are included in this textbook (with a supplementary anthology) and are considered of primary importance because they are the most abstract of the arts (after music) — a fact that we are not always aware of. "Society" has as its objective the development of an open-minded citizen, respectful of other civilizations and cultures, ready to live in a peaceful pluralistic world.

Some other features are also characteristic of the junior secondary cycle. First of all, a computer-oriented approach is being attempted in all disciplines, first in "Language and Mathematics" and in "Nature." While in the primary cycle the only instruments used were calculators, in the secondary cycle personal micro-computers are playing a more and more substantial role. Although we are at the beginning of such an approach, two textbooks on LOGO are already in use in the fifth and sixth grades.

In addition to the three "hard" disciplines we have discussed, related optional activities are available: Translation from Russian, Translation from English, The Naturalist, and Folklore. For all these additional activities special manuals have been published. The children have to choose one of these possible avenues, but can change them after one semester (there is no "tunnel system"); they enjoy this kind of work, where they are the main actors, very much.

There are also two other important disciplines with their own textbooks: "Daily Life" and "Manufacturing." "Daily Life" deals with home economics, especially professional cooking. "Manufacturing" deals with simple crafts and agriculture.

To sum up, the junior cycle is a first cycle, closed from a logical point of view as well as from a chronological one. It is assumed that a subsequent cycle, on a higher level, will follow in the senior secondary cycle. The first clustering of concepts and phenomena arises in the junior cycle, elements of proofs emerge, and a first broad overview of human history adapted to this age is presented. Individual inclinations are given first opportunities to appear and develop.

Remarks on Mathematics

According to the principles of nonexplicitness and of integration, mathematics is introduced gradually among the other themes and topics. The so-called "modern math" approach was abandoned for at least two well-known reasons (not because it proved to be a failure almost everywhere): (1) the contemporary logical structure of mathematics (if such a structure exists at all) does not correspond to the way human beings develop their initial mathematical abilities in numbers and figures; and (2) the concepts of small integers are more fundamental to the human mind than the notion of a set, which requires a higher level of abstraction.

Therefore a more pragmatic and algorithmic approach was chosen, so that children can act, get results, and play, but contemplate and prove later. Notions unusual in Bulgarian general school practice but used in BARGEP are flow charts, simple graphs, coding and decoding and histograms.

What follows is a key-word outline of the math curriculum for the primary cycle:

- 1st year. Addition and subtraction of integers up to 20. Segment, triangle, angle, circle, rectangle, square, cube. Digits versus/and numbers. Sign systems. Order, part of a whole; equal, unequal. Addition and subtraction of numbers with arbitrary length. Multiplying and dividing by 2. Meter, clock. Coding and decoding using simple and frequency codes.

- 2nd year.
  - "Sine qua non" level: Names of numbers up to billions. Multiplication. Prime factors. Multiplication of multidigit numbers with two-digit numbers. Multiplication by 10, 100, 1,000 and 25. Division of multidigit numbers with single-digit numbers. Division by 10, 100, 1,000. Remainder mod 9. Equal and unequal numbers, inequality signs, signs for greater (less) than. Decimal fractions (addition and subtraction). Brackets. Commutative and distributive rules. Assignment of value to a letter-variable. Binary code. Calculators. Angles. Area of rectangle and triangle. Equal figures. Circles, concentric circles. Co-ordinate system.
  - Second level: Punched cards; Tree-graphs; frequency, distribution, histogram; triangular, square and cubic numbers. Magic squares.
- 3rd year. Measures for length, mass, volume, time, area. Money units. Division by two-digit and three-digit divisors. Roman numeration. Percentage. Scale. Common fractions. Classes of common fractions.

Addition, subtraction, multiplication, and division of common fractions. Common denominator. Rectangular parallelepiped, its surface area and volume.

- 4th year.
    - "Sine qua non" level: GCD, LCM. Reducing common fractions. Block-scheme, algorithm, Euclidean algorithm. The four operations with decimal fractions. Negative numbers. Reciprocals. Absolute value. Ordering of the number line. Circle, length, sector; radius, grade. Measuring angles, adding and subtracting angles. Distance from point to line. Sum of angles of a triangle. Regular pentagon. Parallel and intersecting straight lines. Negative powers. Rounding decimal numbers. Symmetry.
    - Second level: Equilateral triangle, regular hexagon and octagon. Coordinates on the Earth's surface. Tetrahedron, cube. Converse figures. Equations. Graph of a function. Symmetry.

It should be noted that no proofs whatsoever are given — they are not needed at this age. One of the novel features of this curriculum is the method used for arithmetic operations with large numbers on a purely algorithmic basis. The process imitates that of a computer and can be stopped at arbitrary points. The geometric concepts aim at developing the child's geometric "Anschauung." No artificial or boring problems are given, nor are things repeated until consolidation. Rather, situations in other fields are used, where the calculation will be needed.

We shall not go into more detail here on math for the junior cycle, which is still in preparation. Its beginnings can be seen in the first textbook. We envisage introducing some elementary Boolean algebra, with set theory, logical and stochastic interpretations, simple axioms and conclusions proved from them, and some attempts to prove geometric facts. There will be more on graphs to serve syntactical analysis. Using microcomputers, more numerical methods and experiments will be introduced. For the time being LOGO is used, but other languages have been considered as possible.

Final Remarks

The results of evaluations made so far on BARGEP are encouraging. It is an experimental project and negative results are also of interest, but as a whole the results are positive. This has been shown by various tests, although in the field of docimetric evaluation much remains to be done.

Computers in Elementary School:
Some Experiments

Renée de Graeve

Université Grenoble 1, France

Introduction

This year the "Elémentaire" group of the Grenoble I.R.E.M. (I for Institute, R for Research, E for Education, M for Mathematics) has been conducting various experiments in teaching geometry with a computer in several primary schools. We have made the hypothesis that the drawing capability of a computer is especially appropriate to an introductory study of elementary geometry for developing intuition, even though a language (Logo) has to be learned. The methods are based on direct intervention in the classroom, and this research requires close cooperation with the teacher. But the present structure does not make such collaboration easy, because the teacher has no release from teaching duties.

Conditions of the Experiments

The Place

The primary school where we are running the experiment does not have a computer, so we have to take the students to another school. As of today, very few French schools have computers. Generally a school is equipped with one or two computers, and it is not very easy to work with such a small number; this is why the computers are often grouped together in one school ("a center of gravity" for three or four schools). Pupils must go to this school in order to work with the computers; they go there by bus during the school day, and they generally like to go because it is like a short break.

The Duration

Every Friday pupils work with the computers for one hour. Once a month I go to their home school and we work without the computers. On those occasions we all do a restatement of questions together. I explain a new theoretical lesson; the children change their rough drafts into good copy, or they write or decode little programs for testing their progress.

The Materials

We work with seven T07s. The T07 is a Thomson home computer: the material is furnished by the Ministry of Education, and so it is always French material. At the beginning of this year Logo was not available on the T07, so we are working with a "mini-Logo"; this is a BASIC program with almost the same graphic primitives as Logo. The drawings are done slowly, and that is suitable for children, but we can't do recursivity, and the parameter procedures are not very good. In each course the children have to complete exercise cards; each lesson is prepared beforehand, except for some exercises or free projects on a chosen topic.

The Individuals

The teacher of the class and an instructor, a professor of mathematics (in this case myself), are present during the lesson. Twenty children, nine or ten years old, comprise this primary-school group. They are pupils of "cours moyen 2e année," called in France "CM2"; it is the year just before secondary school. In this class most of the pupils are immigrant children or belong to the working class, and many of them are disturbed children and have difficulties in school.

The Aims

We have two goals: (1) learning programming through Logo, and (2) learning geometry with the aid of Logo. At first, we use the computers as a tool for modeling geometrical figures, and we do experiments by visualizing properties of these figures. But we also want the children to be involved in these models and experiments, and to be both creative and rigorous.

"Where there's a will there's a way," and the way here is the Logo language, which seems to be responsive to these aims. The children are taught "mini-Logo" so that they can encounter, formulate and solve a number of problems in Euclidean geometry by means of drawings. We think that when the children do programming, they are obliged to analyze and organize the subject of the program, and so what is learned is the substance. Programming is thus a specific situation which allows the children to approach geometry.

Learning Logo

The Contents

In the first lessons we devote all our energies to introducing Logo through the ten "primitives": Forward, Back, Right, Left, Penup, Pendown, Showturtle, Hideturtle, Clearscreen, and Repeat. We introduce these primitives very slowly.

In the first lesson we don't have a computer. The children have to make a drawing on quadrille paper and describe it; then they trade descriptions, and they have to reproduce the drawing. After that they have at least understood the necessity of having a common language for taking one's bearings and having the same points of reference.

The children don't know what an angle is, so in the beginning we work only on quadrille paper and we give them the instructions "RT 90" and "LT 90" for a quarter turn to the right or the left. The turtle is shown on the paper as an arrow: the tail of the arrow is where the turtle is, and the head of the arrow is what it is looking at. In all the exercises we write the start position (arrow number one) and the arrival position (arrow number two) on the drawing; we also have arrow number zero, which is the position of the turtle after a clearscreen. Then we introduce procedures, and we get the children to decompose a figure into elementary figures and thus to write a routine with subroutines. We have just recently introduced parameter procedures, because we finally have Logo on the T07.

The Difficulties and Their Cure

Of course, we have observed a great difference between what the teacher expects and what the pupils are really able to do. Thus, we took note of problems which are sources of mistakes, and we are trying to set up remedial processes. There are few syntactic mistakes, but many semantic ones. For example, the children write "RT 20" instead of "RT 90 FD 20", and they confuse right with left.

The turtle shown as an arrow is difficult, because they see the direction well, but not the position, which is not well-marked. The children also have difficulties in coordinating several elementary operations, and they don't understand that the arrival position is a new start position. They don't understand that the instructions are commands for a change of state; for them instructions produce a drawing, and that's all. We have them put the start and the arrival turtle position on their drawings. This is difficult for them, but it is necessary for comprehension of the concept of state.

To help them understand, they first use an arrow drawn on transparent paper like this: $\frac{L}{R}\rightarrow$ — an arrow where left and right are marked — and they pilot this arrow on their drawings like a car on a road. Then they are asked to do a number of exercises such as:

- Do the drawings corresponding to:

    Repeat 3 [•¹→——↑²], or

    Repeat 3 [•¹→——←² •], and so on.

To do these exercises they make a tracing of the drawing they have to copy and do the final drawing by transferring this tracing.

The Role of the Teacher

The children have to complete exercise cards made up by the instructor. The instructor is careful about the drawing he sets up as a model, and if the children formulate and execute an independent project, he has to enrich and direct this project.

When a pupil asks for help, he is supposed to mimic the turtle's movements with his feet and his body, not with a pencil, and so he finds the corrections himself. The pupil is not penalized for the error, but rather is induced to improve his work. Each pupil works at his own pace and his own level, and before he types his program he works with pencil and paper. When they are at the computer, the children work alone or in pairs. The instructor leaves the pupils free to act and to think up new situations, but he should also feel free to make changes in any of the experiments which he finds inadequate and to provide additional experiments.

What the Children Learn

When a child has to write a procedure, he has to resolve a problem. The problem situation has three characteristics:

1. The goal is to draw a figure
2. The rules of the game, in this case, the syntax of the language (this must be learned specifically)
3. The result is not the goal, but rather a list of instructions which realize the solution.

This is essentially the same as proving a mathematical theorem:
1. The goal is the theorem
2. The rules of the game are the rules of logic
3. The result is a list of deductions which realize the solution.

So when a child has to draw a figure, he has to answer the question, "How is this figure drawn?", and to execute his representation of this figure he does his program. In this way, when he writes programs he learns geometry. On the one hand, programming can help pupils construct new representations by giving them new ideas for the writing of other programs, enabling them to make use of what they have done before. On the other hand, a somewhat ambitious drawing needs modular construction where each element is independent, and thus students learn to decompose a problem into smaller problems.

For example, to test their acquired knowledge after twelve Logo lessons, we ask the children to draw this figure in different ways. It is a comb with ten teeth. The first way that all the children found corresponds to this decomposition: , a tooth.

They argue among themselves about the number of repetitions: is it five, nine or ten? But this is not very difficult to resolve; they count the number on the figure and easily obtain the right result.

The second way they found corresponds to this decomposition: first ⌐‾‾‾‾‾‾‾‾‾⌐₂ , and then iteratively draw 2↓ ↑1 .

This is difficult for them because they don't draw a continuous line; some pupils repeat it nine times and others ten times, but many children don't know where the turtle is at the end of the drawing. I try to get another idea for

113

construction, but the children can't find a new one, and they are tired of this, so I suggest this decomposition: first "edge$_l$" $\quad\begin{smallmatrix}&&^2\\_1&\ulcorner&\end{smallmatrix}$, then iteratively draw a "tee" $\begin{smallmatrix}^1&\rightarrow&^2\\&\top&\end{smallmatrix}$, and at last "edge$_r$" $\begin{smallmatrix}^1&\rightarrow&\\&\lrcorner&_2\end{smallmatrix}$. But I don't mark the arrows; they have to find them.

They thus use four procedures: edge$_r$ and edge$_l$, and with these two procedures they draw a tee, and with these three procedures they draw a comb. Here the number of repetitions poses no problem, even though it is eight. After that we ask them to modify their procedures such that the arrival position and the start position will be the same.

We plan to write up the above procedures with the number of teeth as a parameter; thus we will be teaching arithmetic using the problem of intervals.

A New Approach to Geometry

Angles

In the previous lesson the children defined the procedure "sail" to draw this figure. Now they have to draw the sails of a windmill. How many sails? First they do four, and they have no problem; then they try to draw six sails. The students experiment: they create their drawings and they decide themselves if they are right; they think by themselves and they choose what they must do. Sometimes the students are wrong; for instance, a child hits upon

Repeat 5 [sail RT 71]

and he seems to be satisfied, so I ask him, "Can you find a way to check whether it is right?" He then types "sail" and sees that the turtle does not draw the first sail again, so he says, "Seventy one is too small a number. I'll try seventy two." The exactness and the repetition of the drawings give the children a good illustration of the measure of angles. This gets pupils used to seeing figures where everything is in proportion, and so they learn to measure by eye.

The children proceed by trial and error; they like it and they are involved, and they see that defeats are necessary in order to win. For example, a child writes

Repeat 14 [sail RT 30]

and he says, "It's not right yet, but I have found how to draw a twelve–sail mill." They try to draw more and more sails; for them it is like a cartoon and it is so easy and so beautiful to draw "many" sails. They spare no pains to find the correct angle — always by trial and error.

It is difficult for nine- or ten-year-old children to find the relationship between the number of sails and the angle of rotation, so we ask them a riddle: "Can you find the angle for a seven–sail mill on the first try without the computer?" When we pose this problem, some pupils are aware that if they have found the angle for, e.g., a six–sail mill, they have also found the angle for a twelve–sail mill, a twenty-four–sail mill, and so on, and they say, "If the number of sails is doubled then the angle is divided by two." Although they have the intention of drawing a seven–sail mill, they don't think of dividing three hundred sixty by seven to find the angle of rotation; the pupils don't easily make use of division. Finally, some of them figure out that a complete turn corresponds to three hundred sixty degrees and that the number they are trying to find must satisfy the equation $7 \times ? = 360$. A theorem is thus proved: we have drawn a windmill if the corresponding angles are equal to 360 divided by the number of sails. This illustrates how a mathematical idea is derived by induction from experience and is then proved through formal deduction.

We end this sequence on angles with the drawing of a "trigonometric" circle. In their exercise books the children are given a circle which they are to mark off in increments of approximately ten degrees. The graduation is done by eye and this leads to three different results, depending on the child: a clockwise protractor graduated from 0 to 360, a counterclockwise protractor graduated from 0 to 360, and a right-left protractor with two scales from 0 to 180. Then we give them a LT-RT protractor printed on transparent paper with two arrows: one arrow is fixed and shows the zero point, while the other turns like the hand of a clock and shows the measure of the angle of rotation and also gives the direction (right or left).

Polygons

In this drawing a flag ⚑ is the same as the sail; only the arrival position changes. We ask the children to do the same thing as before. Can they anticipate the different drawings? They try with pencil and paper, but they prefer using the computer. In all the figures they get they can see different polygons in the center of the drawing.

Then they have to draw regular polygons. Here they get good results only by imitating, and we don't explain why the turtle goes back to its start position, but the aim is to show the properties of regular polygons and to explore the formulation of these properties.

Circles

We then illustrate some properties of circles. Circles are constructed in various ways:

1. by rotation of a segment — from this we obtain a disk;
2. with the end points of radii; here the children have to use Penup and Pendown and a cross to mark the points of the circle;
3. with an exterior angle of constant measure; here the children have to draw a regular many-sided polygon. It is difficult for them to see what they have to change in order to get a large circle or a small circle.

What Remains to Be Done

We have just introduced parameters, and we expect to work on similar figures and the generalization of some procedures. We have done a collective project; writing Bon Noel (each team had to write the procedure for one letter). The children also chose to draw a clock. This project is now done, and its realization can be seen on the last page of this article. At the end of the year these experiments will culminate in a test to be given to all the children.

Evaluation

Of course, the efficiency of new methods must be proved, and this is an arduous task. Evaluation is not easy because the pupils are also learning geometry with pencil and paper; what we do merely complements the course, and we don't really know what the contribution of the computer is.

First of all, it is quite obvious that the children like working with the computer very much and want to learn: it is something new for them, and the very fact that they go into the computer room makes them privileged and transforms the image they have of themselves (everything is new — the instructor, the school, the material — and they don't have the feeling that they are doing mathematics).

I think that the pupils are influenced by the efforts and attitudes of the teacher; the change in pedagogical techniques changes their way of learning, their attitudes and their motivations, and they see that they can gain something. But this may disappear when computers are everywhere.

We don't know yet whether the pupils from these classes will be better than others. But we can assert that they have some pictures in their memories — drawings which illustrate some geometrical properties — and that they have a mental representation of mathematical phenomena; they have built this representation themselves, and so they understand it very well.

In addition, computers encourage teacher-pupil interaction. Pupils also learn to participate, to work together, to help each other, and to be curious, inventive and creative; they are not discouraged if they make a mistake, because they now know that errors are necessary to perfect their work to the last detail.

Conclusion

Drawings have always played a prominent part in mathematics. Here it is from drawing that we do mathematics, and we are convinced that with a computer mathematics teaching is less abstract and more effective. But it is certainly important in mathematics for all the facets of a problem to be developed; it is not sufficient to do drawings and figures. The same figure can be exploited in different ways: for example, when we ask the children to draw this symmetrical figure , it is important to point out the key point, which is the center of rotation. This is why pre-service training is important and must be renewed if such experiments are to be generalized.

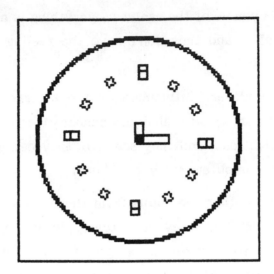

References

Abelson, H. 1984. Le LOGO sur Apple. Cedic/Nathan. Paris.

Berdonneau, C. 1984. Recueil de pratiques pédagogiques autour de LOGO. Agence de l'Informatique.

Bossuet, G. 1982. L'Ordinateur à l'école. Presses Universitaires de France. Paris.

Ferber, J. 1983 (April). "Intelligence artificielle et LOGO." Micro-systèmes 30.

Le Touze, J.C., I. N'Gosso, F. Robert, and N. Salame. 1979 (May). "Apports d'un environnement informatique dans le processus d'apprentissage." (LOGO Project) Ministerie de l'Education I.N.R.P.

Le Touze, J.C. and F. Robert. 1983. "Le système LOGO de commande à cartes. I.N.R.P./C.N.D.P.

Mendelsohn, P. 1985. "L'enfant et les activités de programmation." Grand N 35. C.R.D.P. Grenoble.

Monteil, M.G. 1984. Premiers pas en LOGO. Eyrolles.

Orlarey, Y. 1983 (January). "LOGO: un language d'avenir." Micro-systèmes 27.

Papert, S. 1981. Jaillissement de l'esprit. Flammarion.

Piaget, J. 1972. La représentation du monde chez l'enfant. Fourth edition. Presses Universitaires de France. Paris.

Reggini, H. 1983. LOGO, des ailes pour l'esprit. Cedic/Nathan.

Rouchier, A. 1982 (December). "LOGO et les contenus d'enseignement et de formation." Actes du 1er colloque LOGO: 28-44. Clermont-Ferrand. Edité par l'I.R.E.M. d'Orleans en mars 1983.

Weidenfeld, G., F. Mathieu, and Y.D. Perolat. 1984. LOGO. Eyrolles.

The Introduction of Computers in Elementary and Secondary School in France: Their Effect on Mathematics Teaching

Anne-Marie Pastel

Université de Grenoble, France

The Situation in France

The process of equipping schools with computers is not yet complete, although some schools have been able to purchase what they need. Standard equipment for a secondary school generally consists of 8 microcomputers. The languages usually available are BASIC and LSE; LOGO is also available on some computers. Primary schools, when they have equipment, generally have small "home computers" such as the Sinclair or the Thomson TO 7 and MO 5 or sometimes an Apple II. Often they have bought the computers with their own money because the equipment has not been supplied by the educational system.

Except in a few cases, the schools have not had their equipment for a very long time; the computers are sometimes badly- and/or little-used. In spite of some attempts to introduce computers into the schools (several different plans have been made in the past), the teachers do not know exactly what to do with them. Software is available, but not much of it is really interesting; there is much to be developed in that regard.

Computer science is taught in some technical classes at the secondary level, and also, as an optional subject, in some lycées (grammar schools). Aside from the use of computers in the classroom, many "computer clubs" have been developed.

We should make a distinction between the teaching of computer science and the use of computers for the teaching of mathematics. Most of the teachers using computers are self-taught and have no strong background in computer science. They are often mathematics teachers, but we also try to have teachers of other subjects learn information science because we think it makes for more interesting pedagogy.

Pre-Service Training

There is a problem with pre-service training of teachers in France: there is no specific training for computer science teachers, and most of the teachers of the optional information science classes are self-taught mathematics teachers. So far, pre-service training has not included any training in the use of computers for mathematics teachers. Teachers are selected through competitive examinations. The main competition (CAPES) does not include any test on computers or information science. The "agrégation," a difficult competition which affects only a few teachers, includes for the first time this year an optional test, "The Mathematics of Information Science."

Before these terminal examinations, there are now some elements of information science in the mathematics classes. Last year we instituted a course in Grenoble entitled "Elements of Information Science for the Teaching of Mathematics." This course is part of the "maîtrise", i.e., it is for 4th-year students in the university who are preparing to be mathematics teachers. Within the maîtrise there is an option called "Preparation for Teaching," which consists of two courses: (1) Pedagogical Methods and the History of Mathematics, and (2) Information Science for the Teaching of Mathematics. The latter is divided into three parts:

1. Introduction to Information Science and Computers: elements of algorithms, study of PASCAL and a little BASIC. This part is very classical. The content may vary according to the preparation of the students. (More and more, they will have studied this material previously, but at the moment this is not the case). This part of the course is taken in the first term of the year (October-December).

2. Numerical Mathematics. In this part we study several examples showing how a mathematical problem can be approached in both an algorithmic and a numerical way. The aim is to change the student's outlook on mathematics, which is often very formal and abstract. We want students to acquire some experience in observation, experimentation, conjecture, visualization, and computer simulation. The examples we present are rather classical: convergence acceleration for sequences and series; iterative methods for equations and systems;

difference equations; differential equations (numerical and graphic methods); integration; algorithms for matrix computation; polynomials and interpolation; fast Fourier transforms; and simulation (especially in the area of probability and statistics). An introduction to symbolic systems is also included. This part of the course occupies the second term.

3. Information Science and Teaching. This is the truly new part of the course. It runs during the third term, and includes examples of using information science in the teaching of mathematics:

- pedagogical and didactic aspects
- study of existing software
- study of the LOGO language and its use in primary schools
- writing instructional software, and the pedagogical and information problems associated with it
- expert systems for teaching
- visits to classes where computers are used.

Finally, each student has to work out a personal project: a didactic approach to a mathematical problem and its implementation on a computer (writing a program and technical and pedagogical documents to go with the program).

At the end of the first year of this course, the students expressed the wish that the third part be amplified. This year we are stressing the use of LOGO more, and we are trying to develop with the students some scenarios for instructional software. The students will be required to attend secondary school classes which use computers and to participate in the development of this use. We also envision giving students the basics for evaluating instructional software. Finally, the psychological and cognitive aspects of the use of computers and information science will be studied further.

This year, the course is being given simultaneously in other subjects (geography and physics), and we had several lessons together during the first part of the year. During the second term we worked only with students of mathematics (the subject is not interdisciplinary and the other students have a similar one in their areas). We will work with the students in geography and in physics again during the third term of this year. So the approach is interdisciplinary, and we think it is pedagogically quite interesting.

In-Service Training

Aside from this effort directed towards pre-service training, there are also many opportunities for in-service training in France. The IREMs (I.R.E.M.: Institut de Recherche sur l'Enseignement des Mathématique, or Institute for Research in Mathematics Teaching) have been contributing to the training of teachers for a long time; they have organized many sessions for basic and continued training.

The Ministry has established special programs for three years for training teachers of all subjects in computers. Some teachers are on paid leave for a year to attend the courses. There are 25 centers of pedagogical information science in France, one in each "Académie". Each center accepts about 20 teachers per year. This year a few primary school teachers have also attended at these centers. The instruction is, naturally, interdisciplinary. It consists of a solid base in the characteristics of algorithms, many courses on "information science and education" and programming languages, and some on hardware. The teachers being trained have to complete several projects; in particular, this year they had to produce the information science part of a large instructional program for teaching spelling in primary schools. The teachers also have to complete a project of their own, which is often an instructional program, in the last term of their year.

The training in these centers is not specially oriented towards mathematics (this year in the Grenoble C.I.A.P. — Centre pour l'Informatique et ses applications pédagogiques — there are only 5 mathematics teachers) and the interdisciplinary approach to the use of computers in school makes the teaching at the center very difficult, but also very interesting. Often the teachers carry out their third-term project on the subject they teach, but we have also seen several teachers of different subjects working together on the same program.

The teachers trained there give courses to other teachers, so that by now a substantial number of secondary school teachers have had a small introduction to computer science. These courses are of short duration (50 or 100 hours), but the demand for training is very high and up to now this demand could be satisfied only partially, because there were not enough teachers and not enough money. Today the situation is changing rapidly.

The New Plan

In January 1985 the Prime Minister announced a new plan called "Informatique pour tous" (Information Science for Everyone). This plan demonstrates the government's commitment to progress in computer science in France. The plan has two aims:

1. By June 1986 all pupils will have received an introduction to computer science (at least 30 hours using a computer) by the end of secondary school.

2. All citizens will have access to information science at specially provided sites and it will also be possible for all citizens to use the computers in the schools.

To realize these two goals, 110,000 more teachers will receive special training. Up to now 2,500 teachers have received one year of training and 45,000 have received an introduction, so the number of teachers with computer training will be 157,500 at the end of the year. All teachers are involved, whatever subject they teach. The special training will be given before December 1985; therefore this program will begin during the Easter holidays and will continue during the summer holidays, so that regular school-year courses are not disturbed. The training is oriented mainly towards the use of educational software and data bases in schools. There is not enough time to give the teachers thorough training in algorithms or programming languages; other training sessions are planned for the same teachers in the future. After the initial training we hope that the teachers will be able to use educational programs, that they will know some of the software and will want to introduce it in their own teaching, and also that some of them will be able to give good pedagogical specifications to improve future instructional software. New training for the universities is also envisioned: 300 hours of training for about 800 persons.

"Informatique pour tous" is simultaneously planning how the schools will be equipped.

1. 120,000 new microcomputers will be furnished in 1985 and distributed among 33,000 schools. The future classroom configuration will consist of about ten personal computers (MO 5 or TO 7) connected to a professional computer (SIL'Z III or PERSONA 1600). These "nano-networks" will permit each pupil to have access to a performant system, and they are not too expensive.

2. At the same time, 500 educational programs will be given to the schools.

"Informatique pour tous" is a very ambitious project to introduce computers into primary education, and is also an attempt to introduce information science into real life in France. This plan requires a huge financial effort: 2 billion francs, i.e., about $200,000,000 (France has a population of approximately 55,000,000).

The plan involves only in-service teachers; there are not yet any teachers called "computer science teachers" in secondary schools, nor are there any standardized examinations in pre-service training such as "C.A.P.E.S." or "Agrégation" that exist for the other sciences taught in France.

The Influence of Information Science and Computers on Mathematics

The topic here is not the teaching of information science, but only the utilization of calculators for the teaching of mathematics. (By "calculators" I mean all kinds of pocket calculators and computers.) There are many ways to make use of calculators; one example is the E.A.O. program ("Enseignement assisté par ordinateur": CAI — "Computer Assisted Instruction"). The pupil uses existing lesson programs; the machine asks questions, checks and corrects the pupil's answers; and the pupil works and progresses at his own rate.

CAI can be very useful for pupils who have difficulty with some parts of their studies. Unfortunately, there are few interesting programs for mathematics instruction (important work is now being done in this area). It seems to be difficult to introduce a new concept using only a computer. But it will be useful to make use of calculators before the lesson: computers, too, give us an efficient tool for manipulation and experimentation. Before introducing a new concept one

125

can carry out experiments so that the pupils gain some prior intuition and then are able to make a few conjectures and numerous observations, to visualize complex mathematical objects, and to simulate situations. For example, if we calculate x, $(1 + x)$, $(1 + x)^2$, $(1 + x)^{-1}$, $(1 + x)^{-2}$, for x = 1, x = 0.1, x = 0.01, and so on, we can conjecture that when x is very small $(1 + x)^a \sim 1 + ax$. The Grenoble I.R.E.M. has made experimental use of computers in secondary schools, and it is actually being tried now in the undergraduate biology program of the University (D.E.U.G.B.).

Computers can also be used as a blackboard: in the course of a lesson we can use computers to draw pictures to illustrate the subject and to get an idea about a property. For example, we can view a perspective drawing of the image of any kind of figure using geometric transformations (such as translation, homothety, and rotation). Some graphics programs have been written by teachers working at the Paris VII I.R.E.M. These programs have been used in courses (third and fourth form, i.e., pupils about 13 or 14 years old) to illustrate some of the properties of triangles, some of the properties of parallelograms, and geometric transformations such as symmetry and rotation. Some teachers have used these programs in every geometry lesson, and they find that the experience was good.

Computers can help us to visualize convergence or divergence of sequences and series, and also to draw the graph of a function and the evolution of the graph depending on the parameters. Three years ago we tried to create such a course in the first-year mathematics program at the University of Grenoble. It was an optional course called "Experimental Mathematics," given in the second part of the year. We used pocket calculators in the course because there were not enough computers to do a good job with them. Now we have several new computers in the first-year course, but the difficulties have not disappeared; the undergraduate computers are almost always busy because of the information science courses. For this reason, the professors in the new experimental section of first year are still using pocket calculators along with computers this year.

We can also use computers to simulate random phenomena for experiments in probability, and to do calculus with real data.

It is clear that we have been describing two very different uses of computers in teaching math, and teachers using existing programs will need different knowledge from that required by teachers who want to use computers as a blackboard. The second category need much more knowledge of information science, because there are very few programs for this purpose (and they have not been disseminated to the schools). Some are not too difficult to write, but one needs to know a little about information science to write them.

It may be possible to use some open teaching software: new author programs may permit non-programmers to write short routines to show the pupils certain properties.

I have noticed that almost all the teachers who choose to write an instructional program at the end of the year they spend at the C.I.A.P. — especially the mathematics teachers — write special programs for pupils who are having difficulties; for example, they write programs to "help pupils understand" negative numbers, vectors, angles, or the representations of the points of $R \times R$.

Influence on Notions of Mathematics Themselves

Computers change not only teaching methods but also what is taught; for example, we are not satisfied with an existence theorem but would like to find and exhibit a solution.

The use of computers may also change the curriculum. For example, in secondary school the Newtonian method is now taught for the solution of equations, and the Gaussian method for the solution of a system of linear equations. Perhaps the increased use of computers in French schools will cause greater changes in what is taught. This has happened at the university, where it appears that the teaching of discrete mathematics is becoming more important in the mathematics curriculum.

Instructional Programs

The problem of creating a suitable pedagogical program has always existed. Much research is now oriented in this direction. It is also difficult to find a good pedagogical program because, up to now, programs have been written either only by teachers — pedagogical specialists who have produced only the kind of programs they can write by themselves, and which are thus not very satisfying on the level of information science — or only by professionals, i.e., computer specialists, and thus not entirely suitable pedagogically.

Work nowadays is oriented towards having good pedagogical specifications for a program. Many teachers of a subject have to work together to define the pedagogical interest of a program and give it the required form; if possible, these people should not be computer specialists. Only after the teachers have done that part of the job should they talk with a computer specialist. And finding a language that they both understand and speak is not as easy as one would think. We are trying to plan out the important pedagogical specifications of a problem as precisely as possible, so that the computer science specialist can make the decisions proper to his own science and does not have to make pedagogical decisions.

Perhaps this method will enable us to produce better instructional programs, especially in mathematics, for it is clear that in France really good pedagogical programs for teaching math do not yet exist in sufficient numbers.

Questions Concerning Pre-Service Training

The question of computers and information science in pre-service training is being actively raised in France. One of the main problems is to decide what must be taught to future teachers. Nobody knows what information science will be like 30 years from now; today we are training teachers who will be teaching for 30 or 40 years. It is impossible to give them in advance a body of knowledge for their whole career; the main aim of the training is therefore to give them a foundation, so that they are able to develop and adapt all through their career. This is why it seems to me that it is impossible to solve the problem of pre-service training without taking into account the in-service training of teachers.

Some points about pre-service training:

First of all, of course, the teachers have to learn something about information science. But many questions arise. How far should they go into hardware and the structure of computers? Should they learn only one language or several? Which ones? Is it necessary to have previous training in algorithms, so that they are able to dissociate problems with the algorithm from problems with

128

the language?

At the present time, we do not know what teachers will have to teach about information science. We do not even know exactly what information science will be ten years from now. Therefore, we do not know to what extent the teachers will have to write their own programs themselves. "Expert systems" for teaching are beginning to appear; these will enable the teacher to develop pedagogical uses of the computer according to his desires without programming. Then it will no longer be necessary to know how to program in order to be able to use computers for teaching.

The second point is training in the pedagogical uses of computers. The relationship among the computer, the pupil and the teacher can be dealt with in several different ways. What might the teacher's role be? How much interactiveness and autonomy does the pupil have? Do the pupils write their own programs, or do they use existing programs? Between these two extremes, there are a lot of pedagogical possibilities, and we must improve technology so that the pupils can be interactive.

The same notion in mathematics can be developed in several ways, and it is important that all the interests be served in every situation. That is why appropriate pre-service training must be given to future teachers.

We have still not found a really good way of improving the teaching of mathematics using computers. We are just now seeing that the introduction of information science into teaching, and particularly into the teaching of mathematics, can change our habits and our ways of thinking about and teaching this science. We must find the right solution to this problem, for the world is changing and teaching is changing, and children have to learn to live with all these new technologies.

Mathematics in Swedish Schools

Hans Brolin

University of Uppsala

Compulsory Education in Sweden

The nine-year compulsory comprehensive school in Sweden was established in 1962. It was revised for the first time in 1969. For that reason, the first two curricula[1] are known as "Läorplan for grundskolan 1962 (Lgr 62)" and "Läorplan for grundskolan 1969 (Lgr 69)." The latest reform was decided upon in 1980, and "Lgr 80" has been in force since the 1982-83 school year. Compulsory schooling starts at the age of 7. The municipalities are mandated to offer pre-school to all children at the age of 6, but pre-school is voluntary.

The compulsory school involves two three-year levels of primary education — the junior level (grades 1-3) and the middle level (grades 4-6) — and a third lower-secondary level, the senior level (grades 7-9). In the first two levels instructors teach almost all subjects. At the senior level, teachers generally specialize in two or three subjects. Music, arts, physical education, etc. at the senior level and often at the middle level are taught by teachers who specialize in only one subject.

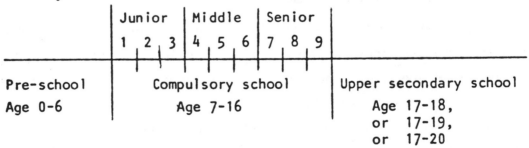

Figure 1. School Structure in Sweden

Almost all children in Sweden attend the public compulsory school. Private schools are very rare in Sweden, though the Education Act takes a liberal attitude towards private persons or groups who want to establish private schools. Less than one percent of all children attend private schools.

The school year starts at the end of August and ends in mid-June. It comprises 40 weeks (Monday-Friday) and is divided into two terms, called the autumn term (roughly August 20 - December 20) and the spring term (roughly January 10 - June 10). The exact dates vary from one school year to another and are not the same in all school districts.

The general goal of the compulsory school is set forth in the first paragraph of the Education Act: "The purpose of the education which society provides for children and young persons is to impart knowledge to the pupils and to develop their skills and in cooperation with their homes, to promote their development into well-rounded individuals and competent and responsible members of society."

The opening sentences of the goals section of the curriculum read: "Compulsory school is a part of society. The curriculum reflects the democratic view of society and man, the implication being that human beings are active and creative and that they both can and must assume responsibility and seek knowledge in order to cooperate with others in understanding and improving their own living conditions and those of other people. The content and the working methods must be designed in such a way that they promote this attitude towards people and society. It is the duty of schools to give their pupils increased responsibility and powers of co-determination as they grow older and become increasingly mature."

The goals and guidelines are formulated in general terms. They are thus not measurable, and they express a direction for the work of schools rather than what schools shall attain.

The syllabuses for individual subjects or groups of subjects also outline the goals for the work in these subjects or areas. They have to be read in parallel with the general goals and guidelines. The goals for each subject are also formulated in rather general terms. Aside from goals, the syllabuses also contain main topics in individual subjects for each of the three levels of the comprehensive school. These, too, are expressed in fairly general terms.

Mathematics in Swedish Schools

Compulsory School

Time Schedule

Lessons per week (one lesson = 40 minutes)

	Junior			Middle			Senior		
	1	2	3	4	5	6	7	8	9
Mathematics	4	4	5	5	5	5	4	4	4

Curriculum (From the curriculum for the compulsory school, 1980)

Mathematics is included in compulsory school instruction because

- it can be used to describe reality and to calculate the consequences of different actions,
- the pupils' mathematical skills must be gradually built up and they must come to realize how those skills can be utilized.

The teaching of mathematics must take as its starting point the experiences and needs of the pupils and must prepare them for their role as adult citizens. First of all, therefore, pupils must acquire the capacity to solve the type of mathematical problems which commonly occur in everyday life. This means that the pupils must gain the following from the instruction they receive:

- a firm command of numerical calculation, with and without aids
- proficiency in mental arithmetic and rough estimates
- above all, a knowledge of percentage calculation, practical geometry, units and unit changes, and descriptive statistics.

Through their school work pupils must also acquire mathematical knowledge and skills which will be useful to them in their studies of other school subjects, in further studies after compulsory school, and in their leisure and working life. In addition to the above-mentioned requirements, therefore, pupils must also acquire a knowledge of the following:

- real numbers
- geometrical relationships

- algebra and the basic theory of functions
- statistics and the theory of probability
- the use of computers and information science

Mathematics teaching must be sufficiently concrete for every pupil to comprehend concepts and understand the use of mathematics in practical situations. Teaching must be arranged in such a way that pupils come to appreciate the need for using mathematics while experiencing the satisfaction of applying the skills they have learned. Mathematics teaching must utilize the pupils' curiosity and imagination and develop their logical thinking. In this way mathematics will become a tool for understanding reality and a source of benefit and gratification.

Main Mathematics Topics

The structure of mathematics is hierarchical. That is, new subject matter is generally based on knowledge of more fundamental subject matter. This must be carefully borne in mind when planning the individual pupil's instruction. A pupil must not start a new topic without sufficient grounding in the preceding topics. Guidelines are suggested by the following division of topics at each major level.

Junior level. Topics which all pupils are to study to acquire a basic knowledge and proficiency during the junior level grades.

Junior and intermediate levels. Topics with which most pupils should become acquainted during the junior level grades and which all pupils are to study to acquire a basic knowledge and proficiency during the intermediate level grades.

Intermediate and senior levels. Topics with which most pupils should become acquainted during the intermediate level grades and which all pupils are to study to acquire a basic knowledge during the senior level grades.

Senior level. Topics the pupils should become acquainted with during the senior level grades. These items are particularly important to pupils intending to go on to lines of upper secondary school or comparable forms of subsequent education involving a great deal of mathematics.

133

Problem Solving

The fundamental goal of mathematics is for all pupils to acquire the capacity to solve the mathematical problems they encounter at home and in the community at large. To solve such problems, one generally has to meet the following requirements:

- One must understand the problem and have a method of solving it.
- One must be able to cope with the necessary numerical calculations.
- One must be able to analyze and evaluate the result and draw conclusions from it.

All these stages in the solution of problems must be covered in teaching. Teaching must include practice in discussing and adopting standpoints concerning both the nature of the problem and the plausibility of the solution, and it must not be confined to one-sided drill on predetermined calculations. Talking mathematics is an important part of teaching.

Problem solving must be included in all main teaching items. Generous scope must be given to practical problems of everyday life.

Problems ought to be selected primarily from the pupils' experience and interests and from the immediate environment, but they should also shed light on social and global problems. Calculations must be closely adapted to the skills of each individual pupil.

Junior and intermediate levels. Problems should be concrete and should emanate from the pupils' experience and immediate surroundings, or from environments jointly constructed in the classroom. Considerable scope must be given to the interpretation and discussion of written problems.

Intermediate and senior levels. Problems at these levels are increasingly taken from environments in the surrounding community which are studied in other school subjects. The selection of problems should also be guided by the pupils' interests and by a focus on future vocational activity.

Senior level. Problems are now also given a theoretical slant and can, for example, include formulae and simple proofs. Examples can be given from occupations where the capacity for solving mathematical problems plays an important part.

Basic Arithmetic

Arithmetic teaching must start with and be rooted in everyday problems and situations of concrete relevance to the pupils. In order to be able to solve mathematical problems of an everyday nature, the pupils need, among other things, proficiency in various methods of calculation. Review and individual diagnosis are particularly essential in arithmetic, because the various modes of calculation are interdependent. The practice of skills must continue individually until each pupil has mastered the subject matter.

<u>Junior level.</u> The concept of numbers is built up by means of laboratory exercises and comparisons of numbers. Natural numbers up to 1,000 are covered in connection with everyday problems leading to addition and subtraction. The concepts of multiplication and division are considered, but the treatment of algorithms should be left until the pupils have acquired a firm command of addition and subtraction. The latter requirement depends, in turn, on a firm knowledge of addition and subtraction tables up to 18. Multiplication tables through five are then learned.

<u>Junior and intermediate levels.</u> The set of natural numbers considered is expanded to 10,000 and decimal numbers, primarily to two places, are dealt with in connection with arithmetic. Multiplication and division are practiced, because they constitute an important basis for mental arithmetic and approximation as well as for learning the corresponding algorithms. Previously learned algorithms are revised and multiplication is learned algorithmically with one factor limited to units. Practical applications of the algorithms thus learned are practiced, mainly in the form of mental arithmetic and rough estimates applied to relevant everyday problems.

<u>Intermediate and senior levels.</u> Arithmetic is limited primarily to natural numbers up to a million and decimal numbers with up to three places of decimals. Previously learned algorithms are revised, and the multiplication algorithm is expanded to include two factors running into several figures. The aim is for pupils to acquire confidence in dividing by at least a single-digit divisor.

Mental arithmetic and approximation are practiced on everyday problems with reference to the algorithms which have been learned. Generous scope is given to practical applications of the four modes of calculations, with and without aids, for example, in connection with wages, costs, unit prices, and so on.

Senior level. Review and practical implementation of the various modes of calculation, using calculation aids where appropriate. Mental arithmetic and approximation.

Real Numbers

Most people have occasion mainly to use natural numbers and decimals. Negative numbers occur less often in practical life, and in the junior grades they are dealt with only in response to pupils' questions and with reference to other subjects. The same applies to squares, square roots and numbers to the power of ten somewhat higher up in the compusory school grades. Numbers such as $\sqrt{2}$ and 3/7, for example, can be rounded off to decimals with the aid of a pocket calculator. It should also be noted that subtraction with negative numbers, like multiplication with two negative numbers, hardly ever occurs in practical situations.

Junior and intermediate levels. Negative numbers occur, for example, in the form of degrees below zero. Fractions such as 1/2 and 1/3 are presented and arranged in order of magnitude.

Intermediate and senior levels. Integers and the commonest fractions are arranged in order of magnitude. Integers are added and subtracted in connection with practical situations.

Fractions with equal denominators are added and subtracted in fraction form. Approximate values of fractions are dealt with. Fractions are added and subtracted after conversion into decimals or decimal approximations.

The concepts of squares and powers of ten, like other topics, are dealt with mainly in connection with everyday problems. Attention is paid to mental arithmetic, approximation and rules concerning plus and minus signs.

Senior level. Integers and fractions are dealt with using all four modes of calculation. Rules concerning brackets and signs are practiced, together with numbers in power form and powers of ten. Attention is also paid to the laws of exponents and to calculation using base numbers. Square roots and their approximations, together with the laws of roots, are studied in connection with geometry and the solution of simple quadratic equations.

Percentages

The concept of percentages occurs in a number of practical situations, e.g., economic deliberations concerning such matters as prices, discounts, interest, installments, and loans and, in the teaching of social and natural sciences, with reference to populations. The pupils should solve problems connected with these and similar situations. In doing so they should acquire proficiency in performing calculations based on percentages and should learn to employ such calculations as a basis for their decisions in everyday situations.

This analytical aspect of problem solving must be given at least the same scope as the calculations, which can often be performed using aids.

Intermediate and senior levels. Percentages are treated in connection with practical problems and other subjects. Shares are calculated in percentage form, and consideration is given to connections between numbers in percentage, decimal and fraction form. Rough calculation and mental arithmetic are treated with the solution of everyday problems, e.g., comparisons of prices, pay changes and discounts.

Senior level. Percentages are calculated when the part and the whole are given. Rough calculation, mental arithmetic and estimates are applied to everyday matters such as loan charges and growth. The pocket calculator is used to work out compound interest, population growth, and so on.

Measurement and Units

Units form an important part of the data employed in most problem solving situations in the home, at work and during leisure. Often one needs to be able to choose the right unit for a calculation and to have mastered the commonest changes of units. In practical situations the unit is seldom given and must instead be worked out in connection with measurement. In teaching, great attention must therefore be paid to measurement using different units and instruments, and to the coordination of this work with the teaching of other subjects, above all home economics, crafts, and science subjects.

Junior level. Length, mass and volume are measured using the units and instruments most commonly occuring in the home. Telling time is practiced, together with specification of time in hours and minutes. The use and denominations of the commonest paper currency and coins are dealt with in connection with practical situations.

Junior and intermediate levels. Units of time are expanded from seconds to years. Writing the date is also treated. Simple changes of units and mental arithmetic are practiced in problem solving with reference to cooking, the calculation of travelling times, prices, etc.

Intermediate and senior levels. The commonest prefixes are studied, as are accuracy of measurement and measuring techniques. Everyday problems are solved to consolidate the commonest units in connection with geometry, speed, local times, trade and currency.

Senior level. Error is calculated in measurement and rounding off into whole numbers. Pupils study units and changes of units in technical and scientific contexts.

Geometry

Geometry teaching must help pupils organize their apprehension of their surroundings. Teaching should start with the home and the immediate environment and tie in with other subjects, especially geography, and crafts. Teaching must also give pupils the capacity to interpret and apply geometric formulas and models.

The capacity for thinking in geometrical models is closely connected with the pupils' development. Most junior and intermediate level pupils have a very limited ability to comprehend and make theoretical use of such concepts as area and volume. The same applies to formal, formula-oriented geometry teaching at the senior level. Therefore, geometry teaching ought to be concrete and applications-oriented, especially at junior and intermediate levels. Furthermore, the arithmetic used in geometrical calculations should be adapted in such a way that the basic geometric ideas will not be obscured by difficulties with calculation.

Junior level. This level treats practical geometry and geometric depictions referring to symmetry and measurement in concrete, everyday situations. The concepts of distance, point, square, triangle, rectangle, and circle are presented.

Junior and intermediate levels. Work at these levels includes exercises to build up geometric understanding. Circumference is presented, together with simple, practical examples of enlargement and reduction. This is done by referring to the local environment via drawings and maps and objects made in crafts lessons.

Intermediate and senior levels. The concept of area is dealt with in everyday situations, starting with examples taken from the immediate environment, such as floor, carpets, walls, lawns, soccer fields, and so on. The concept of volume is dealt with primarily in terms of liters and deciliters. These kinds of calculations should be confined to simple figures, such as rectangles, triangles and cubes. Simple geometric figures are constructed, e.g., rectangles and circles with given dimensions. A number of important concepts should now also be treated in connection with the commonest two- and three-dimensional figures and bodies, e.g., height, base, diagonal, edge, and vertex.

The concept of an angle and the measurement and construction of angles of less than $180°$ are considered.

The everyday use of scale is discussed. The pupils ought to make use of the scales indicated on maps or drawings for simple rough estimates or calculations.

Senior level. The everyday applications and practical uses of geometry are elaborated. Subject matter is taken from the pupils' surroundings and from everyday occupations, e.g., sewing, knitting, painting, and carpentry. The concept of scale is applied in practical contexts.

Furthermore, a foundation should be laid for practical geometry. This instruction calls for great discrimination and careful consideration of the maturity level of the individual pupil. Concepts involving angles in triangles and quadrangles and the intersection of straight lines are dealt with, together with related problems and other common geometrical concepts. The circumference and area of a circle and the area of a polygon are measured and calculated. Congruence and similarity are studied. The construction of geometrical solids and the calculation of area and volume limits of prisms, cylinders, pyramids, cones, and spheres are also investigated.

Some simple geometrical theorems, e.g., the Pythagorean Theorem and the theorem of isosceles triangles, are also considered.

Algebra and the Theory of Functions

This element is of minor importance in everyday life, but all pupils should have some acquaintance with the material. Careful individualization, based on pupils' preferences and capacities, is called for.

Junior and intermediate levels. Simple equalities are solved by trial and error.

Intermediate and senior levels. Simple equations are solved mainly by trial and error and on the basis of problems. The concept of functions is introduced by means of practical experiments. Simple functions in the first quadrant of a graph system are depicted and interpreted. Functional values are calculated by inserting them in formulas from problems referring to everyday life or other school subjects.

Senior level. Graphs are constructed and interpreted using the entire coordinate system. Notation, simplification and calculation of expressions are studied. Parenthetical expressions, the extraction of factors, the rules of quadrature and the conjugate rule are treated with special consideration of the pupils' maturity, interests and needs. First-degree equations, including those with unknowns on both sides and with parentheses and fractions, are covered, as is problem solving using simple equations. Linear functions, especially those indicating proportionality, are examined.

The senior level curriculum also covers linear equation systems and simple quadratic equations, mainly in connection with problem solving and preferably with a graphic solution.

Descriptive Statistics and Theory of Probability

A great deal of the information we receive about the community around us is expressed in the form of tables, charts and probabilities. Instruction is aimed primarily at teaching the pupils to interpret and evaluate this information. It should be integrated with instruction in general subjects and based on real data from the local environment, society and the world at large. Work on this item is particularly suitable for group exercises, and is recommended for projects.

Junior level. This curriculum treats the collection of data and the interpretation of simple tables and charts. Charts ought preferably to be constructed on the basis of empirical investigations.

Junior and intermediate levels. The systematization of collected data is expanded, and the production and interpretation of column charts and graphs are treated as is the concept of mean value.

Intermediate and senior levels. This curriculum covers calculation of averages and the construction and interpretation of pie charts, column charts, etc. It treats frequency and frequency tables in connection with the pupils' own investigations. It discusses critical appraisal of different ways of presenting statistical material, e.g., in advertising, and investigates the probability of events.

Senior level. The senior level curriculum treats classification, histograms, and simple positional measurements, together with relative frequencies and their stability. It also covers the concept of probability and the calculation of probabilities in simple cases.

Examples are given of occupations in which a knowledge of statistics and the probability theory is important.

Information Science

All pupils should be informed about the use of computers in society and the rapid development taking place in this sector. It is particularly important for pupils to realize that the computer is a technical aid controlled by man.

   Senior level. Computer functions are presented, with particular emphasis on the task of the computer program and methods of problem solving. Also discussed are the various fields of data processing in which the rapid pace of technical progress is particularly apparent.

   This topic also offers examples of occupations in which a knowledge of computers is important.

Upper Secondary School

Time Schedule (three-year and four-year lines)

Lessons per week (one lesson = 40 minutes)

Year	1	2	3
Natural Science line	6	5	4
Technology line	6	5	4
Liberal Arts line	5		
Economics line	5	3	3
Social Science line	5	3	3

Time Schedule (two-year lines)

Lessons per week (one lesson = 40 minutes)

Year	1	2
Economics line	3	3
Social line	3	3

Main Mathematics Topics in the Natural Science and the Technology Lines

- Numerical calculations with fractions and decimal numbers
- Some simple theorems concerning the circle and polygons; similarity and some simple geometrical constructions
- The law of sines, the law of cosines, the unit circle, and the additional formulas for the sine and cosine
- Solution of equations and systems of equations and inequalities by graphical and numerical methods
- Proportions; the equations of straight lines
- Limits
- The graphs and derivatives of polynomials, simple rational functions, power functions, exponential functions, logarithmic functions, and trigonometric functions
- Differentiation of a product and a quotient, and the chain rule
- Transformation of algebraic expressions and factoring quadratic polynomials
- Algebraic solution of first- and second-degree equations and systems of linear equations
- The law of exponents and the law of logarithms

The Integral

- The volume of a cylinder and a cone and volumes of solids of revolution
- Calculation of areas
- Probability
- The use of computers and information science. Some simple numerical methods for solution of equations and integrals

At least two of the following alternative items:

- Complex numbers
- Methods of integration
- Differential equations
- Probability, more advanced course
- Vectors
- Series

Some Calculator Projects

The ARK Project[2]

In the spring of 1976 the National Board of Education appointed a group whose task would be to <u>provide an analysis of the consequences of the use of the pocket calculator</u> - known as the ARK group (From the Swedish "Analys av Räknedosornas Konsekvenser). The ARK group's first task was to identify the problems at hand and to launch various projects led by the group's members.

The Problems

Three aspects of the use of the pocket calculator were investigated:

The Pocket Calculator as Numerical Aid

<u>Ages 16-19.</u> The calculator replaces the sliderule and tables. No special problem is seen here.

<u>Ages 13-15.</u> The calculator replaces the sliderule and tables. Problem: What levels of proficiency should now be set for calculations involving decimals and fractions?

<u>Ages 10-12.</u> Problems: (a) If the calculator is approved as an aid, algorithmic calculations will receive less attention. The question is how will this affect:

- the pupil's concept of numbers?
- the pupil's ability to estimate?
- the pupil's ability to solve problems in textual form (word problems)?
- the pupil's own arithmetical capabilities?

(b) What levels of proficiency should now be set for algorithmic calculations?

The Influence of the Pocket Calculator on Current Methodology

In respect to all grades, the following problems have been defined:

a) Should the calculator be employed as an aid in connection with the introduction of certain new concepts?

b) Can the calculator stimulate pupils' interest in problem solving and perhaps improve their capabilities?

c) Can present courses be made more applied and problem oriented?

d) Can practical experiments involving the calculator be introduced into current courses?

The following questions apply only to age group 16–19:

e) Will programmable calculators bring about a new methodology in current courses?

f) Can programmable calculators stimulate an interest in mathematics?

The Impact of Calculators on Course Content

Problem: Can changes in course content be introduced where the calculator is in use?

In respect to all grades there is a need for:

- new ideas
- testing
- formulating and developing new suggestions.

In addition, for age group 16–19, the likelihood of the programmable calculator's influencing future course content should be investigated.

Subprojects

The following sections present a brief account of the most important subprojects undertaken wholly or partly under the auspices of the ARK group.

Ages 10–12

Subproject: IAB (Non-Algorithmic Skills)

The Subproject IAB (Non-Algorithmic Basic Skills) was concerned mainly with the middle level of the comprehensive school (grades 4–6, ages 10–12). Three reports and material for the further training of teachers of this school level have been published.

Non-Algorithmic Basic Skills. The mathematics taught at the middle level of the comprehensive school (ages 10–12) is a complex mixture of skills, abilities and insights at many different levels of abstraction. It is, therefore, very difficult to anticipate the consequences which the introduction of calculators might have on this age group.

The report contains an analysis of such skills and insights which are independent of the means and methods used for routine computations involving the four basic rules of arithmetic (called non-algorithmic basic skills), a test to measure these non-algorithmic basic skills, and an analysis of the outcome of the test.

Children's Ability to Solve Mathematical Problems. The ability to solve routine mathematical problems in the middle age-group of the comprehensive school (ages 10-12) is already a very complicated mixture of many different skills and abilities.

The report describes an attempt to "define" and measure this ability. A test was designed and given to children in grade 6 (age 13). Observations were made and interviews carried out to obtain information on the thought processes of children engaged in solving mathematical problems.

The construction of the test, the test results, and the observations are discussed and analyzed.

Children's Ability to Communicate Using Mathematics. This report describes a pilot study of children's ability to communicate using mathematics. A test was designed to explore and measure some aspects of children's ability to communicate with or about mathematics. The test and its outcome are discussed and analyzed.

The main object of this investigation was to gain insights into children's ways of communication with and through mathematics that would enable the authors to locate the areas in which further research is needed.

Materials for Further Training of Teachers

From the beginning, all those engaged in the ARK Project realized that the very existence of electronic calculators and computers, whether or not they are or will be adopted in school teaching, necessitates a critical analysis of the contents of the mathematics curriculum at all levels and of the methods by which mathematics is to be taught.

A group made up of researchers and practicing teachers set about producing material for further training of teachers at the middle level (for children aged 11-13) which would make teachers aware of the rapidly changing situation and enable them to participate in and influence developments in a conscious and effective way.

The material produced so far is an analysis of the teaching methods and actual content of mathematics as taught to children aged 11-13, independent of the present curriculum. It is to be used in study groups.

Subproject: RIMM

Introduction. The RIMM project was a longitudinal study of the calculator in grades 4-6 (ages 10-12). Its object was to study the consequences of regular use of the pocket calculator where more complicated calculations are required. The following questions were posed: To what extent does regular use of the pocket calculator influence pupils' ability to do mental calculations? Their ability to perform calculations by hand? Their incentive and interest?

In addition, we wished to establish whether pupils' understanding of numbers had improved or deteriorated.

Implementation. For the purpose of this project, specially designed material was created to be used alongside conventional teaching material. The latter excluded the use of pocket calculators. The new material was given a trial run and then revised in accordance with trial results. The revised version was used for the experiment itself, which began with grade 4 (age 10) in 1979-80 and terminated with grade 6 (age 12) in 1981-82. The trial material covered:

- Skills in multiplication tables and mental arithmetic.
- Calculations by hand (fewer in comparison with standard teaching material).
- Everyday problems in textual form requiring a calculator (where word problems were used, the pupils were given practice in estimations).
- Project work where the pupils themselves organized the material and chose the form in which the results were to be presented. Each such project was estimated to require 3-5 lessons.

In all, 8 trial classes and 3 control classes were involved in the experiment.

Method of Assessment. Tests and questionnaires were completed in all classes at the beginning of the autumn term in grade 4 (age 10) and at the end of the spring term in grade 6 (age 12). The test given in grade 6 was set by the IAB group of the ARK project. Classroom observations were carried out and teachers were interviewed.

Results. Test results for grade 4 showed no significant disparity between trial and control classes. As pupils neared the end of grade 6, trial classes showed better results than control classes in

- creative thinking
- ability to estimate orders of magnitude
- estimation
- choice of correct method of procedure for problems in textual form
- ability to extract relevant information in more complex problems.

In other areas covered by the experiment, no tangible differences between trial and control classes were noted. Results provided by the test in mental arithmetic and calculations by hand were interesting. Trial classes did as well as control classes in both tasks.

Questionnaires showed that

- trial classes tended to regard mathematics as being easier than did the control classes after the experiment
- trial classes did not appreciate the sections on estimations which were drilled rather intensively during the experiment
- control classes were considerably more skeptical about the benefits of calculators in school both at the beginning and at the end of the experiment.

Classroom observations of the trial classes provided some interesting results. Four classes were observed for 4-5 consecutive lessons, in general about 3 times per academic year. Among points worth noting:

- many pupils preferred to calculate mentally or by hand even when they had access to a calculator
- on the other hand, some pupils used the calculator to excess, to the extent

of using it for estimations

- many pupils, mainly in grade 4, were nonplussed when the display was filled with figures

- when a new type of problem is introduced, the arithmetic in the introductory text should be sufficiently simple for the pupil to grasp the correct means of procedure. Problems which the pupil can solve mentally are to be recommended in this respect.

From interviews with the teachers it was apparent that they were most enthusiastic about the trial. For one thing, it was felt that the use of the calculator would introduce greater variety into mathematics teaching. We can therefore draw the conclusion that regular use of the calculator in a teaching situation need not adversely affect the pupil's ability to calculate by hand or mentally, as had been feared. We have also established that the increased training in solving everyday problems made possible by the calculator has improved pupils' ability to estimate orders of magnitude. An improvement has also been noted in the pupils' ability to choose the correct method of operation and to extract relevant information from text. In addition, the trial has inspired new ideas for using the calculator. It is hoped that these will lead to an even more effective training of the pupils' problem solving abilities than was possible during the experiment.

Ages 13–15

Subproject: Experimental Mathematics, Ordinary Course

Everyone needs to feel that his/her work is worthwhile. Without this feeling life loses its luster. The situation of pupils who lose interest in school work may be seen in this light. For such pupils, school has no incentives to offer. The school system can then be accused of failing in its duty to help young people develop in such a way that they can find a place in the society we have created. In particular, pupils who are less interested in theoretical subjects have been sadly neglected. These pupils must be helped to change their attitudes if they are to benefit from their school years.

With this in mind, we feel that there is a possibility of bringing about a change in attitudes if mathematical studies can be pursued in a more practical way. Pupils themselves can take problems from everyday life, from the reality which interests them. This will lead to the discovery that mathematics in all its forms surrounds them — an important discovery and one which renders the subject more interesting and exciting. The calculator has proved to be a great boon to those pupils who must otherwise struggle with a calculation, never quite crossing the threshold into discovery. Instead of a situation where young people toil listlessly with endless and meaningless calculations, the opposite is possible: young minds confident of their abilities, expectantly searching out new areas to explore, wanting to know more, to learn. Shouldn't school be more like this?

For several years now we have worked with "Experimental Mathematics with the Help of the Pocket Calculator" and have compiled an "Introductory Pamphlet" which suggests practical problems suitable to start with. We have also created a compendium of practical problems, all of which have been tested in our own classes. We hope that those who find this approach rewarding will be encouraged to create more material in this vein. There is also a program sufficient for a half study-day and the ITV program showing the practical approach and the pupils' response.

Ages 16–19

Subproject: Changes in Methodology in Mathematics for Natural Science/ Technical Courses

The introduction of calculators into the classroom has radically affected the subject of mathematics. It is therefore imperative that both aims and methods in mathematics teaching be reviewed.

To begin with, where the calculator is used as an aid, it is more natural to express numbers in decimal form. In my opinion, all solutions to mathematical problems should as a rule be expressed in decimal form, the precise number being given where appropriate. There will then be more opportunity (and hopefully inclination) to check the plausibility of an answer (nowadays this aspect of problem solving is sadly neglected). Now that we have calculators, problems can and should be constructed in such a way that the numbers do not easily lend themselves to calculations by hand. There is no longer any need to weed out more complex, realistic figures.

In addition, with the advent of the calculator, there is more opportunity to introduce practical steps in conjunction with the teaching of theoretical passages. For example, in connection with limits, series, probabilities, etc., pupils can draw up various kinds of tables. This experience should make the theoretical aspects easier to understand. The pupil should also be given the opportunity to allocate time for writing out as well as working out the calculation. It is hoped that the new emphasis placed on the practical aspects of problems will benefit some of the weaker pupils.

It is worth remembering that the calculator does not perform only arithmetic. The type which should be used in the Natural Science/Technical courses functions as an easily accessible and easily manipulated set of tables. Important areas which were previously regarded as too time-consuming can now be given the attention they deserve, since the pupils can now manage more — and more varied — problems of this kind.

Subproject: Probability and Statistics Using Programmable Calculators

The aim of this project was to investigate how the use of programmable calculators might influence the teaching of probability and statistics in Swedish secondary schools. The project was organized as follows. Preliminary teaching material was written and tried out in a secondary school in Gothenburg (Göteborgs Högre Samskola). On the basis of this experience a complete textbook was written together with supplementary booklets giving programs for different programmable calculators. This textbook has subsequently been used in a large number of classes in various secondary schools in Sweden. Teachers and students have been asked to fill in questionnaires with their responses to the material. The project has shown that programmable calculators can be used in a very effective and, for the students, stimulating way to teach probability and statistics. With the aid of such machines important concepts and methods in these areas can be illustrated and investigated. Random experiments can be simulated, and variability and stability, inherent in such experiments, can be demonstrated in a challenging way. Programmable calculators can also be used for a new methodology in these areas. Probability studied in this way will also provide the students with several non-trivial programming exercises.

Subproject: Project Work With Programmable Calculators and Computers

The aim of this project was to write and test instructions for students who are to participate in mathematical projects using programmable computers or calculators. New teacher guide material was also included in the trials. Instructions have been written for seven projects with the following titles: Nikomachos and Euclid; Fibonacci; Prime numbers; Girl or boy?; Stochastic geometry; Economy; Statistical data analysis.

Subproject: NUMA

From 1976-1980 experiments were conducted in numerical methods in mathematics and physics involving 16-19-year-olds. The main objective of the NUMA experiment was to demonstrate how simple procedures like iteration, step methods, extrapolation, and simulation, can, when used with the appropriate aids, provide extremely effective methods for problem solving. Where programming is concerned, schools could concentrate on either programmable calculators or computers (BASIC), depending on the type of equipment available. The aim has been to improve the pupils' understanding of mathematical concepts rather than to keep them preoccupied with the fine details of programming. The experiment comprised the following areas:

Mathematics
- An introduction to numerical methods in connection with equation solving and integrals.
- Numerical solutions of differential equations.
- Probability theory and simulation.

Physics
- Step methods in mechanics and electricity.
- Step methods in connection with mechanical and electrical vibrations.
- Radioactive decay.

Classes involved in the NUMA experiment were given special problems in the central examinations.

Examples of Questions Posed During the Experiment:
- What type of programmable equipment (pocket calculator or

minicomputer) is most suited for teaching mathematics and physics to pupils aged 16-19?

- What new steps of a numerical nature should be included in mathematics courses for the 16-19 age group? Which areas of physics can be better understood through the use of repeated calculations in small steps?

- Do discrete and step methods create a better understanding of basic concepts in mathematics and physics? For instance, shouldn't numerical work with difference quotients and step methods in applied problems precede the analytical treatment of derivatives, integrals and differential equations?

- Is interest in mathematics likely to be stimulated by the practical steps introduced by numerical methods?

The findings of the NUMA experiment in mathematics are reflected in the latest curriculum and in the textbooks for Natural Science/Technical studies.

Subproject: Applied Statistics for Pupils Aged 18-19

Background. Many mathematics teachers have expressed concern over certain areas in mathematics courses for Civics/Economics students ("a trimmed-down version of the mathematics course taken by Natural Science students, with no attention given to the special needs of those who study Civics or Economics"). Most criticism was directed at the section dealing with mathematical statistics at the end of the pupils' third year. This situation prompted the ARK project group to initiate a trial course where probability theory was for the most part replaced by applied statistics.

Content. Text material for trial purposes was built around newspaper headlines connected in various ways with statistics. This material was to provide the basis for systematic training in the critical analysis of conclusions drawn from statistical data. The material provided examples of inappropriate or shifting definitions, inaccurate measurements, random errors, non-response errors, questionable sampling methods, confusion of cause and effect, and biasing factors. The numerical examples were concerned mainly with non-response errors, the statistical margin of error for percent figures, and the elimination of the influence of biasing factors (method of standardized avarages).

Object. The object of this experiment was to test the attitude of pupils and teachers to the use of a text of this kind in mathematics and also to establish to what extent the pupils can benefit from the contents. Critics have claimed that this type of material covers a difficult area which demands a higher degree of maturity than that possessed by this age group.

Procedure. 8 classes from 6 schools participated in the experiment, which took place at the end of the spring term of 1981. The following courses were represented: 3rd year Civics - 1 class, 3rd year Economics - 1 class, 3rd year Natural Science - 1 class, 2nd year Social Science - 3 classes. The 8 teachers involved cooperated on a voluntary basis. Because of the dissimilarity between the courses, variations in method were permitted. 6 classes read all, or almost all, of the 97-page text over about 20 lessons, with the exception of the 3rd year Natural Science students, who used only 12 lessons. Pedagogical methods also varied. In some cases teachers led the lessons, in other classes pupils worked independently in groups.

Results. It is, of course, hazardous to generalize from the results of such a small-scale experiment. Some pupils reacted negatively to the detailed text section. This was felt to be out of character with mathematics as a subject. On the whole, however, response was favorable. The majority of pupils thought that the course was interesting and that they had gained new insights and skills. Most teachers recommended that the critical analysis and applied statistics be extended over the entire course and were enthusiastic about the possibilities of coordinated study with other subjects — particularly Civics. This experiment has broken new ground by showing that, under the conditions set by the experiment, it is possible to create a course which meets with the approval of the majority of pupils. Perhaps the most important result is to be seen in the new syllabus for mathematics in Civics/Economics proposed for 1984-85.

Some Conclusions and Recommendations
Handling Numbers in Lower Grades

All pupils should be proficient to perform by hand
- multiplication of two-digit numbers

- simple division

In order to note any deterioration and take appropriate measures, tests have been constructed to monitor any changes which may be attributed to the introduction of the calculator.

Teaching Methodology

Mathematics courses should be

- more applied
- more experimental
- more problem oriented
- more directed towards understanding concepts

Impact of Curricula So Far

- Introduction of numerical methods and simulations into mathematics courses for age group 16-19
- Introduction of applied statistics and critical analyses into mathematics courses for age group 16-19

Notes

1. The curriculum is national. It contains general goals and guidelines, time schedules and syllabuses with (a) goals for subjects or group of subjects and (b) main topics for teaching in each subject or group of subjects.

2. A copy of "The ARK Project - A Progress Report for the Period 1975-1983" by Lars-Eric Bjork and Hans Brolin can be obtained from:

Hans Brolin
The Institute of Education, University of Uppsala
Box 2136
750 02 Uppsala
Sweden

An Example of a Long-Life Curriculum Project and Its Underlying Philosophy

M. Bruckheimer and R. Hershkowitz

Weizmann Institute of Science, Rehovot, Israel

Introduction: The School System in Israel

The population of Israel consists of just under 4,000,000 people from a wide variety of cultural backgrounds. About 1,200,000 pupils attend Hebrew-language schools and about 250,000 learn in Arabic.

The school system consists of a 12-year course, free but compulsory only up to tenth grade. Curricula are subject to approval by the Ministry of Education, which is also responsible for the matriculation examination for those completing grade 12. The system allows for considerable flexibility. At the high school level (grades 10-12) there are two parallel systems, the academic and the vocational. There are seven public institutions of higher learning, each of which includes a department for research into problems related to the teaching (and learning) of mathematics.

The original structure of the Israeli school system was an "8 + 4" structure, but a transition to a "6 + 3 + 3" system, that is, including a junior high and a high school, is currently in progress. About half the schools have now been so transformed.

The junior high schools are comprehensive and compulsory. The main goals in their creation have been:

1. integration of different population groups
2. provision of teachers with academic backgrounds in order to make possible an "improved" curriculum, especially for grades 7 and 8.

There is much debate as to whether the first goal has been achieved, but it is clear that the second has been fulfilled at least partially. Thus, 65% of mathematics teachers in junior high school now have the requisite formal academic qualifications, whereas only 25% in the parallel classes still organized in the old two-tier fashion are so qualified.[1]

The reorganization of the school system was taken as an opportunity to develop and implement new curricula, beginning at the junior high school level.

The Israel Center for Science and Mathematics Teaching was set up by the late Professor Amos de Shalit about 20 years ago. The Center has departments in a number of academic institutions, as well as a section in the Ministry of Education responsible for curriculum development and implementation.

Mathematics Curricula

The curriculum changes required by the new junior high schools have in turn led to major changes in both the primary and senior high school systems.

Primary School (Grades 1-6)

A new curriculum encouraging the use of calculators in all grades has recently been adopted. Algorithms such as long division receive much less stress. No attempt is made to use set algebra for introducing natural numbers and the operations of arithmetic. The syllabus for grades 5 and 6 also includes some basic concepts and methods of probability theory. By the end of the 1984-1985 school year, some 100,000 pupils will be practicing arithmetic regularly with the aid of computer terminals.

Senior High School (Grades 10-12)

Academic high schools take only part of the population; the rest go to various vocational schools, whose syllabuses and materials are, in general, traditional.

In the academic high schools and the upper streams in vocational schools, courses are offered at three levels, each with its own curriculum and each leading to an appropriate matriculation examination. Students choose the level they wish to take. The current curriculum is fairly conservative, based on algebra, geometry, and trigonometry. Calculus and probability are optional subjects for students matriculating at the lowest level, but are mandatory, as is analytic geometry, for those taking the examination at a higher level.

Now that a "modern" curriculum has reached general implementation in the junior high schools, it has also become necessary to modernize the senior high school curriculum in order to preserve continuity. Under the new curriculum, which is being implemented gradually, calculus will be taught in Grade 10 at all three levels together with a variety of subjects, including vectors and linear programming. About 30% of the time will be left free for elective courses on special topics such as topology or numerical analysis, including computational laboratories.

Junior High School (Grades 7-9)

Here there are three parallel courses at different levels of difficulty, each with its own textbooks. Students are streamed within each school at the beginning of Grade 7, and are encouraged to move from one stream to another as a result of their experience and achievement.

The curriculum is modern; ongoing gradual change is attuned to students' and teachers' needs and abilities, as well as to external influences. The project at the Weizmann Institute, which is responsible for most of the mathematics activity at the junior high school level, was started some 20 years ago. Later, we will describe some of the work of this project to illustrate the philosophy which guides it.

Mathematics Teachers

The process of changing the curriculum puts demands on teachers. In-service training courses are provided for teachers who wish to participate, and many do (Hershkowitz and Israeli 1981).

Prospective teachers for the elementary schools are trained in teachers' colleges, usually as general teachers. Since they do not always have a deep understanding of mathematics, they have to devote extra effort to cope with the new curriculum. In general, they seek to improve their teaching by in-service training and involvement in experimental programs. These programs include individual instruction, group study, and computer-assisted drill and practice.

All high school teachers are university mathematics graduates and have a relatively sound mathematical background. The academic institutions offer courses for teaching certificates.

The educational background of teachers at the junior high school level and the parallel stream in the "old system" elementary school is very heterogeneous; we will come back to it later.

Students and Mathematics

As mentioned, the population includes a variety of groups with different social backgrounds. About a third of the student population has been defined as "socially deprived." Special projects are needed to enhance the mathematical education of this group, and several such projects exist.

Almost all students take a matriculation examination that includes mathematics, and about 30% of the students take mathematics at one of the higher levels.

A Philosophy of Curriculum Activity

We view the activity of curriculum development as a continuous, long-term process consisting of four interactive components:
1. Creation of learning materials and learning and teaching strategies
2. Their implementation in the classroom
3. Evaluation of the different stages of creation and implementation (which leads to better creation and implementation, and so on)
4. Research that focuses on student and teacher needs, abilities and disabilities, examines results as reflected "in the mirror" of curriculum and teaching strategies, and is used as a basis for planning and replanning the other components.

These components take place in interlocking and ongoing cycles, the aim of which is to improve the conditions, means, processes, and products of teaching and learning. Rather than continue the discussion of this philosophy in abstract, we will try to reflect it in the reality of curriculum activity in a junior high school mathematics project.

The "Revolution," or "First Aid" Stage

Some eighteen years ago, mathematics teaching in Israel consisted of very "traditional" arithmetic: complicated computations at the elementary level and algebra and calculus, with stress on techniques, at the high school level. The political decision to create junior high schools transformed the vision of a few mathematicians and scientists of an updated and improved mathematics curriculum into the official programme for the newly created junior high schools. A mathematics group (and the Science Teaching Department) was formed to make this transformation a reality. A very small team, consisting of a single mathematician and a few mathematics teachers, had to supply materials to the first schools within one month. At that time, most of the schools were located in new immigrant towns whose population had come largely from non-Western countries with different cultures and backgrounds, and had to adjust themselves as quickly as possible to a Western technological society. It may well be that this is one of the factors that kept us from making the naive assumption that the creation of textbooks will, in itself, guarantee the "success" of the program — an assumption made by many curriculum development projects and responsible people in ministries of education. Rather, the project's philosophy had a very pragmatic beginning, in a realization that without efforts and activities by the project team that went beyond the creation of textbooks, the chances of any even moderately successful implementation of the new curriculum were small. An in-school guidance system and in-service teacher training program were organized from the very beginning and reached the vast majority of the target population. The end of this stage in the project came with the completion of the first draft of a complete sequence of textbooks for grades 7 to 9, influenced in content by American and English projects of the 1960's. Separate textbooks were prepared for each of the three ability levels.

While the schools were being reorganized in the three-tier system, the target population grew at an enormous rate. The guidance system and the in-service program were extended to keep pace with demand. The project even became responsible for publishing and distribution of textbooks and other teaching aids. The inclusion of all these functions gave the project a degree of control and flexibility that proved invaluable in many future activities.

The "Creative Implementation" Stage[2]

During the first stage, the project team was driven by the needs of the moment. With the completion of the first stage came the opportunity to move into the driver's seat and plan ahead. Although not all pressures and constraints had been removed, the situation left room to maneuver. For the team members it was clear, first of all, that work must continue; we had not come to any sort of conclusion.

The main characteristics of this new stage were as follows.

1. The core curriculum (textbooks for students and teachers) had been prepared. Activities with teachers had been established (in-service teacher training programs and the guidance system in schools).

2. The rate of growth in the target student and teacher population was enormous. This led to

3. Heterogeneity in the population, which suggested a need for

4. Refinement of evaluation instruments and strategies to gather information on differences between subpopulations, differences in professional background and attitudes of teachers, and differences in the achievement, performance, ability, and motivation of students. This led to

5. Refinement of implementation strategies in order to improve implementation of the curriculum. We have chosen to give these strategies the collective name of creative implementation. As opposed to previous strategies that took the curriculum materials as fixed and acted on teachers and students only through directly course-related activities, the new strategies, although the core curriculum remained relatively untouched, included the creation of supplementary materials for students and teachers designed to improve learning in a variety of situations.

6. There were also some specific changes in the core curriculum arising from such sources as changes in the demands of the discipline and surrounding influences. These changes were accommodated in the curriculum gradually and on an evolutionary basis. For the most part, they have been accompanied by formative and summary evaluation and by intensive activity with teachers.

These general characteristics will become clearer from the following two examples.

Example 1: Activities for teachers. The style of in-service teaching at the beginning of the project was mainly lecturing, followed by exercises taken from student manuals. With time, we adopted the attitude that teachers' teaching strategies could be influenced if we used those very strategies in the in-service courses. So we combined the two elements, teaching subject matter — for which there was and still is a great need[3] — through strategies we hoped the teacher would use in his own classroom. Passive frontal lectures turned into active workshops.

All this was done on the basis of intuition only. However, we reached the point where the number of teachers was so large, their needs so different, and the constraints of money and time so severe as to force us to rethink carefully and plan our activities efficiently for optimal use of resources. To do so, we needed quantitative information about our population and its needs, the effectiveness of our activities, and so on. We began by carrying out a survey (Hershkowitz and Israeli 1981) that served as "patients' files" (Bruckheimer and Hershkowitz 1984) from which we could develop a treatment with hope for a cure. The survey gave us information of the following sort.

- About 80% of all teachers are female;
- Most teachers teach more than one ability stream;
- About 75% of mathematics teachers in grades 7 and 8 in primary school have no formal qualifications whatsoever in mathematics;
- In junior high school this figure drops to about 35%;
- Participation of teachers in summer in-service courses (average duration of two weeks) is summarized in the following chart (figures are rounded).

45%	40%	15%
never participated	1-2 courses	3 or more courses

- Participation of teachers in day workshops (about 10 are organized each year) is summarized in the following chart (figures are rounded).

25%	35%	40%
never participated	1-5 days	6 or more*

* some 10% have attended 11 or more.

- In-school guidance has reached about 50% of junior high mathematics teachers, but less than 15% of primary school teachers (grades 7 and 8).

From the data we were able to identify several major types among our "patients" and their particular needs. Many activities were given impetus and direction by this survey.

1. The need for multipurpose courses on basic mathematics that would be stimulating and motivational, and could provide enrichment for those with stronger mathematical backgrounds, led to a variety of new courses. One of the more original of these is a workshop course on the history of a number of basic mathematical concepts such as negative numbers, irrational numbers, and equations.[4] These courses, designed to give teachers a feeling for the conceptual development of the topic, use original sources whenever possible; teacher activity is guided by questions on these sources. Originally the courses took the form of workshops with tutors. After teachers had gotten all they could out of the worksheets, small group or individual discussions with tutors, and a summary discussion led by the tutor with the entire group, the course participants received detailed answer sheets that often presented further source material and enrichment materials relevant to the work at hand. These materials were also used as the basis of a correspondence course.

Development of these materials was a major activity, comparable to the development of entirely new curriculum materials. But their purpose was to improve the implementation of an existing curriculum by improving teachers' knowledge of mathematics used in the school curriculum and enriching their view of mathematical activity.

It is interesting to note that, in keeping with a general tendency in our work with teachers, the history workshops were accompanied by both cognitive and affective evaluation. Although analysis of the data here is not yet complete, there are indications that some of our intended objectives have been achieved at least in part, but that not all we had hoped for actually happened. When the information is complete, we intend to use the results as a starting point for the creation of further activities. Curriculum research

163

and, in particular, curriculum implementation is still very much at the stage of an intuitive art. What we are trying to do over a long period of intensive activity is to turn such research into an experimental science. We are currently introducing an evaluation component into all our in-service activities to put a sharper focus on whatever problems are identified and to improve our practice. This activity is part of the project's third-stage philosophy, to be described below. We note here only that one of the specific objectives already mentioned — encouraging teachers to use non-frontal techniques through our own use of these techniques in in-service courses — is not an unmitigated success. This is one area that will have to be investigated further and that requires further treatment.

2. The above example of the history course illustrates the implementation of a curriculum through work with the teacher. The following example, although it began in a similar way, ended by affecting the curriculum materials themselves. As happens with so many curricula, certain parts achieve near-zero implementation. This happened to the sixth and final chapter on quadratic functions in the Grade 9 algebra text. A few years ago, we decided to attack the problem and try to implement the chapter. The hypothesis was that if we expressed concern for the chapter and provided the necessary support, we would find teachers willing to adopt it. We therefore arranged one-day courses devoted to quadratic functions.

While developing the teacher worksheets for these courses, members of the team saw the opportunity for a wholly new and much lighter approach to the topic. This approach involved much more activity on the part of the student, thereby shortening considerably the time that had to be devoted to the subject.[5] From beginning to end, our intentions were to implement the existing curriculum; but we allowed the existing curricular material to remain a variable in the process.

Example 2: Activities for students. One of the major activities in this stage of the overall curriculum project was a special sub-project intended to "bridge the gap" between existing mathematics curriculum materials and socially deprived students.[6] What we sought were ways of increasing communication between the three components of teacher - materials - student, or just the components materials – student. There has been a tendency in various parts of the world to create special learning materials for the socially deprived. For a number of reasons, it seems to us more efficient and effective to concentrate on the further development of existing materials, adapting them to the needs and abilities of this population. This is yet another aspect of creative implementation. One pragmatic reason for our adherence to the existing curriculum is that in the future the socially deprived will have to contend with the same criteria and standards as others. To alter or reduce the demands on the socially deprived population in junior high school would only be asking for trouble later on.

During the first year of the subproject, certain fundamental difficulties as well as some major types of successful treatment were clearly identified. As an example, let us consider just one difficulty — the lack of prerequisite knowledge. The original syllabus, like most syllabuses, was based on the assumption that the student enters grade 7 with a certain body of knowledge and degree of technical competence. It was, however, no surprise when it became clear in the first year of the project that this assumption was entirely false.[7] To obtain a quantitative description of the extent of the problem, a needs assessment survey of a sample population was carried out which demonstrated a lack of mastery of even very elementary computational skills. Interestingly, there was a clear and significant correlation between the percentage of students in the school who were socially deprived and the lack of mastery of basic skills.

Remedial programs designed to close the gap in prerequisite knowledge were created. Each of the programs contains a diagnostic stage in the form of a test, which also serves as a handy class evaluation tool, and a treatment stage, consisting of a sequence of small units taking the form of highly motivational individual or group activities. At the diagnostic stage, the teacher is able to decide how to use the materials (e.g., with the whole class or just groups or individuals; the whole subject or just subtopics). In any case, the materials are intended for use in parallel with the main course curriculum so as not to delay or discourage the students. The development of the various remedial programs (in fractions, decimals, geometry, and so on) took place in the classroom, and was followed by a systematic evaluation. The evaluation demonstrated a considerable gain from the use of the materials, and effective retention of this gain after a considerable period of time.

This example is indicative again of our rejection of the naive approach, typical of much of the debate on "new math" curricula, that responds to an apparent failure of the curriculum by holding the syllabus or the materials or some other ingredient of the learning environment to be at fault. The problem is always far more complex and unlikely to be solved by the simple replacement of one curriculum by another, as even a cursory glance at the history of mathematics education in the last hundred years shows. Devoting more creative effort to implementing what already exists may yield better results. We would contend that we have some evidence in support of this thesis.

The Third Stage: A Deeper Look

The creative implementation of the second stage continues, but at least part of the team are now devoting themselves to a new but complementary stage. The rationale for this stage is given in the following (Carpenter, 1980):

> It has been observed that most research in cognitive development is only incidentally connected with the learning of mathematics,...the objective for mathematics educators should not be to verify some aspect of a general theory of cognitive development; rather we should attempt to identify how the theories and techniques of cognitive development can be applied to deal with issues that are significant for the teaching and learning of mathematics.

Thus, our intention in this stage is to look closely and understand better those learning processes involved in the mathematics curriculum for which we are responsible. We wish to study actual interactions between teacher, student, and materials and use this information to enhance creative implementation activities as well as the learning processes of this stage itself.

The needs assessment surveys from the previous stage are replaced by more sophisticated evaluation of the implementation activities by means of applied cognitive research and research on the interaction between teacher and student.

Example 1: Cognitive research. In this area, we have been looking at the way students recognize geometrical objects and the factors which influence this recognition. The existence of common cognitive paths, canonical objects, and the

166

influence (or lack of influence) of verbal definitions all have a direct bearing on the redesign of curriculum materials and work with teachers. In addition, in parallel studies with teachers there is strong evidence of a correlation between teachers and students, which is of particular importance in our work with teachers.[8]

Another example of our work in this area is a multi-year study of ways of encouraging and evaluating student activity at the upper end of the cognitive spectrum. This has led to materials designed to extend the student's cognitive achievement as well as teacher strategies for encouraging and evaluating such activity (Zehavi et al. 1985).

Example 2: Interaction between student and teacher. The cognitive research in geometry discussed earlier touches on this area in its study of teacher responses to identical items given to students.

Another example is that of student achievement tests. These were first given to teachers, who were asked to predict the success rate of their own students. The responses were later compared with the actual success rate of those students, and the results represented graphically. Examples of over- and under-estimation were used in subsequent discussions with the individual teachers, and anonymously as part of in-service workshops on classroom evaluation.[9]

Not all this research is original, nor was it meant to be. It differs, however, from much educational research in its context and impact. The very fact that the curriculum project team members are engaged in this sort of research as an ordinary part of their activities sharpens their practice, and makes them more aware of the facets and processes of the learning activity, the existence and possible sources of misconceptions, and so on. Research experience — just as much as the results of that research — has an immediate and ongoing impact on curriculum materials and activities with teachers.

Conclusion

In this paper we have described briefly what we regard as the major features of the philosophy underlying the curriculum project for which we have been responsible. (Further information can be found in the items cited in the notes and references.) We see the major, interlocking features of our curriculum work as long-term, evolutionary, intensive, and comprehensive. We regard curriculum development as an ongoing, long-term activity. It may take as much as an entire generation before reasonable implementation of certain aspects is achieved. The bewildering variety of rapidly changing curricula characteristic of the past 25 years in Western countries seems to have had less than optimal impact: curricula sometimes appear as transient as "shooting stars." Teachers are unable to absorb, internalize, and transmit curricula undergoing rapid change in content or teaching strategies. Therefore we have adopted a gradual evolutionary approach to curriculum development, in which relatively minor adjustments are made to content or teaching strategy, while the major corpus of the curriculum remains unchanged. Such evolutionary phases are accompanied by intensive implementation activity, involving teacher courses, workshops, in-school guidance, and so on. All of this activity is accompanied by formative and ongoing (rather than summative) evaluation of both materials and activities, together with a spectrum of applied research ranging from research on general educational concepts to very subject-specific projects. These features give the work of the department a comprehensiveness rare among curriculum projects.

Postscript

It should not go unnoticed that microcomputers have not been mentioned anywhere in this paper. This should not be taken to imply that we are not working on their integration into the junior high school mathematics curriculum — we are. But this work is being undertaken in line with our general philosophy of evolutionary development, adequate teacher preparation, and careful development of materials. (Not all that glitters is gold, and not all that motivates is educationally sound — as much computer software currently available easily demonstrates.) We reject the overenthusiasm of some of our colleagues, and deplore the political and commercial pressures and the financial incentives that may cause the failure of yet another promising educational innovation.

Notes

1. See Hershkowitz and Israeli (1981). This survey (in Hebrew), carried out in 1981, contains data on teacher qualifications, distribution, in-service activity, and so on.

2. More extensive descriptions of this stage can be found in Bruckheimer (1979) and in Hershkowitz (1979b).

3. See Fresko and Ben-Chaim (1985). This study shows, among other things, the existence of a continuing need for strongly content-based in-service programs. It also highlights areas in which the present programs are inadequate.

4. The materials for these workshops have been translated into English. Single copies are available free from the Department. See also Arcavi, Bruckheimer and Ben-Zvi (1982).

5. A full description of the approach can be found in Hershkowitz and Bruckheimer (1985).

6. More extensive descriptions of this stage can be found in Bruckheimer (1979) and in Hershkowitz (1979b).

7. See Hershkowitz (1979a). A more extended documentation of the same topic can be found in Israeli and Hershkowitz (n.d.).

8. This research is discussed in a sequence of papers including Hershkowitz and Vinner (1983, 1984); and Vinner and Hershkowitz (1980, 1983).

9. See, for example, Buhadana and Zehavi (1984); Zehavi and Bruckheimer (1981); Zehavi and Bruckheimer (1983); and Hershkowitz and Zehavi (1985).

References

Arcavi, A., M. Bruckheimer, and R. Ben-Zvi. 1982. "Maybe a Mathematics Teacher Can Profit from the Study of the History of Mathematics." For the Learning of Mathematics 3: 30–38.

Bruckheimer, M. 1979. "Creative Implementation." In Proceedings of the Bat Sheva Seminar on Curriculum Implementation and Its Relationship to Curriculum Development: 43–49.

Bruckheimer, M. and R. Hershkowitz. 1984. "In-Service Teacher Training: The Patient, Diagnosis, Treatment and Cure." In Proceedings of the Bat Sheva Seminar on Preservice and Inservice Training of Science Teachers: 135–140.

Buhadana, R. and N. Zehavi. 1984. Bringing Research Constructs to the Teacher Through Relation of Expectations and Difficulties. Departmental Technical Report.

Carpenter, T.P. 1980. "Research in Cognitive Development." In Research in Mathematics Education: 146-206, R.J. Shumway, ed. Reston, VA: NCTM.

Fresko, B. and D. Ben-Chaim. 1985. "The Impact of Inservice Teacher Education on Subject Matter Confidence and Competency." Teaching and Teacher Education 1: 317-324.

Hershkowitz, R. 1979a. "Entry Behaviour: The First False Assumption." In Proceedings of the Third P.M.E. Conference: 108–112.

Hershkowitz, R. 1979b. "Case Study of Creative Implementation." In Proceedings of the Bat Sheva Seminar on Curriculum Implementation and its Relationship to Curriculum Development: 143–147.

Hershkowitz, R. and M. Bruckheimer. 1985. "Deductive Discovery Approach to Mathematics Learning — or — In the Footsteps of the Quadratic Function." International Journal of Mathematical Education in Science and Technology 16: 695-703.

Hershkowitz, R. and R. Israeli. 1981. Who Is the Mathematics Teacher in Grades 7-9. Department Technical Report.

Hershkowitz, R. and S. Vinner. 1983. "The Role of Critical and Non-Critical Attributes in the Concept Image of Geometrical Concepts." In Proceedings of the 7th P.M.E. Conference: 223-228.

Hershkowitz, R. and S. Vinner. 1984. "Children's Concepts in Elementary Geometry - A Reflection of Teacher's Concepts." In Proceedings of the 8th P.M.E. Conference: 63-69.

Hershkowitz, R. and N. Zehavi. 1985. "Research Leading to Novel Classroom and Inservice Activities." In Using Research in the Professional Life of the

Teacher, T. Romberg and D. Dessart, eds. Proceedings of ICME 5: 196-205.

Israeli, R. and R. Hershkowitz. n.d. _A Study of Achievement in Basic Skills and Its Application to the Development of a Remedial Programme._ Departmental Technical Report.

Vinner, S. and R. Hershkowitz. 1980. "Concept Images and Common Cognitive Paths in the Development of Some Simple Geometrical Concepts." In _Proceedings of the 7th P.M.E. Conference_: 177-185.

Vinner, S. and R. Hershkowitz. 1983. "On Concept Formation in Geometry." _ZDM_ 83(1): 20-25.

Zehavi, N. and M. Bruckheimer. 1983. "A Case Study of Teacher Thinking and Student Difficulties." In _Proceedings of the 7th P.M.E. Conference_: 395-401.

Zehavi, N. and M. Bruckheimer. 1981. "A Method of Analyzing Tests Using the Teacher's Predictions." _Journal for Research in Mathematics Education_ 12: 142-151.

Zehavi, N. and M. Bruckheimer. 1983. "Teachers' Expectations Versus Students Performance: A Guide to Some Inservice Activities." In _Proceedings of the Bat Sheva Seminar on Preservice and Inservice Education of Science Teachers_: 573-579.

Zehavi, N., M. Bruckheimer, and R. Ben-Zvi. 1985. "Qualitative Evaluation of Mathematics Activity and Its Relation to Effective Guidance." _Educational Studies in Mathematics_ 16: 27-40.

Mathematics in Junior and Senior High School in Japan: Present State and Prospects

Tatsuro Miwa

Institute of Education, University of Tsukuba

Introduction and Overview

Mathematics has provided an indispensable method for guiding science and technology in a democratic society. Accordingly, it is emphasized as a vital part of formal education, especially in developed countries like the U.S. and Japan. If we look at its present state, however, it is doubtful that school mathematics meets society's expectation.

In this paper, the author will describe the present state of mathematics education in Japanese junior and senior high schools, focusing on students and on ways to improve high school mathematics education in Japan. First, student achievement in mathematics is presented by showing the results of the Second International Mathematics Study conducted by IEA (International Association for Evaluation of Educational Achievement) and the Achievement Survey Tests of the Ministry of Education. From these tests we can see that the achievement level of Japanese students is very high compared with that in other countries, and that many Japanese students are able to solve hard problems. Next, student attitudes toward mathematics are presented by summarizing the results of the IEA Mathematics Study, this author's survey, and teachers' observations. These measures indicate that student attitudes toward mathematics are not very positive, and a considerable number of students are passive, merely waiting for the answers to be given by teachers. In addition, there are great differences in student achievement within a school or even within a class. We believe that while achievement is a product of yesterday's effort, attitude will determine tomorrow's achievement, and we are therefore not especially optimistic about the future.

In Japan today, the school mathematics curriculum is uniform up to the tenth grade. Within this framework measures are being taken to improve mathematics education in senior high schools, in areas such as teaching methods and mathematical content. Several methodological questions are discussed, the most important one being how to get students to participate actively in mathematics classes. In the area of mathematical content, problems relating to applications of mathematics and slow learners are discussed. But the most essential thing is the reconstruction of the philosophy of mathematics teaching and the framework of the mathematics curriculum in senior high school. A good example of this is the Core and Option Modules mathematics curriculum.

The Japanese Educational System

As a preliminary step, we will present an outline of school education in Japan. Japan's school system is basically a 6-3-3-4 system, i.e., six years of elementary school, three years of junior high school, three years of senior high school (or four years for part-time courses) and four years for college or university. Other types of schooling are offered as needs arise, e.g., two-year junior colleges, technical colleges, special training schools, and so on. In the Japanese system, elementary and junior high school education are compulsory and free, and children 6-15 go to the schools designated by their district boards of education, with a few exceptions for those who go to private schools or special schools. Enrollment in compulsory education has now reached 99.99%.

After graduating from junior high school, students take entrance examinations for admission to senior high schools. According to 1984 statistics, 94% of the children in the corresponding age group are enrolled in senior high schools. Looking at the breakdown of students in terms of curriculum, about 70% are in general education courses and 30% are in specialized or vocational education courses.

Enrollment in universities and junior colleges was 35% of the corresponding age group in 1983. Selection is made on the basis of an entrance examination. The entrance examinations for prestigious and famous universities are very difficult, and mathematics is usually one of the key subjects. School curricula (from elementary school to senior high school) are based upon the "Course of Study," the standard syllabus which is set by the Ministry of Education, while details and teaching methods are left to the teachers.

Mathematics is compulsory in elementary school and junior high school. In senior high school, all first-year students (tenth grade) have to take Mathematics I, but in the subsequent years mathematics is optional. The objectives and contents of the mathematics curriculum are determined in the Course of Study and are the same for all students up to the tenth grade. Textbooks used in the schools are compiled by commercial publishers according to the Course of Study and must be examined and approved by the Ministry of Education before being used in schools.

Student Achievement in Mathematics

Let us look at student achievement in mathematics by examining the results of the IEA Second International Mathematics Study and the Achievement Survey Tests administered by the Ministry of Education.

The IEA Second International Mathematics Study

First we will present the achievement of Japanese students in the IEA Second International Mathematics Study. The test was carried out in Japan on Population A in February 1981 and Population B in November 1980. Population A consisted of thirteen-year-olds, students in the first year of Japanese junior high schools (seventh graders), and Population B consisted of students in the terminal year of the secondary education system — in Japan, the third year of senior high school for full time students (twelfth grade) — who are studying mathematics as a substantial part of their academic program. The following table is a summary of the results, which were available to this author as a member of the specialist subcommittee of the National Committee for the Second International Mathematics Study. In this table the figures under "Score" are the average percentage of correct responses to the test items by content area, and the numbers under "Opportunity to Learn" represent the average percentage of students who had already learned the material of the test items in each content area. "International" is the mean value for the twenty participating countries.

Table 1

Mean Achievement Scores and Opportunity to Learn

Population A

Content Area	Score		Opportunity to Learn	
	Japan	International	Japan	International
Arithmetic	60%	50%	85%	73%
Algebra	60	43	83	67
Geometry	57	41	51	46
Probability & Statistics	71	55	75	50
Measurement	69	51	95	74
Total	62	—	75	62

Population B

Content Area	Score		Opportunity to Learn	
	Japan	International	Japan	International
Sets, Relations, & Functions	80%	62%	95%	73%
Number Systems	72	50	82	76
Algebra	75	57	100	82
Geometry	58	42	85	61
Analysis	69	44	94	77
Probability & Statistics	72	50	83	63
Finite Mathematics	76	—	99	66
Total	69	—	91	—

We next give examples of IEA test items and the scores of Japanese students.

Outline of Test Items   Percent of Correct
                                              Responses

- 2/5 + 3/8   88.9%
- 5x + 3y + 2x – 4y   73.7
- Soda costs a cents for each bottle, but   61.5
  there is a refund of b cents on each empty
  bottle. How much will Henry have to pay
  for x bottles if he brings back y empties?
- If segment PQ were drawn for each figure   73.9
  shown below, it would divide one of the
  figures into two congruent triangles.
  Which figure?

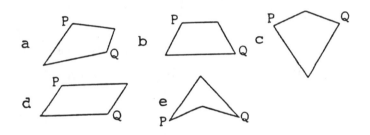

-   48.9

The triangles shown above are congruent.
The measures of some of the sides and
angles are as shown. What is x?

- The length of the circumference of the cir-   75.3

cle with center at 0 is 24 and the length of
arc RS is 4. What is the measure in degrees
of the central angle ROS?

Population B

- If X and Y are sets, then what is   90.5%
  $(X \cup Y) \cap (X \cap Y)$?

- Which of the operations defined below is   77.2
  commutative?
  a. $x * y = x + xy$ b. $x * y = x - y$
  c. $x * y = x(x + y)$ d. $x * y = xy(x + y)$
  e. $x * y = x^2 + xy^2 + y^4$

- If $\dfrac{x - 1}{x^2 + 3x + 2} = \dfrac{P}{x + 1} + \dfrac{Q}{x + 2}$,   89.9

  then find P, Q.

- In a Euclidean plane the coordinates of   43.8
  a moving point M at time t are x = 2 sin t,
  y = 2 cos 2t - 1. Find the path of the
  point M.

- The curve defined by $y = x^3 - ax + b$ has   86.8
  a relative minimum point at (2,3). Find the
  value of a and b.

- Evaluate   54.3
  $$\int_0^1 \frac{dx}{x^2 + 5x - 6}.$$

- A set of 24 cards is numbered with the   87.2
  positive integers from 1 to 24. If the
  cards are shuffled and if only one is selected at random, what is the probability

that the number on the card is divisible
by 4 or 6?

Ministry of Education Achievement Survey Tests

In February 1981, achievement tests were administered by the Ministry of Education to one percent samples (about 17,000 children in each grade) of the fifth- and sixth-graders in elementary school. The objective of the tests was to assess the students' achievement with respect to the new Course of Study which was introduced in April 1980. In 1982 other achievement tests were administered to one percent samples (about 16,000 students in each grade) of junior high school students. The objective was the same as that of the elementary school tests. Some results for sixth-graders in elementary school and students in the third year of junior high school (ninth-graders) are given below. The elementary school children's results may be regarded as a baseline for junior high school, and those of the junior high students for senior high school.

Sixth-graders in elementary school

Outline of Test Items	Percent of Correct Responses
• 5/6 x 4/9	91.8%
• The following graph illustrates the relation between the length x m and weight y g of a wire.	

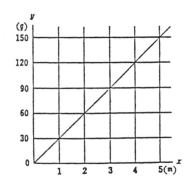

(1) Express the relation between x and y.	54.8

(2) Find the length of this wire which
   weighs 450 g.   82.5%

● When we substitute a positive number into □   63.0
   of the following expressions, the greatest is

   a. $\square \times \dfrac{1}{2}$ b. $\square \times 1\dfrac{1}{2}$ c. $\square \div 1\dfrac{1}{2}$ d. $\square \div \dfrac{1}{2}$

Ninth-graders in junior high school

● Calculate $\sqrt{600}$, given that $\sqrt{6} = 2.449$ and   67.1
   $\sqrt{60} = 7.746$.

● Solve the equations:
   $x^2 - x - 20 = 0$ 72.8
   $x^2 + 5x + 3 = 0$ 62.3

● In triangle ABC, the lengths of AB and
   BC are 10 cm and 20 cm respectively, and
   the degree of angle ABC is 90. Point P
   moves on side AB from A to B and point Q
   moves on side AC from A to C, preserving
   line PQ perpendicular to AB.

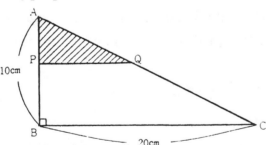

   (1)   If the length of AP is 3 cm, find the area of
         triangle APQ.   72.1

   (2)   If the length of AP is x cm, find the area
         of triangle APQ.   46.5

● In the circumcircle of triangle ABC, the bisector   53.5
   of angle BAC meets side BC at D and arc BC at E.

179

Prove that triangle ABE is similar to triangle
ADC.

The results given in the IEA Second International Mathematics Study and in the Ministry of Education Achievement Survey Tests bring out several points.

a. Generally speaking, the achievement of Japanese junior and senior high school students is high compared with that of other countries. Especially in the IEA study, Population B scored very high because many students are required to solve hard problems in senior high school mathematics. They may be in the top group of students from all participating countries.

b. One of the reasons for this high performance is that the Opportunity to Learn (OTL) is very high compared with other countries. For Population A, the OTL is already high except for geometry, and for Population B the OTL is even higher. This is due to the fact that the Course of Study establishes a high nationwide standard in Japan.

c. The results of the Ministry of Education Achievement Survey tests are compatible with those of the IEA study. In elementary school, children perform well in calculation and in mathematical understanding. This is also true in junior high school.

d. About half of the ninth-graders could express quantitative relations using letters (variables) and could write geometrical proofs. Since we have about 94% of junior high school students going on to senior high school, as mentioned above, senior high school teachers should bear this fact in mind.

Student Attitudes Toward Mathematics

Student attitudes toward mathematics are treated in the IEA Second International Mathematics Study, in this author's survey, and in the teachers' observations.

The IEA Second International Mathematics Study

Some of the responses of Japanese students to the questionnaires are shown below. In Table 2, the numbers in the upper line are the results for Population A and those below in parentheses are the results for Population B. The percentage of neutral answers (undecided) is omitted.

Table 2
Responses to Questionnaires

Outline of Items on Questionnaires (Mathematics as a Process)	Response Pattern of Students			
	Disagree Strongly	Disagree	Agree	Agree Strongly
• There is little place for originality in solving mathematics problems.	5% (11	26% 43	13% 13	2% 2)
• There are many different ways to solve most mathematics problems.	2 (1	6 3	56 61	15 22)
• Learning mathematics involves mostly memorizing.	13 (28	33 43	11 6	2 1)
• Mathematics is a set of rules.	4 (4	14 19	29 33	6 6)
• Mathematics helps one to think logically.	2 (3	5 7	34 40	8 9)

(Mathematics and myself)	Disagree Strongly	Disagree	Agree	Agree Strongly
• I really want to do well in mathematics.	1% (1	2% 2	37% 39	44% 44)
• My parents really want me to do well in mathematics.	1 (1	1 2	36 34	31 18)
• If I had my choice I would not learn any more mathematics.	11 (15	30 37	10 12	7 7)
• No matter how hard I try I still do not do well in mathematics.	11 (12	32 42	9 7	4 3)
• I usually feel calm when doing mathematics problems.	7 (6	22 27	10 10	2 2)

From these results we can see that many students recognized that there are various methods for solving mathematics problems and the mathematics helps to foster logical thinking, but some of them did not recognize the potential for originality in solving mathematics problems, and they viewed mathematics as a set of rules. This suggests that many students might have the idea that there are rules in mathematics which will be given by the teachers and that they can solve mathematics problems simply by applying suitable rules. Thus, for them mathematics seems to be something which they do not have to think out by themselves.

The most remarkable responses are those relating to the students' own desire to do well in mathematics and to their feelings that their parents expect this of them. About 80% agreed and only 3% disagreed. But 13% of Population A and 10% of Population B agreed that they do not do well in mathematics no matter how hard they try. On the other hand, 17% of Population A and 19% of Population B agreed that they would not learn mathematics if not compelled to. This is one of the problems of attitude among Japanese students.

Student Attitudes Towards the Application of Mathematics

It has been noted that senior high school students in Japan seem to have little interest and less confidence in the application of mathematics. In October and December of 1982 this author carried out a survey of eighty students in the first and second years of a senior high school which is known for having bright students. The following results were obtained:

- two thirds of the students agreed that mathematics is useful;
- half of them agreed that mathematics is useful only because of the service it offers to other subjects and because of its immediate applicability;
- as examples of the application of mathematics, students listed everyday use, everyday statistics, economics, betting, computers, and so on; the limited kinds of examples named show that students have little experience in applying mathematics to solving real problems.

These results were confirmed when the same survey was taken in another senior high school. This demonstrated that some students, especially those bright ones in the school mentioned above, do not realize that mathematics is useful in solving various applied problems: they may think that it is not useful for anything more important than solving the problems presented in mathematics classes and that only the content of the given mathematics lesson can provide them with the keys for solution.

Teachers' Observations on Student Attitudes

Some of the observations made by senior high school mathematics teachers at a meeting of teachers in December 1984 are presented below.

- The individual differences in student ability and achievement in a given school or even in a single class are considerable.

- Some students are studying mathematics involuntarily and their lack of concentration in mathematics class is quite noticeable; some students chatter during class and do not pay attention to the teacher.

- The number of students who do not participate spontaneously and who passively wait to be supplied with the answers without thinking for themselves is increasing.

- The number of students who are looking only for the answers without going through the processes of reasoning is increasing.

- The number of students who do not prepare for class and who do not finish their homework is increasing.

- The number of students who are not very good at making calculations is increasing.

It seems that these observations contradict the results of the Second International Mathematics Study shown in the IEA Second International Mathematics Study, but we can assume that both are correct by taking note of the following considerations (leaving aside the fact that teachers always expect much of their students).

- The International Mathematics Study was carried out on Population A, consisting of seventh-graders who had just entered junior high school with fresh hopes, and Population B, consisting of twelfth-graders who were studying mathematics as a substantial part of their academic program in senior high school. The teachers cited, by contrast, teach mathematics in senior high schools of varying degrees of excellence and with varying specializations, where many of the students are not necessarily diligent or talented in mathematics.

- Since the social climate demands uniformity in education, the mathematics curriculum up to the tenth grade is the same for all students, and differential tracking is rather limited. Therefore, the differences in ability and in willingness to learn are rather great even within a single class, and this may make the classroom atmosphere not very conducive to learning mathematics.

We can extract several points from "Student Attitudes Toward Mathematics" above with regard to student attitudes toward mathematics in junior and senior high schools.

a. Many students want to do well in mathematics and feel that their parents also hope they will, but in reality many of them are not very diligent about learning mathematics and do so passively.

b. Some students do not realize the power of mathematics in applied work and see mathematics as exercise for solving given problems only.

c. Individual differences in ability and willingness to learn are great even within a single class. This makes the classroom atmosphere unconducive to learning mathematics.

Generally speaking, student achievement in mathematics is very high and we can be proud of it. We consider that achievement is the product of yesterday's effort on the part of students, teachers, parents and educational administrators. However, it should be noted that by "achievement" here we mean only the results of paper-and-pencil tests, for it is mainly through paper-and pencil tests that we measure the degree of mathematics knowledge acquired and mathematical skills mastered. Such knowledge and skills are important, and are especially so in a competitive situation like an entrance examination. In mathematics education, though, it is more essential to develop students' mathematical intelligence and to foster mathematical thinking. For this purpose student attitudes towards mathematics are crucial. We recognize that at the moment student attitudes towards mathematics learning are not very positive; since it is these attitudes that determine tomorrow's performance, we are not very optimistic about the future of school mathematics in Japan.

Measures Being Taken to Improve Mathematics Education

We would like to point out that one of the features of the mathematics curriculum in Japan is uniformity or "non-differentiation." In elementary school and junior high school, which nearly all children attend from the age of 6-15, the mathematics curriculum is determined by the Course of Study and is the same for all students, i.e., there is no curricular provision for differentiation among students. This is partly because of the great expectations for education on the part of the parents and partly because of the social tendency against apparent differentiation in the populace. At one time in the past, compulsory and optional mathematics courses were set up in the third year of junior high school (ninth grade), but almost all students chose to take both of them, and thus the plan to differentiate among students was abolished. In senior high school, Mathematics I is compulsory for all students in the first year (tenth grade), irrespective of their specializations, and its content, too, is the same for all students. In the following grades mathematics is optional, but in the general course, in which about 70% of senior high school students are enrolled, more mathematics is required, e.g., Basic Analysis, Algebra-Geometry or Mathematics II, and so on. Several ways of improving mathematics education have been devised under these conditions. Measures relating to teaching methods and mathematical content for senior high school will be discussed below.

Teaching Methods
Classes Formed According to the Ability and Achievement of the Students

As mentioned above, there are considerable differences in ability and achievement among students in a given school. Classes formed according to the ability and achievement of the students are being instituted in many senior high schools, but practices vary from school to school according to their respective situations. A common type is presented here as an example. Students are divided into several groups according to their ability and achievement for mathematics lessons, regardless of the homerooms to which they belong for their other subjects, and they study the appropriate material and exercises with teaching methods suited to them. It is important to note that the subject matter to be learned is fundamentally the same as that in the Course of Study, and the differences are only in the depth and/or the extent of the content, the degree of difficulty of the exercises, and whether the presentation is more abstract or more concrete. The division of the students into the different groups is not an easy matter, and it often causes psychological and sociological frustration both to students and to their parents. The teachers try to persuade both students and parents to accept this measure, since it is relatively successful in improving mathematics education.

Teaching the Slow Learners

About 94% of the corresponding age group is enrolled in senior high school, but not all the students are well-prepared in mathematics. Guiding these students is one of the urgent tasks for senior high school teachers. New materials for slow learners are being devised and small after-school classes are often held, but these classes are not very effective. This may be because slow learners are not willing to devote extra time to learning mathematics after class and/or because the materials are not necessarily suited to their needs. The reasons which stand behind why some students are slow learners should be correctly understood by the teachers. This will be mentioned further in "Mathematical Content."

Utilizing Educational Technology

Audiovisual technology is helping to improve mathematics education; however, some senior high school mathematics teachers have strong confidence in their board-and-chalk methods, and audiovisual facilities are not widely used. Japan is famous for its electronics industry, but electronic technology is not very common in education. Pioneer work is now being done, e.g., microcomputers are being introduced into senior high schools, and some teachers report that their use in mathematics classes has achieved success in inspiring and motivating students to learn mathematics.

Motivating Students in Mathematics Class

The most important problem in mathematics classes is motivating students to learn mathematics and letting them do mathematics actively by themselves. Many teachers have been making efforts to develop their own methods for this purpose. We mention briefly two interesting general methods. One of these is the "Open Approach" developed by Professor Nohda. This is a systematized approach to mathematics teaching, directed towards fostering students' creativity. Under this approach, breadth of mathematical activity is assured in the presentation stage by devising problems which are not only open-ended but also rich in development, and student initiative is emphasized in the solution stage by enabling students to come up with diverse processes and expressions according to their aptitudes. Trials of this approach have already been successfully conducted in several areas of mathematics, and we hope it will start to be used in a greater range of contexts in senior high school mathematics. The second method is the "Developmental Treatment of Mathematical Problems" worked out by Mr. Sawada and others. In this developmental treatment, after solving the problem given, students are encouraged to formulate new problems by modifying some parts of the given problem through discussion of the method and the solution. The students and the teacher select a new class problem from among those problems which the students have formulated. The students work together on solving this common problem, and then finally they solve their own problems. The developmental treatment is effective in fostering mathematical thinking and in enhancing students' interests, and it is helpful for improving their attitudes toward mathematics.

Mathematical Content

There are many mathematical subject areas and materials that are useful in motivating students to study mathematics actively and eagerly. The teaching methods described in "Teaching Methods" should be used in conjunction with suitable subject matter and materials. In the proceedings of each annual conference of the Japanese Society for Mathematical Education, thirty to forty reports of this kind are published. Only the present author's contributions will be dealt with here. One in particular concerns the application of mathematics to other areas. As mentioned in "Student Attitudes Towards the Application of Mathematics," Japanese students have little experience with genuine applications of mathematics; thus, it is important to set up a learning unit on mathematical modeling in senior high school and to teach these techniques to students. An experimental instructional unit along these lines was devised by this author. The theme selected for this unit was population growth, because the application of mathematics to biology had not yet been introduced into senior high schools; because it was thought that students might be interested in this material; and because the required mathematical techniques were suitable. An experiment with this unit was carried out successfully in the second year class of senior high school (eleventh grade). This experiment allows us to be optimistic about overcoming the difficulties that cause students to have little interest and even less confidence in the application of mathematics.

Another one deals with students' ability to use letters (variables). Being able to operate freely with letters is the most important prerequisite for senior high school mathematics. In the case of slow learners, however, the situation is not very propitious. We maintain that grasping the idea behind the use of letters is not separate from doing the same thing for numbers, and that this notion is very closely connected to the understanding of numbers, especially rational numbers. The enhancement of the understanding of and ability to manipulate letters on the part of students is based upon a full understanding of rational numbers. The notion "rational number" contains many-sided and abstract concepts. Among these, the most important are the notion of ratio contained in the very concept of rational number and the notion of multiplication by rational numbers. Students, especially slow learners, often encounter difficulties in comprehending multiplication by fractions, although multiplication by integers does not trouble them. This multiplication depends crucially on the notion of ratio.

It is thus urgent to develop written materials for enhancing the understanding of rational numbers which focus on the notions of ratio and multiplication by fractions. Such materials have been produced in Osaka by a team of senior high school teachers directed by Professor Takenouchi and the present author.

Need for a New Framework for the Mathematics Curriculum

In order to be responsive to the great diversity in aptitude, achievement and willingness to study mathematics among senior high school students, and to meet the need for a modern methodology for teaching mathematics, a new framework for the mathematics curriculum in senior high school is required. It should not be uniform for all students, but should rather be adaptable to the needs, interests and aptitude of the individual student. Moreover, it should reflect modern thinking in mathematics, and would be appropriate for the young people who will live and play an active role in the twenty-first century. We believe that the Core and Option Modules mathematics curriculum worked out by Professors Fujita, Terada and Shimada is worthy of consideration.

References

Educational Affairs Study Group. 1978. Education in Japan. Foreign Press Center, Japan.

Fujita, H. and Terada, F. 1984. "The Present State and Current Problems of Mathematical Education in Senior High Schools of Japan." Speaking draft for AG4 and project presentation, ICME 5.

Fujita, H., F. Terada and S. Shimada. 1984. "Towards a Mathematics Curriculum in the Form of the Core and Option Modules." Manuscripts for the project presentation, ICME 5.

Hashimoto, Y. and T. Sawada. 1984. "Research on the Mathematics Teaching by Developmental Treatment of Mathematical Problems." In Proceedings of ICMI-JSME Regional Conference on Mathematical Education, T. Kawaguchi, ed.

Japanese Society of Mathematical Education. 1982, 1983, 1984. "Memorandum of the 64th Annual Meeting," "Memorandum of the 65th Annual Meeting," "Memorandum of the 66th Annual Meeting." In Journal of Japan Society of Mathematical Education, Supplementary Issues of vol. LXIV, LXV, and LXVI (in Japanese).

Kawaguchi, T. 1980. "Secondary School Mathematics." In Studies in Mathematics Education 1, R. Morris, ed. UNESCO.

Ministry of Education. 1982. "Monbusho." Printing Bureau, Ministry of Finance.

Ministry of Education. 1984a. "The Report of Comprehensive Survey of the State of Enforcement of the Course of Study - Elementary School, Arithmetic" (in Japanese). Ministry of Education.

Ministry of Education. 1984b. "Statistics of Education 1984." Daiichi Hoki Shuppan (in Japanese).

Ministry of Education. 1985. "The Report of Comprehensive Survey of the State of Enforcement of the Course of Study - Junior High School, Mathematics" (in Japanese). Ministry of Education.

Miwa, T. 1984. "Mathematical Modelling in School Mathematics." In Proceedings of ICMI-JSME Regional Conference on Mathematical Education, T. Kawaguchi, ed.

The National Institute for Educational Research. 1984. "Mathematics Achievement of Secondary Students" (manuscript). The National Institute for Educational Research (in Japanese).

The National Institute for Educational Research. 1981, 1982. Mathematics Achievement of Secondary Students - Second International Mathematics Study - National Report of Japan 1, 2. The National Institute for Educational Research (in Japanese).

Nohda, N. 1984. "The Heart of 'Open-Approach' in Mathematics Teaching." In Proceedings of ICMI-JSME Regional Conference on Mathematical Education, T. Kawaguchi, ed.

Takenouchi, O. and T. Miwa. 1980. "The Structure of the Formation of Mathematical Concepts and Abilities," Part 1 and 2. In ICME 4 Abstracts of Short Communications.

The Present State and Current Problems of Mathematics Education at the Senior Secondary Level in Japan

Hiroshi Fujita

University of Tokyo

Introduction and Overview

In this paper we will review the present state of mathematics education in Japan, mainly at the senior secondary level, and, in particular, the mathematics curriculum. At the same time we will analyze and identify current problems in mathematics education in Japanese high schools which must be taken into account in any effort to reform our curriculum. Some of these problems seem to be common to countries with developed or developing technology.

We will then describe some results of our study of how to solve these problems. In the process, we will state our philosophy about the goals of mathematics education in Japanese high schools and propose some appropriate strategies for addressing the problems in the context of the Japanese educational system and social environment.

Here are the essential points of our philosophy and our strategies.

i) Cultivation of "mathematical intelligence" should be emphasized as the goal of mathematics education in the high schools of Japan. "Cultivation of mathematical intelligence" includes fostering sound "mathematical literacy" for the majority of students, as well as the development of deep mathematical potential among the brighter (moderately gifted) students.

ii) A flexible and efficient curriculum should be designed. To increase flexibility we should try to organize a "core and option modules" type of curriculum. To increase the efficiency of the curriculum, more honest efforts should be made to delete obsolete material and/or obstacles from the traditional and new math components.

iii) The introduction of active use of calculators and computers is unavoidable.

Calculators should be used to assist students in understanding mathematical notions and methods through numerical work. Computers and other information technologies should be introduced to create more opportunities for enriched mathematical experiences for students, and also to facilitate personalized learning by students without increasing the burden on teachers.

This paper is based on projects undertaken by our study group in mathematics education, and is a revised version of the papers read last summer in Adelaide by the present author and his collaborator, Professor F. Terada of Waseda University, at the session on curriculum of Action Group 4 of ICME 5 and at a project presentation there. The study group mentioned above, composed of about 30 university professors, several educators and a few high school teachers of mathematics, was originally founded by the late Professor Y. Akizuki twenty years ago, and was recently presided over by the author of this paper. Financial support for the study group and, hence, for this work, was provided by a Grant in Aid for Scientific Research (Project numbers 00538031, 58380037, 57380030) from the Ministry of Education of Japan.

The Present State of Japanese School Mathematics and Social Environment

The system of school education in Japan is strongly centralized. The curricula of all subjects are controlled by the Ministry of Education through its Course of Study, from elementary school up to (senior) high school, while the content of university education is essentially left to each university on the principle of academic freedom.

Thus, the mathematics curriculum, as well as the curricula of other subjects through the 12th grade, is a kind of national curriculum which must be taught uniformly in all schools, including private schools.

It should also be noted that in Japan any change in the mathematics curriculum is a national measure which exerts inevitable influence upon all schools and all students at once. In the past, the Course of Study has been revised approximately every 10 years. When it is to be revised, scholars (university professors), educators and teachers are asked to cooperate in serving on a committee of some twenty members (in the case of mathematics) set up by the

Ministry and organizing a new curriculum.

It may be peculiar to Japan, but commercially published textbooks must undergo examination by the Ministry and must be given official approval before they are adopted for school use. The judgments of the textbook examiners are usually fair and reasonable; however, some authors become frustrated when required to rewrite a presentation or to replace exercises they regard as well-suited.

Generally speaking, the competence and the performance level of mathematics teachers in junior and senior high schools are rather high. All of them are graduates of 4-year universities with majors in mathematics education or in mathematics. The competition to get a job as a high school teacher is stiff. Although there is some need for retraining mathematics teachers due to rapid innovations in the curriculum, most mathematics teachers are well-qualified and respected in society; at any rate, there is no shortage of mathematics teachers in secondary education at the moment. When compared with other countries which are experiencing such shortages, the favorable situation in Japan may be due to the small difference in salary between company employees and school teachers, and to the attitude of some graduates of mathematics departments, who prefer the stability, intellectual independence and social status enjoyed by high school teachers.

Now we proceed to problems arising from social factors. The Japanese public is aware of the growing importance of mathematics and mathematics education in a society with modern technology. Parents encourage their children to study mathematics — sometimes because they think that a productive life and good job opportunities in such a society require a better background in mathematics, and sometimes because they simply believe that mathematics is a crucial subject in the entrance examinations for respected high schools and prestigious universities. Indeed, mathematics is required even of applicants to law school, and as part of the entrance examinations for national universities (like the University of Tokyo, Kyoto University, and Osaka University), where applicants must exhibit, among other things, a mastery of formal operations in differential and integral calculus and a basic knowledge of elementary linear algebra (vectors and 2 x 2 matrices). Therefore, it is not necessary to persuade Japanese society or individual Japanese that mathematics education is important. The issues are limited to what we should teach in mathematics and how we should teach it.

One of the most serious problems of mathematics education in Japanese high schools (10th grade - 12th grade) is the result of a sociological factor, namely, the very high rate of continuation to high school education (94.2% of the age cohort). To begin with, Japanese by nature respect education, and, generally speaking, they believe that the possibility of obtaining a good job — and, consequently, satis-factory social status — depends on the level and quality of the education the individual receives. Moreover, due to improvement in the financial situation, all families can afford higher education for their children. Thus, higher education is on the rise, and parents, particularly mothers, are enthusiastic about sending their children to prestigious universities and popular departments.

The extraordinarily high rate of advancement to high school means that the aptitude of the average high school student is close to that of the whole age group. These are no longer elite children, and high school teachers see a greater range of aptitude among the students in their classes.

On the other hand, there are social tendencies against apparent differen-tiation in education. No mathematics course for low-achievers can ever become popular if it deviates from the compulsory curriculum through 10th grade. Students and their parents reject explicit remedial courses, preferring one single track.

Recent Curricula

Here we will sketch the preceding and the current mathematics curricula for Japanese high schools, which were set by the Ministry of Education in 1968 and 1978, respectively, and put into effect three years after they were announced. Before doing so, however, a few words about the Japanese school system are in order.

The Japanese school system is a 6-3-3-4 system and is quite similar to the American system, simply because it was reformed under strong American influence during the post-World War II period. Elementary School Education for 6 years and Lower Secondary Education (middle school education, or junior high school education) for 3 years are compulsory and free.

The rate of advancement in the compulsory portion is 99%. The 3 years of Upper Secondary Education (senior high school education) are not compulsory. The advancement rate, as mentioned above, is currently about 94%, while in 1954, 1964 and 1974 it was 50.9 %, 69.3% and 90.8%, respectively. Incidentally, 64% of the children who are newly enrolled in elementary school have had one or more years of kindergarten education. At the other end, 32% of high school graduates enroll in 4-year universities or 2-year colleges (see Appendix A for further statistics).

The 1968 curriculum followed a policy of modest modernization of high school mathematics, and it introduced in quite simple fashion some notions of modern mathematics, e.g., the notion of 2 x 2 matrices and the notation of informal set theory. These approaches were accepted and regarded as efficient in the classroom. However, as a version of the modernization curriculum, the 1968 curriculum still placed its emphasis on the systematic approach, on conceptual structure and abstract reasoning. These features, which are popular among research mathematicians, turned out to be far beyond the average student's aptitude and mental maturity. In fact, it was nearly impossible for many students to exercise their ingenuity when they were forced to follow abstract exposition of the theory. Furthermore, little effort was made to eliminate obsolete materials from the curriculum, although this was necessary to make room for new topics. Thus, the content appeared to be a simple union of traditional materials and new math topics, which made it broad, but shallow (see Appendix B for an outline of the 1968 curriculum).

When the current curriculum was designed in 1978, it aimed at remedying those defects of modernization which had been pointed out; for instance, to respond to overloading, the 1978 curriculum reduced the overall content to some degree and introduced selective courses, which are intended to allow more intensified study of particular topics according to student preferences. We appreciate the fact that the 1978 curriculum has improved the situation, although we do not condone the reduction in standard credits allotted to mathematics (from six units to four in 10th grade); but again the change was too limited. As can be seen from examination, the curriculum is still too wide-ranging in scope and too hard to follow for the majority of students, while at the same time stifling the development of the brighter students. Some detailed information on the content of the 1978 curriculum is given in Appendix C. No official statistics concerning the number of students who take each of the optional subjects are available yet; however, we can roughly estimate percentages of those students within the age cohort from the following data on the number of copies of textbooks sold in 1983:

Mathematics II	675,000 (copies)
Algebra/Geometry	871,000
Basic Analysis	1,038,000
Probability/Statistics	387,000
Differential and Integral Calculus	339,000

Thus in the 10th grade all the students (about 1.5 million) study compulsory Mathematics I. Then in the 11th grade 40% of the students take the easiest of the elective subjects (Mathematics II) and terminate their study of mathematics with it, while the other 60% take Algebra/Geometry or Basic Analysis. Algebra/Geometry contains trigonometry and an informal treatment of differential and integral calculus. Only 25% of the students continue mathematics in the 12th grade, taking Probability/Statistics and/or Differential and Integral Calculus. Most of these students intend to enter 4-year universities with majors in science, engineering, medicine, economics, or other fields of applied science which require mathematics.

The mathematics background of students entering senior high schools can be extracted from the information given in Appendix D, which is an excerpt from the official curriculum for lower secondary education. For example, we note that the Pythagorean theorem, conditions for congruence and similarity of triangles, linear functions and simple quadratic functions, straight lines in coordinate geometry, factoring of quadratic expressions, and simple cases of probability are taught in middle school (i.e., in lower secondary education).

We cannot underestimate the influence of the university entrance examinations when considering the quality of and motivation for learning mathematics on the part of high school students. Recently, about 26% of all high school graduates have been accepted each year by 4-year universities. At the time of the entrance examination applicants are divided either into specific subgroups which correspond to their prospective majors or into two broad groups corresponding to the science division and the literature division, depending on the university. If we classify all freshmen enrolled in universities (i.e., successful applicants) into the latter two groups, the ratio is 33% (science division) to 67% (literature division), i.e., 140,000 to 280,000 in numbers of students. Although a considerable portion of the age cohort is accepted by universities overall, the competition on the entrance examinations is very stiff since the number of popular and prestigious universities is limited. Mathematics is a crucial part of the exam not only for those applying to the science division, but also for those applying to the literature division (in the case of national universities). If we examine the mathematics problems from the entrance examinations for the better universities, we see that they require full mastery of high school mathematics as well as intuitive and logical strength in problem solving. (English translations of some of those problems are given in Appendix E.)

Thus, many students who want to be successful on the entrance examinations study mathematics far beyond the usual level of classroom teaching or of the textbook. Some work with sophisticated reference books and some attend evening classes at preparatory schools. Many of those applicants who fail the entrance examinations enter preparatory schools for an extra year of study in order to try again the next year. About half of the freshmen at the University of Tokyo, for instance, have had this experience. Such extreme eagerness to pass the entrance examinations for preferred universities can produce serious distortion of the normal and human atmosphere of secondary education. In fact, both children and parents believe that the competition starts at an earlier age. In Tokyo, for example, ambitious children in primary schools attend evening classes or Sunday classes offered for them by preparatory schools in order to be successful on the entrance examinations for the best private high schools of the 6-year system, where intensified instruction is given in a systematic way. Ultimately they want to be well-prepared for the entrance examinations to the universities they prefer. On the other hand, however, it would be fair to note that the scholarship — particularly the problem-solving ability — of the brighter students is enhanced noticeably by their efforts and focus on the entrance examinations.

At this point we should mention the The Joint First Stage Achivement Test for national and public universities. The JFSAT is run by a government agency, the Center for University Entrance Examinations, and all applicants for national or public universities are required to take it before they attempt the secondary examinations given individually by each university at a later time. Each year about 400,000 students take the JFSAT. Their examination papers are scored by a "mark sheet and computer" process. The problems on the JFSAT are, in principle, taken from the topics of the compulsory subjects in the senior high school curriculum. In mathematics the problems have been algorithmic and simple (partly because of the restrictions imposed by the mark sheet form), and most students who intend to enter the better universities do not regard mathematics as a key subject on the JFSAT, since it requires only speed, care in computation, and processing. Most of the university professors criticize the JFSAT as being responsible for a tendency toward superficial mastery and, hence, a lowering of the overall level of scholarship of university-bound high school graduates.

Each university decides how to weight the scores on the JFSAT in making its admissions decisions. For instance, in the case of the University of Tokyo, the ratio of the weights of the JFSAT to its secondary examination is 20/80%, but most universities set the ratio at 40/60% or 50/50%.

Finally, we note that the scope of the high school curriculum is not affected by the university entrance examinations, although preparatory study by university applicants is more sophisticated than in the usual implementation of the curriculum. In other words, all universities (and the agency in charge of the JFSAT) must base the entrance examinations on the Course of Study (the national curriculum).

Policy and Strategies for Future Reform

In this paper we are concerned mainly with reform of the mathematics curriculum for Japanese high schools. However, it would be appropriate to start with a consideration of ultimate goals, general policy, and feasible strategies for mathematics education.

We propose first that, as a goal of mathematics education at the senior high school level in countries like Japan, more emphasis be placed on the cultivation of mathematical intelligence than on covering a wide range of mathematical knowledge. Needless to say, the acquisition of sound fundamental knowledge is indispensable for the development of mathematical intelligence, although under many circumstances an efficient way to foster the students' mathematical intelligence is to encourage so-called "problem-solving activities." However, we would like to assert that even in those latter cases, problem solving is a means of cultivating mathematical intelligence and is not in itself the goal. Obviously, the aspects of mathematical intelligence include the ability to recognize the mathematical structures of various phenomena, to reason logically and correctly, to make heuristic considerations, to make penetrating mathematical insights, to exercise ingenuity in solving problems, and, above all, to understand what is important and what is not.

The level of mathematical competence necessary to live and work in a modern technological society is much higher that that needed in a traditional one. Furthermore, in order to maintain and develop its scientific and technological level, such a society must have a considerable number of students of higher aptitude study mathematics to an advanced level. From this point of view, what those gifted students should develop in their high school study of mathematics is a reliable mathematical intelligence on which their future specialized studies can be based. We should note that even if we assume an optimal range in the repertoire of high school mathematics, the knowledge students can acquire in high school is so restricted and primitive that it is by itself inadequate for meaningful applications to problems encountered in their university studies and future professions. In particular, in this age of scientific and technological progress, students should be provided with the ability to discern and master the mathematics necessary for problems they encounter on their own. Incidentally, we hesitate to approve the optimistic introduction of realistic scientific and technological applications into high school mathematics as materials for problem solving.

On the basis of our belief in mathematical intelligence, we would like to set the following concrete goals for Japanese mathematics education in the foreseeable future.

i) The majority of students should acquire sound "mathematical literacy."

ii) A significant number of students with high aptitude (3–8% of the age cohort, i.e., 45,000–120,000) should be given ample opportunity to develop their mathematical intelligence and to prepare for university study and, ultimately, their future professions.

By "mathematical literacy" we mean a standard level of competence in mathematics required for an intelligent citizen in a high-technology society. Citizens should have a sound comprehension of useful mathematical notions like functions, solutions, derivatives, probabilities, etc; however, they are not expected to carry out rigorous mathematical proofs by themselves. Concerning the second goal, we have in mind not geniuses who may turn out to be eminent mathematicians, but simply those students who are gifted to the extent that they are qualified for higher education according to traditional standards.

Under the social pressure against formal differentiation in education referred to earlier, the two goals i) and ii) are not compatible with each other if we adopt a curriculum which is both uniform and demanding, and if we stick to the traditional method and style of classroom instruction. Nevertheless, we hope to find a solution through the following strategies:

a) designing a flexible curriculum composed of a core and option modules;

b) introducing inspiring problems which assist self-study and future development on the part of students;

c) using calculators, computers and other technological aids to enrich students' mathematical experience and facilitate students' learning at their own pace.

Details of proposals a) and b) have been discussed elsewhere (e.g., Fujita et al. 1984). We note here only that the core part of the curriculum should contain solid material of central importance which must be chosen carefully from among traditional and modern topics. Obsolete material should be deleted to reduce overloading. Hurdles in the early stages must be removed to minimize the mathematics anxiety experienced by the majority of students. The core should be learned by the majority of students in keeping with goal i), while the optional parts of the curriculum should be rich and diverse in topics and levels, so that gifted students can choose what they study according to their aptitude and background.

Calculators, Computers and Other Information Technologies

In the current curriculum, calculators and computers are explicitly treated only in Math II, although use of these computational aids is recommended wherever appropriate in the teaching of all topics.

It is our belief that the free and full use of calculators should be encouraged in classroom teaching as well as in individualized learning, not only to lighten the computational burdens encountered in problem solving, but also to facilitate numerical and activity-oriented approaches to mathematical notions in traditional and new math topics. This will help particularly in promoting both confidence and understanding of mathematics for the majority of students. In Japan, it is not a financial problem for each family to provide their children with their own calculators, since a simple calculator costs the equivalent of two lunches or 8 cups of coffee. Standardization of calculators for school use will most likely be a difficult problem.

The present curriculum is obviously deficient in introducing computers into mathematics education. Math II mentioned above is not normally taken by students bound for a university. Many mathematics teachers are reluctant to introduce computers into mathematics education, partly out of conservatism, and partly because they do not regard teaching about computers or computer literacy as a proper topic for mathematics. In general, it is a painstaking effort for teachers to give experimental instruction on topics outside the national curriculum, particularly to students who are seriously concerned about the entrance examinations. On the other hand, the availability and popularity of personal computers in Japan is remarkable. Many students are already enthusiastic about personal computers, especially computer games. Some pupils in primary schools and junior high schools even compile their own programs for computer games. It is said that 15-20% of pupils or their families in big cities own personal comuters. Some teachers already report that the performance of a number of pupils has deteriorated in basic subjects and/or that the pupils have become isolated from their classmates because of their fascination with personal computers.

It is thus an urgent matter for us to try to find out how and when a simple and systematic introduction of computers into mathematics education should be made. Under the Japanese system such questions are never trivial. We have to revise the national curriculum, equip all schools with computers for teachers' use as well as for pupils' use, train (or retrain) a considerable number of teachers, and get universities to cooperate in including some topics on computers in their entrance examinations. All this must be done simultaneously as a national measure. In addition, for the time being, it is equally important to develop the extensive use of computers and other information technologies in order to create more opportunities for individualized learning which will allow students to be autonomous and to proceed at their own pace in learning mathematics, and which will supplement classroom instruction without increasing the burdens on teachers. With these kinds of aids, high school teachers can encourage underachievers to recover their pace and thus help all the students develop their mathematical intelligence to the utmost. In this vein, Professor F. Terada and his colleagues have succeeded in inventing an exemplary system, the THE system, which combines microcomputers and optical videodiscs and offers ample opportunity for interactive and individualized learning of mathematics (see F. Terada et al. 1984). A report on the present status of THE system is being given by Professor Terada at this conference.

Appendix A

Some Statistics on Secondary Education in Japan

I. Lower Secondary School

Total number of pupils (three grades)

1960	5,899,973		
1970	4,716,833		
1980	5,094,402		
1982	5,623,975	1st year	1,884,912
		2nd year	1,885,554
		3rd year	1,853,509

Pupils per class   38.0
Pupils per full-time teacher   20.9
Total number of lower secondary schools   10,879

II. Upper Secondary School

1) Total number of students (three grades)

1960	3,239,416		
1970	4,231,542		
1980	4,621,930		
1982	4,600,551	1st year	1,486,038
		2nd year	1,541,107
		3rd year	1,539,691

2) Total number of schools in 1982 = 5,213

National	17
Local	3,954
Private	1,242 (23.8%)

3) Percentages of students by course

General	69.7%
Agriculture	3.4
Technical	10.0
Commercial	12.1
Fishery	0.4
Home Economics	3.1
Health	0.6
Others	0.7

III. Universities and Junior Colleges

1) Total number of universities in 1982 = 455

National	95
Local	34
Private	326 (71.6%)

2) Total number of junior colleges in 1982 = 526

National	36
Local	51
Private	439 (83.5%)

3) Total number of university students (4 years) in 1982 = 1,817,650

Female	405,127	(22.3%)
National	425,141	
Local	52,632	
Private	1,339,877	(73.7%)

4) Total number of junior college students (2 years) in 1982 = 374,273

Female	335,992	(89.8%)
National	16,389	
Local	19,768	
Private	338,122	(90.3%)

5) Percentages of university students by fields of specialization in 1982

Humanities	13.9%	Social Sciences	39.7%
Sciences	3.2	Engineering	19.4
Agriculture	3.4	Medical & Dental	4.3
Home Economics	1.8	Education	7.8
Arts	2.6	Others	3.9

IV. Admission and Graduation

1) New entrants

	Kindergarten	Elementary School	Upper Secondary School	University
1970	1,011,640	1,621,635	1,381,997	333,037
1980	1,299,741	2,055,699*	1,623,069	412,437
1982	1,177,305	1,865,573	1,474,789	414,536

* Peak of the second wave of the post-war baby boom.

2) Enrollment and Advancement rate

	Enrollment rate for Lower Sec. Sch.	Advancement rate for Upper Sec. Sch. (Female)		Advancement rate to Univ. & Jr. Coll. (Female)	
1950	99.91%	42.5%	(36.7%)		
1960	99.93	57.7	(55.9)	10.3%	(5.5%)
1970	99.89	82.1	(82.7)	23.6	(17.7)
1980	99.98	94.2	(95.4)	37.4	(33.3)
1982	99.98	94.3	(95.5)	36.3	(32.7)

Appendix B

Topics included in mathematics subjects for senior high schools in the 1968 curriculum (the previous curriculum).

Math I (Compulsory, to be studied in 10th grade)

1. Numbers and Expressions:
   real numbers, the number line, polynomials (ring operations), rational expressions.
2. Equations and Inequalities:
   quadratic equations, equations of a higher degree, the factor theorem, inequalities.
3. Plane Figures and Equations:
   plane coordinates, the equation of a straight line, the equation of a circle, domains defined by inequalities.
4. Vectors:
   vectors in a plane, linear operations with vectors, the components of vectors.
5. Functions:
   simple functions, quadratic functions, fractional linear functions, exponential and logarithmic functions, inverse functions.
6. Trigonometric Functions:
   trigonometric ratios, radians, trigonometric functions, the sine theorem, the cosine theorem.
7. Probability:
   permutations and combinations, probability, conditional probability.
8. Mapping, Sets and Logic:
   mapping, the composition of mapping, 1-to-1 mapping, onto-mapping, inverse mapping, propositions, the truth set, $\forall$ and $\exists$, union, intersection, complement.

Math II A (Elective subject, to be studied in 11th grade by non-academically oriented students)

1. Matrices:
   matrices, multiplication of 2 x 2 matrices, linear transformation in a plane.
2. Differentiation and Its Applications:
   limits, the derivative, differentiation of polynomials, the tangent, maxima and minima, velocity.
3. Integration and Its Applications:
   indefinite integrals (of polynomials), definite integrals, area and volume.
4. Probability and Statistics:
   random variables, binomial distribution, normal distribution, the sample mean, statistical inference.
5. Electronic Computers and Flow Charts:
   outline of electronic computers, programming, flow charts, arrays and files.

Math II B   (Elective subject, to be studied in 11th grade by academically
              oriented students)

1. Coordinates and Vectors in Space:
    coordinates in space, the equation of a sphere, vectors in space, the inner
    product of vectors, planes and straight lines in space.
2. Matrices:
    matrices, multiplication of 2 x 2 matrices, inverse matrices, systems of
    linear equations, linear transformation, rotation, the addition theorem for
    sine and cosine, groups.
3. Sequences:
    finite sequences, arithmetic and geometric series, mathematical induction,
    the recurrence formula, the binomial theorem.
4. Differentiaton and Its Applications:
    limits, the derivative (of polynomials), the tangent, maxima and minima,
    velocity.
5. Integration and Its Applications:
    indefinite integrals (of polynomials), definite integrals, area and volume,
    movement on a line.
6. Axiomatic and Plane Geometry

Math III (Elective subject, to be studied in 12th grade by academically oriented
          students mostly bound for science divisions of universities)

1. Limits of Sequences:
    the convergence of series, limits, infinite series.
2. Differentiation and Its Applications:
    the limit of a function, continuity, differentiation of quotients,
    differentiation of inverse and composite functions, the derivative of
    trigonometric, exponential, and logarithmic functions, higher derivatives,
    the tangent, the mean value theorem, decreasing and increasing, maxima
    and minima, convexity of a curve, velocity, acceleration, parameter
    representation of curves, approximation.
3. Integration and its Applications:
    indefinite integrals, change of variables, integration by parts, formulas of
    integration, the definite integral, area, volume, the length of an arc,
    approximation of definite integrals, differential equations.
4. Probability and Statistics:
    random variables, probability distribution, binomial distribution, normal
    distribution, the mean, variance, standard deviation, statistical inference,
    the sample mean.

Appendix C

The Mathematics Curriculum for Upper Secondary Schools in Japan

revised by the Ministry of Education in 1978,
effective from 1982

Excerpt from the General Provisions of the
Course of Study for the Upper Secondary Schools

Part I. The Mathematics Program for General Students

1. The overall subject "Mathematics" is composed of several subjects whose titles
    and associated credits (other than "Science Mathematics" and "Integrated
    Mathematics") are shown in the following table.

Subject	Standard Number of Credits
Mathematics I	4
Mathematics II	3
Algebra/Geometry	3
Basic Analysis	3
Differential and Integral Calculus	3
Probability/Statistics	3

(1 credit is 35 lesson-hours or 35 fifty-minute periods of study throughout the
year.)

I. The Objectives of Mathematics

II. Subjects

I. Mathematics I

1. Objectives

2. Content

    (1)   Numbers and Algebraic Expressions
    a.   (i)   numbers and sets
         (ii)   integers, rational numbers, real numbers
    b.   Algebraic expressions
         (i)   polynomials
         (ii)   rational expressions
    (2)   Equations and Inequalities
    a.   Equations
         (i)   quadratic equations

(ii)   simple equations of a higher degree
(iii)   simultaneous equations
b.   Quadratic inequalities
c.   Algebraic expressions and proofs
(Terms / Symbols)
     discriminant, imaginary number, i, complex number
(3)   Functions
a.   Quadratic functions
b.   Simple rational and irrational functions
(Terms / Symbols)
     inverse function
(4)   Geometric Figures
a.   Trigonometric ratios
(i)   sine, cosine, tangent
(ii)   the sine theorem, the cosine theorem
b.   Plane figures and equations
(i)   points and coordinates
(ii)   equations of straight lines
(iii)   equations of circles
(Terms / Symbols)
     sin, cos, tan

3. Remarks about the Content

(1)   In (1)a(i), the fundamental facts of sets should be included.
(2)   In (1)b, the index should be extended to include zero and the negative integers.
(3)   In (2)a(ii), use of the factor theorem should be limited to solving third- and fourth-degree equations with simple numerical coefficients.
(4)   The content of (2)a(iii) should be limited to simultaneous equations in two variables of the first and second degree.
(5)   In (2)c, necessary conditions, sufficient conditions, contraposition, etc. should be included.
(6)   In (3)b, functions such as $y = \dfrac{ax+b}{cx+d}$ and $y = \sqrt{ax+b}$ should be included.
(7)   In (4)b, simple cases of the regions represented by inequalities should be included.

II. Mathematics II

1. Objectives

2. Content

(1)   Probability and Statistics
a.   Permutations /combinations
b.   Probability
c.   Statistics

(Terms / Symbols)

       nPr, nCr, factorial, n!, complementary event, expectation, standard deviation

(2)   Vectors
  a.   Vectors and their operations
  b.   Applications of vectors
(3)   Differentiation and Integration
  a.   The meaning of differential coefficients
  b.   Derivatives and their applications
  c.   The meaning of integration

(Terms / Symbols)

       limit value, lim, indefinite integral, definite integral

(4)   Sequences
  a.   Arithmetic sequences
  b.   Geometric sequences
(5)   Various Functions
  a.   Exponential functions
  b.   Logarithmic functions
  c.   Trigonometric functions

(Terms /Symbols)

       power root, $\log_a x$, generalized angle

(6)   Computer and Flow Charts
  a.   Functions of electronic computers
  b.   Algorithms and flow charts

3. Remarks about the Content

(1)   The content of Mathematics II (1)–(5) forms a base for what is treated in "III Algebra/Geometry", "IV Basic Analysis" and "VI Probability/Statistics". The reader should refer to the "Contents" and "Remarks about the Contents" for each of these subjects in order to gain a clearer understading of the mathematical level of Mathematics II.

(2)   When actually teaching Mathematics II, the instructional content of the class should be appropriately selected from among the above areas (1)–(6). In so doing, consideration should be given to the pupil's ability and to the number of school hours assigned to the class.

(3)   In (6), actual experience preparing programs for the computer, running them, and analyzing the results should be included.

III. Algebra/Geometry

1. Objectives

2. Content

(1)   The Conic Sections
  a.   The parabola
  b.   The ellipse and the hyperbola

(2)	Vectors in a Plane
a.	Vectors and their operations
b.	The inner product of vectors
c.	The applications of vectors;
	equations of straight lines and circles, etc.
(3)	Matrices
a.	Matrices and their operations
b.	The inverse matrix
c.	Linear transformation and mapping

(Terms / Symbols)

$$A^{-1}$$

(4)	Figures in Space
a.	Points, straight lines and planes in space
b.	Coordinates in space
c.	Vectors in space;
	include equations of straight lines, planes and spheres

3. Remarks about the Content

(1)   In (1), the equations of the conic sections should be limited to normal form, and rough sketches and foci should be included.

(2)   Matrix multiplication in (3)a, should include matrices up to 2 x 2.

(3)   The content of (4)a should be built around the relations of parallelism and perpendicularity and the theorem of three perpendiculars.

IV. Basic Analysis

1. Objectives

2. Content

(1)	Sequences
a.	Simple sequences;
	arithmetic sequences, geometric sequences, etc.
b.	Mathematical induction

(Terms / Symbols)

$$\Sigma$$

(2)	Functions
a.	Exponential functions
b.	Logarithmic functions
c.	Trigonometric functions
	(i) the generalized angle and the circular measure method
	(ii) trigonometric functions and their periodicity
	(iii) addition theorems for trigonometric functions

(Terms / Symbols)

power root, $\log_a x$

(3)　　　Variation of Values of Functions
a.　　The meaning of the differential coefficient
b.　　Derivatives and their applications
　　　　(i)　derivatives of sums, differences and products multiplied by
　　　　　　real numbers of functions
　　　　(ii)　the tangent, increase and decrease of the values of a function,
　　　　　　velocity, etc.
c.　　Integrals and their applications;
　　　　　　indefinite integrals, definite integrals, area, etc.
(Terms / Symbols)
　　　　limit value, lim

3. Remarks about the Content

(1)　　　The content of (1)a should be limited to the sum of an arithmetic
　　　　sequence, the sum of a geometric sequence and the sum of the
　　　　sequence $\{n^2\}$.
(2)　　　In (2)b, computation using logarithms is not included.

V. Differential and Integral Calculus

1. Objectives

2. Content

(1)　　　Limits
a.　　The limit of a sequence
b.　　The limit of the value of a function
(Terms /Symbols)
　　　　convergence, divergence
(2)　　　Differential Calculus and Its Applications
a.　　Derivatives
　　　　(i)　differential calculus of the product and quotient of functions
　　　　(ii)　differential calculus of composite and inverse functions
　　　　(iii)　derivatives of trigonometric functions
　　　　(iv)　derivatives of exponential and logarithmic functions
b.　　Applications of derivatives
　　　　the tangent, increase and decrease of values of a function, velocity,
　　　　acceleration, etc.
(Terms / Symbols)
　　　　natural logarithm, e, the second derivative, point of inflection
(3)　　　Integral Calculus and Its Applications
a.　　Integral calculus
　　　　(i)　the meaning of integration
　　　　(ii)　integration by substitution and integration by parts in simple
　　　　　　cases
　　　　(iii)　the integration of various functions
b.　　Applications of integration
　　　　(i)　area, volume, length of path, etc.

(ii)   the meaning of differential equations, solving equations such as dy/dx=ky

3. Remarks about the Content
(1)

(2)   In (2), the mean value theorem should be approached intuitively and used in such a way as to make clear the relation between changes in the value of a function and its derivative.

(3)   In (2)a(ii), simple functions such as $y=x^k$ (k is a rational number), $y=\sqrt{ax+b}$ and $y=\sqrt{ax^2+b}$ should be studied.

(4)   In (3)a(ii), integration by substitution should include functions such as ax+b = t or x = a sin θ. Integration by parts should be limited to those cases where only a single application is required, relating to simple functions.

VI. Probability/Statistics

1. Objectives

2. Content

(1)   The Arrangement of Data
a.   Distribution of the variable
b.   Representative values and measures of dispersion
(Terms / Symbols)
     variance, standard deviation, Σ
(2)   The Number of Cases of Possibilities
a.   Permutations / combinations
b.   The binomial theorem
(Terms / Symbols)
     nPR, nCr, factorial, n!
(3)   Probability
a.   Probability and its basic laws
b.   Independent trials and their probability
c.   Conditional probability
(Terms / Symbols)
     complementary event, exclusiveness, independence, dependence
(4)   Probability Distribution
a.   Random variables and their probability distribution
b.   Binomial distribution, normal distribution
(Terms / Symbols)
     expectation
(5)   Statistical Inference
a.   Population and sample
b.   The idea of statistical inference
(Terms / Symbols)
     estimation, test

3. Remarks about the Content

Item (5) should be limited to gaining an understanding of statistical inference through concrete examples.

III. The Construction of Teaching Plans and Remarks on the Content of Each Course

1. In designing teaching plans, the following should be considered.
   (1) Mathematics I is required for all pupils in the 10th grade, but the other mathematics subjects are optional. Mathematics II, Algebra and Geometry, Basic Analysis, and Probability and Statistics should, as a rule, be studied after Mathematics I.
   (2) Differential and Integral Calculus should, as a rule, be studied after Basic Analysis.

2. In teaching, the following should be considered.
   (1) When pupils study subjects beyond Mathematics I, the teacher must take into account the relationships between the content of the various subjects and also the systematization of the curriculum.
   (2) (Terms / Symbols) given for each subject of II are included to clarify the extent and scope of the material treated in the subject. In teaching, these terms and symbols should be closely related to the content of the subject.
   (3) Wherever possible, pupils should be encouraged to use computers or other mechanical aids to computation.

The Mathematics Curriculum for Lower Secondary Schools in Japan

revised by the Ministry of Education in 1977,
effective from 1981

Partial Excerpt from the General Provisions of the
Course of Study for Lower Secondary Schools

1. Mathematics is a required subject in each grade.

2. Standards for the total number of school hours in each
   grade are prescribed as follows:

7th grade (12 years old)	105
8th grade (13 years old)	140
9th grade (14 years old)	140

   Notes:
   1) A school hour is 50 minutes.
   2) Lessons in mathematics must be given over at least
      35 weeks each year.

Mathematics

I. Objectives

II. Objectives and Content in Each Grade

1st Grade
1. Objectives
2. Content
A. Numbers and Algebraic Expressions
(1)   To help pupils deepen their understanding of the properties of integers.
    a.   Expression of an integer as the product of prime numbers.
    b.   Properties of divisors and multiples.
(2)   To help pupils understand the meaning of positive and negative numbers,
and compute with those numbers.
(3)   To help pupils develop their abilities to express relations and rules in a
formula through the use of letters, and to calculate simple expressions.
(4)   To help pupils understand the meaning of an equation, and solve
a linear equation.
(5)   To help pupils understand approximate value, and use it properly
under various circumstances.
(Terms / Symbols)
    natural number, factor, the greatest common measure, the least common

multiple, sign, absolute value, term, coefficient, similar term, $\leq$, $\geq$

B. Functions

(1) To help pupils deepen their understanding of a function through studying the relationships between phenomena, paying attention to two variable quantities.

    a. Change and correspondence.

    b. Variable and domain.

(2) To help pupils understand that a functional relation can be expressed by means of a table, graph, formula, etc., and investigate the characteristics of change and correspondence by using those representations.

C. Geometric Figures

(1) To help pupils think about geometrical figures through various manipulations, and deepen their understanding of figures in space.

(2) To help pupils develop their abilities to construct figures that meet given conditions.

    a. To construct basic figures such as the bisector of an angle, the perpendicular bisector of a line segment, a perpendicular, etc.

    b. To consider a figure as a set of points that meet certain conditions, and to construct the figure.

(3) To help pupils develop their abilities to measure a figure.

    a. The length of an arc and the area of sector.

    b. The surface area and volume of a cylinder, a conical solid, and a sphere.

(Terms / Symbols)

body of revolution, arc, chord, π, $/\!/$, $\llcorner$, $\angle$

3. Remarks about the Content

2nd Grade

1. Objectives

2. Content

A. Numbers and Algebraic Expressions

(1) To help pupils apply a simple formula using letters and make the computations, and the four fundamental operations.

(2) To help pupils develop their abilities to find further quantitative relationships between phenomena, and to express and apply such relationships in a formula using letters.

(3) To help pupils understand the concept of inequality and solve a linear inequality.

(4) To help pupils understand the significance of setting up simultaneous equations or inequalities and their solutions, and the ability to solve them.

    a. The significance of a linear equation in two variables and its solution.

    b. Solving simple simultaneous linear equations or linear inequalities.

B. Functions

(1)   a. Some phenomena may be described through the use of linear functions.

    b. A linear equation in two variables can be considered to express the functional relations between the two variables.

(2) To help pupils understand the characteristics of linear functions and

develop their abilities to use them.
    a.   The equation of a straight line and the graph of a linear function.
    b.   The ratio of change of a linear function is a constant.

C.   Geometric Figures
(1)   a.   Displacements by parallelism, symmetry, and rotation.
    b.   The properties of parallel lines.
    c.   The conditions of congruence for triangles.
(2)   To help pupils clarify the concepts of similarity of figures and to develop their abilities to consider the properties of figures by using the conditions of congruence or similarity for triangles.
    a.   The meaning of similarity and the conditions of similarity for triangles.
    b.   The properties of the ratio of segments of parallel lines.
    c.   Properties of triangles and parallelograms.

(Terms / Symbols)
    opposite angle, interior angle, exterior angle, center of gravity, R, $\triangle$, $\equiv$, $\backsim$

D.   Probability and Statistics
(1)   To help pupils collect data in accordance with their objectives, and to arrange the data by using tables and graphs, and thereby to ascertain the trends of the data.
    a.   The meaning of frequency distribution and methods of examining histograms.
    b.   The meaning of relative or cumulative frequency.
    c.   The meaning of mean value and range.

(Terms / Symbols)
    frequency, class

3. Remarks about the Content

3rd Grade
1. Objectives
2. Content
A. Numbers and Algebraic Expressions
(1)   To help pupils understand the square root of a positive number and be able to use it.
(2)   To help pupils expand and factor simple expressions.
    a.   Multiplication of a simple linear expression.
    b.   Expansion and factoring of an expression by use of the formulas indicated below:

$$(a + b)^2 = a^2 + 2ab + b^2$$
$$(a - b)^2 = a^2 - 2ab + b^2$$
$$(a + b)(a - b) = a^2 - b^2$$
$$(x + a)(x + b) = x^2 + (a + b)x + ab$$

(3)   To help pupils understand quadratic equations.
    a.   Quadratic equations and their solution.
    b.   Solving quadratic equations by using factoring and the formulas for solution, etc.

(Terms / Symbols)

radical sign, rational number, irrational number, $\sqrt{}$

B.   Functions

(1)   a.   Various phenomena and functions describing them.

         b.   Functions which are in direct or inverse proportion to x^2.

         c.   The ratio of change of functions.

(2)   To help pupils think about the relations of correspondence between the elements of two sets and deepen their understanding of the meaning of function.

         a.   Sets and functions.

         b.   The domain and range of a function.

C.   Geometric Figures

(1)   To help pupils deepen their understanding of the properties of a circle.

         a.   Circles and straight lines.

         b.   The relationship of the angle of circumference to the central angle.

(2)   To help pupils understand the measuring properties of figures.

         a.   The Pythagorean theorem and its application.

         b.   Measurement of height and distance.

         c.   The relationship between the ratios of length, area, and volume in similar figures.

(Terms / Symbols)

     tangential line, point of tangency

D.   Probability and Statistics

(1)   To help pupils understand probability by paying attention to the frequency obtained through large numbers of observations or trials.

         a.   The meaning of probability.

         b.   Computing probability in simple cases.

(2)   To help pupils understand that the trends in a population can be estimated from a sample.

3. Remarks about the Content

(1)   In A(3)b quadratic equations with real numbers as their solution should be taught. For factoring as a method of solution, the available formulas should be limited to the ones indicated in A(2)b.

(2)   In D(1) only events which can be easily classified with the aid of tree diagrams, etc., should be dealt with.

(3)   The content of D(2) should be limited to the level of dealing with experiments and observations.

III. The Construction of Teaching Plans and Remarks on Content in Each Grade

Appendix E

Sample Mathematics Problems for Entrance Examinations to National Universities (Secondary Examinations) and Private Universities in Japan

I. Full Set of Problems, University of Tokyo, 1982

6 problems for the science section (students bound for the Faculties of Science, Engineering, Medicine, Agriculture, and Pharmacy). Time: 150 minutes. Four problems for the literature section (students bound for the Faculties of Law, Economics, Humanities, and Education). Time: 100 minutes.

<u>Science Section</u>

(S1) Let f be a linear transformation of the xy-plane defined by the matrix $A = \begin{pmatrix} a & b \\ c & d \end{pmatrix}$, and suppose that there exists a point P different from the origin which is mapped to P itself by f. Prove that there exists a straight line ℓ which does not go through the origin and which is mapped into (or onto) ℓ by f.

(S2) Let T be a regular tetrahedron and S be a sphere of radius 1. Suppose that each edge of T is tangential to S. Find the length of an edge of T, and find the volume of the solid portion which is exterior to T and, at the same time, interior to S.

(S3) A and B are two moving points in the xy-plane; A moves along the subarc lying in the first quadrant of the unit circle with radius 1 and center at the origin 0, while B moves on the x-axis in such a way that the length of the segment AB is kept equal to 1. Moreover, the point C of intersection of the segment and the circle is different from either A or B. Under these circumstances, answer the following questions.
  (1) Determine the range of $\theta = \angle AOB$.
  (2) Express the length of BC in terms of θ.
  (3) Denoting the midpoint of OB by M, determine the range of the length of CM.

(S4) A point P moves along the curve y = sin x in the xy-plane from left to right as indicated in the figure below. Assume that the speed of P is a constant value V (V > 0). Find the maximum of the magnitude $|\vec{a}|$ of the acceleration vector $\vec{a}$ of P. The speed of P here means the magnitude of the velocity vector $\vec{v} = (v_1, v_2)$ of P, and $\vec{a}$ is given by $\vec{a} = (\dfrac{dv_1}{dt}, \dfrac{dv_2}{dt})$, t being the time variable.

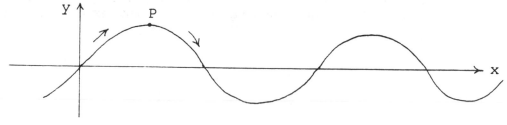

(S5) Find the volume of the solid in xyz-space which is the set of all points with coordinates x, y, z subject to

$$0 \le z \le 1 + x - y - 3(x - y)y, \qquad 0 \le y \le 1 \quad \text{and}$$

$$y \le x \le y + 1.$$

(S6) A die is initially placed with its 1-spot face at the top. Then the position of the die is changed by repeated rotations. At each rotation, the angle of rotation is 90 degrees and the axis of rotation is a straight line connecting the centers of a randomly chosen pair of square faces parallel to each other. Assume that any possible choice of axis is equally likely, and that for each chosen axis either of the two possible directions of rotation is equally likely.

Let p_n, q_n and r_n denote the probabilities that the 1-spot face is located on the top, on one of the lateral sides and on the bottom just after the n-th rotation, respectively.

(1) Find p_1, q_1, r_1.

(2) Express p_n, q_n, r_n in terms of p_{n-1}, q_{n-1}, r_{n-1}.

(3) Find $p = \lim_{n \to \infty} p_n$, $q = \lim_{n \to \infty} q_n$, $r = \lim_{n \to \infty} r_n$.

Literature Section

(L1) A and B are two fixed points in a plane, and the length $\overline{AB}$ of the segment AB is equal to $2(\sqrt{3} + 1)$. Now consider three points P, Q and R which move in this plane keeping the relations $\overline{AP} = \overline{PQ} = 2$ and $\overline{QR} = \overline{RB} = \sqrt{2}$. Find the domain S in the plane which is the set of all points Q can reach, and find the area of S.

(L2) A, B and C are three variable points on the curve $y = x^2$ in the xy-plane, in the order of their x-coordinates (the x-coordinate of A is the smallest). While they move, the difference between the x-coordinates of A and B is kept equal to **a** (**a** is a positive constant), and the differenc between the x-coordinates of B and C is kept equal to 1. Express the x-coordinate of A in terms of **a** for the case where $\angle CAB$ attains its maximum, and find the value of **a** such that $\angle ABC$ becomes a rectangle when $\angle CAB$ attains its maximum.

(L3) Consider the 4 roots of the equation $x^4 + ax^2 + b = 0$, where a and b are integers. We are given approximate values of -3.45, -0.61, 0.54 and 3.42 for these 4 roots, where the absolute value of the error of any of these approximate values does not exceed 0.05. Give the numerical values of the exact roots to two decimal places.

(L4) Let

$$A = \begin{pmatrix} \frac{1}{2} & \frac{1}{2} \\ 0 & 1 \end{pmatrix}, \quad B = \begin{pmatrix} 1 & 0 \\ -\frac{1}{2} & \frac{1}{2} \end{pmatrix}.$$

Starting with $P_0 = (1.1)$, we generate sequences of points P_0, P_1, P_2, ... in the xy-plane using the following procedures, where (x_n, y_n) stands for the coordinates of P_n.

a) If $x_n + y_n \geq \frac{1}{100}$, then (x_{n+1}, y_{n+1}) is given either by

$$\begin{pmatrix} x_{n+1} \\ y_{n+1} \end{pmatrix} = A \begin{pmatrix} x_n \\ y_n \end{pmatrix} \quad \text{or by} \quad \begin{pmatrix} x_{n+1} \\ y_{n+1} \end{pmatrix} = B \begin{pmatrix} x_n \\ y_n \end{pmatrix}.$$

b) If $x_n + y_n < \frac{1}{100}$, then (x_{n+1}, y_{n+1}) is given by $\begin{pmatrix} x_{n+1} \\ y_{n+1} \end{pmatrix} = A \begin{pmatrix} x_n \\ y_n \end{pmatrix}$.

In this way a number of sequences of points is generated.

(1) Find all possible points P_2, and show them graphically.

(2) Express $x_n + y_n$ in terms of n.

(3) How many possible points P_n are there?

II. Selected Problems

(1) Kyoto University, 1979. Science Section

Find polynomials $P_1(x)$, $P_2(x)$ and $P_3(x)$ whose coefficient of the highest degree term is equal to 1 and which satisfy the following conditions, respectively.

(1) $P_1(x)$ is linear and $\int_{-1}^{1} P_1(x) C dx = 0$ for any constant C.

(2) $P_2(x)$ is quadratic, and $\int_{-1}^{1} P_2(x) f(x) dx = 0$ for any polynomial f(x) with degree equal to or less than 1.

(3) $P_3(x)$ is cubic, and $\int_{-1}^{1} P_3(x) f(x) dx = 0$ for any polynomial f(x) with degree equal to or less than 2.

(2) Kyoto University, 1979. Science Section

P is a moving point in the xy-plane. At time t, the coordinates (x,y) of P are given by

$$\begin{pmatrix} x \\ y \end{pmatrix} = e^{-at} \begin{pmatrix} \cos bt & -\sin bt \\ \sin bt & \cos bt \end{pmatrix} \begin{pmatrix} c_1 \\ c_2 \end{pmatrix},$$

where a, b, c_1 and c_2 are constants with $a > 0$, $b > 0$. The point $C(c_1, c_2)$ is assumed to be different from the origin $O(0,0)$.

(1) Show that the angle between the velocity vector of P and the radial vector OP is constant.

(2) After starting from $C_0 = C$, the point P intersects the segment OC for the first time at C_1, for the second time at C_2, ..., for the k-th time at C_k. Show that the length of the arc between C_k and C_{k+1} of the orbit of P forms a geometric sequence, i.e., that the length of $\overset{\frown}{C_0 C_1}$, $\overset{\frown}{C_1 C_2}$, ..., $\overset{\frown}{C_k C_{k+1}}$, ... is a geometric sequence.

(3) Kyoto University, 1980. Literature Section

Suppose that $f(x)$ is a polynomial of degree n and $n \geq 2$. Show that

$$f(x) + f(x + 1) - 2\int_0^1 f(x + t)\,dt$$

is of degree n - 2.

(4) Kyoto University, 1980. Literature Section

Let S be a set $S = \{a_1, a_2, ..., a_n\}$ of n distinct positive numbers with the following property: if we take any two different elements a_i, a_j from S, then either $a_i - a_j$ or $a_j - a_i$ belongs to S.

Show that by a suitable rearrangement of the order, a_1, a_2, ..., a_n can form an arithmetic progression.

(5) Waseda University, 1982. Faculty of Science and Engineering

Let R be the image of each of the figures in (1)–(3) below by the linear transformation $\begin{pmatrix} x' \\ y' \end{pmatrix} = \begin{pmatrix} 1 & 3 \\ 2 & k \end{pmatrix} \begin{pmatrix} x \\ y \end{pmatrix}$. Determine for each case whether R will be a point, a straight line or the whole plane. If R is a point or a straight line, then write its coordinate or equation.

(1) straight line : x + 2y = 0
(2) straight line : x + 2y – 3 = 0
(3) whole plane.

(6) Waseda University, 1982. Faculty of Science and Engineering

Suppose that the point P(x, y) in the xy-plane moves on the unit circle whose center is at the origin. Find all the points where $z = x^3 + y^3$ attains its maximum. Also give the maximum value of z.

(7) Waseda University, 1982. Faculty of Science and Engineering

The coordinates x, y and z of a variable point P in space are regarded as functions x(t), y(t) and z(t) of time t $(0 \le t \le \frac{1}{2})$ and are subject to the following conditions:

(i) $\frac{dx}{dt} = z - y$, $\frac{dy}{dt} = x - z$, $\frac{dz}{dt} = y - x$

(ii) $(x(0), y(0), z(0)) = (1, -1, 0)$,

(iii) $x(t), y(t), z(t) \le 0$.

(1) Find $f(t) = x(t) + y(t) + z(t)$.

(2) Find $g(t) = (x(t))^2 + (y(t))^2 + (z(t))^2$.

(3) We denote the angle between the vector $(1, -1, 0)$ and the vector $(x(t), y(t), z(t))$ by $\theta(t)$. Express $\sin \theta(t)$ in terms of z(t).

(4) Find $\theta(t)$.

(8) Waseda University, 1982. Faculty of Education

Let α and β be two planes in xyz-space such that α contains $(0, 0, 0)$, $(1, 1, 1)$, $(0, 1, -1)$ and β contains $(0, 0, 0)$, $(-1, 1, 1)$, $(1, 0, 1)$. Find the point of intersection of the plane β with the straight line which passes through P(a, b, c) and is perpendicular to the plane α.

(9) Waseda University, 1982. Faculty of Education

Let f(x) and g(x) be differentiable functions defined on $0 < x < \pi$, and suppose that f and g satisfy the conditon $f'g = \sin x + \cos s$, $fg' = -\sin x$, $f(\frac{\pi}{2})g(\frac{\pi}{2}) = 1$, $\lim_{x \to 0} g(x) = -1$.

Determine f(x) and g(x).

References

Fujita, H. and F. Terada. 1983. "Mathematics Education in a Society with High Technology." Proceedings of ICMI-JSME Regional Conference on Mathematical Education, 55–59. Tokyo.

Fujita, H. and F. Terada. 1984a. "Mathematical Education in a Society with Modern Technology. A Case Study in Japan." Project Presentation, ICME 5, Adelaide.

Fujita, H. and F. Terada. 1984b. "The Present State and Current Problems of Mathematics Education in Senior High Schools of Japan." AG4, ICME 5. Adelaide.

Fujita, H., F. Terada, and S. Shimada. 1984. "Toward a Mathematics Curriculum in the Form of the Core and Option Modules." AG4, ICME 5. Adelaide.

Iyanaga, S. 1981. "Mathematical Education in Japan." J. Sci. Educ. 5: 467–477. Japan.

Terada, F. and H. Fujita. 1981. "The Present State and Current Problems in General Mathematics Education at Japanese Universities." Proceedings of the Seventh International Conference on Improving University Teaching, 327–336. Tsukuba.

Terada, F., H. Fujita, and S. Shimada. 1984. "Learning Mathematics Through the New Media and the THE System: An Exemplary Videodiscs Learning System." Project Presentation, ICME 5. Adelaide.

SMP 11-16: The Most Recent Work in Curriculum Development by the School Mathematics Project, and Its Relation to Current Issues in Mathematical Education in England

John Ling

SMP, Westfield College, London, England

The School Mathematics Project (SMP) is one of the longest-lived mathematics projects in England, and in terms of range and output, the largest. Its published material at present covers the age range 7-18, not counting a handbook for teachers of children aged 5 to 7. The title "Project" is now slightly misleading, suggesting as it does something of a unified nature. During the 24 years of its existence views have changed as circumstances (and personnel) have changed, but the overall aim of the SMP remains that of improving the quality of mathematical education through the writing of material for use in schools.

The SMP started in the early 1960's. It was one of a number of projects in England that set out to reform school mathematics at the secondary level (ages 11-18). The original writers were concerned about the large and widening gap between the mathematics being taught at school and the subject as treated at university, and about the complete lack of reference in school to the growing range of newer applications of mathematics, including computing. On the first of these issues — the school-university divide — the approach adopted by the original SMP writers differs somewhat from the approach of certain contemporaneous American projects. Whereas American projects tended to incorporate into school mathematics the abstract and axiomatic approaches current in universities, the SMP way was to provide rather more concrete and informal approaches to topics that would receive a formal treatment later, at university. (The SMP approach to geometry — through practical work on mappings — was dismissed in a review by Carl B. Allendoerfer as being merely physics.) On the applications side, the SMP, along with other similarly motivated projects, introduced such topics as statistics, probability, linear programming, networks, and computing into the school syllabus.

The SMP was, as I have said, one of a number of initiatives that set out to reform the school syllabus. In terms of its take-up it has been the most successful, and the financial rewards of its success, most of which were retained by the trust that administers the Project, have been invested in a series of projects of which the new SMP 11-16 course is the most recent.

The first SMP material was written for the most able 20-25% of pupils in the age range 11-16. The SMP expanded subsequently into the 16-18 age range and later into the 7-11 age range, but in this paper I shall be concerned only with developments affecting pupils in the 11-16 age range.

The success of the original course for the top 20-25% of pupils led to a similar rethinking of the mathematics offered to pupils in the "middle" band of abilities (from the 20th down to the 60th percentile). The content of the resulting course for these pupils was derived by simplification of and omission from the original course for the top 20%, and was embodied in books A to H, which became very popular indeed. For a time this series was the single most widely used maths textbook series in England.

A further development was occasioned by the growth of mixed-ability teaching in the early years of secondary school. As a partial response to this situation, a set of individual workcards, intended particularly for the lower end of the target audience, was written to cover broadly the same ground as that covered in books A to D of the main school series.

When electronic calculators became readily available, the SMP was a pioneer of their use in schools, and special material exploiting the use of the calculator was written to supplement the existing textbook series.

These developments take us to about 1975. Throughout this period, SMP work in the 11-16 age range consisted of modifying the original course to meet different circumstances (such as the growth of comprehensive schools, of mixed-ability teaching, and the use of calculators), but not substantially changing the overall content. In 1976 the SMP held a large conference to discuss what it should do next in the 11-16 age range. By this time there was widespread criticism of the SMP books, which appeared to be suitable for a much narrower ability range than had been envisaged originally. SMP decided that work should start on a new course whose content would be worked out afresh. The course would set out to cater for the needs of about the top 85% or so of the ability range. The work of writing and testing the new course, which is called SMP 11-16, has, not surprisingly, turned out to be the SMP's most ambitious undertaking.

In the foregoing brief account of the SMP's work in the 11-16 age range, I have made brief references to external developments in education to which the SMP responded. I do not propose to say anything more here about these developments and their effect on SMP's pre-1976 material. Rather, I will turn to more recent developments which are related more directly to the new SMP 11-16 course.

The Cockcroft Report (Cockcroft 1982) is the most significant recent public document on school mathematics in England, and its comments on existing practices and its recommendations for action have become the starting point for discussion of future policies for mathematical education. The Cockcroft Committee was convened in 1978, at a time when the "issue" of the day was the alleged inability of many school leavers applying for jobs in industry to cope with basic mathematics. The Committee's brief was a broad one — to report on and make recommendations concerning the teaching of mathematics in schools. It did make a thorough and useful investigation of the school-employment interface, in the process accumulating much information about the mathematical skills actually needed in various occupations. The report, published in 1982, refused to underwrite a back-to-basics movement (but by then the issue had faded somewhat from the public mind). On the contrary, its views, which related to the whole field of mathematical education in schools, were generally welcomed by teachers as enlightened and providing a sound basis for future action.

The report drew attention to the fact that for many pupils their experience of mathematics was an experience of failure, and that the public examinations they sat were often simply opportunities to show how little of the work they had covered in school had been understood. The Committee thought that many children were following syllabuses that were too large and that they were required to cover the ground too fast. For the least able 40% or so of pupils, the Committee produced a "foundation list" of mathematical topics that would set a worthwhile and attainable goal.

The SMP 11-16 team, which had started work in 1977, and which by 1982 was well into its program of writing and testing, welcomed the Cockcroft Report, which gave public recognition to many of the views that had united them as a team on the project. This concurrence between the team's outlook and that of the Committee was not surprising: an innovative project naturally attracts those whose criticisms of current practice are deepest. What the report has done for projects such as ours is to enable us to approach bodies like examination boards, whose cooperation we need, in the knowledge that what we are trying to do is in line with public policy.

I shall now describe the SMP 11-16 project. At some points I will make reference to the Cockcroft report and to the recent decision to restructure the 16+ examination system, both of which have a direct bearing on the later years of the course.

The SMP 11-16 curriculum development team has a number of broad aims which are closely related to one another. We believe that most existing courses for secondary schools in England go too fast and try to cover too much ground for many children. There is much evidence for this view in the findings of research projects, such as CSMS,[1] and it is also supported by the findings of the Cockcroft Committee, as mentioned above.

We believe that children need to be allowed time to become familiar with mathematical ideas in a concrete way. If they are expected to work at a relatively abstract level too early, they will lose their grasp of the meaning of what they are doing. A major aim of the course is to avoid the sense of meaninglessness that many pupils feel in mathematics lessons, and which many adults recall in later years. So we have set out to present mathematics concretely. In the early stages children make models and use simple apparatus to find out things for themselves. We make extensive use of illustration to allow pupils to enter imaginatively into situations that embody mathematical ideas and contexts in which people use mathematics. This aim of relating mathematics to the "real world" applies throughout the course, and there is a constant emphasis on the applications of mathematics.

We have been concerned with making the syllabus for each level of ability of a realistic size. We think it better to have a secure grasp of a limited range of mathematics than to have a shaky hold on a wider range. The course includes investigative and practical work, which encourages pupils to think out for themselves how to use the mathematics they know.

The writing team is large, numbering around 40 or so, mostly teachers working on the project in their free time and attending weekend meetings. At first I was the only person employed full-time on the work, as team leader. Later, writers who showed particular talent were seconded to the project for limited periods, and as the testing program got under way and production schedules tightened, the full-time staff was increased to a maximum of five (including myself) during the period 1981-84.

The writing team was initially organized into five groups, working in the topic areas of number, algebra, graphs, space and statistics. The "chairmen" of these groups were members of a central planning team. The subdivision of the team by topic areas had the advantage that contributors were able to concentrate attention on the particular difficulties encountered by children in each topic area. In the early stages some members of the CSMS team (referred to above) participated in the work of some of the writing groups.

The groups started work in January 1978, and some experimental material was tried out on a small scale in the writers' own schools. Eventually it was decided that the most appropriate form for the material being written was short topic booklets. Throughout the production of the material — both testing and published versions — the SMP 11-16 team has been as closely involved in the presentation and design of the printed materials as in the content. The trial version was produced by the Cambridge University Press to the same standard as published material, and testing began in about 40 schools in September 1980. The feedback from the testing schools to the writers is through written reports, visits to classrooms by members of the project team and, in some areas, local meetings of teachers with team members. At present (March 1985) the testing program is in its fifth and final year. The material for the first two years of the course (ages 11 to 13+) has already been revised and was published by the Cambridge University Press in 1983-84. Revision of the later material for publication is under way.

The course materials fall into two parts. Part 1 comprises material for the first two years of the course (ages 11 to 13+), and Part 2 covers the last three years (ages 13 to 16+).

The material in Part 1 consists mainly of 16- and 8-page topic booklets. They are organized in four successive levels. Extension booklets are added to levels 2, 3 and 4, forming levels 2(e), 3(e) and 4(e).

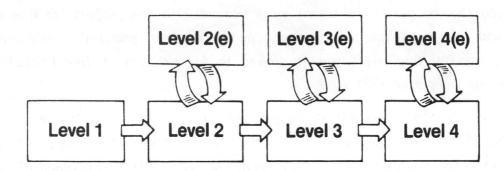

Pupils need a satisfactory understanding of work at one level before proceeding to the next. But the levels are not linked to chronological age: more able pupils will not need to work through all of level 1 and by the end of two years will have completed level 4, while less able pupils may still be working on level 2.

The booklets are designed to enable pupils to work from them directly at their own pace, but they also provide scope for group work and teacher-led activities.

The material does not depend on a particular way of grouping pupils; classes may be mixed-ability or not as the school chooses.

The extension booklets are not intended for a precisely defined upper-ability range: most extension booklets follow on from one of the main course booklets. A pupil should only tackle an extension booklet if he or she has coped well with the main course booklet to which it is linked.

Practical work with simple equipment is an essential part of the course and some special learning aids — including several games — have been devised. Some work involves the use of a calculator, but there is also number work for which calculators are not to be used.

There are altogether about 120 booklets in the material for years 1 and 2, together with review books, teacher's guides and a collection of mathematical investigations with special problem cards for very able pupils.

In most schools that use the booklets, children spend much, if not most, of the time working individually, or in pairs, or in small groups. There is thus no "common pace" for a class, so children in the same class will be working on different topics at the same time. This is an unfamiliar situation for those teachers who have hitherto done most of their teaching "from the front", addressing the whole class at once, and we recommend that such teachers should initially limit the number of different booklets in use at any one time. In practice, schools have developed different ways of organizing the use of the material to suit themselves.

The distribution in the classroom of booklets, worksheets and equipment — the "service activity" — has to be smooth-running if the teacher is not to become totally immersed in giving things out. The children themselves have to be trained to take responsibility for as much as possible of the service activity. This, together with the fact that they are usually given some degree of choice as to which piece of work to tackle next, gives the children a sense of involvement in, and responsibility for, their own work.

Part 2 of the course covers years 3, 4, and 5 (ages 13 to 16+). The material in part 2 consists mainly of books for class use. At the start of year 3, pupils are allocated, provisionally at first, to one of three series of books: Y (yellow - upper ability), B (blue - middle ability) and G (green - lower ability). The starting assumptions of each series are defined roughly in terms of coverage of material in the booklet scheme. To allow for "late developers" initially allocated to the B series, this series branches after two books. The upper branch is designated R (red). There are extension books added to the Y series for pupils of very high ability. The G series includes booklets for individual use and resource packs of cards containing investigations, games and puzzles.

The overall structure of Part 2 of the course is set out below.

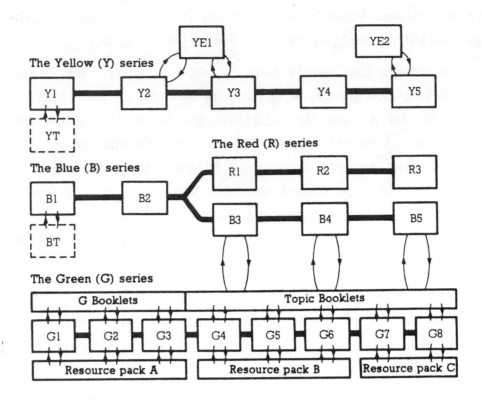

The books continue to pursue the aim of the course to present mathematics in relation to other areas of life. This is particularly, but not only, true of the content of the B (middle) and G (lower) series. The B series excludes many topics at present included in courses in England for the middle ability range, such as vectors, matrices, and functions, and concentrates instead on a smaller range of widely applicable elementary mathematics. The syllabus for the G series corresponds very closely with the "foundation list" recommended by the Cockcroft Report and referred to earlier.

The pupils at the lower end of the ability range in the age range 14 to 16 present the greatest problems for teachers. Their plight has been exacerbated in England by the absence of any public assessment scheme that might provide at least an external goal for their efforts. It is the view of the writers of the G material that it is pointless to give these pupils the same diet of elementary mathematics again and again which they then fail to master again and again. On the contrary, it is possible for such pupils to make progress and to gain in competence and get a sense of achievement from gaining in competence. Because the system of public examination does not extend to the lowest 40% or so of the ability range — the range for which the G series is intended — we have had the opportunity to develop an assessment scheme of our own for these pupils. The main component of this scheme is a series of graduated tests taken in school by pupils as and when they are ready for them. These tests are supplemented by oral and practical tests. An examination board has agreed to award special certificates on the basis of these tests. The certificates are at three stages, and a pupil receives a certificate as soon as he or she reaches the level of overall competence required by one of the stages. The subject of graduated tests is one which is of great interest in England at present, and the government has selected three schemes, including our own, for special study.

The Y, B, and R series are designed to prepare pupils for the "official" public examination system. A pilot scheme of assessment has been designed for the schools that have been following the draft version of SMP 11-16. Here we have broken new ground by including coursework assessment along with the traditional written examination papers for all candidates. The need for other instruments of assessment besides written papers arises from the fact that there are important kinds of work in the course which cannot be adequately asssessed by written examination papers. These include investigative work and work requiring the candidate to collect and select appropriate data from the real world; and, generally speaking, work that requires time for thought and execution. The inclusion of coursework assessment in an examination scheme for the whole of the top 60% of the ability range is a novelty, and the pilot scheme is being treated as a feasibility study and being carefully monitored.

In this scheme, a pupil's coursework assessment program extends over two years. The tasks to be done are devised by the SMP team and marked, following guidelines, by teachers in the schools.

Some examples of coursework tasks are given below.

1. You have a piece of plywood 150 cm by 300 cm.

    You have to design a dog kennel that could be made from the piece. Try to make your kennel as large as you can.

Make a scale drawing to show how the parts of the kennel have to be cut out of the plywood. Give the measurements.

Draw a sketch or sketches to show what the finished kennel will look like. Write the measurements on the sketches.

2. Some daily newspapers give a lot of their space to photos, others not so much. Choose two newspapers with different page sizes and different numbers of pages. Find out how much of the space in each paper is taken up by photographs. Write a report in which you compare the two papers. Include one or more diagrams or charts to illustrate what you found.

3. There are many different ways of arranging twelve 1 cm squares edge to edge (leaving no "holes" surrounded by squares). For example:

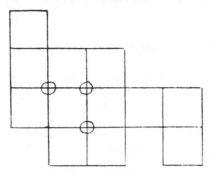

In this example the perimeter is 20 cm, and there are 3 points where four squares meet together (marked by circles). Draw other arrangements of twelve squares and see if you can find a connection between the perimeter and the number of "four-square" points.

The main technical problem in this scheme arises from the need to standardize the assessment over a large number of schools. However, the main practical problem is the amount of sheer hard work teachers have to do. It is in the nature of the work being assessed that marking is not a straightforward business. Nevertheless, it is now known that the new system of 16+ examinations that will be in effect from 1988 on will include coursework assessment, and the SMP 11-16 course is in this respect in the forefront of developments in the field of assessment.

Although the way in which the new course is related to the public examination system may be of little general interest outside England, I mention it because the examination system has a far-reaching backwash efffect on the teaching of mathematics, and if the aims of a curriculum development project are at variance with the demands of examinations, it is the latter that will hold sway. This explains why the SMP has always tackled curriculum and examinations together, as being (in England at any rate) the only way to ensure that curricular ideals have some chance of being realized in practice. This, and a continued concern with in-service training, have always been and still remain features of the SMP's work.

The publication of SMP 11-16 will not be completed until 1987, but already SMP is turning its attention to the continuation of what has been begun.

As SMP is funded from the royalties on its published material, its continued health depends largely on its "success" as measured in commercial terms. The present time is one of great financial stringency in English education, and schools have difficulty finding the money to buy new materials.

The take-up of the course has nevertheless been very encouraging indeed, and there will be a continuing need in the future for in-service work involving teachers using the new course. Local "user groups" are already being formed, and these will be encouraged from the center. One important area which lies open for exploration by such groups is the use of microcomputers in relation to the course.

SMP plans to devote greater attention than it has done in the past to monitoring the way in which the material is used in schools that decide to adopt it. (I must emphasize that a school's decision to follow the course is its own, and our material competes in the marketplace with other materials.)

A full-time 11-16 development officer is to be appointed, and in three local authorities SMP is arranging to provide part of the funding for the appointment of advisory teachers who will spend the "SMP" part of their time monitoring the use of SMP 11-16.

However, ours is not a course which is suitable for every school. Running it requires a maths department which is willing to plan as a team, is well-led, and keeps its practices as a team under continual review. We make no attempt to hide the organizational complexity. There will be schools that will take on the course and handle it badly. We have tried to be realistic in our assessment of what schools are capable of, but if we attempted to produce something which is workable in all circumstances, however unfavorable, we would have to abandon our role as an innovative project altogether.

So far the SMP 11-16 team has been heartened by the many reports of children's increased enthusiasm for mathematics in schools using the material, which is of course what the team set out to achieve.

In conclusion, I would like to extend my best wishes to the UCSMP, with which my own project already shares some initials, and offer to share whatever of our experience may be relevant to your situation.

Note

1. Concepts in Secondary Mathematics and Science, based at Chelsea College, University of London. See Hart (1981).

References

Cockcroft, W.H. 1982. Mathematics Counts. Report of the Committee of Inquiry into the Teaching of Mathematics in Schools. London, Her Majesty's Stationery Office.

Hart, K.M., ed. 1981. Children's Understanding of Mathematics 11-16. London, John Murray.

Post-"New Math" Since 1963:

Its Implementation as a Historical Process

Tamás Varga

National Pedagogical Institute, Budapest, Hungary

The aim of this paper is to present a case study: the continuing reform of math education in Hungary, a small country in east-central Europe.

In the title, 1963 indicates the year of birth of this reform process in a sense still to be explained. The year of its conception was 1962, the year of an International Symposium of Mathematics Learning in Budapest.

Post-"New Math" since 1963 sounds contradictory. The New Math movement dates back to the late fifties and early sixties, and lasted more than a decade. It was followed by the Back to Basics Movement. Neither of the two was, in my judgment, a convincing success. Now we are in the Post-"New Math" era. The University of Chicago School Mathematics Project has been a major stimulus in this regard. My reason for using such a presumptuous title was the pleasant surprise of finding that the underlying principles of the UCSMP mostly match those of our project in Hungary — or rather, what was a project in 1963 but an implemented curriculum by 1985.

Not that we were able to reach all our declared goals in two decades. Implemented does not mean attained or realized! There is nothing to boast of in having our curriculum implemented. As a matter of fact, it was regrettable, as you will see below.

Yet it is true that because of a lucky concurrence of circumstances we were able to avoid some of the major pitfalls of the New Math Movement and reach — in this sense — the Post-"New Math" Era two decades earlier. The favorable circumstances include an early acquaintance with persons, or at least their writings, who could foresee and foretell those pitfalls.

The 1962 Symposium was instrumental in this respect, which is why I fixed 1962 as the year of conception of our reform. This is so despite the fact that Hans Freudenthal — one of those who could and did foretell the pitfalls — was not present: a clerk at UNESCO (co-sponsor of the Symposium) had stricken his name as a "notorious quarreller."

Along with Freudenthal, Figure 1 lists others whose ideas contributed to shaping a post-"New Math" curriculum during the early sixties — one which did not need any substantial readjustment or turn-about later on.

R.B. Davis (USA)	D. Page (USA)
Z.P. Dienes (UK)	G. Pólya (USA)
H. Freudenthal (Holland)	W.W. Sawyer (Canada)
A.Z. Krygowska (Poland)	W. Servais (Belgium)

Figure 1

Note that the persons listed here — and many others could be included — take rather different positions on particular issues in math education, and I could not stand behind any of them as one whose position exactly matches ours.

Before touching upon some Hungarian geography and the history of our curriculum, I might illustrate what I mean by UCSMP's post-"New Math" character that I am so much in accord with by some quotations from a UCSMP document on K-6 curriculum development.

The UCSMP, as I read,

1. "is directed toward the general student population" (the New Math tended to forget the non-mathematically inclined, even if it declared otherwise);

2. "intends thoroughly to cover arithmetic" (yes, arithmetic remains central in grades K-6, even if the very meaning of the word arithmetic undergoes substantial changes, as it did before the advent of calculators and still more since);

3. "and build firm foundations in geometric, proportional and probabilistic thinking by grade six" (the list could be — and later is — extended by adding terms such as logical, algebraic, relational, functional, combinatorial, statistical... but the basic idea is clearly this: rather than offering juxtaposed courses, the emphasis is on the learner who nurtures

geometric, proportional, and probabilistic ideas along with arithmetical ones, and the school has the responsibility of assisting him or her in further developing and integrating such ideas into some sort of mathematics as a whole, of developing, in fact, a holistic way of thinking beyond mathematics as well).

4. "One of the most dramatic differences is that several topics are started earlier in the grades than in traditional approaches. By doing so, the topics are carried along throughout the grades and adequately formed over several years as opposed to the present approach of delaying topics, possibly past germination periods, and trying to cover the topics too rapidly."

In commenting on this last point, let me cite a story illustrating how difficult it is to make such a position acceptable even after decades of pilot work. A few months ago a live program was broadcast on one of our two national television networks: an educational program with sixty-six participants who could raise thorny questions (and vote on others). One of the sixty-six, a teacher of mathematics in the intermediate grades, took up the discussion and announced how absurd the newly implemented mathematics curriculum was for grades five to eight: instead of being confined to traditional topics it tended to cover much of what had earlier been assigned to grades nine to twelve. Apparently, in his thinking, as among many of his colleagues still today, juxtaposition is not one of the possible ways (perhaps obsolete) of building up a mathematics curriculum, but the only possible or logical way of doing so. Twenty-two years were not enough for us to challenge such a deeply rooted belief in quite a number of militant defenders of a contrary view.

Let me now continue with some geography and history, as promised.

I come from Hungary, and here in Illinois I feel very much at home. Not only because this is my sixth or seventh visit to this state since 1967, when I spent two weeks in Chicago with the Madison Project, but because this state reminds me of Hungary, with its climate, its cornfields, its single large city (though Budapest has not much more than two million inhabitants) along with many small ones. Like Illinois, Hungary has a total population somewhat over ten million. Moreover, the two have roughly the same area; even their shapes are somewhat similar (though Hungary is longer east to west than north to south).

You may or may not know that quite a number of world-famed mathematicians and scientists were born Hungarian. (See the two noninclusive lists, Figures 2 and 3). This fact may not be totally independent of some good traditions in Hungarian math and science education. I might also add that during the twenty-five-year history of the International Mathematical Student Olympiad, the Hungarian team has almost always been among the three highest scoring teams.

Mathematics

J. Bolyai (Non-Euclidean Geometry)

P. Erdös (Number Theory, Graph Theory, etc.)

P. Halmos (Set Theory, Measure Theory, etc.)

J.G. Kemeny (Probability, Computers)

I. Lakatos (Philosophy of Mathematics)

J. von Neumann (Logic, Geometry, Computers, etc.)

G. Pólya (Complex Analysis, Probability, Heuristics)

F. Riesz (Functional Analysis)

Figure 2

Sciences

G. Békésy (Acoustics, Nobel Prize 1961)

D. Gábor (Holography, Nobel Prize 1971)

G. Hevesy (Isotope Chemistry, Nobel Prize 1943)

T. Kármán (Aero- and Hydrodynamics)

A. Szent Györgyi (Biology: Vitamin C, Nobel Prize 1937)

L. Szilárd (Nuclear Physics)

E. Teller (Nuclear Physics)

E. Wigner (Quantum Theory, Nobel Prize 1963)

Figure 3

Both facts have mainly to do with the math and science education of the gifted, especially at the secondary level. This is where we have really good traditions: we have had mathematical student competitions and student journals since the end of the nineteenth century, interrupted only by the two world wars. Promoting mathematical mass education is a different problem, however, and has been our main concern for more than two decades.

In this paper I have repeatedly mentioned the 1962 Symposium on Math Education. For me it opened up new dimensions. In writing a background paper before the meeting, I needed, obtained, and read the most recent relevant literature in English, French, German and Russian. This endeavor took me six months, along with my courses in math education at Budapest University. Two weeks of discussions followed, mainly during the academic breaks. Most importantly, after these two weeks the highest ranking educational authorities were ready to authorize educational experiments in mathematics, even quite wild ones which earlier would have been out of the question. I had to devise a curriculum, which I managed to do during the next school year. The experiment — or rather, pilot work — started in two first grade classes of a school in Budapest with two volunteer teachers.

In subsequent years teachers from the same school, and later from other schools and cities as well, volunteered to teach similar courses in their own classes. In a decade the number of classes grew to about two hundred, 0.5% of the classes in the eight grade schools all over the country. The number of classes from year to year grew in a geometric progression. The quotient of the progression was

$$\sqrt[10]{\frac{200}{2}} \approx 1.58.$$

This is the pattern of an organic growth. The number of classes joining the scheme each year was proportional to the number of classes within the scheme.

As with contagious diseases, when the sources of infection increased (the increased possibility of visiting classes), the chances of "catching the disease by personal contact" increased proportionately.

Boundary conditions such as maturation tend to slow down this kind of growth after a while; the value of the function approaches an upper bound which in the present case could not be higher than the number of classes in the country (in the order of forty thousand). Applying this extremely simplified model to the present case, one could expect every — or nearly every — class in the eight Hungarian grade schools to be involved some time early in the next century. A promising scenario.

Attempts to speed up a natural, organic growth rarely give satisfactory results, however, and unfortunately such attempts were made. High-level decisions required that new curricula be developed and implemented for every school subject in grades 1-8. Our project was found worthy of serving as the basis of a new mathematics curriculum for the entire country.

We who were involved in the project sensed the danger, but could not argue strongly enough. After all, those responsible for other school subjects were even less prepared. "Let other schools — every school in Hungary — enjoy the fruits of your work," the enticement sounded.

As a compromise, we molded a milder version of the curriculum adopted in our pilot schools and advised that it be implemented gradually: first in 5% of the first grades (the number actually became 8%); in the following year 15% of the new first grades, as the pathbreakers came to the second grade; and so on, with 50, 80, and 100% of the first grades in subsequent years.

At the time of this writing (1984-85), 100% of the seventh and about 80% of the eight graders "enjoy the fruits of our work" — or do they? You may have guessed that not all of them do.

The gradual implementation was intended to simulate organic growth, but it did not work well, for various reasons. First, the pace was too quick. Second, it was usually not volunteer teachers who decided when to start with the new curriculum, but their supervisors or higher authorities. Many of those who had been suspicious of our pilot work before it had official recognition hastily jumped on the bandwagon as soon as they found it had. (This made the originally planned 5% become 8%.) Thus even in the early stages of implementation, and increasingly thereafter, many teachers were obliged to follow the new curriculum even though they would have preferred to keep the old, which they felt at ease with. And so the good reputation of our project met its fate. We became compulsory, as a state religion or state ideology. The landlords — the educational authorities — converted the masses, and the latter had to follow the new faith, a practice bequeathed from the late Middle Ages. Cuius regio, eius religio: "whose region — his religion."

It often happened that when the inspector appeared, the teachers would come out with some kind of activity which showed that they were true followers. In other cases the teacher was happy with the new curriculum, but the inspector or the head teacher was not, and tried to discourage the teacher. The cuius regio, eius religio principle worked both ways, and still does.

Teacher training — both initial and in-service — should serve to remedy the situation, but its effectiveness is not noteworthy. Much depends on the trainers, of course. Imparting new knowledge is relatively easy. Difficulties arise when trainees are supposed to unlearn obsolete concepts, abandon familiar views, change habitual practices, or — most difficult though most important of all — change attitudes. I mean, for instance, accepting children as fellow-learners whose ways of thinking, silly as they seem, merit serious attention: not a standard attitude on the part of Hungarian teachers. I will come back to this point and illustrate it with an example.

Before doing so, I would like to sketch some features of our curriculum as it emerged in our pilot classes. I will stress the word "emerged." It has been, in fact, the result of cooperation between the teachers who did the actual daily work and the leaders who were regularly present and often took over the teachers' role.

The content and the structure of the curriculum were influenced by two aims: first, to approximate the ways children of the given ages think and feel; and second, to remain close to the traditional curriculum, except where this would contradict the first point.

Concerning the first, we make efforts not to reduce the content to mere arithmetic. After all, children spontaneously develop other kinds of mathematical ideas, about space, relations, chance, and so on, even before they go to school. It is the responsibility of the school to nurture the existing germs of such ideas in children's minds. With this in view, instead of asking when to teach geometry or

243

probability, and so on, we would ask the question: <u>what</u> kind of mental food would serve best in grade one, then in grade two, and so on, to nurture those seeds and let them bud and blossom. This is, if you like, an application of Jerome Bruner's famous principle to school situations.

This can, of course, lead to disastrous results unless the principle is moderated by common sense and by a constant consideration of children's immediate responses as well as the long-range effects on their development. The traditional curriculum and teachers' habitual practices may not meet the requirements of our rapidly changing world, but they have their riches: in the course of time teachers have accumulated a great deal of common sense. This again is also a question of attitude: without the highest respect for the teaching profession and individual teachers, how could we expect teachers to respect the children?

Instead of describing in detail how these two complementary endeavors shaped our <u>intended</u> curriculum, let me give you two examples. They will shed some light, I hope, first on the <u>implemented</u> curriculum, as it is put into practice by teachers, and secondly on the <u>attained</u> curriculum, the effect it has on children. Both examples will reflect the difficulties we encounter, but favorable auspices will not be totally lacking.

The first example is a story based on a problem in our math workbook for third graders. Here is the problem:

- Somebody tells a joke on Monday to five persons. The next day, Tuesday, each of the five tells the joke to six other persons. Each of the latter tells it to seven persons on Wednesday. How many will have heard it on Wednesday?

In an interview published in one of our national newspapers a mathematician complained about this problem, mentioning his nephew's dilemma. The boy had received it as homework, and found three different solutions, depending on the interpretation of the text:

a. Five persons heard the joke on Monday, five times six or 30 on Tuesday, five times six times seven or 210 on Wednesday. The answer is 210.

b. In another interpretation the answer is 5 + 30 + 210 or 245: those who heard the joke on Monday or on Tuesday <u>will have heard it</u> on Wednesday, together with the 210 who heard it precisely that day.

c. The person who told the joke to the first five persons must have heard it

244

previously — unless he invented it — so the answer is 246.

The nephew was desperate. "If I come up with any of these solutions," he said to his uncle, "the teacher may have in mind another solution and she will make a fool of me before the class because I could not find the real solution. The whole class will laugh at me!"

The point of the story is the inference drawn by the mathematician. He said, "Problems in math workbooks should be more carefully worded so as to exclude different interpretations."

For my part, I like giving children problems which they can interpret in different ways. Finding different interpretations is a first step toward inventing problems on their own, or toward mathematizing an open situation. Such activities are at least as important as solving ready-made problems, if learning mathematics has any goal beyond itself. In some cases — test items, contest problems — unambiguous wording may be a virtue. There is a watershed here: either you believe that the main goal of mathematics problem solving is to enable pupils to solve further, more difficult problems in order to pass tests and win contests, or you see something, maybe a great deal, that can be achieved beyond this. In the first case you will see no point in ambiguous problems or open problem situations; in the second case you will.

In our example the boy really was in trouble because of the ambiguous problem: not because of the problem itself, but because it was given by an authoritarian teacher. Let me argue in favor of the mathematician interviewed: as long as there are authoritarian teachers in the schools, pupils may get in trouble unless problems in the workbooks are unambiguously worded. But should we ban such problems to the detriment of some major goals of mathematics learning?

There is no easy solution to this dilemma. The policy we have adopted is not to ban ambiguous problems, but to convey to teachers how to make use of them. Authoritarian personalities will not change overnight, but are they really what we suppose them to be? Maybe the teacher of the boy in our story only needs an encouragement, and she will accept different answers to our problem, depending on the children's interpretations.

She may even find it fun to suggest further interpretations. 246 seems to be the greatest number. But is 210 the smallest? What does Figure 4 tell you?

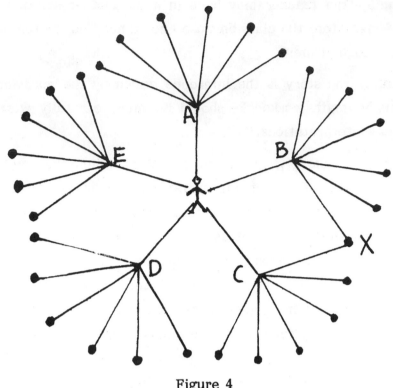

Figure 4

On Tuesday somebody, X, hears the joke from two different persons B and C out of the five who heard it the day before. He is polite and does not repeat it to those who have heard it already. So Monday 5, Tuesday 29; and on Wednesday? 29 times 7. Could it be less? Still less?

Children compete in bringing the number down and down further still. They use their imagination — the right hemisphere of their brains if you like — and they use the other hemisphere, too, in producing and defending their answers. Whether they come down to the number 7 (each of them listening to the joke Wednesday as many times as there are persons who heard it Tuesday) is almost irrelevant. The main interest is not in the minimum solution, but in a succession of good, better and still better solutions. The situation will probably make children laugh. (And not at the expense of a classmate!)

Let me turn to my second example, the one which throws light on the attained curriculum. Fifth graders struggled with the following problem, just one item in a battery of items, some time in May (age at the beginning of the school year: 10+).

- The area of a rectangular flowerbed is 36 m^2. Surrounding the flowerbed along its edge is a rope, with knots at one meter distances everywhere. How many knots are there on the rope?

In an earlier version of the problem, which proved to be more difficult (guess why), tulips were planted along the edge at one meter distances.

The interesting finding with this test item was this: in the majority of classes very few pupils or nobody at all could solve it, or even leave some trace of an attempted solution. But in a minority — though quite a number — of classes most pupils, including those with very low marks, found one or more solutions, or at least worked on it in a way which made sense, even if they miscalculated the answer. (See Figure 5 for some sketches.)

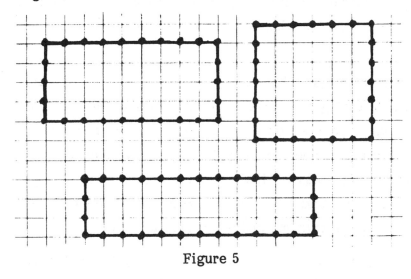

Figure 5

The social background of the pupils in the first and in the second type of classes showed no differences that could explain the finding. Further inquiry indicated a single reason: the "good" classes were "good" because their teachers had gotten pupils used to experimenting with concrete materials such as colored rods, or geoboards, or just making drawings on their own in order to figure out solutions to problems. The classes with low average scores on this item were sometimes quite good in routine problems and calculations. At this item they were blocked probably because the problem was not in their stock of knowledge. They could not recall having ever met any similar problem, so they did not do anything.

The above test item required some sort of originality, perhaps even creativity, on the part of the pupils. Every non-routine problem does. Many teachers become indignant at such test items. They feel that it is unfair at an evaluation to ask a question for which the answer has not been taught, or to give a problem for which the solution method has not been provided previously. I cannot say they are altogether wrong. But they work for a different set of goals. All I can say is that their set of goals has to be extended. The world we live in requires that. They will find that they cannot prepare their pupils by including flowerbed problems among those they teach them to solve. Next time still other non-routine problems will appear in the tests, unexpected ones. That is what makes them non-routine. The only remedy will be to get pupils involved in using their imagination, not just their memory. In other words, to develop their creativity, if only in a modest sense.

Many teachers will ask, "At the expense of skills?" Well, along with, not at the expense of. The slogan "Do non-routine work and the skills will come" is self-deception. In our project we were not cautious enough to avoid this kind of self-deception completely. Some of our textbooks, workbooks, maybe even some teachers manuals overstress originality at the cost of drill. This lack of balance tends to discredit the sound principle. Overstressing one extreme may contribute to reinforcing the other extreme instead of fighting it down.

In our example, children who do draw rectangles on grid paper but do not know the shortcut way of finding the areas and therefore count the tiny squares instead will probably fail in solving the problem. In life, too, not only in the classroom, they will be rather helpless without a minimum equipment of mathematical skills.

This equipment will be reduced. My father learned to extract the cubic root of numbers. I did not, but I learned to extract the square root. My children did not: they learned to find it in a table. They spent much time learning long division. My grandchildren still learn it, but their children probably will not. Or they will only in order to understand how it works, for no practical purpose. Calculators are here, and will stay.

Yet the reduction of the stock of required skills does not mean less emphasis on those skills which remain necessary. Some skills — think of estimation — will even gain in importance.

Creativity, yes — but skills too. This is the essence of the message I have brought you from a little country on the confines of Eastern Europe.

Action and Language — Two Modes of
Representing Mathematical Concepts

Elmar Cohors-Fresenborg

University of Osnabrück

Abstract

This paper reports on research and curriculum development which has been carried out by the research group "Foundations of Mathematics and Mathematics Education" at the University of Osnabrück during the past 10 years, and recently at the Forschungsinstitut für Mathematikdidaktik, Osnabrück.

The paper presents curriculum materials which give children aged 9 to 14 insight into computer operation and automatic processes from a mathematical perspective. The main idea is that constructing an algorithm is more organizing a sequence of actions than expressing a mathematical structure in a formal computer language. Our methodology is to create a suitable microworld in which the basic concepts can be introduced, so that later explanation of more sophisticated concepts in real computer application situations has an adequate foundation.

Our research on pupils' behavior when dealing with algorithms shows that it is useful to build a conceptual framework that distinguishes between the level of cognitive structures and the cognitive strategies that are applied to them. On the level of cognitive structures, we distinguish between the language-oriented "predicative" and the action-oriented "functional" way of thinking, and on the level of cognitive strategies between the pre-structuring "conceptual" way and the more interactive "sequential" way. The existence of these differences could explain some problems that have accompanied the introduction of the New Math in schools. These differences could also help explain the varying opinions on suitable computer languages and approaches for schools (e.g., the interactive approach of Logo versus the structural approach of Pascal) and at the university level (e.g., between the functional structure of Lisp and the predicative structure of Prolog.

The research into curriculum development and children's cognitive strategies on which this paper concentrates emphasizes a virtually nonverbal approach to mathematical concept formation. But this approach is complemented by a project in which 13-year-old pupils are first introduced to the role of language in the axiomatic approach of mathematical concept formation; a text in an unknown language must be deciphered. In this approach we start from language and must proceed to meaning, while computer programming may be regarded as a problem of expressing a given meaning in a particular language. This introduction to the axiomatic approach to mathematics enables children to understand the role of mathematical modeling in understanding reality.

The appendix to the paper contains a brief description of the West German school system and teacher education.

Introduction

Mathematics teaching should be based on a philosophy of mathematics. There are different philosophies about the nature of mathematics, and therefore, there is more than one idea about the nature of mathematical thinking. It is an historical and psychological fact that mathematics is concerned with the abstraction and formalization of concepts and the relations between them. These concepts, which are useful in describing our knowledge and our theories of reality may either represent actions or express predicates (formulated in a particular language).

We know from research on the foundations of mathematics that (insofar as mathematical theories are enumerable) the following conceptions are equivalent as a basis on which to build mathematics: the abstraction of actions to (recursive) functions and the abstraction of language to the concepts of formal logic or set theory.

But this does not mean that in the teaching or learning contexts both approaches are equally useful, independent of the social contexts or the individual preknowledge and knowledge structure of the pupils and the teachers.

It seems to us that the mathematics curriculum reform of twenty years ago, known as the movement to New Math, was based merely on the preference of language over actions. This clearly was a result of the development of (pure) mathematics in this direction, the Bourbakism. But we believe that the general increased popularity of science (not only natural sciences) formulated as theories in a scientific language led to an increased acknowledgement of language-oriented approaches to scientific concept formation in school systems all over the world. The movement to New Math in schools often degenerated into a movement towards verbal-oriented concept formation.

Our recent research on children's cognitive strategies (see the "Research on Cognitive Strategies in Algorithmic Thinking" section), although so far concerned only with some algorithmic aspects of mathematical thinking, gives some hints that they have an action-oriented approach to concept formation and mathematical thinking which may be independent of the currently favored verbal approach.

This could provide an abstract explanation why there is so often a demand for more computational skills in schools, and also why rather many children like traditional competitions in mental arithmetic at the late primary level. Organizing the sequence of computational operations is what these children like, and not necessarily the subject of arithmetic. The traditional curriculum contained training in organizing skills — connected with arithmetic — and abilities in the handling of formal objects (in algebra), which is not necessarily bound psychologically to arithmetic. It is possible that the computer fever, which we recognize in many pupils and adults, has the same roots. One example may elucidate this: a pupil weak in mathematics, asked by a television reporter why he likes computer programming, replied that here, for the first time, he has a slave who is doing what he was instructed to do. The increasing availability of computers for children may give some of them a new challenge in constructing mathematical models (in the form of computer programs) and in mathematical thinking, where they can use their sequential action-oriented concept formation.

The fact that the balance between action and language (the two modes in which concept formation, knowledge, and thinking are represented in school) must be reconsidered may be seen from the title of Aebli's book on cognitive psychology (Aebli 1980-81) Denken: das Ordnen des Tuns (Thinking: The Ordering of Doing).

In this paper, we report on the research and curriculum development which has been carried out by the research group "Foundations of Mathematics and Mathematics Education" in the department of mathematics and computer science at the University of Osnabrück and the "Forschungsinstitut für Mathematikdidaktik" (Research Institute for Mathematics Education), Osnabrück.

The work may be grouped as follows:

- Development of a didactic aid, <u>Dynamic Mazes</u>, for introducing fundamental ideas of automatization and computer programming at the primary level (grades 3–6).

- Development of a software system, <u>Registermachine</u>, for introducing a microworld to help pupils understand computer programming, the design of computer languages, and the complexity of algorithms.

- Development of curriculum for introducing the concept of functions on the basis of algorithms.

- Empirical research on cognitive strategies of pupils (in grade 7) when they construct or analyze algorithms on different levels of representation.

- Development of curriculum to introduce the fundamental ideas of the axiomatic approach in mathematics.

Understanding Automatization and Computers at the Primary Level
Using <u>Dynamic Mazes</u>

Introduction

Through our efforts to enable elementary school pupils to approach the problem of automatization (in a play situation), to develop switching networks, and to prepare for computer programming, we found it necessary to develop material which linked aspects of thinking and organizing with actions. It seems further necessary that the pupils' organization should not be confined to conceptual structuring, but should be expressed as an ordering of actions. In 1974 we began to develop the building-block kit <u>Dynamic Mazes</u>, which has since been tested as instructional material in extensive experiments with children from 8 to 12 years of age. The concept of toy train networks is used as an approach to the development of algorithms, and this gives the instructional material a unique flair. Because of the mechanical functioning of the network (instead of an electronic realization with chips or on a computer screen), the pupils are able to develop their ideas sequentially while constructing the network or traveling through it. Our classroom experiments have shown that in addition to introducing the mathematical background of automata and computers, these building blocks improve the pupils' ability and increase their pleasure in problem solving (Cohors-Fresenborg 1978).

Building Blocks

We start by considering the network of a toy railway, in which one train travels in one direction and all switches are controlled by contact rails. The network had (probably several) "Begin" stations (input I) and "End" stations (output O). This simple railway network can be seen as a network of automata: the switches (Figure 1) are the automata, linked together by the rails; the traveling train is the information which determines the condition of the network. The switch (including both contact rails) is a simple automaton with a memory, which can accept one of two possible states S: left l and right r. Passing through entrance 2 (or 3) changes the state into the left (right) position. The mathematical behavior of this building block is described by its automaton table (Figure 2).

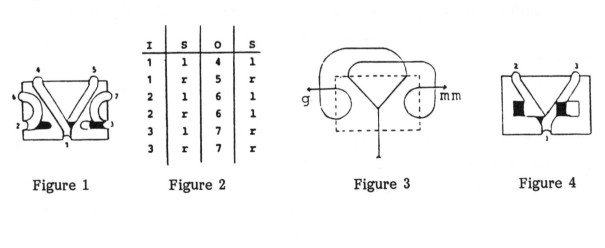

I	S	O	S
1	l	4	l
1	r	5	r
2	l	6	l
2	r	6	l
3	l	7	r
3	r	7	r

Figure 1   Figure 2   Figure 3   Figure 4

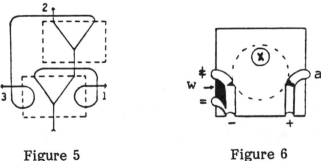

Figure 5   Figure 6

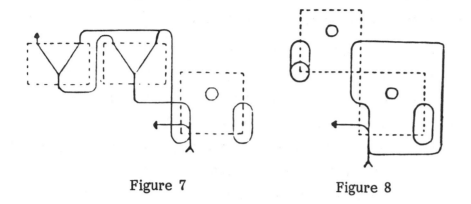

Figure 7   Figure 8

The other important building block is the counter (Figure 6). It has two entrances + and –, and three exits a, =, $\neq$. Inside is a cogwheel on which the numbers from 0 to 10 are written. If one passes through entrance + (and comes out at a), the number shown in the window x is increased by one. If one enters at entrance –, the number is decreased by one if it was greater than zero, and exit $\neq$ is reached. If the number was initially zero, it remains unchanged but exit = is reached, because by counting from 1 to 0 the wing W is turned by the cogwheel to block exit $\neq$ and open exit =. The kit contains two switches, counters and flip-flops (Figure 4), and 92 rails (straights, curves, crossings, junctions). All building blocks are fixed on a pinboard.

Examples

With the switch (or flip-flop), networks can be constructed to simulate counting machines, as used in vending machines for stamps, cigarettes, or tickets.

Figure 3 shows such a network, which simulates a two–coin vending machine. (Exit mm indicates, that more money has to be put in, exit g initiates the output of goods.) But the same network also simulates a sorting machine for two types of bottles, A and B, if these bottles are coming in on a conveyor belt in the sequence A B A B. This network is mathematically equivalent to a flip-flop (Figure 4): the two exits are reached alternately. If this sorting machine has to sort a sequence of numbered bottles, it sorts it into odd and even numbered bottles. This idea may be generalized to a three-bottle sorting machine or a counter mod 3, and so on. Those bigger sorting machines mod n check whether numbers are divisible by n, or find the remainder after the division. These examples also show the material's connections to the normal mathematics curriculum.

Using the counters (Figure 6), networks can be built for doing addition, subtraction, multiplication, division, etc. These counting networks (Figure 8 shows a network for addition) may be regarded as equivalent to computer science flow-charts. Figure 7 shows the idea for networks which translate numbers from the decimal system into the binary system and vice versa. According to Bruner's theory, there are three levels of representation: constructing a network (enactive), drawing (iconic), and describing a network (symbolic) by automaton table (Figure 2).

Even at the primary level it is possible to discuss with children in what sense an automaton, given as an automaton table (e.g., Figure 2), and a network (e.g., Figure 3 or Figure 4) are isomorphic, or what it means that the two-coin vending machine and the two-bottle sorter have the same mathematical model. It is an opportunity to begin with reflection on the range of mathematical concepts and models at the primary level. Our experience is that it is exciting to observe the children's discussions in these lessons.

Practice and Research

From our classroom experiments we have obtained a sequence of problems, which have been published in 9 booklets (Cohors-Fresenborg et al. 1979). In our experience, pupils even as young as 10 can work through these booklets by themselves. In a field study (Cohors-Fresenborg 1978) with about 540 pupils at the age of 10-11, the results of the test on the problems described above were compared with the scores on tests of intelligence and anxiety. Later we conducted case studies of the pupils' problem-solving behavior. Our action-oriented approach to introducing the concepts enables even deaf pupils to achieve astonishing success in working with the Dynamic Mazes (Cohors-Fresenborg et al. 1982b). Our material has also been used in practice and research in other countries. One study is reported in Lowenthal and Marcq (1982). With respect to sex differences, we might have assumed boys to be more successful in this field than girls. But the field study and the later case studies have shown no

256

differences between boys and girls in their problem-solving ability in this area. Evidently, if girls have suitable didactic material, they are much more creative in mathematics and technolgy than is normally assumed.

Registermachines as a Preparation for Pascal

Introduction

We have developed a method of introducing a microworld to help pupils understand computer programming, starting with Grade 7. It not only takes into account our research on algorithmic concept formation, but also is intended to introduce an understanding of the basic mathematical structure of programming languages. For this purpose we have developed the Registermachine, a model computer on which the pupil can visually follow an actual program step by step as it runs (Figure 10). The Registermachine (RM) is realized by a software system on a microcomputer (e.g., an Apple).

Rodding (1968), pursuing ideas first mentioned by Minsky (1961), developed the mathematical concept behind the Registermachine. The programming language is recursively defined throughout the processes of linking and iteration for the basic elementary operations: forward and backward counting. We developed this concept further to include the possibility of subprograms (Cohors-Fresenborg et al. 1982a). This expansion of the language implies that recursive procedures may be employed.

This mathematical concept has been built into an instructional unit which consciously applies Bruner's ideas on the spiral construction of concepts. Our main idea — confirmed by our research on algorithmic concept formation by pupils — is that developing a computer program primarily means organizing a sequence of elementary actions to be executed by the computer.

In the following we introduce the major components of our concept which we have tested in several classroom experiments. Details can be found in the first part of the textbook for pupils in grades 7-9, Registermachines and Functions, which is supplemented by a very detailed Teacher's Handbook (see Cohors-Fresenborg et al. 1982a).

From Computing Networks to the <u>Registermachine</u>

The first task in the sequence is to construct an algorithm for addition using forward and backward counting. Pupils learn that such an algorithm can be represented at various levels, and they become aware of the possibilities:

- as a series of simple actions with matchsticks
- as a computation network
- as a flow chart or "structogram"
- as a program in the language for the <u>Registermachine</u>.

A detailed analysis of the advantages of the different levels and the representation of the concept of variables is presented in Cohors-Fresenborg (1986).

Our observation in classroom experiments and our case studies on algorithmic concept formation by pupils have shown that for most pupils the major problem in constructing an algorithm is not expressing the algorithm in the programming <u>language</u>, but rather organizing the algorithm as a series of elementary actions (for example, the addition or subtraction of single matchsticks). This compelled us to choose an <u>action-oriented</u> approach and to make the pupils aware that their organization of elementary actions can be expressed on other representational levels. This can be seen in the following example, an algorithm for addition.

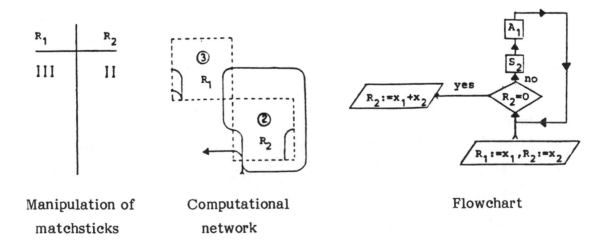

Manipulation of   Computational   Flowchart
matchsticks   network

258

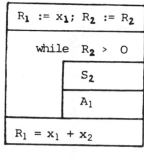

$$(_2\ S_2\ A_1\ )$$

Structogramm   RM program

Figure 9

The computation networks are constructed with the bricks (mainly the counters) of the <u>Dynamic Mazes</u> described in the previous section. This shows that the problems of constructing computational networks for addition, subtraction, multiplication, and division may straddle the curriculum at the end of the primary level and the beginning of the secondary level.

Already in the first lesson of our course we make use of the full mathematical complexity of the language: the <u>linking</u> of elementary operations and <u>iteration</u>. Pupils learn to manage the RM system in just a few minutes.

The RM (see Figure 10) has one line for the main program and several lines to define subprograms. If a subprogram is called during the execution of a program, and during the execution of this subprogram another subprogram (or in case of a recursive procedure, the same subprogram) is called, the first 3 subprograms occur in lines 2-4. If the depth of the calling hierarchy of subprograms is four or more, then line 5 displays the subprogram which is currently being executed. A cursor moves from character to character (elementary operations or names of subprograms). The speed can be varied.

These five lines of programs are followed by two lines to show the actual value of the 8 registers, which change during the computation. The last line contains a counter for the number of steps (CS, see next section) which the RM has executed, as well as any comment about errors.

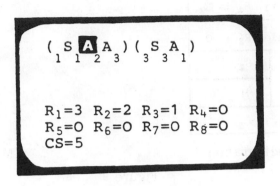

Figure 10

Complexity of Algorithms

The number of necessary steps during execution of an RM program is a measure of the complexity of the algorithm (step-function). We use this idea to introduce the concept of multivariable functions in a natural way. A function which describes the number of steps of an algorithm is ideally suited to encourage pupils to create efficient programs through competition.

A basic technique used in constructing efficient programs and formulating their step-functions is the rules of term-reordering. At this stage children may regard that part of pure (and sometimes boring) mathematics as a powerful tool in constructing efficient algorithms. Our classroom experiments have shown that the concept of step-function strengthens the functional understanding of algebraic terms. It also sharpens the ability to express functional ideas by algebraic terms.

The Syntax of the RM Language

The concept of a subprogram is introduced as an abbreviation for programs which have already been written and appear again and again in more complex programs. The notation $F_3 \langle 1,2 \rangle$ for a subprogram which calculates the function f is derived from the usual notation $x_3 = f(x_1, x_2)$. We should stress here that pupils first conceptualize registers 1 through 9 by analogy to the compartments of a register box, in which matchsticks are either added or taken away. The treatment of these registers within the programming language identifies them as a model for variables (in the sense of Pascal).

Pupils recognize very early that variables (in the mathematical sense) should be employed in the subprograms for the register numbers. This simplifies the further utilization of the subprograms. In syntactic terms, the pupils create local parameters. After they have expanded the language in order to realize their algorithmic ideas on the RM, we introduce in a special chapter the notion that a programming language has a grammar.

We discuss why the grammar must have been installed in the computer: otherwise, the computer could not "understand" the intended "meaning" of a program.

Pupils are astonished to learn that a grammar is an object of mathematical reasoning. This offers interesting ramifications for discussing with pupils the nature of mathematics and mathematical thinking. It also permits cooperation with language teachers on the role grammar plays in formal and natural languages.

At the end of this chapter we get the following summary of the RM language:

1. Correct <u>RM Programs</u>
   1.1 a) A_i, S_i $(1 \leq i < 9)$ are correct programs.
      b) Correct abbreviations (for subprograms) are correct programs.
   1.2 a) If P and Q are correct programs, then so is the <u>linking</u> PQ.
      b) If P is a correct program, then so is its <u>iteration</u> $(_iP)$ for $(1 \leq i \leq 9)$.

2. Correct <u>abbreviations</u> consist successively of:
   2.1 one character as a name;
   2.2 one footnote at the name, which may be a number from 1 to 9 or a character,
   2.3 a) the sign "<";
      b) numbers or characters for registers, separated by a comma;
      c) the sign ">."

Examples of abbreviations: $M_3\langle 1,2\rangle$ or $A_i\langle i,j\rangle$.

3. A correct (main) program is <u>ready to run</u> if all abbreviations which occur in the program are completely defined in subsequent lines.

4. A <u>complete definition</u> of a subprogram consists successively of:

4.1 one correct abbreviation;

4.2 the sign "=";

4.3 a correct program.

Example: RM program for multiplication $(x_1, x_2, 0, 0) \rightarrow (x_1, 0, x_1 \cdot x_2, 0)$

a) <u>iterative</u>

$M_3<1,2>$

$M_3<1,2> = (_2S_2A_3<1>)$

$A_j<i> \quad = (_iS_iA_h) (_hS_hA_iA_j)$

b) <u>recursive</u>

$M_3<1,2>$

$M_3<1,2> = (_2S_2 \ M_3<1,2> \ A_3<1>)$

$A_j<i> \quad = (_iS_iA_h) (_hS_hA_iA_j)$

The definition of the concept "correct program which is ready to run" is recursive for the following reasons:

- the definition of "correct RM program" contains the concept "correct abbreviation";

- the definition of "ready to run" contains the concept "complete definition";

- the definition of "complete definition" contains the concept "correct RM program."

While executing a recursively defined program, the <u>Registermachine</u> displays the construction of the recursion cellar on the screen.

Transition to Pascal

We intend for the microworld <u>Registermachine</u> to help pupils understand the fundamental ideas and concepts of algorithms and programming languages. In a model-world we have discussed the syntax and semantics of programming languages, the efficiency of algorithms, and structured programming. Experiments have shown that pupils in grade 8 achieve an astonishing understanding of the conceptual framework. We have been able to reach this level because we have consistently adopted a spiral curriculum based on an action-oriented approach to the creation of algorithms, and employed metacognition so that from the beginning pupils understand that they are involved in a process of concept <u>formation</u>

RM	PASCAL

$M_3\langle 1,2\rangle$	```PROGRAM MULT;```
	
	```VAR R1,R2,R3: INTEGER;```

$A_J\langle I\rangle =$
 	
$(_I$
 	
S_I
 	
A_H
$)$
 	
$(_H$
 	
S_H
A_I
A_J
$)$
	```PROCEDURE ADD (VAR J,I: INTEGER);```
	
	``` VAR H: INTEGER;```
	``` BEGIN```
	``` H:= O;```
	
	``` WHILE I > O DO```
	``` BEGIN```
	
	``` I:= I-1;```
	
	``` H:= H+1```
	
	``` END;```
	
	``` WHILE H > O DO```
	``` BEGIN```
	
	``` H:= H-1;```
	``` I:= I+1;```
	``` J:= J+1```
	
	``` END;```
	
	``` END;```

$(_2$
 	
S_2
$A_3\langle 1\rangle)$
	```BEGIN```
	``` WRITE ('R1=');```
	``` READLN (R1);```
	``` WRITE ('R2=');```
	``` READLN (R2);```
	``` R3:=O;```
	
	``` WHILE R2 > O DO```
	``` BEGIN```
	
	``` R2:=R2-1;```
	
	``` ADD (R3,R1)```
	
	``` END;```
	
	``` WRITELN ('R3=',R3)```
	```END .```

Figure 11

and not merely learning a programming technique.

After this preparation on _Registermachines_, the pupils are ready to be introduced to the language Pascal, including the concept of procedures. But all variables are still of the integer type. In the following section we will use the program $M_3 \langle 1,2 \rangle$ for multiplication (see Figure 11) to demonstrate how our RM language corresponds to Pascal: In $M_3 \langle 1,2 \rangle$, the numbers 1,2,3 (abbreviations for R_1, R_2, R_3) are global _variables_; the characters j, i in $A_j \langle i \rangle$ are formal parameters, and the character h in the definition of $A_j \langle i \rangle$ corresponds primarily to a local variable. Before executing $A_j \langle i \rangle$, the RM substitutes the global variables 3 and 1 for the formal parameters j, i, and the local variable is replaced by a global variable which does not occur in the subprogram and has the value zero at that time. (In our case it is the number 4.)

Introduction of the Concept of Functions

One of the main topics in secondary school mathematics is the introduction of the concept of the function. The difficulties that pupils have with functions may be considered from both the syntactic and semantic points of view. On the one hand, they have problems using the notation properly, especially in handling of variables; on the other hand, it is quite difficult to understand that a value _depends_ on several variables and that the mathematical concept of the function actually describes this dependency in a precise way. For this reason it is difficult to find the proper mathematical expressions for word problems. In these problems the understanding of the function is combined with the understanding of a text in natural language. Complex functions occur in a situation which is difficult to understand.

In our approach to introducing functions (see Cohors-Fresenborg et al. 1982a) we have chosen a situation which is easy to understand, but for which the teacher may create as difficult a description as he wishes: the complexity of a RM program as a function of the inputs, the step-functions (see the "Complexity of Algorithms" section). This approach has the following advantages: the use of variables is quite natural, because they are names for the content of registers; the goal of predicting the computation time of programs (in the case of competitions) is an interesting and concrete problem for many pupils; and the computer can be used to check the answer. The concept of the function is used in a naive way and is not an object of reflection. But the pupils learn to handle the "language" of functions of several variables.

Functions do become the object of reflection in a second approach. At the beginning of our course Registermachines and Functions there is an assignment to write programs for given mathematical problems in addition, multiplication, etc. If the problems are complex, such as $(2x)^2 + 2x$, the technique of structured programming uses the unreflected idea that such a problem (function) may be regarded as composed of simpler problems. The question formulated for the pupils is whether a set of simple functions can be invented, together with simple rules for constructions comparable to the techniques of linking and iteration. This leads to the concepts of composition and recursive definitions of functions. The technique of decomposing given arithmetic terms into simpler terms can be developed by solving problems which ask for a quick algorithm. Often reorganizing a term leads to better algorithms, e.g., $(2x)^2 + 2x$, $4x^2 + 2x$, $2x(2x + 1)$. Which idea includes the best algorithm?

At the end of this chapter pupils understand that there are two different languages in which a functional dependency may be described: programming language or the traditional mathematical representation as function terms. They learn that theories about constructing function terms are very useful tools for solving the applied problem of writing efficient programs. It leads to the insight that giving a formal definition or writing a program are two sides of a coin, and that turning the coin over often helps solve a problem.

Our approach to introducing functions enabled pupils in grade 7 to handle the notation of functions in a way teachers did not believe possible at that age. Our explanation for our success is that besides the algorithmic approach, a very powerful tool is extended reflection on a meta-level about the process of concept formation. The pupils became aware why it is convenient to use symbols.

The last question that we have discussed with pupils is whether there exist functions which in principle cannot be computed by computers. By proving that the stop-problem cannot be resolved, we have probably hinted to the pupils that the question of how the use of language creates meaning is an exciting and very difficult one in understanding the nature of mathematical concepts.

Research on Cognitive Strategies in Algorithmic Thinking

Although there is a major movement to introduce computers into schools throughout the world, there has been very little research into the thinking processes of pupils when dealing with computer programs. It is our opinion that such research is necessary as a background for decisions in curriculum development.

Our investigations since 1981 have been concerned with some aspects of these problems. First we have looked at the question of whether different representations of algorithmic concepts influence concept formation by pupils. From our classroom observation we were led to the hypothesis that pupils' ability to analyze algorithms may differ from their ability to construct them. In a third step, we are now investigating whether pupils have different cognitive strategies when constructing and analyzing algorithms.

To investigate such questions we could use our experience in developing curriculum materials. With the Dynamic Mazes and Registermachine it is possible to give even younger pupils rather difficult algorithmic problems after only brief preparation. Because we are interested in the thinking processes of the pupils, we have chosen the methodology of clinical interviews. In the last four years we have observed about 80 pupils in grade 7 of a Gymnasium.

Individual pupils had four hours to solve problems of two different types: constructive and analytical tasks. The problem-solving sessions were videotaped.

In a constructive task the pupils had to create a RM program (e.g., addition, subtraction). The pupils were free to choose the form in which they would like to represent the algorithm as they were constructing it: as matchstick manipulation, as computational network, or directly as a program for the RM (see Figure 9). One finding of our research is that the main difficulty in writing a program is not the computer language, but rather the problem of how to organize the elementary actions. A deeper analysis of the importance and advantages of the different forms of representation is given in Cohors-Fresenborg (1986).

In an analytic task the pupils had to analyze a given RM program. There were two different types of tasks: in one the pupils were asked which function was computed by a given program; in the other, they were asked which was the step-function of the program (see "Complexity of Algorithms" section). We have found that competence in analyzing is independent of competence in constructing, in both directions. We measured success by counting how many hints pupils needed before they got the complete solution.

A deeper and more precise analysis of pupil behavior led us to the hypothesis that there are at least two different cognitive strategies for dealing with our tasks of constructing and analyzing algorithms. To describe this we have established two concepts: conceptual and sequential strategy. An important role in classifying pupil behavior is played by the way the pupils use language to describe their aims and the process of solving.

Pupils who prefer a conceptual strategy begin by structuring the given problem. For this purpose they use mathematical concepts. They try to build up a conceptual framework incorporating their preknowledge about previous problems and their solutions.

Pupils who prefer a sequential strategy are goal-oriented, but they start toward a first solution before they have completely structured their ideas. They develop their ideas in a dialog with the material and in an interaction with their partial solutions.

In trying to describe pupils' cognitive processes by a theory, we have found that the concepts "VMS" (visually moderated sequences) and "frame," as used by Davis (1984), offer a suitable system for explaining some pupil behavior.

Our investigations will continue in the coming years. A first overview is reported in Cohors-Fresenborg and Kaune (1984). A report on the first study is provided by Kaune (1985).

The process of analyzing pupils' behavior and constructing hypotheses is ongoing. A powerful new theoretical framework is presented by Schwank (1985, 1986). She distinguishes between the cognitive structures (predicative versus functional) in which thinking processes are expressed, and the cognitive strategies

(<u>conceptual</u> versus <u>sequential</u>) which determine the process of planning and solving the problem. If a person is said to prefer <u>predicative</u> (versus <u>functional</u>) thinking, this means that the internal conceptual representation into which a given problem is translated is built up by <u>predicates</u> or relations (versus <u>functions</u> or actions). In the context of our experiments, a pupil who prefers <u>predicative</u> thinking views the problem in such a way that certain <u>relations</u> between start and goal have to be established. A pupil with <u>functional</u> thinking imagines that a sequence of <u>actions</u> has to be invented.

One question for further research is whether the level of cognitive structure (predicative/functional) is independent from that of cognitive strategies (conceptual/sequential). Recently Marpaung (1986) was able to show in his study with Indonesian pupils (for overview see Cohors-Fresenborg and Marpaung 1986) that there exist predicative/sequential and functional/conceptual combinations in addition to the other two (predicative/conceptual and functional/sequential), which probably occur more often. Our research in this field is still in the beginning stages. A comparison of Cohors-Fresenborg/Kaune (1984) and Schwank (1986) will reveal how these ideas are developing. The research will continue. We are convinced that our conceptual framework — although developed in the area of computer education — is useful for explaining some kinds of pupil behavior and difficulties in other fields of school mathematics.

How to Introduce Axiomatic Mathematics

In the previous sections we have discussed the relationship between actions and language with respect to the role language plays in the representation of algorithms. But mathematical concepts are not just formalizations of <u>actions</u>. A second root of mathematics is the description of <u>relations</u> between objects and the formulation of <u>statements</u>. The axiomatic point of view is a fundamental part of modern mathematics. It is the basis for mathematical modeling, i.e., from a philosophical point of view, the axiomatic foundation of mathematics is the reason for its applicability.

We believe that it is important that pupils have an insight into the axiomatic method in order to gain a wide understanding of mathematics. Therefore, we have developed a curriculum element for pupils in grade 7 of the Gymnasium (in our state of Lower Saxony, the Gymnasium begins with grade 7). Our aim was to create a didactic situation in which these pupils could understand that a single axiomatic system can be interpreted in different ways, that the words by which the variables are expressed have no meaning in the common sense, and that therefore knowledge from outside the system may not be used in proofs.

Often teachers introduce these ideas together with Euclidean geometry. But the problem is that pupils cannot understand why obvious facts have to be proved. They do not understand, for example, that the words "points" and "line" have no meaning, that they describe variables. In our analysis we became convinced that one reason is that it is too difficult to abstract from manipulation of familiar "geometrical" objects that pupils already know; another reason is that the words "point" and "line" have meaning for pupils, and it is simply too strange for them to ignore that meaning.

In our approach we made use of pupils' familiarity with problems in which someone needs to know the translation of a text given in a foreign language. We combined this idea utilizing the fact that pupils in grade 7 learn a great deal of ancient history and that they like tales about decoding old texts (e.g., in cuneiform characters); we created a project called "Sentences from the Desert and Their Interpretation."

The textbook "Sentences from the Desert" (Cohors-Fresenborg and Griep 1985) again creates a microworld in which we introduce the concepts "variable," "axiomatic system," "interpretation," and "model." The following story is presented to the pupils:

During excavations in the desert an archeologist named McDonald finds some papyrus scrolls with sentences in a language which he has recently decoded. But there are words in these sentences which he is still unable to translate. He therefore fills in three artificial words and tries to understand what the text could mean. His translation:

I. There exist at least two different bres.

II. For every two different bres there exists exactly one ket, such that both bres dety these kets.

The pupils' imagination is virtually unlimited. They try to invent meanings for these unknown words. They use information reported in the story that the scrolls were found in a temple. They learn that not every idea leads to a possible translation, and that one can prove that some translations are wrong even though nobody knows the correct translation.

During the lessons the teacher offers more sentences from the desert:

III. For every ket there exists at least one bre which does not dety this ket.

IV. For every ket and for every bre which does not dety this ket there exists one ket, such that the bre deties this ket and both kets are silukan.

V. Two kets are silukan, if there does not exist a bre which deties both kets.

The children discern that ket and bre must be nouns, dety a verb, and silukan an adjective. They realize that there are more possible translations for sentences I and II than for all five together. They learn to systematize the "proofs" about possible translations. They detect a relation between the number of bres and kets. There are very impressive discussions among the pupils about how it is possible for them to know these numbers when they do not know what the words mean. They discuss whether these lessons properly belong to mathematics or to language; the teacher may ask what the pupils mean by the word "mathematics." We were very impressed by visits to such lessons.

The reader may have discerned that sentences I–IV are axioms of plane finite incidental geometry and that V is the definition of "parallel." At the end of this course the pupils are led by the teacher to realize that this is one possible interpretation. We cannot report on this project in detail, but we would like to mention that chapter five of the textbook creates a microworld in which a teacher can explain what problems occur in the relation between contracts and laws on one hand and the intended meaning on the other. This could be a starting point for interdisciplinary cooperation between math and social science teachers.

A future project will test whether our goal of giving pupils some insight into the idea of axiomatic mathematics can be useful when the axiom of groups (of mappings in geometry and of whole numbers in algebra) is introduced. We are convinced that the "Sentences from the Desert" will serve as a suitable microworld in which the methodology of pure mathematics can be explained to schoolchildren.

Summary

The curriculum materials we have described in the second, third, and sixth sections are linked by the common idea of explaining fundamental mathematical concepts in a suitable microworld. In sections two and three we have given a survey of two curriculum materials we have developed to give younger pupils a fundamental understanding of automatic processes and computer programming. The starting point was the idea that organization of actions is fundamental for this field. In this approach, the language is only the last link in the chain of concept formations. Our research (section five) leads us to suppose that many pupils represent those concepts in their memory in primarily nonverbal form, like riding on a bicycle, for example. These pupils build their conceptual framework using representations of actions (functional structure). But there are pupils who need language-oriented concepts for structuring a problem before they can invent the desired algorithm. They build their conceptual framework using representations of predicates. On the level of cognitive strategies we have found at least two different strategies used by the pupils. Some prefer planning first (conceptual strategy), while others prefer a more interactive style (sequential strategy). Obviously, a computer language like Pascal supports conceptual strategy, but is disadvantageous for pupils who reason sequentially.

It is important to emphasize that it is not our aim to build up a constructive approach to mathematics, like Papert (with his turtle geometry and Logo). For us it is important to prepare the mathematical symbolism and to give pupils a deeper understanding of variables and the use of functions. We believe that an important goal of school mathematics is to make pupils aware of the existence of common models for different phenomena and the idea of the isomorphism of concepts. Therefore, the reflection phases in all four of our courses are essential.

If one looks at this aspect of our development of curriculum materials, the last example fits in seamlessly: we are convinced that it is important to present both variants to the pupils, both the algorithmic/action-oriented and the axiomatic/conceptual approaches to mathematics. Therefore, we are introducing the three curriculum elements described in sections three, four, and six into our grade 7 mathematics project.

If we compare the role that language plays in algorithms and axioms, we may remark on the following difference. Our research has shown that algorithms can be invented by pupils and represented in memory non-verbally; algorithms can exist independent of language. But in an axiomatic approach, there first exists a text written in a formal language; mathematical reasoning must then invent a meaning that fits.

Appendix

Some Remarks on the German School System and Teacher Education

Germany is a federal republic of 11 states. Cultural affairs such as schools and universities are the responsibility of the states. Therefore there are numerous differences in detail. The following remarks are intended to characterize the general situation. Most of the schools and nearly all universities are run by the state.

Pupils begin their school life at the age of 6 with four years in primary school. After this the parents and teachers decide at what level pupils go to secondary school. On the average, the upper 30% go to Gymnasium, the next 35% to Realschule and the last 35% to Hauptschule. But there are also comprehensive schools. The Hauptschule ends with grade 9 or 10; the Realschule with grade 10; the Gymnasium with grade 13. Every pupil who has successfully graduated from the Gymnasium is allowed to enroll at a university without an entrance examination. Pupils who leave school after grade 10 have to go to a part time technical college for another two years during their apprenticeship.

Mathematics is compulsory for all pupils in all grades (the one exception: grade 13 in the Gymnasium). On the average, the schedule includes 4 hours of mathematics per week.

The teachers for all types of schools study at the university level. The teachers for primary schools and Hauptschule study two or three subjects for 3 1/2 years (in some states mathematics is compulsory for primary teachers), teachers for Realschule study 4 years, and teachers for the Gymnasium study two subjects for 5 years. After passing their final university examination (which is controlled by the states), they must attend a special teacher training college for 1 1/2 years, where they are trained in the methodology and didactics of their subjects. After this they have to pass a second examination in teaching ability.

Although even primary school teachers study particular subjects, the reality in school is that because of the shortage of mathematics teachers, mathematics is very often taught by teachers who did not study it.

References

Aebli, H. 1980-81. Denken: das Ordnen des Tuns. Stuttgart: Klett-Cotta.

Cohors-Fresenborg, E. 1978. "Learning problem solving by developing automata-networks." Revue de phonétique appliquée 46/47.

Cohors-Fresenborg, E. 1986. "On the Representation of Algorithmic Concepts." In Pragmatics and Education, F. Lowenthal, et al., eds. New York: Plenum Press.

Cohors-Fresenborg, E., D. Finke, and S. Schütte. 1979. "Dynamic Mazes." Osnabrücker Schriften zur Mathematik, U 11-19, 11A - 19A (9 booklets with programmed lessons; in English); U 21 (teacher's handbook in German). University of Osnabrück.

Cohors-Fresenborg, E., M. Griep, and I. Schwank. 1982a. "Register machines and Functions - A Textbook Introducing the Concept of Functions on the Basis of Algorithms." Osnabrücker Schriften zur Mathematik, U 22 (textbook, in English); U 22 LE (overview and solutions to the exercises, in English), U 25 (teacher's handbook, in German). University of Osnabrück.

Cohors-Fresenborg, E., and H. Strüber. 1982b. "The Learning of Algorithmic Concepts by Action - A Study with Deaf Children." In Language and Language Acquisition, 95-106, F. Lowenthal, et al., eds. New York: Plenum Press.

Cohors-Fresenborg, E. and C. Kaune. 1984. "Sequential Versus Conceptual-Two Modes in Algorithmic Thinking." In Proceedings of the 8th Conference for the Psychology of Mathematics Education, 261-267. Sydney.

Cohors-Fresenborg, E. and M. Griep. 1985. "Sätze aus dem Wüstensand und ihre Interpretationen - Ein Textbuch für Schüler zur Einführung in die axiomatische Auffassung von Mathematik," Schriftenreihe des Forschungs- instituts für Mathematikdidaktik 2. Osnabrück. (English translation in preparation.)

Davis, R. 1984. Learning Mathematics: The Cognitive Science Approach to Mathematics Education. Norwood, New Jersey.

Kaune, C. 1985. "Schüler denken am Computer." Osnabrück: Forschungsinstitut für Mathematikdidaktik e.V.

Lowenthal, F. and J. Marcq. 1982. "How Do Children Discover Strategies (at the Age of 7)?" In Proceedings of the 6th International Conference for the Psychology of Mathematics Education, 287-292. Antwerpen.

Marpaung, Y. 1986. "Profile indonesischer Schüler beim Umgang mit Algorithmen und ihre Analyse." Osnabrück: Forschungsinstitut für Mathematikdidaktik e.V.

Minsky, M.L. 1961. "Recursive Unsolvability of Post's Problem of 'Tag' and Other Topics in the Theory of Turing Machines." Annals of Mathematics 74: 437-455.

Rödding, D. 1968. "Klassen rekursiver Funktionen." In Lecture Notes in Mathematics 70: 159-222. Berlin: Springer.

Schwank, I. 1985. "Zum Problem der Repräsentation algorithmischer Begriffe." In Axiomatische und algorithmische Denkweisen im Mathematikunterricht der Sekundarstufe I des Gymnasiums, I. Schwank, ed. Osnabrück: Forschungsinstitut für Mathematikdidaktik e.V.

Schwank, I. 1986. "Cognitive Structures of Algorithmic Thinking." In Proceedings of the 10th International Conference for the Psychology of Mathematics Education: 404-409. London.

The material Dynamic Mazes and the software system Registermachine are available from Forschungsinstitut für Mathematikdidaktik, Postfach 18 47, D - 4500 Osnabrück, West Germany.

Address:
Professor Dr. Elmar Cohors-Fresenborg
FB Mathematik/Informatik
Universität Osnabrück
Postfach 44 69
D-4500 Osnabrück
West Germany

Part 2: General Principles, Curriculum Design, and Instructional Strategies

Mathematics Starting and Staying in Reality

Hans Freudenthal

University of Utrecht, The Netherlands

In a survey and analysis of current Dutch arithmetic textbooks, A. Treffers distinguished the following approaches:

a) mechanistic

b) empiristic

c) structuralist

d) realistic.[1]

Though I am not fond of classifying, I will, for a moment, adopt this classification — but not exactly in the sense Treffers intended — and apply it to teaching and learning mathematics in school in general. I will do this by identifying the philosophy or ideology behind each approach.

The mechanistic approach to mathematics education regards man as a computer-like device programmed by drill to perform arithmetic and algebraic operations, and to distinguish types or patterns of word and application problems in order to solve them. When Treffers drew up his classification scheme, the great majority of Dutch arithmetic textbooks were characterized by this approach. (It has since lost ground to the realistic approach, which will eventually dominate, or so I hope.) To be sure, even in most of the mechanistic textbooks the introduction of new operations and concepts has been preceded and accompanied by explanations using material and visual aids. But the attention to insight has been too superficial to be taken seriously, a mere intellectual alibi for the sake of conscience and a smoke screen to hide the ideology. Although this approach has proved inefficient everywhere for teaching mathematics for applications, it might regain its momentum in the future if computer-aided instruction is pursued too single-mindedly.

The ideology behind the empiristic approach is the utility of mathematics. This is a respectable ideology as long as the focus on utility is not too narrow. The empiristic method has deep roots in British utilitarian instruction. By providing the learner with tools for concrete activities, it strives to respect and account for the child's everyday environment as seen from an adult's perspective, rather than to broaden it by rational means.

In the ideology behind the structuralist approach mathematics is viewed, and taught, as affixed to a grand superstructure which gives it overall organization. Historically, the structuralist view has ancient roots. It reached its zenith in education when curriculum developers and textbook authors tried to adapt Bourbaki to classroom teaching by means of false concretizations of seemingly simple but, in reality, far too general abstract ideas.

The ideology behind the realistic approach is that mathematics, from both the historical and individual perspectives, starts in reality. Mathematical structures are not presented to the learner so that they might be filled with realities (or, rather, pseudo-realities). They arise, instead, from reality itself, which is not a fixed datum, but expands continuously in man's individual and collective learning process. Natural number arose, and continues to arise, from counting objects, actual or imagined; gradually, number itself becomes part of this same reality, together with numerical operations and relations. It is the art of education to guide this process and make it conscious to the one who undergoes it, so that it can be reinforced. The paramount expression of this growing consciousness is verbalization, but the feasibility of verbalization should not be the criterion for deciding what should be taught and learned. People — even mathematicians — are accustomed to knowing and performing more mathematics than they can verbalize.

Instead of number, let me consider the next most powerful mathematical tool — ratio and proportionality, which to my mind preceded and still precedes number in the development of mankind and the individual.

First let me recount a little story from a diary I kept for years on a boy who was then six years old. After a period of sunny days and blue skies, he again sees clouds and says, "It will rain." I tell him, "These are very high clouds, from which no rain falls. Rain clouds are low and dark." He asks, "How high are these clouds?" I exaggerate, "Ten thousand meters." "And rain clouds?" he asks. "A thousand meters." The boy points to the ground and says, "So if we are here and this (he indicates a height of about thirty centimeters) is rain clouds, then this (indicating about one meter above the ground) is no rain clouds."

A year and a half later I find another story in the diary which I have told so often that I am tired of repeating it. It involves measuring the height of a tower and its clock dial by putting a cheek against the top of a low wall and sighting over the end of a vertical forty-centimeter stick. Rather than repeating the entire story here, I will instead reproduce the boy's drawing (Figure 1), which is not beyond reproach, and a sketch (Figure 2) which demonstrates a proposal he himself made afterwards for a practical improvement in this method. I should add that the original method was not his own invention, but after I had shown him how to put his cheek against the wall and sight along the stick to the tower, he immediately grasped the idea.

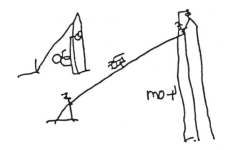

Figure 1

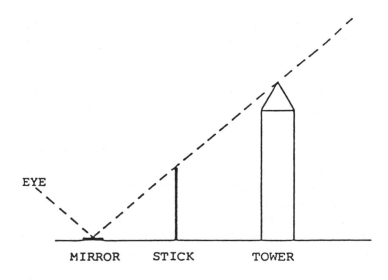

Figure 2

It obviously represents a huge increase in sophistication to proceed from the first observation, at age six, to the second, a year and a half later. But let us analyze these anecdotes more carefully. The first reports a spontaneous act. I cannot tell whether he invented at that precise moment the trick of comparing two large heights (or other dimensions) proportionally by marking off spaces with his hands, or whether he already knew it and applied it. In any event, the degree of verbalization was remarkable: "If we are here and this is rain clouds, then this is no rain clouds." It approaches the acuity of the classical "this is to that as this is to that." This spontaneity was absent in the second case, and there was no attempt at verbalizing. All he did was mentally substitute the stick for the tower and calculate corresponding distances and heights. He reported this action in a drawing, but he certainly did not say anything resembling the acute verbalization in the first story. To be sure, the second situation was more involved, and with paper and pencil at hand there was no urge to verbalize.

Rather than flooding you with more observations, let me quote a passage from one of my books:

> There can be no doubt that children recognize early the different sizes of objects, and their being larger or smaller. It is equally certain that they can handle similarity as an operational equivalence. I would even go so far as to assert that congruences and similarities are built-in features of that part of the central nervous system that processes our optical perceptions. The immediate reidentification of objects after a rotation (of the object or the perceiver) and after a change of distance presupposes something in the brain like a computer program for the elimination of this kind of mapping — it is a riddle to me what such a program looks like; its existence, which I do not doubt, is like a miracle to me.

> At a young age a child recognizes drawings and models of animals, furniture, cars, bicycles, ships as images of these objects — it does not matter on which scale, and whether they are pictured side by side on different scales. "How big is a whale really?", a child can ask, convinced that the picture, except for the scale, is faithful. Well, sometimes whales are sketched by drawings in one line, but even the difference between a photograph and a characteristic sketch is grasped early.

> Without any hesitation, children accept that objects at the blackboard are drawn ten times as large as on the work sheet, that the number line at the blackboard has a unit of 1 dm compared with that of 1 cm on the work sheet. They accept number lines where the same interval means a unit, or ten, or hundred, side by side. Children would, however, immediately protest structural modifications that violate the similarity of the image:

what is mutually equal in the original
should be mutually equal in the image,
which implies the invariance of internal ratios, characterizing mappings as similarities.

(Freudenthal 1983:190)

If ratio and proportionality precede number developmentally, then why does the latter overtake the former? The answer is obvious: number (cardinal and ordinal number with all its appendages) is easily verbalized, to the extent that it is eventually integrated into everyday language. Ratio and proportion are different. Their verbalization looks so much more difficult that educators do not even consider trying to make students conscious of them, even on a non-verbal level. When attention is finally devoted to ratio and proportion in traditional education, it is most frequently on the level of full verbalization or even formalization, and what is still more important, instruction comes much too late to take advantage of the student's built-in ability to handle ratio and proportion intuitively and operationally. In my opinion, this explains why most people never learn to handle ratio and proportion mathematically.

Ratio and proportionality can be made conscious in two seemingly opposite ways, visually and numerically; accordingly, they can be verbalized and formalized in the language of geometry and arithmetic. In between there is a gray area of applied ratio and proportion: Is this mixture of lemon juice and sugar sweeter than, or as sweet as, that one? Is this green mixture of blue and yellow more bluish than that one? Is this country more densely populated than that one? Is this person slimmer than that one? Is this brand of a certain commodity cheaper than that one? Is this event more probable than that one? We could go on indefinitely. Questions like these and the corresponding answers can be visualized graphically and can be made precise numerically, but we would like to connect the geometry and arithmetic of ratio and proportionality in a more fundamental way, or even integrate them with each other, by linking them to the same aspects of reality. If I speak of mathematics starting in reality, I mean as much reality as feasible, and starting again and again as reality extends, rather than incidentally.

Learning is not a linear process, but for the sake of teaching we are obliged to discern and organize learning strings, the loose ends of which may be seized

283

again later and extended. It was precisely in learning and teaching ratio and proportionality that L. Streefland (1984–85) stressed the necessity of intertwining learning strings at an early stage, thereby anticipating future, even distant objectives. In fact, anticipatory learning (or prospective learning, as I call it) has become a very natural feature of elementary instruction. The four basic operations of arithmetic are no longer taught in their logical order and as separate topics, but rather, in good teaching, with as much anticipation of each other as possible. This principle should be applied to ratio and proportionality as well.

As I have asserted, the ratio–proportionality string starts earlier than the number concept string. It starts in a context of similarity, but it remains a loose end for a long time (or even forever), disregarded in instruction even when the teaching of similarity begins formally. This loose end deserves to be seized early and intertwined with the constantly growing number concept string. This is certainly not an easy task, but teaching units such as those developed by IOWO collaborators H. ter Heege and E. de Moor (1978) and research, particularly by L. Streefland (1984–85), are quite encouraging. Nor is it an impossible task. First, one must look for real world situations where the intuitive geometrical aspect of ratio and proportionality is smoothly fitted to the more rational arithmetic aspect. To be sure, becoming conscious of both aspects and verbalizing and eventually formalizing them can require long, intertwined learning processes.

The most valuable didactic tool that elementary arithmetic has borrowed from higher mathematics, as I have pointed out many times, is the number line, from its most concrete to its more abstract realizations. It seems to me that we have not yet sufficiently exploited its didactic power, certainly not with respect to ratio and proportionality.

In preparing this paper, I was reluctant to indulge in extensive theoretical effusions. Let me illustrate what I mean in a more concrete way, with the caveat that this illustration is a mere shadow of the much richer contexts I have in mind.

The teacher displays a closed string (Figure 3) with repeating groups of three white and four black beads. How many beads of each color could be on the string?

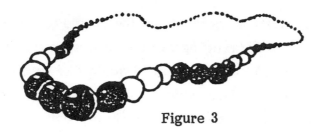

Figure 3

This question can be answered at various levels of arithmetic ability with increasing skill and knowledge of the basic operations. The students can start with simple addition:

White	3	6	9	12
Black	4	8	12	16

and if the string seems long enough, bold, enterprising children might continue:

30	60	300	600	660	663	657
40	80	400	800	880	884	876

At the stage of addition, this is prospective learning for multiplication, perhaps even for division, and at all stages it is prospective learning for ratio and proportion.

But what use is it? Unlike with traditional methods, ratio is not introduced abruptly at a specific moment, but rather it is prepared and informally shaped over a number of years before being made explicit. It does not come as a bolt from the blue, but as a ripe fruit falls from a tree — and so do fractions, which have been prepared over the years along with ratio. Of course, strings and beads make up only one of the innumerable contexts that prepare ratio. Another is price tables: three items cost four dollars, so how much does another quantity cost? Or weight tables: three items weigh a total of four pounds, so how much would six items weigh? Or taste tables: three spoons of syrup and four spoons of water — what tastes the same? Or three of my paces equaling four of your paces — what will happen if . . .? The proportionality table, eventually extended to fractional entries, is a powerful didactic tool. Its discreteness is almost obliterated in the

285

string of beads, and is completely overcome by the double-scaled number line, an equally powerful didactic tool for relating two magnitudes, e.g., time and distance or volume and weight, in a realistic context. I would like to do more justice to similarity in two and three dimensions, which is an early childhood experience, yet is so dreadfully neglected in teaching that I simply cannot draw on enough practical teaching experience intertwining the similarity string with other learning strings.

So far, I am afraid I have described mathematics starting in reality too theoretically and in much too poor a context. Let me now try to give a more concrete and richer context, without slipping into the other extreme.

I will briefly describe a six-lesson IOWO unit for grade 3, "The Giant's Greeting" (ter Heege and de Moor 1978), based on a sequence of illustrations.[2] A classroom window is open. The blackboard shows an enormous handprint. There must have been a giant in the classroom. Giants are tall. But how tall was this giant?

"Look at my hand." The teacher puts her hand on the giant hand-print, which appears to be four times as large as hers. The teacher is measured. A string is cut off to a length four times as great as the teacher's height. The children write a letter to the giant on the blackboard. "This is your height," they write. The next day the giant has answered the letter.

It is difficult for a giant to live in the world of people. Why is it difficult? Where can a giant go and where can't he go? How many sandwiches will he eat? The teacher's shoe is measured on a piece of paper. There are various reactions as to how big the giant's foot is. Finally, the students try to fit 16 of the teacher's footprints together to get one of the giant.

A baker has found strange footprints in his garden. He calls a reporter from the local newspaper, who takes a picture. Now the children discover human footprints, the giant's footprints, and intermediate footprints, which they interpret as belonging to the giant's son. The reporter writes about it in the newspaper.

The giant notices the story. But real giants' newspapers are larger. How large should they be?

Again footprints are found, this time in the snow. The children compare them by proportionality tables. They want to bake a cake for the giant. What size should it be? Now it is cold, and the giant undoubtedly needs a mitten. What size should it be? How long would it take to knit it? How many balls of wool would be needed? There is plenty of opportunity for using proportionality tables.

You may ask me, "Is this reality — a giant's world?" Yes, it is. It is just as much reality as any good fiction. And this is good fiction for third-graders, just like Gulliver in Lilliput, which we adapted for use in grade 5. On the other hand, traditional word and application problems are very bad fiction.

I have displayed one snapshot out of an extended curriculum, one stage in a process, which tries — albeit still unsatisfactorily — to intertwine long learning processes by means of real situations.

I have provided only a few details, but space limitations prohibit even this restricted degree of detail if I am to attempt a broader view of mathematics starting and staying in reality. I can only state that even in teaching the algorithms of addition, subtraction, multiplication, and division, we refrain as much as possible from posing bare numerical problems. The algorithms are taught following a method we call progressive schematization, integrated with reality, which is a marvelous guide in the labyrinth of formal procedures. In teaching long division, in particular, empirical research has shown this method to be far superior to traditional methods (Rengerink 1983).

At the secondary level, in grade 7, pupils from various primary schools merge together to form classes in which the pupils differ enormously in arithmetic understanding and skill. Most of them hate arithmetic, which they believe they are too stupid to learn. One way out is to teach geometry: not formal, axiomatic, algebraized geometry, but visual geometry, characterized by such unit titles as "Light on Shadow," "Can You Look?," and "Regular Surfaces."[3] "Turnpikes" is another source of mathematics, including some arithmetic. Let me briefly demonstrate a few pages from a unit developed at SLO in Enschede entitled "Graphs and Connections" (SLO 1984). Two pages are about "Lightbulbs" (Figure 4).

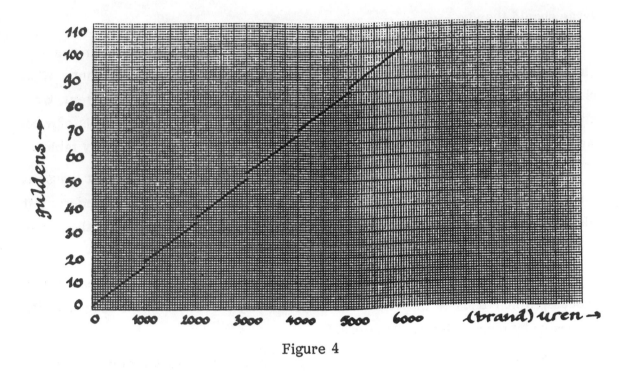

Figure 4

An ordinary lightbulb which costs 2 florins, consumes 1.5 cents worth of electricity per hour of operation, and lasts 1,000 hours is described in the text and illustrated with a graph. A different lightbulb is available, and it costs as much as 15 florins, consumes 4 florins worth of electricity per 1,000 hours, and lasts 5,000 hours. The students must draw the corresponding graph and discuss it, answering a large number of questions related to the graphs.

If you consider that "Lightbulbs" is insufficiently motivating in the reality of seventh-graders, let me present a sequence entitled "Cycling" (Figure 5a). In the Netherlands, almost all children of this age go to school by bicycle. This sequence deals with riding from Losser to Enschede (about 10 km). Four rather global graphs of the route are to be related to stories about the bike trips, one of which should be filled in extemporaneously. The next graph (Figure 5b) is more precise — it is easy to imagine the variety of questions that can be asked about it. The same material is then repeated, but now more specific questions are asked dealing with a girl who first rode faster than a boy, but nevertheless arrived at school later — the data are specified — so what happened? On the next sheet (Figure 5c), a child taking a bus to school enters the picture, and many questions can be asked connected with various aspects of the situation. The last sheet announces that the highway will be closed off on a particular day at a certain time, and so the bicyclist has to change his strategy.

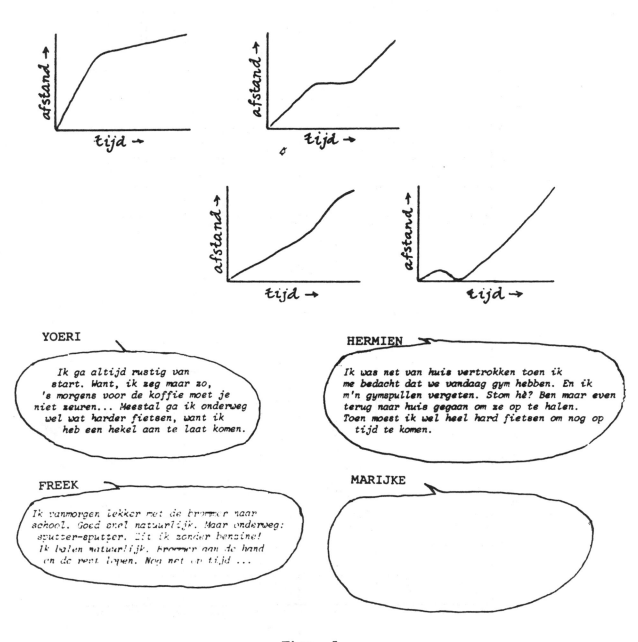

Figure 5a

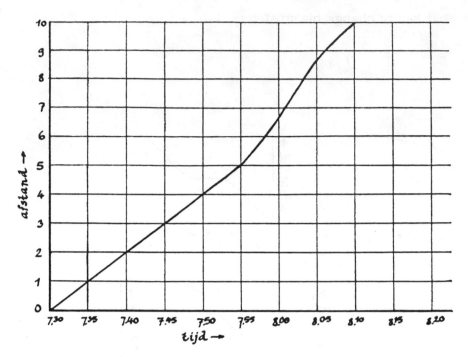

Figure 5b

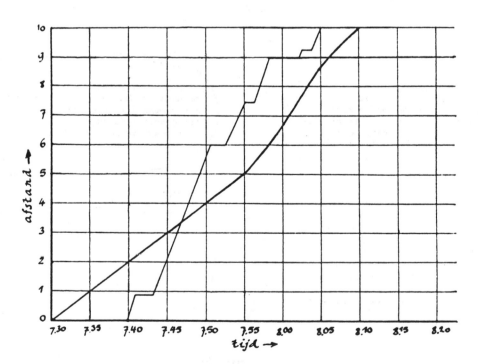

Figure 5c

290

This unit "Graphs and Connections" has been the subject of profound empirical research in small, mixed-ability groups.[4] I have studied the protocols of this research; duly analyzed, they are a rich treasure of knowledge about logic — not formal logic, but the way students think and reflect their own thoughts and those of their peers in reality.

A marvelous ninth-grade book on trigonometry and vectors has been developed around one theme: flying with hang gliders, gliders, and planes.[5] Books for grade 10 on "Exponents and Logarithms" and "Calculus"[5] have been developed which start in reality and never lose contact with it. Unfortunately, all this material still exists only in Dutch, but I hope that at least part of it will be translated into English.

I will conclude by discussing the only material which also has an English version, "Matrices"[6], but first I must briefly describe our educational system.

Primary education, up to the age of twelve or thirteen, is essentially uniform, although individual schools may differ greatly from each other. Secondary education is diversified, though attempts at unifying it are being made. Most secondary schools have a four-year course of study. Five-year schools prepare students for polytechnical institutes and colleges for primary and low level secondary teachers. Six-year schools prepare students for the university. In the upper two grades of six-year schools mathematics is not compulsory. Recently mathematics in these grades has been fundamentally reformed. The main feature of the reform is a brand new curriculum for future university students of social sciences, economics, and humanities (rather than mathematics, natural sciences, and technology). Besides calculus and statistics, this curriculum includes such subjects as computers, linear programming, and matrices. I will now give a quick demonstration of how "Matrices" has been interpreted.

We have an imaginary island with a number of villages and three roads (Figure 6a). Where should a school be built? The geographic situation is simplified into a road graph (Figure 6b), which in turn is translated into a connectivity matrix (Figure 6c); then the relation between the graph and matrix is generalized. Distance matrices are introduced (Figure 6d) and are applied; they are followed by direct road matrices (Figure 6c).

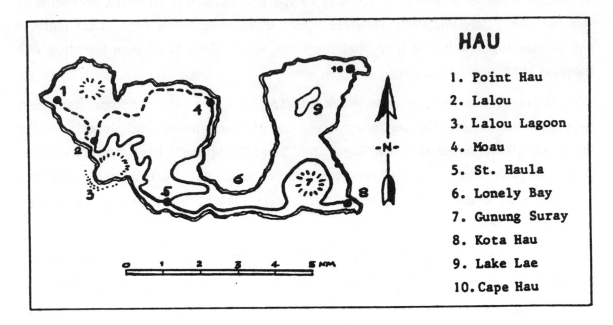

Figure 6a

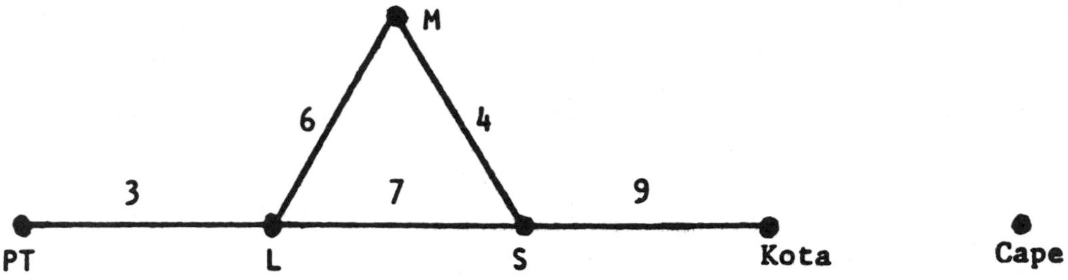

Figure 6b

	FROM				
	K	R	N	O	L
K	0	1	1	0	0
R	1	0	1	1	0
TO N	1	1	0	0	0
O	0	1	0	0	1
L	0	0	0	1	0

K = Katwijk; N = Noordwijk; R = Rijnsburg;
O = Oegstgeest; L = Leiden

These are all towns in The Netherlands.

Figure 6c

292

FROM

$$
\begin{array}{c}
\qquad\quad K \quad R \quad N \quad O \quad L \\[4pt]
TO \quad
\begin{array}{c} K \\ R \\ N \\ O \\ L \end{array}
\left(
\begin{array}{ccccc}
0 & 3 & 4 & 5 & 7 \\
3 & 0 & 4 & 2 & 4 \\
4 & 4 & 0 & 6 & 8 \\
5 & 2 & 6 & 0 & 2 \\
7 & 4 & 8 & 2 & 0
\end{array}
\right)
\end{array}
\qquad \text{or} \qquad
A =
\left(
\begin{array}{ccccc}
0 & 3 & 4 & 5 & 7 \\
3 & 0 & 4 & 2 & 4 \\
4 & 4 & 0 & 6 & 8 \\
5 & 2 & 6 & 0 & 2 \\
7 & 4 & 8 & 2 & 0
\end{array}
\right)
$$

Figure 6d

FROM

$$
\begin{array}{c}
\qquad A \quad B \quad C \\[4pt]
\begin{array}{c} A \\ B \\ C \end{array}
\left(
\begin{array}{ccc}
0 & 0 & 1 \\
1 & 0 & 2 \\
1 & 2 & 0
\end{array}
\right)
\end{array}
$$

Figure 6e

A matrix for shoe stock gives rise to matrix multiplication in conjunction with price matrices. Matrix multiplication is also applied to connectivity matrices. Then comes a great leap forward: a migration matrix and the conclusions that can be drawn from it. The concluding chapter is ecological. We see the hooded seal, its population trend analyzed with Leslie matrices, and finally a computer printout showing the population structure under various conditions of hunting.

To date the new mathematics curriculum for the upper two grades of university-preparatory schools has been a great success. The number of students electing mathematics as an exam subject is increasing sharply, especially among girls, and the exam results are highly satisfactory. But this is just its impact to date; dangers loom large on the horizon. In the Netherlands, instruction is determined entirely by the way its results are evaluated, i.e., by the final examinations. There is a tendency to replace traditional examinations by batteries of quadruple-choice tests, which are less expensive to score. Presently exams are of the essay type. If the advocates of multiple choice exams gain the upper hand, the new mathematics starting and staying in reality will degenerate within a few years into meaningless junk, as dead as the traditional approach.

In the meantime, serious efforts are being made to extend the reform to other types of schools and, gradually, to the lower grades — but these efforts are again menaced by examination dogmatism.

Notes

1. Meanwhile he has elaborated his theory. See Treffers (1986).

2. The illustrations shown during the lecture have been omitted here. Compare ter Heege and de Moor (1978).

3. These units can be obtained at Vakgroep OW & OC, Tiberdreef 4, Utrecht. The first is available in English under the title "Shadow and Depth."

4. By Rijkje Dekker and Paul Herfs. Obtainable at SLO (see SLO 1984).

5. By J. de Lange. Obtainable at the address mentioned in (3).

6. Obtainable at the address mentioned in (3).

References

Freudenthal, Hans. 1983. Didactical Phenomenology of Mathematical Structures. Dordrecht, Reidel.

ter Heege, H. and E. de Moor. 1978. Wiskobas Bulletin 7 (5–6):23–42. (In Dutch.)

Rengerink, J. 1983. De staartdeling. (In Dutch.) Obtainable at Vakgroep OW & OC, Tiberdreef 4, Utrecht.

SLO (Stichting voor de Leerplanontwikkeling). 1984. Hoe langer hoe meer. Enschede: SLO.

Streefland, L. 1984–85. "Search of the Roots of Ratio." Two parts. Educational Studies in Mathematics 15:327–430 (part I); 16:75–94 (part II).

Treffers, A. 1986. Three Dimensions — A Model of Theory and Goal Description in Primary Mathematical Education: Wiskobas. Reidel, Dordrecht.

Reflections on Mathematics Education

Felix E. Browder

University of Chicago

In the present meeting, we heard some stirring speeches which gave us a number of very diverse precepts about the philosophy that must guide our efforts to remold mathematics education if our efforts are to bear fruit. We were told that we must preserve the romantic flavor of mathematics by emphasizing insight and discovery, that we must practice realism by emphasizing the reality of the materials presented to mathematics students at every level of their learning, and that the most essential ingredient for achieving success in our effort to reform mathematics education is a mixture of high ideals in our objectives with low cunning in their practice. In my view, these are all deep truths (though obviously each is a partial truth) and my subsequent remarks are intended to illuminate them on the basis of historical comparisons and examples.

There is an important point I have not mentioned above, the international character of the present Symposium. The fact that this Symposium is international makes it an especially significant event in the ongoing effort to reconstruct American mathematics education. Comparing our thoughts and experiences with those of leading representatives of research and reform in mathematics education in other countries can only help overcome our parochialism and show that there are possibilities and forms of action quite different from those we are used to. Just as historical comparisons make us less provincial in terms of our time, comparisons across national boundaries make us less provincial in terms of the peculiarities of our national experience and educational institutions.

I referred earlier to E.H. Moore as embodying both a powerful devotion to mathematical research and all phases of education in mathematics. There is no doubt whatever that Moore's model in this union of roles was Felix Klein in Göttingen. There is no doubt of this because of the direct personal connection of Moore with Klein, on whose behalf he organized the first international meeting of mathematicians at the Columbian Exposition in Chicago in 1892. Klein was the principal luminary at this meeting, which he attended as the Commissioner of the

German Emperor to the Columbian Exposition, a meeting which connoisseurs refer to as the Zeroth International Congress of Mathematicians. Klein subsequently gave a series of 32 Colloquium Lectures in Evanston which appeared in book form as the first publication of the American Mathematical Society and was subsequently translated into 23 languages. A final trace of Klein's connection with Moore can be seen in the otherwise mysterious fact that in 1899, E.H. Moore was awarded an honorary doctoral degree by the University of Göttingen.

Felix Klein played a decisive role in the academic and educational structure of the Germany of his time in several completely separate roles. He was a mathematical visionary with an over-arching concept of the development of the mathematics of his time; his celebrated Erlangen Program tied geometry to the study of groups of transformations. He was the academic dictator of most of German mathematics and science, with the power to make academic appointments in these universities under the control of the Prussian Ministry of Education. Finally, he was the most ardent and energetic reformer of pre-university mathematics education in Germany. Some aspects of this last role can be seen from his well-known volumes, Elementary Mathematics from an Advanced Standpoint.

Klein is an unusual but important example of a research career of the highest distinction combined with a major contribution to the development of mathematics education at basic levels. The great French geometer Monge from the Napoleonic period is another example. We have yet another example at the present Symposium in Hans Freudenthal, a research mathematician who has made extraordinary contributions over a number of decades to algebra, analysis, topology, and the foundations of mathematics, and in very recent decades equally extraordinary contributions in his leadership of the reform process in mathematics education in the Netherlands and on a world level. We can point to other examples, such as Andrei Kolmogorov in the Soviet Union, a mathematician who has played just as energetic a role in the transformation of Soviet mathematical education. That men of such remarkable intellectual distinction could become so intensely involved in the concrete problems of mathematics education at its basic levels not only testifies to the importance of these problems but also challenges invidious stereotypes of mathematics education. What these examples prove is that the problems of mathematics education are not only of great practical seriousness; they attract and deserve serious intellectual notice and concern as well.

This point deserves special stress in the American context. Despite the fact that a small elite of the American school population reaches high levels in its mathematics and science education at the pre-college level (a very small elite indeed), as our colleague Izaak Wirszup has very eloquently and persistently pointed out, the basic level of the great mass of American students in primary and secondary schools is extremely poor in mathematics by international standards. Some reasons for this situation have been the pervasive contempt for intellectual skills propagated through dominant educational philosophies, negative attitudes toward learning among students and throughout the population, and the continuous weakening of standards, status, and conditions among the body of teachers in the schools. As we are all well aware, we have been deluged with a flood of official reports in the past several years discussing the questions. We are also all well aware that announcements of this kind provide no solutions, automatic or otherwise, to the problems they are publicizing.

The fundamental difficulty to be faced is not just the enormous scale of the resources needed for an effective solution. Rather the problem can be found, I believe, in the massive structure of influence and power we call the educational establishment. To paraphrase a political slogan of the 1960's, the establishment is not part of the solution, it is most of the problem. The United States has some very gifted and insightful scholars and researchers in education; it certainly has the world's strongest research establishments in mathematics and the sciences by far. Yet neither the educational scholars nor the scientists have any appreciable influence on the thrust of educational policy and decision. Who does? The answer is very simple: The caste of professional educational administrators and ex-administrators. With a few exceptions, they are not teachers or scholars or scientists (and have not been for most of their careers). Those in positions of power seek only to justify their power; those out of power have the authority of failure, to give an ironic usage to a famous phrase of F. Scott Fitzgerald. They have an aura of greatness because of the gigantic magnitude of their failures.

I can cite two personal experiences which left me with a strong firsthand impression of the nature of this establishment. The first was membership in a Committee of the National Research Council set up to report on problems of national scientific and technical education. This Committee included two members of the genus _educational establishmentarian_ as I have defined it, both devoted to proving at any cost that all the works of the establishment were right and good and we live in Paradise. One was particularly devoted to making sure there would be no international comparisons, since international comparisons allow at least a minimal check upon the complacency of those who hold power or have held it. Many witnesses appeared before this Committee, and of them, a number of similar establishmentarians, including the dean of one of the leading schools of education, equally devoted to the thesis that nothing needed to be done. In a second experience, I participated as an observer in a meeting of the National Academy of Education held on this campus and devoted to a discussion of the flow of recent reports on the status of the nation's schools. The meeting had two parts. In the first part, summaries of such reports were presented, mostly by establishmentarians led by several former Commissioners of Education. The establishmentarians all agreed that there were no real problems that could not be solved by a happier rhetoric, with a small admixture of higher teacher salaries. The second day, all the establishmentarians had vanished and the facts were presented, belying every one of their assertions.

If this picture is accurate, and I don't think anyone familiar with the situation will deny its essential accuracy, how can we hope for meaningful improvement in the basic situation? Let me suggest there is indeed hope, and try to point out where this hope lies. I shall do so by drawing two kinds of historical analogies.

The first is related to the category of _low cunning_ to which I referred above. In the context of educational strategies, this phrase was used to refer to the artful employment of detailed strategies expressly designed to fit well with special features of an experimental situation. There is another usage to the phrase, closer to the practical reality of the uses of power and influence in educational politics as in any other variety. Let me turn to the examples already cited of Monge and Klein, two educational reformers of extraordinary influence.

How did Monge get his influence? Through his connection with Napoleon. How did that connection arise? Through Monge's use of his influence in the Academy of Sciences to get Napoleon elected to the Academy as a mathematician (in a slot understood to be filled by political figures) at a time when Napoleon was a rising but not yet established figure. How did Felix Klein get his academic and educational power in Prussia? Through his personal connection with Altdorff, later von Altdorff, a clerk in the Ministry of Education whom Klein helped raise to key posts, and finally to complete control of the Prussian universities.

Low cunning in this sense is an invariable part of the political process. Indeed, it is the only real skill of the establishmentarians. What I suggest is that their opponents ought to be willing and able to play by the same rules if the opportunity arises.

The second historical analogy is broader and much more important. Our present educational situation in the United States can be summarized in the following schematic way: Our primary and secondary educational system is basically inadequate and even deplorable in many of its most important characteristics; our collegiate educational system is adequate though showing signs of wear and tear; our system of graduate education, at least in the sciences, is the envy of the world. Being used to these phenomena (which in the present form go back only to the end of the Second World War), we have the tendency to think of the best universities as research universities and to see them as the engines of progress in the educational system as a whole. We also tend to extrapolate this system back into the historical process and to see this interrelation of the different levels of education as a permanent historical fact. This is far from the case, however.

The educational tradition from which our own institutions and tradition derive arose in Western Europe in the 12th and 13th centuries in the system of urban universities in Italy, France, England, and Spain, for the most part under the sponsorship and control of the Church. This system encompassed education from beginning to end, since it included the equivalent of at least a secondary education, and there were always students who persisted in the theology courses for decades. During the 13th century, it attained great vitality and intensity of intellectual purpose, which made this period a paradigmatic model for educational reformers of the 1930's and 1940's, men such as Robert Hutchins and his collaborators. But those who praise the intense vitality of 13th century scholasticism fail to point out that almost all the positive traits they discern vanished very quickly in the 14th century. What did them in were two endemic diseases of systems of education, conservatism and corruption. The vitality of 13th century scholasticism arose from the tension within this tradition between the religious impulse based upon the Christian Revelation and the awakened powers of intellectual inquiry fortified by the translations of the corpus of scientific and philosophical writings of the ancient world. These two impulses conflicted but their tension generated a powerful source of intellectual vitality. From the viewpoint of ecclesiastic and academic orthodoxy, this tension was dangerous since it might lead to heresy, and in a number of important steps, it was suppressed.

Educational systems are vital when they combine two functions, that of discovering new truths and that of transmitting truths, both new and old, to the rising generations of youth. In each society, the educational system is given the task of formation of these rising generations, to develop their skills and talents, but also in one way or another to fit these skills and talents into the framework of the existing society. Academic creativity arises from the possibility of molding the young into new forms which will be adequate for new situations. Academic conservatism arises from the pressure to maintain what already exists and resist the pressure of change. It becomes especially strong when the function of the educational system moves to an emphasis on preserving the status quo of social influence and power or of religious orthodoxy. From the middle of the 14th century till the beginning of the 19th century, the primary emphasis of the European university system was academic conservatism hostile to innovation or creativity in almost any form, designed to buttress the system of social status of the nobility and gentry of which the university population, teachers and students, was almost exclusively composed. The corruptions which this system engendered were the corruptions of the surrounding society, but corruptions made much more acute (as in the case of the Church itself) by the contrast with the ostensible ideal purposes of the institution.

301

Thus, in all arenas of intellectual and cultural advance, whether humanism, the Scientific Revolution and its consequences, or the advance of new conceptions of the world in scientific or philosophical form, the universities formed the chief wall of resistance. We think of Newton or Galileo as models of the university professor as scientist, but they were nearly unique cases. A more representative case, one which is especially apposite for a meeting on the reform of mathematics education, is the great educational reformer of 16th-century France, Pierre de la Ramee (Peter Ramus), Professor of Mathematics at the College de France, murdered in the St. Bartholomew's Day Massacre in Paris in 1572. The College de France itself had been founded by Francois I because of the Sorbonne's opposition to humanism, a stand which had led to the prohibition of the teaching of Greek. Ramus' death is symbolic of the more extreme perils which can be implicit in the process of reform. A wave of clerical reaction had led to the appointment of another professor of mathematics at the College de France who was religiously orthodox (Ramus had become a Protestant) and a true educational establishmentarian. He knew very little mathematics. Ramus, always a difficult person, was revolted by the appointment and published a public protest against his new colleague's mathematical incompetence. The latter retaliated in his own way. When the massacre of Protestants took place, Ramus was the object of attention of a special Death Squad composed of mercenary soldiers with no perceptible tinge of religious fanaticism, who cut him down in the top floor of his own college and threw his corpse out the window. Thus was the cause of academic conservatism and mathematical incompetence vindicated in 1572.

Let me say in parenthesis that (with the appropriate modification for vast differences of context) many of Ramus' emphases in education, and especially in mathematics education, have a number of major points of affinity with themes presented at this Symposium, especially the realism which Hans Freudenthal emphasized in his lecture. It is indeed fortunate that the advance of civilization has preserved the latter from the attentions of academic conservatism as they were applied to Ramus.

It would be too lengthy a project to review in any detail the intellectual decadence into which the university system of Western Europe sank in the period in question. One case might suffice, that of Oxford University, a case which is particularly well-documented. Here we have the testimony of any number of prominent ex-Oxonians from the 18th century, men such as Adam Smith, John Locke, or Edward Gibbon. Their testimony is savage. With only a few exceptions, there was little teaching or learning, and independent thought was discouraged. We might distrust their testimony as based upon special interest; Locke's books were banned at Oxford, for example. Rather than trust such disgruntled dissenters, no matter how distinguished, let us instead take up the testimony of someone who may be characterized as a gruntled conformist, John Henry Newman. In his celebrated book, The Idea of a University, a book which propounds as one of its major principles the thesis that no new truths ought to be investigated at an ideal university, Newman remarks of a situation that was typical at Oxford in generations very close to his own. He recalls an Oxford in which no teaching or active tutoring was carried on, and certainly no scholarship or research to speak of then. He asks if such a situation, involving the collegiate cohabitation of a group of young men in happy solitude, is better or worse from his perspective than a non-residential university in which teaching, learning, and scholarship are actively pursued. His conclusion is that residence (with all its important social and acculturating features) is decisively more important than teaching and learning. This is the final verdict of a certain concept of academic conservatism, which from the point of view of educational as opposed to social ideals, can only be described as decadent.

Starting at the beginning of the 19th century, a wave of massive reform began in the university systems of Western Europe, mainly in response to the new social and political pressures generated by the French Revolution and the Napoleonic campaigns across Europe. Leaving aside France, the most conspicuous symbol of this process of transformation was the foundation of the University of Berlin in 1809 with a Charter written by the celebrated humanist Wilhelm von Humboldt. This Charter committed the University to the principles of freedom of teaching and research as its central values. This symbolic act, which had enormous consequences in the later development of universities in Germany and throughout Europe, took place under circumstances that illustrate quite vividly the importance of action at a decisive moment. The Prussian universities had traditionally been under the control of the police, under the Ministry of Interior, and such notable figures as Kant had been chastised by Royal Edict as freethinkers whose ideas were impinging upon the religious orthodoxy of the realm. The Prussian defeat by Napoleon at the Battle of Jena in 1806 gave rise to the Stein Ministry, a brief liberal interlude in Prussian history. Humboldt was made head of the department of public instruction in the Ministry of Interior, staying in that position for less than a year and a half, during which time he promulgated his history-making Charter. He took the post of minister to Vienna in 1810, and left Prussian politics in 1815 after the removal of the liberal ministry. He was strongly influenced by the writings of Kant as well as Schleiermacher, a theologian who had responded strongly to the attack on Kant. After Humboldt's departure from office, the first rector of the new University was the philosopher Fichte, a ferocious verbal opponent of the intellectual freedoms espoused in the Humboldt Charter. Yet the structure set in place by Humboldt did not buckle, even in the hands of extremely unsympathetic practitioners.

We may ask why the wave of massive reform of the decadent University structures continued and became ever stronger as the 19th century proceeded. It did not really affect the old universities in England till the middle of the century. Everywhere, it faced the determined and often savage opposition of the academic establishmentarians, the beneficiaries in very personal terms of the status quo and the controllers of the levers of power inside the universities. The answer is very apposite for the reform problem we face in precollegiate education. In their decadent or 'conservative' state the universities did not serve the societies of which they were a potentially important organ. When these societies faced massive political pressures from sharp commercial and industrial competition, the leaders of each society began to question the appropriateness of the status quo. How elegant the universities might appear as they rotted could no longer serve as an excuse.

This is what we can and must count on in our present educational problems. Societies like the U.S. cannot tolerate decadence in their educational systems, no matter how powerfully entrenched, or how skillfully the educational establishment tries to camouflage the problem. In this context, we should note that mathematics education plays a particularly strategic role. We need not be mathematics educators or mathematicians to recognize the central role of mathematics in primary and secondary education as an indispensable prerequisite in terms of skills and concepts for the successful learning of a large proportion of the disciplines and professions needed in modern society. This has become transparently true in view of the dominant role of the computer as an instrument for organizing all the processes of advanced societies. Indeed, effective use of the computer and other skilled tools in social processes depends upon the level of intellectual maturity and understanding achieved by those who live in advanced societies.

The Western European tradition of which we are an offshoot inherited from the Greco-Roman tradition an interesting set of categories which govern the basic structure of education, usually referred to as the seven liberal arts These seven branches of learning are composed of two groups, the verbal arts of logic, grammar, and rhetoric, and the mathematical arts of arithmetic, geometry, astronomy, and music. Though the Platonic tradition from which this clas-sification came tended to emphasize the importance of the mathematical arts (the quadrivium) over the verbal arts (the trivium), since the time of the Greeks, educational practice in the primary and secondary schools has given precedence to the verbal arts. In fact, if not in name, the central art of educational practice turned out to be rhetoric, which in practical terms may be defined as the art of using persuasive speech and writing to convince others of what may be so (or, for that matter, what isn't so). I venture the suggestion that in most periods of educational history, rhetoric has been the chief instrument and focus of academic conservatism (except in revolutionary periods, when it turns into the chief tool of revolution). It is usually not so simple a matter to turn the mathematical arts (which by a very simple extension of terms could include the mathematizable natural sciences in place of astronomy and the more objective structures of the arts in place of music) into apologetic instruments for any status quo. Usually the mathematical arts are neutral in such matters and even exhibit a dynamism from their close relation (especially in the most recent times) to the rapid development of the scientific and technological infrastructure of modern society.

If we accept this hypothesis for our present situation, we then have very good grounds for believing that the reform of mathematics education in the primary and secondary schools can well become a powerful lever for reforming the entire educational structure. Criteria for judging the desirability of reform as well as the importance of reform in terms of overall social interest are also far clearer. This, I believe, explains the curious emphasis of some opponents of fundamental change on a greater role for rhetoric in education as the only kind of reform that is really necessary. We can reply simply without worrying appearing one-sided or seeming to be only pleading our own special interests. The major arena of educational reform in America lies in the domain of mathematical skills and understanding.

Let me conclude by repeating the exhortations of earlier speakers that we go forward in reforming mathematical education as vigorously as possible, with a strategy that combines high ideals, romance, realism, and low cunning.

The Role of Applications in the Pre-College Curriculum

Peter J. Hilton

State University of New York at Binghamton

Across the country, and beyond the borders of the United States, the cry is being heard that we mathematicians should be concerning ourselves more with applications of mathematics, both in our research and in our teaching at all levels. It is argued that we have been overemphasizing mathematics as an autonomous discipline at the expense of mathematics' usefulness: its role in science, in engineering, and in the conduct of modern society. Some put it crudely — there is too much "pure mathematics," too little "applied mathematics." The issue is rendered more crucial by its relation to the critical problem of making mathematics courses more accessible to our students and more relevant to their perceived needs and interests without any sacrifice of standards or integrity.

Many mathematicians and mathematics educators have devoted considerable effort to coming to grips with these problems, at various levels. There is the report of the NRC Committee on Applied Mathematics Training (National Research Council 1979); a panel of CUPM, under the chairmanship of Professor Alan Tucker, has produced sample curricula with a decidedly "applied" flavor; there is a joint MAA-SIAM Committee considering undergraduate and graduate courses; there are the recommendations of the PRIME 80 Conference; there is the NCTM Agenda for Action; and much else.

Now it is plain that to do good work in applied mathematics it is necessary, though not sufficient, to be a good mathematician. Thus it is naive to suppose that simply increasing the emphasis on applications in our teaching will make things easier for students. It is true that a very traditional course in applied mathematics may be an easier option, but such courses are partly fraudulent, for they are based on the notion that to apply mathematics to the real world, we presume that the real world problems have already been stipulated and require precisely the piece of mathematics just learned, no more and no less. To truly teach students how to apply mathematics is undoubtedly a difficult task — difficult in conception, difficult in execution. One component is to instill a comprehension of the strategy of problem-solving, but I warn the reader in the last section of this essay against supposing that there is a royal road to success in problem-solving which somehow circumvents the needs to learn a lot of mathematics. George Polya would never make such a mistake!

From my own observations of applied mathematics in action, I draw two related conclusions which do not seem to be part of the common wisdom. First, there is no natural division of mathematics into applicable mathematics and mathematics sui generis — all parts of mathematics are potentially available for applications. This point is the more conspicuous today with the new applications of discrete mathematics and the ubiquitous presence of the computer. Second, published papers in applied mathematics differ from published papers in pure mathematics not in the manner of treatment or the rigor of the arguments but in the real-world motivation for the mathematical problem. Thus, if our students are to genuinely learn how to use mathematics, they must acquire a broad and deep understanding of mathematics and a positive attitude toward applications through their education. This conclusion is in harmony with that of the NRC Committee report referred to above, in which both points are made with great emphasis, and in which, in keeping with the conclusion that mathematics, pure and applied, is a unity, a plea is made for a broad-based undergraduate major in the mathematical sciences, giving all students the opportunity and encouragement to acquire familiarity with the way mathematics actually is applied. The implications for pre-college mathematics education are obvious.

If this is to be the case, what might be a reasonable working definition of applied mathematics? Let me attempt one, based on a definition of applied analysis given by Kaper and Varga.[1] I propose the following:

> The term applied mathematics refers to a collection of activities directed towards the formulation of mathematical models, the analysis of mathematical relations occurring in these models, and the interpretation of the analytical results in the framework of their intended application. The objective of applied mathematics research activity is to obtain qualitative and quantitative information about exact or approximate solutions. The methods used are adapted from all areas of mathematics; because of the universality of mathematics, one analysis often leads, simultaneously, to applications in several diverse fields.

In this paper I will first make some general remarks about the nature of applied mathematics and mathematical modeling. I will then suggest that the methodology of applied mathematics is not, in fact, as distinctive as it at first appears; that, in fact, it has much in common with processes of abstraction and generalization that go on within mathematics in general. These remarks will lead us to the conclusion that, by modifying our approach to the curriculum in certain ways that give expression to the unity of the mathematical sciences, we may prepare students to use mathematics to solve problems both from within and outside mathematics. The sterile antagonism one sometimes finds today between pure and applied mathematics — and pure and applied mathematicians — would be eliminated by abandoning these labels and reverting to the notion of a single discipline, mathematics. The paper closes with a section on the special role of geometry and a section analyzing current approaches to problem-solving.

The Nature of Applied Mathematics

In this section I discuss what applied mathematics means in practice, based on the definition of Kaper and Varga. As far as the general methodology of applied mathematics is concerned, I will not feel obliged to limit the discussion to the secondary level, but when I come to discuss pedagogical implications I will limit my examples to mathematical material accessible to the secondary student.

What can be said in general of the process of mathematical modeling? The following schema seems to reflect the methods which mathematicians have successfully exploited.

$$\text{Real-life problem} \xrightarrow{\text{(1)}} \text{Scientific model} \xrightarrow{\text{(2)}} \text{Mathematical model}$$

Step (1) occurs whenever a mature science is involved. However, in the "soft" social sciences one may proceed directly from a real-life problem to a mathematical model. Such a process is dangerous, because it is within the scientific model that one locates the measurable constructs about which one theorizes, quantitatively and qualitatively, within the mathematical model. Thus the direct passage from the problem to the mathematical model, while often intellectually exciting, is open to the objection that one may well be using sophisticated mathematical tools to reason about extremely vague concepts involving very unreliable measurements.

Let us assume that a scientific model is articulated. The model will typically consist of objects (observables, constructs) and laws (physical, chemical, biological) about the behavior of matter in the form of liquids, gases, and solid particles. The selection of the appropriate scientific model, step (1), may be called constructive analysis; for example, in the study of energy systems (fission reactors, combustion chambers, coal gasification plants), the laws express the rates of flow of mass, momentum, and energy between the components of the system.

Step (2) consists in choosing a mathematical model for the analysis of the scientific system. The mathematical model is, by its very nature, both more abstract and more general than the scientific system being modeled. Thus the conservation laws for an incompressible viscous fluid lead to the Navier–Stokes equation, which is an evolution equation; but we may also derive equilibrium equations leading to the study of bifurcation phenomena. Again, the study of non-linear wave mechanics leads to the Korteweg–de Vries (KdV) equation for the behavior of long water waves in a rectangular channel.[2] Here the theory predicts solitary waves that interact but emerge unchanged — the so-called "solitons" of modern theoretical physics. In these examples, the original constructive analysis leads to the next stage of mathematical analysis (here, qualitative analysis) within the mathematical models. Typically, this stage consists of proofs of the existence of solutions, together with a study of their uniqueness, stability (sensitivity to changes of parameter), and behavior over large time intervals (asymptotic analysis).

Quantitative analysis, however, also plays a key role in today's applied mathematics, due largely to the general availability of high-speed computers. Sophisticated numerical methods have been developed; asymptotic and perturbation methods are widely used. One particularly important new tool of quantitative analysis that might be mentioned here is the "finite element" method, invented by Richard Courant and rediscovered by engineers who saw its potential once computers had become available. The method itself is, of course, undergoing improvement and refinement.

In several of the best applications of mathematics, we see evidence that our scheme is incomplete in one significant respect — the process of elaborating a mathematical model may well be <u>iterated</u>. Thus our scientific model may lead to a first-order differential equation that can be interpreted as a dynamical system and embedded in the theory of vector bundles or, more generally, fibre bundles. The existence of a solution is then translated, first into an integrability problem, and then into a cross-section problem to which the techniques of obstruction theory may be applied. It would be very misleading to think of the process of abstraction and generalization as a single-stage procedure; by the same token, it is a mistake to think of an area of mathematics as ineffably "pure" simply because all its direct contacts are with other areas of mathematics.

It is plain that if our students are to be able to apply mathematics effectively, they must gain some understanding and experience of the art of mathematical modeling. I do not recommend a special modeling course; rather, the modeling process should be explicitly discussed when an application of mathematics to a scientific problem is in question. I would recommend that the discussion include the following issues: the selection of a suitable problem; the development of an appropriate model; the collection of data; reasoning within the model (qualitative analysis); calculations (quantitative analysis); reference back to the original problem to test the validity of a solution; modification of the model; and generalization of the model as a conceptual device. Moreover, the entire modeling process must be set against the background of a strong computer capability (actual or assumed).

It is my claim that applications do not have to be separated from the rest of mathematical activity in order to emphasize these processes. I will suggest, in fact, that good "applied" mathematics and good "pure" mathematics have a great deal in common, and that this complementarity should be reflected in the undergraduate and pre-college curricula. The question I wish to consider now is this: how special to applied mathematics are the techniques and procedures described in this section? In one sense they are certainly special, for if we start off with a "real world" problem and apply mathematical reasoning, we are <u>ipso facto</u> doing applied mathematics. For the question to make any sense, we must allow the original problem to be be a mathematical problem. I would then claim that the process of abstraction characteristic of the schema described above also occurs in work within mathematics itself. Let me give an example, admittedly involving a relatively trivial piece of mathematics, but a piece accessible to secondary school students. Here and below, we allow ourselves to consider the closely related processes of abstraction and generalization.

Suppose it is observed that $5^6 \equiv 1$ mod 7. This may be verified empirically — we compute $5^6 - 1$, obtaining 15,624 and check that 15,624 is exactly divisible by 7. This argument is compelling and convincing, but unsatisfactory; we do not feel with this demonstration that we understand <u>why</u> the assertion is true.[3] The situation is ripe for generalization — let's make a mathematical model!

We conjecture, and then prove, that if p is any prime number and if a is prime to p, then (Fermat's Theorem) $a^{p-1} \equiv 1$ mod **p**. Notice that this is a generalization, not an abstraction, because we are still talking about rational integers. However, we may not feel entirely satisfied with the generality of Fermat's Theorem. We could proceed in one direction to Euler's Theorem, or we could regard Fermat's Theorem as itself a special case of Lagrange's Theorem that the order of a subgroup of a finite group divides the order of the group. This latter development involves abstraction as well as generalization, for we are now discussing abstract groups, postulating in our abstract system a single binary operation only (whereas the integers admit two, both of which were involved in our original demonstration that $5^6 \equiv 1$ mod 7), an operation that need not even be commutative. (At the secondary level, this further generalization might only be described, without aiming at student mastery.)

One feature of this example is the iteration of the modeling (generalizing) process; as we have said, this is also frequently a feature of applied mathematics. On the other hand, let me immediately admit that there is also a difference between modeling a non-mathematical problem, and the modeling we did here. In our example, we obtained incontrovertible proofs of our original congruence, and of related congruences, whereas in applied mathematics the mathematical reasoning can at best establish a scientific assertion as a good working hypothesis, a good approximation to the truth. I give a second "mathematical" example to show that this difference is not so absolute.

There is a beautiful numerical process, based on our base 10 enumeration system, called <u>casting out 9's</u>. What is involved here, in mathematical terms, is the canonical ring projection $\theta: \mathbb{Z} \rightarrow \mathbb{Z}/9$; we use the residue ring $\mathbb{Z}/9$ because it is particularly easy to compute θ. Now we may say that θ provides a means of modeling identities in $\mathbb{Z}$ by means of identities in $\mathbb{Z}/9$. This is a good checking procedure, because it is far easier to do calculations in $\mathbb{Z}/9$ than in $\mathbb{Z}$. However, we cannot <u>prove</u> identities in $\mathbb{Z}$ by modeling them by true identities in $\mathbb{Z}/9$; we can only <u>disprove</u> them by modeling them by false identities. Here we are dealing with the important mathematical process of <u>simplification</u> (with preservation of structure), a process that exhibits a strong similarity to the simplification involved in modeling a real-world situation.

If we are to treat abstraction, generalization, and simplification as key processes within mathematics itself, we find ourselves inevitably led to give prominence to the essential unity of mathematics. In practical terms this means insisting far less on the autonomy and apparent independence of the various mathematical disciplines, and emphasizing instead their real interdependence. This poses severe curricular design problems, but I believe that perhaps the most important desideratum is the breadth of view of the instructor.

Examples of interaction between different mathematical disciplines abound at all levels. Thus we use topology in the foundations of real analysis (a continuous function from a compact metric space to a Hausdorff space is uniformly continuous); we use algebra to do topology (the fundamental group); we use complex variable theory to do algebra (the fundamental theorem of algebra); we use algebra to do geometry (the syzygy method for proving Desargues' Theorem in the coordinatized plane); and we use linear algebra to study systems of linear differential equations (eigenvalues and eigenvectors). These and other examples show how one mathematical situation is modeled by another. It is my contention that this "applied perspective" should be adopted in mathematics education — and not merely for the worthy reason that this will help the student become familiar with the different ways of doing applied mathematics. I give one example of natural interaction between disciplines in the next section, but let me describe now an interaction between geometry, algebra, and combinatorics accessible at the secondary level which comes up in studying a problem that even couched in the language of "pure" mathematics has a distinctly applied flavor.

313

This problem is discussed in detail elsewhere (Hilton and Pedersen, in press); I note here that it is particularly apposite today in light of the considerable emphasis on discrete mathematics at both the pre-college and college level.

Consider (minimum) paths on the integer lattice in $\mathbf{R}^2$ from (a, 0) to (m, n); it is not difficult to show that there are $\binom{m+n-a}{n}$ such paths. Now assume $m \geq n$ and count the number of paths that meet (and perhaps cross) the line $y = x$. By setting up an ingenious one-to-one correspondence, it may be shown that this is just the number of paths from (0, a) to (m, n), or $\binom{m+n-a}{m}$. Thus there are $\binom{m+n-a}{n} - \binom{m+n-a}{m}$ paths that do not meet the line $y = x$. We may now calculate how many paths from (a, 0) to (m, n) meet the line $y = x$ without crossing it. Using another ingenious one-to-one correspondence, that between the set of paths from (a, 0) to (m, n) that do not cross $y = x$ and the set of paths from (a+1, 0) to (m+1, n) that do not meet $y = x$, we see that the number of paths from (a, 0) to (m, n) that meet the line $y = x$ without crossing it is $\binom{m+n-a}{m} - \binom{m+n-a}{m+1}$.

This example highlights the difficulty of designing a curriculum featuring discrete mathematics: the methods of discrete mathematics are far more varied and unsystematic than those of traditional courses in algebra and calculus. Mathematics professors at our universities and colleges who enthusiastically believe the problem of inadequate preparation of many students for the core calculus curriculum can be solved by simply replacing that curriculum with one with a strong emphasis on discrete mathematics (Ralston 1983, Hilton 1984) ought to take heed. There is a strong case for such a replacement, but it cannot rest on such naive optimism.

Our examples — and our arguments — leave us with an important question. Are there essential differences between the methodologies of "pure" and "applied" mathematics? This is, in my view, a very interesting subject for research, with strong implications for the teaching of mathematics. My own thinking is still at a fairly primitive stage on this question, but let me offer one fairly obvious example of an essential difference.

Suppose we are modeling some physical phenomenon and produce a differential equation with certain boundary conditions. Suppose further that we prove that this mathematical system has no solution. The effect of this discovery is to discredit the model — we must have oversimplified (say, by linearizing), or we must have neglected some aspect of the physical situation that was, in fact, highly relevant to the dynamic process we were modeling. However, if we model a mathematical situation, the connection between the situation and the model is far closer. There is still, as in applied problems, the very difficult art of choosing a good model (e.g., a useful generalization); but if the problem in the model has no solution, then the original problem has no solution, either. This remains true whether we are generalizing or simplifying in constructing our model.

The Special Role of Geometry

The place of geometry in the curriculum is, today, a special concern and a special problem. Students are arriving at our universities and colleges woefully ignorant of geometry and seriously lacking in any geometric intuition. These failings undoubtedly contribute to the difficulties they experience with the regular calculus sequence. Among the upper division courses we also find that courses in geometry are under-subscribed (along with certain other "traditional" offerings). Indeed it may happen that the only geometry course with an adequate enrollment is a course designed for future high school teachers — a course usually failing to do justice to geometry as a living branch of mathematics.

I would recommend that the geometric point of view figure prominently in virtually all secondary and undergraduate mathematics courses. It is a point of view that allows one to conceptualize more easily; besides, geometry is a wonderful source of ideas and questions. Geometry, in this informal sense, may be thought of as partaking of the quality of both pure and applied mathematics. In truth, of all the branches of mathematics, it is the one closest to the world of our experience.

It is probably not realistic to recommend an attempt to revive the study of geometry for its own sake through courses devoted exclusively to the discipline. But geometry is very "real" to the students; it provides questions whose answers must be sought in the disciplines of analysis and algebra. Without geometry, these latter disciplines must often seem to the students to be answering questions which they could never imagine themselves asking.

Let me give one example to illustrate how geometry and algebra may be made to interact in a completely natural way at the secondary level; this example will be discussed in much greater detail in a forthcoming book (Hilton and Pedersen, in press).

We start with the equation

$$S: x^2 + y^2 + 2gx + 2fy + c = 0$$

of a circle in $\mathbb{R}^2$. Given two circles $S_1 = 0$, $S_2 = 0$, it is natural to ask what might be the significance of the equation $S_1 = S_2$. This clearly represents a straight line. Moreover, if the circles intersect in P and Q, then P and Q obviously lie on this straight line since $S_1 = S_2 = 0$ at P and Q. Thus $S_1 = S_2$ is the equation of the common chord of the two circles. Notice that we have found this equation without knowing the coordinates of P and Q; it would be a horrendous task to discover them!

Now we ask what if the circles do not intersect. We then look for an interpretation of the quantity $S(x_0, y_0) = x_0^2 + y_0^2 + 2gx_0 + 2fy_0 + c$ and show that it is the square of the length of the tangent from $P_0(x_0, y_0)$ to the circle $S = 0$, provided P_0 lies outside the circle. This is sometimes called the _power_ of P_0 with respect to the circle. Thus points of $\mathbb{R}^2$ that have the same power with respect to the two circles lie on the straight line $S_1 = S_2$. We have effortlessly proved the theorem that such points do lie on a straight line. But what significance can we attach to $S(x_0, y_0)$ if P_0 lies inside the circle? We recall the geometrical theorem (probably proved by appeal to similar triangles) that the power of P_0 is $P_0 S \cdot P_0 T$, where $P_0 ST$ is a chord through P_0 intersecting the circle in S and T. This suggests the hypothesis that, even if P_0 is inside the circle, the power of P_0, defined as $S(x_0, y_0)$, is still $P_0 S \cdot P_0 T$. However, we must regard $P_0 S$ and $P_0 T$ as _signed_ distances to obtain a negative power when P_0 lies inside the circle. We may prove our hypothesis by showing that $P_0 S \cdot P_0 T$ is independent of the choice of chord (again, appealing to similar triangles), and then by calculating $P_0 S \cdot P_0 T$ when ST is a diameter. We may now say that the locus of equipotent points is, indeed, a straight line.

The story does not, of course, end here, since mathematics never ends naturally with an answer — all answers are stimuli to further questions. A typical further question might be, what is the significance of the equation $S_1 = \lambda S_2$ as λ varies? Here again we find it instructive to consider both the case where the circles intersect and the case where they don't.

Have we strayed too far from applied mathematics in describing this sequential segment of potential mathematical curriculum? I argue not, since the methodological principles exemplified above may, if desired, easily be cast in the language of the relationship between "problem" and "mathematical model." Moreover, students will not feel that the discussion is taking them out of the real world, especially if the obvious generalization to three dimensions is then pursued as a line of inquiry.

Let me close this section with what might be called an argumentum ad hominem. The Soviet Union is never accused of neglecting real-world applications of mathematics. Yet geometry figures as a prominent and highly significant part of the curriculum in each year of Soviet secondary education. Study of Mikhail Postnikov's delightful book (Postnikov 1982) shows, moreover, how to create a mutual enrichment of algebra and geometry. Contrast the single year devoted to geometry in the traditional U.S. curriculum. A year is spent doing algebra without geometry, followed by a year doing geometry without algebra. If, subsequently, some work is done in the name of analytical geometry, it tends to consist of attempting to sweeten the pill of some very dull algebra by giving it a coating of geometric language — seldom is any genuine geometry done. We who endeavor to teach undergraduate students the rudiments of the calculus know the consequences of this misguided pedagogical strategy.

Problem-Solving

I would like to close this essay with a few remarks on problem-solving as a curricular or pedagogic device. The clamor for applications finds its echo at the pre-college and even undergraduate level in a strong plea (endorsed by the PRIME 80 Conference and the National Council of Teachers of Mathematics, as a key element in their platforms) for greater emphasis on problem-solving.

It is not in question — and should always have been obvious — that the principal reason for learning mathematics is that it enables one to solve problems. If certain programs have neglected this proposition, then to that extent they are seriously defective. However, what is emerging from all the clamor for problem-solving is usually something quite different in nature from a simple forceful recommendation to keep in mind the reason for learning mathematics. For the advocates of problem-solving seem to be arguing that we should be teaching problem-solving as an alternative to our traditional approach. Good pedagogical strategy should be "problem-oriented," they argue; and, if problem-solving is taught effectively, we need not trouble the students to absorb the "theory" that has hitherto proved a stumbling-block to them. An example of this attitude is to be found in the publishers' puff for a (very good) book on applied combinatorics that reads, "Its applied approach gives your students the emphasis on problem-solving that they need to participate in today's new fields. Rather than focus on theory, this text contains hundreds of worked examples with discussion of common problem-solving errors..."

Not for the first time, I must insist that this false dichotomy between the building and analysis of mathematical structure on the one hand, and problem-solving on the other, is dangerous. If problems are to be solved mathematically, the mathematical model must be chosen. Either it must already be available to the would-be problem-solver, or he (or she) must be capable of developing it by the modification of existing mathematical models. Thus the investigator absolutely needs to know, understand, and be able to discriminate between different mathematical structures.

There are, it is true, certain problem-solving strategies and precepts which it is worthwhile enunciating explicitly. But these cannot serve as a substitute for a knowledge of mathematics. It would be truly calamitous if the belief were to spread that difficult quantitative problems could be solved merely by becoming proficient in the field of problem-solving, allying a knowledge of a few general principles with "sound common sense." This is just a sophisticated restatement of the old egalitarian fallacy.

There is a further reason, less obvious at first, why it would be a mistake to concentrate so exclusively on problem-solving. For it is implicit in the concept of problem-solving that the problem has already been formulated; it is further suggested that it is then a matter of selecting the best mathematical model and successfully exploiting it. Thus the question of how we formulate good questions is totally ignored — and this is an essential question in scientific work. Moreover, the problem-solving approach ignores the fact that it is often the mathematical concept and the mathematical result which suggest the promising question. Frequently it is an advance in mathematics that enables us to see the true nature of a scientific problem more clearly and to pose the significant questions (this is eminently true, for example, of electromagnetic theory and, more recently, the theory of solitons). Thus the essential two-way flow between mathematics and science is lost in an exclusively problem-solving mode of instruction.[4]

None of this, of course, is to gainsay the stimulus of attempting to solve interesting problems on the understanding and doing of mathematics. The UMAP modules can be extremely valuable as a component of a rich mathematical education. It should not be thought, however — and here I believe I have understood the intentions of the UMAP editorial board — that these modules are to concentrate on applied topics. It is easy to supply a list of mathematical topics suitable for modular treatment. A brief sample, drawn from both college and pre-college levels, might include maximum and minimum problems treated by various methods; thought-provoking paradoxes; linearization and linearity in mathematics; computational complexity; the geometry of three-dimensional polyhedra; topics in combinatorics; algebraic curves; and the classical groups. However, as these topics should suggest, the modules do not replace the systematic study of mathematics; they stimulate, enrich, and enliven it. Ultimately, we can only serve the purpose of mathematics education, in all its variety, by inculcating both the ability and the will to do mathematics and to use it. If, as some enthusiasts for a more applied curriculum and for more problem-solving rightly claim, the ability without the will leads to sterility, it is also true that the will without the ability leads to frustration. We avoid both these unpleasant consequences by teaching all of mathematics as a unity, emphasizing its unique generality and its immense power.

Notes

1. See Kaper and Varga (1980). I am most grateful for the opportunity to see this paper, from which I have drawn many ideas.

2. More precisely, the KdV equation describes the propagation of waves of small amplitude in a dispersive medium.

3. We are thus in the unfortunate situation so typical of our students! They are compelled to accept but do not truly understand.

4. Problem-solving may be characterized as "going from question to answer"; but scientific and mathematical progress often consist of going from answer to question.

References

Hilton, Peter. 1984. Review of The Future of College Mathematics, Anthony Ralston and Gail S. Young, eds. Amer. Math. Monthly 91(7): 452–455.

Hilton, Peter and Jean Pedersen. In press. Bridging the Gap. Reading, MA: Addison-Wesley.

Kaper, Hans G. and Richard S. Varga. 1980. Program Directions for Applied Analysis. Applied Mathematical Sciences Division, Department of Energy.

National Research Council. 1979. The Role of Applications in the Undergraduate Mathematics Curriculum. Washington, DC: AMPS.

Postnikov, M.M. 1982. Lectures in Geometry. Semester I. Analytic Geometry. Moscow: Mir Publishers.

Ralston, Anthony and Gail S. Young, Jr., eds. 1983. The Future of College Mathematics. New York: Springer.

Curricula for Active Mathematics

Hugh Burkhardt

Shell Centre for Mathematical Education

University of Nottingham, England

This paper reviews the difficulties of achieving large-scale curriculum change and relates them to the aims of UCSMP. Illustrations are drawn from various projects and activities at the Shell Centre, and from other groups in Britain and elsewhere. The main conclusions are that:

1. Planned curriculum change is always difficult to achieve.

2. The kind of process objectives essential for "active mathematics" place exceptional demands on the teacher.

3. Materials must be provided that are at least as supportive as those used for familiar types of teaching.

4. Large-scale change is possible only with careful empirical development of imaginatively exciting draft material in representative classrooms; if the material does not achieve its objective in "random" classrooms taught by the target group of teachers, it does not succeed.

5. Purpose-built structured classroom observation methods are an important element of such curriculum development.

6. New possibilities, including the use of the calculator and, particularly, the microcomputer, have real potential in helping teachers and students achieve classroom learning activities (discussion, non-routine problem-solving, and open investigation) that have long proved inaccessible to all but a few teachers.

7. The "levers" of large-scale curriculum change vary from society to society; they are not easy to discern, though we need as many of them working together for us as we can get. The virtues of gradual, modular change over an all-embracing global approach should be carefully considered.

Dynamics of Curriculum Change

My personal interest in the aims of UCSMP will be known to some of you (Burkhardt 1981). I have been working on the teaching of mathematical modeling, and particularly the tackling of real problems from everyday life, for more than two decades; it is this work that brought me to the Shell Centre. This paper is based on the experience and experiments of my colleagues and myself, and parallel work at other centers. Its aim is to help develop a realistic view of the problems and possibilities involved in introducing curriculum changes of this kind so as to provide a firmer basis for planning our efforts.

There has been a tendency to assume that we can plan curriculum changes and achieve those plans in a reasonably straightforward way — that feasibility is not a problem. I believe that the evidence of past experience suggests that this is far from true. The system we aim to affect (children learning and teachers teaching in classrooms in schools in the community) is largely unaffected by the efforts of innovators, and the changes that have taken place are usually a travesty of the aims of those who initiated them.

For example, let us look back at the reform movement of 25 years ago in mathematical education — "new math," "modern mathematics," and so on. Comparison of the initial aims agreed upon at conferences such as this with the pilot schemes in a few exceptional schools and the classroom reality of today show the contrasts vividly. For example, in England the applications of mathematics occupied a central place in the original design; in most of the major courses that emerged they are not to be found. Similarly, new mathematical concepts were introduced, but often with none of the pay-off that motivated their inclusion, because the serious examples envisaged initially proved too difficult for most students and were replaced with trivial ones.

The basic reason for our failure in the past is that the detailed innovations that were proposed could not possibly work for the people for whom they were envisaged. Now how could I make such an outrageous assertion, when such splendid things were achieved in these developments? Two crucial factors were given inadequate attention: teacher target groups and realistic empirical development.

Over the last few years it has come to be more widely accepted that the problems of mathematical education are complex and difficult and that progress is unlikely to come simply from trying to implement the most attractive-seeming suggestions put forward by people, however wise or experienced in the field. The various "solutions" advocated and enthusiastically taken up in various past generations are seen, decades later, largely to have failed in their goals and their claims. The need for tougher, more effective methods of tackling the central problems is clear; in particular, there is an obvious need for ways to develop and test hopeful ideas and to learn much more rapidly their strengths and limitations. The Shell Centre's approach over the last few years has been to develop methods of working that integrate curriculum design flair with a classroom development procedure based on empirical research techniques, some of them new. (Shell Centre 1984a; Burkhardt and Fraser 1982). The aim is to devise practical teaching approaches that are attractive and accessible to the average teacher, and at the same time to acquire a deeper understanding of the fundamental dynamics of learning and teaching. Such an approach, though well recognized as effective in engineering, for example, is relatively new in education.

The core of the approach is empirical study in representative classrooms, the "laboratories" in which our ideas are developed and tested. As in other more rapidly advancing fields, there is always a need for better strategies for probing systems of interest and capturing data of importance to their understanding. We have found, for example, that the microcomputer and purpose-built notations are, respectively, surprisingly powerful for these complementary aims. The study of teachers and what they can be helped to achieve is a necessary complement to the study of children's learning.

The range of performance in mathematics teaching is likely to remain wide, and realistic programs for change must build into their development both these aspects and other key factors involved in the education system.

Levels of Curriculum Development

It is useful to distinguish four levels of progress in our understanding of learning and teaching and our understanding of how to develop materials and procedures to make them more effective. I call them (Burkhardt 1982) the L level (how children learn); the T_1 level (teaching possibilities); the T_2 level (realizable teaching); and the C level (large-scale curriculum change). Each level underpins the later ones though each presents quite different problems and requires quite different methods, some of them new, for successful implementation. In particular, the crucial distinction between T_1 ("there exists a teacher, usually in the development group, who can help children learn better in this way") and T_2 ("what the 'target group' of teachers will achieve with the 'target group' of pupils, with whatever help in materials and other support we can be realistically expected to offer them") is crucial. Most curriculum development work has been at the T_1 level, producing useful materials and personal development for the authors and some fellow enthusiasts. T_2 level work is rare; in most development any "trialing is often badly structured, with minimal feedback done late in the day when the material is effectively fixed in all its essentials. T_2 developments may also be aimed at a small band of teachers, though most people vaguely hope their target group is a substantial proportion of teachers. This seems to us where the greatest potential benefit, and difficulty, lies. Achievement will obviously look much more limited than at the T_1 level; the range of performance, as well as style, of teachers is as wide as in any other highly skilled occupation. From a curriculum development planning point of view, the essential fact is that guesses do not work at the T_2 level, which has to involve development and trials with a representative sample of the chosen teacher target group.

The C level (a useful double entendre) involves all the other factors that affect whether teachers at large change their classroom practices (materials and activities) in the way envisaged. There is some evidence that an empirical approach, trying out the hopeful with rich structured feedback, is essential here too. It is likely to involve far more powerful "levers" on the curriculum than simply making materials available with added exhortation. Such levers are hard to find and can be used only responsibly when the evidence of qualitative enhancement is clear at the T_2 level; in our own work (see later) the projects with the JMB and Granada TV are aimed along these lines. Important factors for success at the C level are that the suggested innovation should: (1) make the teacher's job easier; (2) make it more fun; (3) tackle a problem they already know they have; and (4) have some public support behind it. Many reforms fail on all four counts; they are not easy to meet but should never be ignored if one is serious about the C level.

The emphasis throughout this paper is on <u>qualitative curriculum enhancement</u> is important; at least at the present stage, is it worth bothering teachers with the trauma of change if it only helps them do what they presently do somewhat more effectively or easily? There are essential components of learning mathematics that are entirely missing from most classrooms, particularly those involving substantial non-fragmented tasks and student activity that is autonomous rather than imitative. These are my focus here, though the methodological arguments are more general.

Realistic empirical development is simply the requirement that any "package" (teaching materials, outside support, and so on) be developed and shown to work in realistic circumstances and with people representative of the target groups. There is a need, of course, to begin with the exceptional and innovative in order to develop draft ideas and draft materials (though even here we continually remind everybody that they are designing for "the second worst teacher in their department" or some other readily identifiable figure). The development of the draft materials, however, must be done in realistic circumstances without special support that cannot be provided on the scale ultimately intended. Very often in the past these sensible constraints were ignored in the desire to get better apparent results from the children; it was assumed that proper dissemination and in-serivce training would close the gap and bring ordinary teachers into the promised land that the development group's work seemed to offer. It didn't — and couldn't — happen. Those parts of the innovation that were taken up by the commercial (i.e., successful-on-a-large-scale basis) sector were rarely a reasonable reflection of the original aims. Such people do not take the time to do empirical development work — it is, surely, the job of the innovators to find ways of producing materials that work. It will mean reducing their apparent objectives to increase their actual achievements — a useful heuristic throughout mathematical education.

Classroom observation is the other key factor I want to stress. We find it absolutely essential to use purpose-built research-based methods if we want to get feedback on draft materials in action adequate for their revision stage by stage (Burkhardt and Fraser 1982). It is also invaluable for advancing fundamental understanding, and incidentally for teacher training.

These methods represent a partly new hypothesis on more effective ways of promoting qualitative curriculum enhancement on a wide scale. They contain the means for their own evaluation over a period of just a few years, rather than the usual two decades. Alternative approaches should do as much to ensure they are swiftly tested. "Fail fast - fail often" is a motto we think is valuable.

I shall illustrate these methods, which were first developed in our work through the ITMA Collaboration on the potential of the microcomputer as a teaching aid in mathematics, by outlining our most recent project: the TSS program. The following is from the "official announcement" of the program.

The Testing Strategic Skills Program

For the last four years the Joint Matriculation Board (JMB), England's largest public examinations board, and the Shell Centre have been working together to explore and develop a new, modular approach to syllabus change. The first module of this program (Shell Centre 1984b) is now in action in schools in preparation for the Board's 1986 0-level examination in Mathematics. This first module, "Problems with Patterns and Numbers," illustrates what is intended, and the associated materials are available from the Board. Now, therefore, seems a good moment to bring this approach to the attention of a wider audience.

The key principles of the TSS program are:
1. Explicit acceptance by an examining board of the responsibilities placed upon it by the very strong influence its examinations have on curriculum in the schools where they are given.
2. A cautious realistic approach, based on empirical development in the classroom, to the introduction of new examination tasks which will encourage curriculum changes that will be widely recognized as desirable.

The essential elements in implementing this approach are:
1. Gradual Change. We think in terms of changing the basis of just one question in any year, corresponding to about 5% of the syllabus, or three weeks' teaching. Why this modular approach? Everybody, including teachers, finds it difficult to change substantially what they do, and this is particularly true when rather different methods of working are involved. Thus sweeping syllabus reforms have often had limited success, either

because they have been confined to details of subject content — the easiest changes to absorb. The logo of the TSS program is a tortoise, chosen for its well-known race-winning potential — though in this case that has yet to be proven.

2. <u>Effective Support.</u> Teachers are very busy people, and most base their teaching on well-defined schemes of work, even in areas with which they are familiar. In asking them to move into less familiar topic areas or modes of classroom learning activity, it is important to offer them something equally supportive, which is the rationale for the development of the module teaching package. Of course, teachers will do things in their own way in their own classroom, and there will always be a range of approaches and materials for teachers to choose from, or to develop themselves. The module package is offered as one proven possibility.

3. <u>Empirical Classroom Development.</u> It is not possible to guess reliably what the effects of any significant change will be, whether in the classroom or in the examination. As a result, the outcomes are often quite different in important respects from the original intention. In the TSS program, these dangers are much reduced by an empirical approach in which all the elements of the module are developed and tested until they work in the way intended with a broad sample of teachers and children, representative of those who take the examinations concerned. It is here that the Shell Centre for Mathematical Education plays the central role: a research and development center of the University of Nottingham, it has developed new methods for this task, with particular emphasis on structured classroom observation of teachers and pupils in action. These methods are crucial for the refinement of the materials, and so have been designed to work in a robust way for teachers with a wide range of styles and approaches.

The Module Package consists of three components:

- <u>Specimen examination questions,</u> with marking schemes and sample (not "model") pupils' answers.
- <u>Sample classroom materials</u> to support about three weeks' teaching to prepare children for the new types of questions. The materials include pupil work-

sheets, notes for the teacher, and extracts from pupils' work.

- Support materials that help teachers adapt to new aspects of classroom activity or teaching style — and extend them to other appropriate parts of their teaching with other classes, where there are no equally supportive classroom materials. This component can be viewed as a do-it-yourself inservice course.

The first module, "Problems with Patterns and Numbers," introduces some non-routine problem-solving into 0-level Mathematics. Mathematics has proved a good subject in which to explore this approach, partly because the Cockcroft Report identified some widely agreed and important changes, including the need for problem-solving as a regular element within the curriculum. It is interesting that his module required no addition to the existing JMB syllabus — tackling unfamiliar problems has long been included — but adequately reliable ways of examining such skills have not previously been established. Some deletions from the syllabus have been made to make time for the new classroom activities that are a principal aim of this approach.

"The Module Book" describes all the various components. The module package also contains masters so that teachers can make photocopies of the worksheets, checklists, and other printed materials for use in the classroom. The support materials include two unusual elements. First, they include a videotape of various teachers using the materials in the classroom, with a booklet of "Video Notes" to guide the viewer through this "evidence" and suggest questions they may like to consider. Second, the materials contain a disc with five microcomputer programs (for Apple or BBC Model B) with two booklets of teaching notes for their use; research has shown that a microcomputer used as a "teaching assistant" is a great support to the teaching of problem-solving. The response from schools has been remarkably positive, mirroring the views of the 30 representative schools that took part in the final trials.

The syllabus module is described in some detail in the Board's Revised Syllabus Leaflet RS86(k), part of which is included here as Appendix B. Details of the complete teacher's package or the Module Book, which can also be obtained separately, are available from the Shell Centre, or in the U.S. from Dale Seymour.

The second module is entitled "The Language of Functions and Graphs." It aims to develop children's fluency in "speaking mathematics" and in translating information about the real world from one form to another. These are important skills that are seriously lacking even in many able pupils who can manipulate mathematics fluently. This module is aimed at the new GCSE examinations as well as the 0-level examinations which it will replace, and is now in advanced trials in about 30 schools. Further modules are in the course of development. Other Boards have also expressed an interest in joining in this approach.

Outlook

It is too early to judge the potential of this new approach, both cautious and ambitious, as a way forward in curriculum change and in the improvement of examinations. The evidence so far suggests that it offers advantages and avoids some of the problems of the more traditional approaches. The development effort involved is substantial but, because it is concentrated on "growth points," it is probably less in total than when a global syllabus reform is undertaken.

Societies with a different educational system structure may find this use of assessment as a key "lever" (intentionally for good rather than accidentally for ill, we hope) surprising, or even distasteful. Do not reject it without thought, unless you are sure you have better levers.

Some Current Concerns in the United Kingdom

Mathematical education in Britain is still dominated by the Cockcroft Report (Cockcroft Committee 1982). This survey of problems and directions of progress was very well received, partly because it articulated clearly a professional consensus that had grown over the previous few decades. This had been illuminated by a number of detailed surveys of classroom activity and pupil performance, notably the Primary and Secondary Surveys of Her Majesty's Inspectorate and the Chelsea CSMS study of facility levels on a wide range of tasks for 12-15 year olds (HMI Secondary Survey 1977; Hart 1981).

It is interesting to note that Scotland (and Northern Ireland) were not covered by the Report. England and Wales have a highly decentralized system where curriculum responsibility probably lies ("probably" because of our national affection for unwritten constitutions) with the headteacher (principal) of the individual school; nonetheless, the HMI Secondary Survey found <u>uniformity</u> the overwhelming impression in the 334 schools (and about 10,000 hours of mathematics teaching) they observed (HMI Secondary Survey 1977). Scotland, on the other hand, has a centralized curriculum similar to the French model; nonetheless, there is a great deal of experimentation going on. The conclusion to be drawn, which I think is echoed in Australia, the U.S. and elsewhere, is that effective determination of the curriculum is not as simple a matter as formal structures usually suggest.

The Cockcroft Report is a comprehensive review, well worth reading for the questions it raises. Its most quoted paragraph, 243, focuses on learning activities:

"243 Mathematics teaching at all levels should include opportunities for

- exposition by the teacher;
- <u>discussion</u> between teacher and pupils and between pupils themselves;
- appropriate <u>practical work</u>;
- consolidation and practice of fundamental skills and routines;
- <u>problem-solving,</u> including the <u>application</u> of mathematics to everyday situations;
- <u>investigational work</u>.

In setting out this list we are aware that we are not saying anything which has not already been said many times and over many years. The list which we have given has appeared, by implication if not explicitly, in official reports, DES publications, HMI discussion papers and the journals and publications of the professional mathematical associations. Yet we are aware that although there are some classrooms in which the teaching includes, as a matter of course, all the elements which we have listed, there are still many in which the mathematics teaching does not include even a majority of these elements."

The absence of the activities underlined (by us) is not surprising — these are more demanding on the teachers than the prevailing exposition-plus-exercise in three senses:

- mathematically, because a variety of different tracks through the problem or argument must be handled, some of them unanticipated.

- pedagogically, because diagnosis and "minimum correction" are needed, different for each child
- personally, because the teacher will at times be sailing unchartered waters, and will not know all the answers.

Members of the Cockcroft Committee did not anticipate this unusual focus of "public" interest, which I believe reflects a central concern about the inability of children to deploy the mathematical skills they learn, either inside mathematics or in tackling realistic problems of concern to them from their everyday life and work. Incidentally, the Committee was unable to find a single employer who would articulate to them the belief, supposedly widely held, that employers are critical of the mathematical competence of the children who come to them from school.

Much of what is now going on reflects this concern with real-life applications. Perhaps the largest single "movement" is centered on providing "staged assessment" for pupils in the 11-16 age range, with particular attention to lower achievers. Some of this work springs directly from the Cockcroft Report, which emphasized the important influence public examinations have on the curriculum and recommended (paragraph 521) that fundamental changes in examinations should "enable candidates to demonstrate what they do know rather than what they do not know" and "should not undermine the confidence of those who attempt them."

British examinations at all levels have tended to focus on discriminating effectively between able students by asking difficult questions and expecting relatively low levels of response: typically only about half the class will get over 50%, and in the CSE examinations, which are aimed at the 20-60% ability range (from the top) at age 16, raw marks of 25% will often secure a pass. This is quite efficient as assessment but disastrous from a curriculum point of view, for the students know how little they have done. The U.S. tradition does not suffer from this particular defect; neither are examinations such an important factor, although matters of student confidence and the curriculum influence of the narrow range of technical skills measured in standardized tests give grounds for concern.

The spectrum of those involved in this "graded assessment" movement could hardly be wider. By far the greatest number of groups are from many of the 100+ local education authorities (LEAs) who see a need to provide schemes of assessment of this kind — usually by yesterday. I do not see how anybody can develop the really supportive materials that teachers will need to do something significantly new on this timescale and across this range. There are various other groups. One is based on SMP. The School Mathematics Project has been the most commercially successful of the "modern math" projects of the 1960s with, at one stage, over 60% of all 11-16 year old students using these books. They are now in the course of a major revision that attempts to take on board some of these aims, particularly the applications ambition that they failed to implement the first time round.

Another group is based on the NFER team that devised the material for the national monitoring project of the Assessment of Performance Unit. A third group is based on a collaboration between the Chelsea College CSMS team and the (largest) Inner London Education Authority. But the first in the field in England was OCEA, an ambitious collaboration between an exam board and a consortium of LEAs. The Scots began this work in response to the Munn and Dunning Reports several years earlier.

It is my view that no group has yet shown it can realize its aims; the tendency to revert to testing mathematical skills, needless of their deployment, is remarkable.

This movement (in which we are heavily involved — see the following section) has tremendous potential for curriculum good or harm — with the balance of probability on the latter. If the effect is to provide more technical math tests for one group of children who until now have not had their curriculum narrowed by public examination (with a further narrowing of emphasis for the rest), that will be a serious leap backwards. All the groups are eager to avoid this and make the test directly related to the use of mathematics in the real world. But they underestimate how difficult this is, and many of the draft items are the same mixture as before, with a thin coat of realistic paint. A "nursing mathematics" example (believe it or not!):

332

- A nurse has a boyfriend who goes jogging every morning; he goes 6 miles in 40 minutes; how fast does he jog?

There is no doubt of the genuine wish to do better, but whether we are equal to the task is yet to be seen. This is obviously centrally relevant to UCSMP aims; I hope we shall have useful and positive results over the next year or two to contribute to that program.

Another important initiative in the UK is an increase in in-service training — which here and there occupies a status akin to peace, motherhood, and apple pie. Our own researches suggest that most in-service training (INSET) has no observable effect on the classroom behavior of teachers. This is not surprising, since teaching is — and must be — a fluent sequence of largely automatic responses of a kind unlikely to be changed by the sort of civilized discussion between fellow professionals that constitutes most in-service courses.

We have shown that it is possible to have direct effects, but that courses are designed differently (and, of course, we develop them empirically until we know they work!). The methods are still wide open but involve more emphasis on technical teaching exercises and are generally activity-based. We have something to learn from coaching of athletes and musicians in this regard. Also, we have found a single microcomputer used as a "teaching assistant" to be immensely powerful in stimulating and supporting style changes in teaching (see below).

A second Cockcroft recommendation has the vivid nickname of "bottom up." History is full of examples of curriculum reform started by exceptionally gifted teachers with bright children. They have often produced delightful results in terms of classroom activities and student performance across the range of skills from the technical to the strategic. Typically, the subsequent attempt to "let every child benefit" involves two processes which I call "dilution" and "corruption." I hardly need describe them in detail. In dilution examples are watered down to make them accessible to the broader target group of students (and teachers) — often to such a point of triviality that the justification for introducing the topic disappears (matrices in some of the "new math" schemes provides a vivid illustration). Corruption arises when the intended classroom approach goes beyond the style range of many of the target group of teachers, who continue to handle the new material in the old ways. The "bottom up" approach seems likely to have a greater curriculum stability. If a scheme works in the way intended with less able pupils (and teachers), it is generally easier to make it more challenging and richer for abler pupils (dilution is avoided). The avoidance of corruption requires the development of material until it provides the necessary support for the teacher target group concerned.

Numeracy and the TSS Approach

The Shell Centre is also working with the JMB on a feasibility study of testing "numeracy"; we have deliberately chosen a title different from mathematics to avoid having to take responsibility from the very beginning for the whole math curriculum. The word "numeracy" is employed with different meanings (often, unfortunately, to mean simply the ability to do arithmetic), but we have in mind the original meaning, parallel with literacy, of being able to deploy your mathematical and other skills in tackling problems of concern from everyday life. Since work is in progress and the outcomes unsure, I think it best to present here a stark statement as we saw it a year ago, with some notes of commentary.

Definition

Numeracy is the ability to deploy appropriate mathematical and other skills in an effective manner in tackling problems of concern, or situations of interest, as they are met in everyday life in all its aspects.

There is no suggestion that, for the foreseeable future of a few years, work on numeracy will constitute the whole of the mathematics curriculum for any child.

Aims

To develop a scheme of graded assessment, with associated classroom materials and other support as may be needed, that:

- will promote the numeracy of children at all levels of ability found in normal schools, and
- will provide evidence of these achievements in a variety of practical contexts.

It is worth stressing that this is a difficult problem and that little progress has been made towards its solution; such problems tend to yield only to imagination in ideas combined with hard-headed realism in developing and testing them. The usual failure path leads to tests of arithmetic, with a thin cosmetic coat of realism, providing children with dubious skills they cannot actually deploy.

Development Criteria for Inclusion in the Scheme
Specificity

Each "module" of assessment will have specific classroom objectives and criteria of performance that reflect the aims of the scheme directly; any other objectives will be secondary and not be reflected in the criteria of performance. The target groups of pupils and of teachers for whom the "module" is intended will also be specified.

This may be controversial. Many people will be inclined to relax these constraints; in particular, to allow criteria based on performance in pure mathematical tasks such as arithmetic, perhaps set ilustratively in practical contexts. There are strong arguments against this, including:

1. Successful performance at such tasks is no indication of the ability to deploy the skills involved.
2. These skills can be adequately tested in the appropriate context of their applications.
3. The mathematics curriculum, at least between the ages of 8 and 16, is unbalanced by the narrow focus on pure mathematical techniques and concepts which ignores their deployment and practical use almost entirely. This background suggests that, if techniques alone can gain explicit credence within "numeracy," the new demands of tackling real problems are likely to be evaded.

It is clear that the further development of pupils' mathematical skills as well as their deployment will figure in classroom activities designed to prepare children for the assessment and correspondingly in teaching material developed to support the module (see issue below). Successful developments in modern languages have adopted this approach, testing linguistic skills only through their deployment by the pupil in simulated realistc contexts.

Validity

The assessment will be based on valid, useful tasks (useful in the sense of the definition of numeracy; valid in the sense that success in each such task represents a worthwhile achievement in numeracy). In particular, fragmented tasks that are only "potentially useful" will not be assessed.

Feasibility

In the course of its development, any "module" must be shown to work, i.e. to lead to classroom activities and assessment performance as specified for the module, with a representative sample of that target group of pupils and the target group of teachers for whom the module is intended.

Standards in establishing feasibility are often low, the reality of classroom activity, assessment tasks and performance on these tasks being a travesty of the stated objectives and often of no intrinsic value to the children concerned. But reliable and straightforward methods of a fairly obvious kind do exist. They are based on reasonably systematic observation of an appropriate sample of teachers and pupils working on the "module" in reasonably realistic classroom conditions. For anything significantly new, such feedback is essential for the development of objectives, tasks, criteria, and materials from draft ideas to a valid coherent working form. Such observation cannot be done by the particular teacher; each role — teacher and observer — requires full attention. Efficient observation requires a small amount of training. We at the Shell Centre have developed techniques both of observing and of training that seem to be effective.

Some Issues of Importance

1. The focus of "modules." Should it be on strategic skills for a specific type of real problem (e.g., "How to plan a ...") or a particular area of practical concern (e.g., planning a holiday)? The roles of fact, and of fantasy? A wide range of possibilities needs to be imaginatively explored; experiment will provide answers.

2. The relationship between numeracy work and the rest of the mathematics curriculum. Practicalities of all sorts suggest that numeracy will be seen as a rather different aspect of mathematics (with different priorities, different skills,

different demands) that will grow slowly in the light of experience and the range of good quality material available. It will loom larger for some children than for others.

3. The role of technique, and the way technique/concept-centered work in pure mathematics is best related in the classroom to the context-centered focus of numeracy, is an important question to explore. Again, the priorities of numeracy are essential, and in some ways are the reverse of those in the mathematics curriculum. They are the priorities of a "tool-kit" in use, e.g., recognition of the potential uses of each tool, mastery in their use (rather than just acquaintance), flexibility of use, the recognition that simplicity brings reliability, and so on.

4. How can "levels" be determined? The obvious line is to make this norm-referenced. When a criterion-referenced module has been developed and evaluated, its level is fixed by the proportion of children at an appropriate age who can pass it, after a specific learning program. There are obvious technical problems here, including linkage between different ages. A similar approach can map a set of results onto a 16+ scale, if that is desired.

5. The status of different modules. It seems highly desirable to many people that any scheme should provide an open framework within which different groups can contribute. A "modular" system is one approach. There will continue to be a need for "board-based" and "school-based" offerings; they will each continue to have advantages and disadvantages educationally and in the eyes of customers. The framework should allow these things to co-exist and support each other as simply and as naturally as possible.

The essence of the validation process lies in the establishment of feasibility (see above). The target groups involved (one class, one school, or a random sample thereof) and the care with which the responses are observed will vary widely. The reporting of results, provided some are carefully moderated and labeled as such, can allow the reliability of the others to be roughly judged by the user through a comparison of pupil results on more or less carefully moderated modules.

Modular or Global Change?

One of the interesting issues to arise in discussion at the Chicago

Conference concerned this question. The standard approach to syllabus reform is a global one — a fundamental rethinking of the principle underlying the curriculum concerned and the development of appropriate teaching materials and procedures to realize the new goals. This approach has the virtues of consistency and coherence. It can also tap the stream of enthusiasm that motivates change to help it succeed. The course textbook that is the key vehicle for a curriculum in the U.S. must be replaced by one related to the new approach, and thus naturally accommodates a global reform. How, therefore, can I suggest a gradual, modular approach — an alternative that lacks all these virtues?

There are a few key principles, actually at a higher level than those underlying any particular curriculum design, that seem to me to encourage gradual change. My advocacy is not stronger than this since noone, as far as I am aware, has a method of planned large-scale curriculum change in the U.S. or U.K. contexts that has been shown to work reliably. (This is not a statement that things are better in other countries, but an avoidance of that far-from-simple issue for the moment.)

- It has proved to be extremely difficult to achieve planned changes in the curriculum; normally, they happen only on a small scale or are grotesquely distorted in the course of realization. The slope we face is, therefore, uphill and steep.

- Everybody finds change in established habits of work extremely difficult. Retraining requires a lot of time and "drill" as well as reflection, particularly if it is a matter of skills used at an "automatic response" level. Change is a threat. To advocate the need for change is to imply fundamental criticism of current performance, and therefore of current performers. (Technology helps by manifestly changing the game.) Any new challenge carries for those who attempt to face it some fear of inadequacy or failure.

- We do not know enough to predict the behavior of teachers and students under substantially changed circumstances, so any global rethinking of the curriculum must be largely speculative, requiring empirical feedback for its refinement. (Such feedback is a feature of effective stable systems in any field.)

The gradual approach concentrates on the <u>direction</u> of change. The kinds of new concepts and learning activities we want can be clearly articulated. How much of each and how far we want to go becomes a matter of learning from feedback as change begins. The new "balanced diet" is the end point of the development, not the beginning.

The virtues are clear:

- Noone has to change more than a small proportion of what they are doing, so that change can receive the attention (from developer and from user) that is so necessary for a <u>qualitative shift</u> of any kind. The implied criticism of current practice is slight, as is the worry of not being able to cope.

- We the developers don't have to do the whole job, using most of our energy simply to reproduce the current curriculum with minor variants; we can concentrate time, talent, and money on the growth points and key areas of change.

- The pace is under control too. The system is stable; if this year's module fails to meet our standards, we can defer it.

- The user gets bite-sized chunks of novelty to concentrate on and even ordinary human teachers can enjoy that!

Catastrophe theory enables us to see more clearly how small changes can bring a system to a point where it undergoes a qualitative shift apparently spontaneously. Gradual change can lead to global change, surprisingly quickly, hence the tortoise logo for our TSS project. Observations have been made of the classroom behavior of those teachers (still atypical) who have learned to broaden their style range to promote discussion, non-routine problem-solving, and other learning activities characterized by autonomous student behavior. These observations show that there comes a point when this approach pervades their teaching, carrying over into other parts of the curriculum.

Then the new global text is needed, though by now the slope towards change is downhill, the ingredients are there, the audience is prepared, and the task is mainly one of tidying and tuning. (The mixed metaphor is not dead, though the real glories are past — "The British Lion walking hand in hand with the floodgates of democracy.")

Building the Infrastructure

What I have described is an "engineering" approach to innovation in mathematical education. It seems to me that we are not primarily social scientists, simply trying to understand more, rather, we are trying to make a system work better. However, just as Bell Labs is a telephone company (or several telephone companies) which sees it as essential for its purposes to pursue fundamental studies in solid state physics, information theory (which "they" invented), and many other basic fields, so we believe that enhanced fundamental understanding and the development of devices related to it are essential components of progress. Let me illustrate with some examples. I have already referred to the results of the CSMS project of our friends at Chelsea College (Hart 1981). This was a study of learning (L level) which gave us, astonishingly for the first time, a detailed map of the facility levels of a fairly broad range of mathematical tasks over the 12–15 age range. Some explanation of the cognitive basis of their results was also inferred. All sensible curriculum development now takes these findings into account; they may not be immutable, but they are certainly not ignorable.

At the T_1 level, I should like to mention our work on the possibilities of a "diagnostic teaching" approach, which is led by Alan Bell (Bell 1981, 1982a, 1982b, 1983, 1984). Its starting point was the discovery that effective mathematicians at all levels are characterized, not by accurate recall of algorithms, but by "debugging skills," such as the ability to monitor, detect, and correct errors in their own procedures. The teaching materials for a particular topic aim to begin with a rich exploratory situation containing various items of information and an invitation to consider what further information can be found out from what is given. Following this initial exploration, there is a focus on a few particular questions that contain important conceptual obstacles. The questions are deliberately posed to allow misconceptions to come to the surface, if they exist, and thus create a conflict that can be discussed and resolved. The third phase of the teaching cycle consists of exercises with built-in feedback of correctness. The new awareness reached during conflict–discussion is put into practice in a situation where pupils know immediately if they have made an error, and can reconsider their response.

340

The approach is unconventional in that most teachers avoid discussion of errors, preferring instead the "correct restatement approach"; but the results show the superiority of diagnostic teaching. A "responsive" technique, however, places additional demands on the teacher; materials over a range of topics that aim to make this technique widely accessible to teachers (T_2 level) are now being developed and we shall see how it goes.

Our first serious work at the T_2 level was through the collaboration of Investigations on Teaching with Microcomputers as an Aid (ITMA 1985) with the Plymouth group led by Rosemary Fraser, who is now at the Shell Centre. This was primarily a study of the effect of using a single microcomputer serving as "teaching assistant" on the classroom dynamics, but I believe it led us to important advances in our understanding of teaching and teacher behavior (Fraser et al. 1983) and in our methods of working (Burkhardt and Fraser 1982; Beeby, Burkhardt and Fraser 1980) which proved crucial in the C level developments of the TSS program outlined above. Let me present a few points:

We already have evidence that the potential of the microcomputer for helping teachers enhance student learning presents a tremendous opportunity for curriculum enhancement (Burkhardt 1984; Burkhardt 1985). The effects on the dynamics of the classroom can be profound, but they are often subtle; for this reason there is a great deal still to do before we have even a broad understanding of what can happen in the various modes of computer use of the kind listed in the background paper.

I shall illustrate the sort of thing that may be expected by describing one application that has been developed and studied in some detail, and proved to be particularly rich — the use by a teacher of a single microcomputer programmed as a "teaching assistant." I do so for various reasons: it is less familiar to most people, it brings out some general points about the overwhelming importance of the people involved, both teacher and pupils, and of the dynamics of their interaction, and it is particularly relevant to schools as we know them because it seeks to enhance the performance of a teacher working with a group of children in the classroom in the normal way. It also only requires one microcomputer per class rather than one per child.

This mode of use, first emphasized by Rosemary Fraser, has been shown to have remarkable effects in leading typical teachers in a quite unforced and natural way to broaden their teaching style to include the "open" elements that are essential for teaching problem-solving. Since this is a crucial aim that we have been trying to achieve for at least thirty years with little or no effect, it is a valuable result. It is worth explaining briefly why these effects come about (see Fraser et al. 1983). First, the computer is viewed by students as an independent "personality." It takes over for a time a substantial part of the teacher's normal "load" of explaining, managing, and task-setting, key roles played by every mathematics teacher. The computer takes these over in such a way as to lead the teacher into less directive roles, including crucial discussion with children on how they are tackling the problem in which s/he provides guidance only of a general strategic kind — counseling if you like.

Several of the outcomes of this study have become integral parts of our broader curriculum development methods, and of the materials themselves:

- The "design principles" listed below in "Program Design and Use. Some Principles" are useful for all teaching materials.

- SCAN (Systematic Classroom Analysis Notation; see Beeby, Burkhardt and Fraser 1980) is the skeleton of our "observation kits," permitting observers who were not there to discuss what took place in a lesson in terms that are credible to the actual observer who wrote the record, and in sufficient detail to guide the revision of the material. Appendix A gives a brief look at SCAN; other equally purpose-built notations are on the way.

- We shall be holding a working conference under the heading "Classroom Observation Methods for Curriculum Development and Research," in Nottingham on 1-5 September 1986 to share experiences. Do ask us for an invitation if you are active in this area, or want to be.

- The "roles analysis" (Fraser et al. 1983), even its crude form outlined above, is a robust guide to the design and use of materials and for in-service teacher training. Rosemary Fraser has noted that "role imitation leads to higher-level learning activities; rule imitation leads to exercises, often done wrong."

These important general outcomes arise, I believe, because the microcomputer is an outstanding "probe" of the classroom dynamics, producing credible, controllable perturbations of the kind that, for any system in any field of investigation, show us much more about the dynamics than mere observations of the unperturbed system could possibly do. SCAN represents the other key aspect of research design — selective, cost-effective data capture. When compared with more successful fields of endeavor, mathematical education has devoted relatively little effort to improving methods in this sort of way. Like X-rays in medicine, or in molecular biology, new technology and new ideas can help crucially. Even more important, we cannot afford to be chained by "off-the-peg" paradigms of research and development. If we are to make an impact on the problems we confront, we must design and justify more effective and appropriate lines of attack.

Program Design and Use. Some Principles

The systematic observation of a total of some 200 mathematics lessons in 17 classes, with and without the microcomputer, also brought out some results useful from a program design point of view. We can only state, not justify them here.

1. Pupil activity centering. The designer should set his attention on the pupil learning activities he wishes to promote, and not on the micro and its screen images, which are only secondary stimuli for these activities. This almost self-evident but widely ignored principle has a less obvious corollary, namely:

2. Omission design. The most important aspects should be actively produced by the pupils in their minds and on paper or some other personal medium under their direct control; so these aspects should not normally be shown by and on the micro.

3. Ambition. Design material that at least promises a qualitative impact in its educational benefits. There are very few ($\sim$100) program teaching units have any effect on the whole 5-18 curriculum ($\sim$50,000 hrs). It is an effort for teachers to use them, and this situation is likely to persist. The mediocre is not worth the effort.

4. Access. The "access level" measures the balance between ease-of-use and power; any program should be accessible to its clearly identified target

audience, any extra power-plus-complexity being made available in a way that protects those without extra expertise from disagreeable experience and the dangers of disillusionment.

5. <u>Teacher appeal.</u> If the teacher doesn't like it, the children won't get it.

From these a number of more specific points emerge:

a. <u>Willingness.</u> The program should allow pupils and teachers to feel they are learning and teaching in their own way, and not seem forced on them by the specifics of the program. (So many programs officiously dominate the whole process producing a claustrophobic user reaction; only psycho-motor games have strong enough appeal to ignore this factor with impact.)

b. <u>Activity centering.</u> Since there is an enormous range and number of topics in the curriculum but relatively few important learning activities, real educational benefit is more likely to emerge if program design concentrates on promoting learning activities that are valuable and elusive, such as problem-solving, open investigation, and discussion, rather than teaching particular topics or providing yet more practice.

c. <u>Follow on.</u> Look for ideas that will lead to further activities without the computer, and link up with other topics and subjects.

d. <u>Recognizable usefulness.</u> If the teacher sees that the unit is recognizably useful in the present curriculum, particularly in awkward areas, she is likely to try it — and be led to discover other, more profound benefits.

e. <u>Showing possibilities.</u> A unit will be particularly worthwhile if it shows teachers new possibilities for their teaching that can carry over into other lessons, and if it supports them while they acclimatize to this new experience. This in-service training role may initially be the most important contribution of the micro.

f. <u>Supporting role shifts.</u> One important way of implementing (e) is through role shifting and role sharing. The program exemplars in this discussion should help designers to see the ingredients, such as clear scene-setting, solid unobtrusive task-setting and management, and a task with a rich and varied strategic problem-solving demand.

g. <u>Teacher-added material.</u> Often teacher-produced materials added to a

344

program point the way forward to further learning activities of value that do not require computer support. Designers should have this aspect constantly in mind. It may be of even greater value to design programs that allow teachers to incorporate and retain material in the program system itself, though the additional complexity in use should not be ignored.

Appendix A

SCAN – An Outline

To give the reader some flavor of SCAN (Beeby, Burkhardt and Fraser 1980), we present a sparse SCAN record of a short lesson segment, together with the transcript of the dialogue and a table summarizing the notation.

LESSON IDENTIFICATION: C

Resources Used	Activity	Events/Episode Summaries	
PMC	W1 ⊃	o / cf / ch / iα2 ‖ co pch / eβ1 / qαiν / eα1 ‖ co	1
		pch / cf / iα1 / iα1 / e / pa / cf ‖ co pq / eαi ‖ co	2
		pq / e ‖ co i ‖ co ch / qβ2o / qβiν	3
		qβ2o / ∧ ✓ ‖ co o / cf ‖ CH / i ‖ I m ‖ F	4

The dialogue described in the first two lines was as follows. We hope that our description of the system will help the reader follow its illustration on this brief example.

Teacher: //Yes, that's nice/ but what does it mean? What is it?   o / cf

Pupil 1: I plot this against this./   ch

T: What are you going to write? A title. The most important thing about any sort of chart is a title.//   iα2 ‖ co

Pupil 2: Do I have to write that down like that?/   pch

T: You've already got it like that in base 5. Now you've got to write it down as an ordinary number,/ so how many stars have you got?   eβ1

P2: I've got 13./   qα1✓

T: So that's what you write down. That's in base 10.//   eα1 ‖ co

Pupil 3: Is this right? Is that how you do it?/   pch

T: That's right. That's right./
P3: And um...

cf

T: Three — draw stars for the following
numbers. 9. That's right./
Now change these two into base 5 numbers,
by putting...

ιαι

P3: I just put nine little crosses?/

ιαι

T: Yes — you split it into a five and
however many are left over./

e

P3: And I divide that into four lots of
five and one, two, ... four./

pα

T: That's right.//

cf || co

Pupil 4: I don't see what it means here. It
says "Take 9 in base 10." What does that
mean?/

pq

T: That means that 9 is already in base 10, and
you've got to draw nine stars.

eαι || co

P4: Oh, is that all I've got to do?//

Table VII SCAN 1M Code Sheet

RESOURCES (R =)

TM - teacher produced material
PMB - printed material (books)
PMC - printed material (cards)
C - computer
OHP
BB

ACTIVITY LEVEL

E - Exposition teacher to whole class (E_w) or group of 5 or more (E_n)

D - Dialogue, teacher to group < 5

W_n - Pupil work, in groups of n

PP - Pupil-pupil dialogue (use t to prefix teacher remark)

T - Teacher Initiated

P - Pupil Initiated

MISCELLANEOUS

t - teacher slip
c_x - computer slip
z - major tactical change

EPISODE LEVEL (SUMMARY)

D - Defining
I - Initiating activity
CO - Coaching
B - Explaining (new material)
C - Confirming
R - Revising
SS - Searching successfully
SU - Searching unsuccessfully
CN - Conversing
F - Facilitating
AR - Arguing (resolved)
AU - Arguing (unresolved)
CP - Competing

Ⓡ - resource in use
Ⓡ̶ - use ended

EVENT LEVEL

Social, Organizational, Procedural

g - gambit
m - managerial
ch - question for checking
v - vote
p - observation
w - withdraws statement
l - leaves discussion

Associated with Content

q - question of content, A-repeated
a - assertion
e - explanation
x - giving example
cc - conclusion
b - being
s - suggestion
i - instruction/initiation
cf - confirmation
r - rejection
k - correction

FATE OF QUESTIONS

✓ - correctly answered
v - partly correct
x - answer incorrect
h - consistent hypothesis offered
ĥ - inconsistent hypothesis offered

QUALIFIERS

Prefix p for pupil remark

Nature of Activity or Depth of Remark

α - recall, single fact, single act no processing involved

β - exercise of straight forward nature, putting together several facts or acts

γ - extension of previous work involving new ideas

Situation or Level of Guidance

1 - highly structured, close direction, small number of choices

2 - some guidance offered but requires connection of facts rather than selection

3 - minimum guidance, open, investigatory

o - no pupil response
δ - teacher does not take response
hc - hypothesis confirmed
hr - hypothesis rejected
∴ - time to think
( ) - off line
___ - same question

348

Appendix B

Effective Support for Innovation – The TSS Approach

Here we give the official outline for schools of the first Module Package (Shell Centre 1984b); it was described in the official in-service guide to the new GCSE examination in mathematics for 16-year-olds as "a gentle introduction to investigative work" (i.e. non-routine problem solving).

The second module "The Language of Function and Graphs" is similar in structure.

Problems with Patterns and Numbers: A Description of the Module

This Module aims to develop the performance of children in tackling mathematical problems of a more varied, more open and less standardised kind than is normal on present examination papers. It emphasises a number of specific strategies which help such problem solving. These include the following.

. try some simple cases
. find a helpful diagram
. organise systematically
. make a table
. spot patterns
. find a general rule
. explain why the rule works
. check regularly

Such skills involve bringing into the classroom a rather different balance of classroom activities than is appropriate when teaching specific mathematical techniques: for the pupils, more independent work and more discussion in pairs or groups, or by the whole class; for the teacher, less emphasis on detailed explanation and knowing the answers, and more on encouragement and strategic guidance.

The Module is not concerned with any narrowly defined area of content or mathematical technique within the existing syllabus. Because the strategic skills it focuses on are demanding, it concentrates on the simpler techniques which most pupils will have mastered (e.g. using numbers and discovering simple patterns), while giving credit to those who bring more sophisticated techniques (e.g. algebra) to bear on the problems.

Problems with Patterns and Numbers: Five Specimen Examination Questions

These specimen questions indicate the range of questions that is likely to be asked. The questions actually set may be expected to differ from those given here to about the same extent as they differ from each other.

The marking schemes, which are given in part (d) of this leaflet, are designed to give credit for the effective display of strategic skills, in particular for demonstration of the following.

understanding of the problem
organising the attack on it systematically
finding results, particular and general
explaining the approach and the results

The distribution of marks between these headings will vary with the problem, but all carry significant weight. Further specimen questions with marking schemes, illustrated with actual pupil answers, are given in the Module Book.

Specimen Question 1
Skeleton Tower

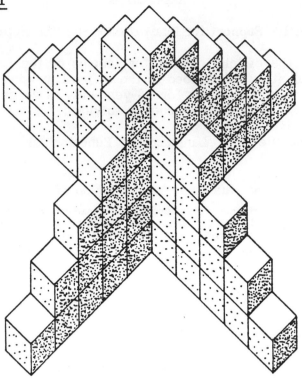

(i)   How many cubes are needed to build this tower?
(ii)   How many cubes are needed to build a tower like this 12 cubes high?
(iii)   Explain how you worked out your answer to (ii).
(iv)   How would you calculate the number of cubes needed for a tower n cubes high?

Specimen question 2
The Climbing Game

This game is for two players.

A counter is placed on the dot labelled "start" and the players take it in turns to slide this counter up the dotted grid according to the following rules:

At each turn, the counter can only be moved to an *adjacent* dot *higher* than its current position.

Each movement can therefore only take place in one of three directions:

The first player who slides the counter to the point labelled "finish" wins the game

(i) This diagram shows the start of one game, played between Sarah and Paul.

   Sarah's moves are indicated by solid arrows (———▶)

   Paul's moves are indicated by dotted arrows (– – – ▶)

   It is Sarah's turn. She has two possible moves.

   Show that from one of these moves Sarah can ensure that she wins, but from the other Paul can ensure that he wins.

(ii) If the game is played from the beginning and Sarah has the first move, then she can always win the game if she plays correctly.

   Explain how Sarah should play in order to be sure of winning.

351

Problems with Patterns and Numbers: Two Specimen Marking Schemes
Specimen question 1
Skeleton Tower

(i) *Showing an understanding of the problem by dealing correctly with a simple case.*

Answer: 66

2 Marks for correct answer (with or without working)

Part Mark: Give 1 mark if a correct method is used but there is an arithmetical error.

(ii) *Showing a systematic attack in the extension to a more difficult case.*

Answer: 276

4 Marks if a correct method is used and the correct answer is obtained.

Part Marks: Give 3 marks if a correct method is used but the work contains an arithmetical error or shows a misunderstanding (e.g. 13 cubes in centre column)

Give 2 marks if a correct method is used but the work contains two arithmetical errors/ misunderstandings.

Give 1 mark if candidate has made some progress but work contains more than two arithmetical errors/misunderstandings.

(iii) *Describing the methods used*

2 marks for a correct, clear, complete description of what has been done providing more than one step is involved.

Part Mark: Give 1 mark if the description is incomplete or unclear but apparently correct.

(iv) *Formulating a general rule verbally or algebraically*

2 marks for a correct, clear, complete description of method.
Accept number of cubes$=n(2n-1)$or equivalent for 2 marks. Ignore errors in algebra if the description is otherwise correct, clear and complete.

Part Mark: Give 1 mark if the description is incomplete or unclear but shows that the candidate has some idea how to obtain the result for any given value of n.

Specimen question 2
The Climbing Game

(i) *Showing an understanding of the rules of the game by systematically dealing with the various possible moves.*

    1 mark   for indicating that Sarah can force a win by moving to point A
         or   for indicating that she could lose if she moves to point B.

    2 marks for a correct analysis of the situation if Sarah moves to point A including the consideration of both of Paul's possible moves.

    Part mark: 1 mark for an incomplete or unclear analysis.

    3 marks for considering the situation if Sarah moves her counter to point B and making a correct analysis.

    Park marks: 2 marks for an analysis which is complete but omits to consider one of the two possible moves for Sarah from point A or C.

                1 mark for a more partial analysis.

(ii) *Formulating and explaining a winning strategy for the game*

    4 marks for a clear, complete and correct explanation.

    Part marks: 3 marks for incomplete or unclear but correct explanation.
                 Up to 3 marks can be given for the following:
                 1 mark for recognition of symmetry
                 1 mark for evidence of a systematic approach
                 1 mark for correctly identifying some winning and/or losing positions above line m
           or 2 marks for correctly identifying some winning and/or losing positions below line m (or above and below line m)

353

Sample Classroom Materials

These offer some resources by which pupils can be prepared for the questions on the examination. All the materials and suggestions are offered in the explicit recognition that every teacher will work in their own classroom in their own individual way. A Module Book, *Problems with patterns and numbers*, is available for sale to assist teachers who prefer not to develop their own teaching resources; the material which follows is taken from the Module Book.

The aims of the material are to develop and give pupils experience in

solving easy unfamiliar problems.

a number of specific strategic skills, with practice in trying these strategies on a range of harder problems, many of which respond to them.

reflecting on, discussing and explaining in writing both their approach to the problem, and their discoveries.

The classroom material is organised as three Units (A, B and C, each of which is intended to support roughly one week's work), together with a collection of problems providing supplementary material for the quicker student, or for revision. Through the three Units, the guidance provided to the pupils is gradually increased so that by the end they are facing challenges similar to those that the examination questions will present.

Unit A consists of a series of worksheets based around a set of problems, which aim to teach a number of powerful problem solving strategies, and show their "pay off".

Unit B gives the pupil less guidance, now in the form of "checklists" which contain a list of strategic hints. It is intended that these "checklists" should only be offered to pupils who are in considerable difficulty, or later as a stimulus for reflective discussion. The problems in this Unit respond to similar strategies to those introduced in Unit A, but begin to vary in style. In particular, one task involves the strategic analysis of a simple game.

Unit C is built around three tasks which differ in style, but which again respond to similar problem solving strategies. No printed guidance is offered to pupils, but the teacher has a "checklist" of strategic hints which may be offered orally to pupils in difficulty.

The Problem Collection provides supplementary material at any stage for those pupils who move rapidly—though because many of the problems in the Units are open and allow various extensions, able pupils will continue to find challenges in them if encouraged to do so.

In this Module all pupil materials are "framed" and it is assumed that calculators will be available throughout.

Notes for the teacher in each Unit provide specific teaching suggestions. Inside the back cover of the Module Book is a Checklist of suggestions on managing and coaching open learning activities. These were found by the teachers involved in the trials to be particularly helpful. As was emphasised earlier, all the teaching suggestions are offered in the recognition that every teacher will work in their classroom in their own individual way. The trials of the material established that teachers found it helpful to have explicit detailed suggestions from which they could choose and modify. All the material contained in this leaflet and given in the Module Book has been used in a representative range of classrooms and has proved to be effective in developing the skills that are the concern of this Module.

The following extracts from Unit A and Unit B show the style of the materials that have been developed.

An Item from Unit A

A1   ORGANISING PROBLEMS

The Tournament

A tournament is being arranged. 22 teams have entered. The competition will be on a league basis, where every team will play all the other teams twice—once at home and once away. The organiser wants to know how many matches will be involved.

Often problems like this are too hard to solve immediately. If you get stuck with a problem, it often helps if you first **try some simple cases.**

So, suppose we have only 4 teams instead of 22.

Next, if you can **find a helpful diagram**, (table, chart or similar), it will help you to **organise the information systematically.**

For example,

Avoid this   Be organised   or, better
               like this   still, like this

1

By now, you should be able to see that our 4 teams require 12 matches.

* How many matches will 6 teams require?
  How many matches will 7 teams require?
  Invent and do more questions like these.

* **Make a table** of your results. This is another key strategy . . .

Number of Teams	4	6	7			
Number of Matches	12					

* Try to **spot patterns** in your table.
  Write down what they are.
  (If you can't do this, check through your working, reorganise your information, or produce more examples.)

* Now can you **use your patterns** to solve the original problem with 22 teams?

* Try to **find a general rule** which tells you the number of matches needed for *any* number of teams. Write down your rule *in words* and, if you can, by a formula.

  Check that your rule always works.

* **Explain** why your rule works.

2

Now use the key strategies . . .

Try some simple cases

Find a helpful diagram

Organise systematically

Make a table

Spot patterns

Use the patterns

Find a general rule

Explain why it works

Check regularly

to solve the problems on the next page . . .

3

Mystic Rose

This diagram has been made by connecting all 18 points on the circle to each other with straight lines. Every point is connected to every other point . . .
How many straight lines are there?

Money

Suppose you have the following 7 coins in your pocket . . .

　　　　1p, 2p, 5p, 10p, 20p, 50p, £1

How many different sums of money can you make?

4

A1 Organizing Problems

The aim is to introduce pupils to specific strategic skills which are often helpful when attacking unfamiliar problems.

Length of time needed 1 hour (approx.)

Suggested Presentation

1 In order to involve the class, describe the problem of the "tournament" orally, and if possible, relate it to a tournament with which the pupils are themselves familier.

2 Now either issue the booklet and ask the pupils to work through it individually or in pairs, or, alternatively, ask the class to describe how they would set about solving the problem, and follow up a few of their suggestions before starting the booklet.

3 Invite the pupils to tackle at least one of the two problems at the end of the worksheet. The "Mystic Rose" generates a quadratic sequence closely related to the "tournament" sequence while the "Money" situation generates an exponential sequence. This is a more difficult problem, and is studied in further detail in A2: do not allow children to struggle with it too long before moving on to A2. Note that in the Mystic Rose, the number of lines does not depend on the equal spacing of points around the circle, (although the symmetry of the pattern does).

4 While the pupils work, try to help by giving *strategic* hints and encouragement rather than by leading them through the problem with specific, detailed instructions. For example, "Have you tried a simpler case? How can you make it simpler?" Is much better than "Try using 2, 3, then 4 points round the circle". (Only give more detailed help if strategic hints *repeatedly* fail). Examples of strategic hints are given inside the back cover of this Module. Pupils will need much encouragement in these early stages, until they can make these strategies part of their own toolkit. When arithmetic slips occur, point out their mistake. Technical skills like these are not our major centre of concern. When pupils are following a line of reasoning that is unfamiliar to you, or apparently unfruitful, allow them to pursue it and see what happens. They may surprise you!

5 Towards the end of each problem, encourage the pupils to generalise their results giving a rule in words or perhaps even an algebraic formula. Expressing a pattern as an algebraic formula is a very difficult mathematical activity. Pupils may be greatly helped towards it if they are encouraged to write their rules in words and explain them, as an intermediate step, rather than to attempt to translate the patterns directly into algebra. It is easy to underestimate just how difficult this is: below we give some examples of responses of pupils to the problem of the "tournament".

I can spot a pattern in my table. If you square the top number, then you take the top number away from the result, which gives you the bottom number in the table. So 4 x 4 = 16 - 4 = 12. (which is correct).

The pattern I notice is that the number multiplied by itself and then minus itself = the bottom number.

This is because if you make a square of the no. of teams and then cross

eg. 3 teams =

out the 'leading diagonal' which represents the matches played between similar teams, the no. of matches which you have to cross out will equal the no. of teams that there are.

$22 \times 22 = 484$. $484 - 22 = 462$ $x \times x - x = y$.

---

6 teams need to play 30 matches
7 " " " " 42 "
8 " " " " 56 "

Instead of drawing the box you could use a formula $6 - 1 = (5)$ $6 \times 5 = 30$
So $T - 1 = X$, $T \times x = m$
$= T(T-1)$

$. 22 (22-1)$
$= 22 (21)$
$= 462$ matches

The rule is so because although all the teams play each other home and away, they don't play themselves and so for each team playing you have to subtract 1 match as they can't play themselves.
e.g. 7 teams playing
$7 \times 7 = 49$ matches
7 teams can't play themselves so $49 - 7 = 42 = 42$ matches.

357

6   Emphasise the importance of *checking* that the rule always works. Take the opportunity to show that a variety of rules work and can be developed into equivalent algebraic expressions.

7   A suitable conclusion to the lesson may be to discuss how useful the pupils found the strategies, and also if they have discovered any new ones which could be added to the list. For example, the pupil below has discovered the strategy of generating further cases from simpler cases rather than by starting afresh each time:

Four teams would require 12 matches
Five teams would require 20 matches
Six teams would require 30 matches
Seven teams would require 42 matches

Number of teams	2	3	4	5	6	7	A
Number of matches.	2	6	12	20	30	42	B

THE "FIRST TO 100" GAME

This is a game for two players.

Players take turns to choose any whole number from 1 to 10.

They keep a running total of all the chosen numbers.

The first player to make this total reach exactly 100 wins.

Sample Game:

Player 1's choice	Player 2's choice	Running Total
10		10
	5	15
8		23
	8	31
2		33
	9	42
9		51
	9	60
8		68
	9	77
9		86
	10	96
4		100

So Player 1 wins!

Play the game a few times with your neighbour.

Can you find a winning strategy?
*   Try to modify the game in some way, e.g.:
     — suppose the first to 100 *loses* and overshooting is not allowed.
     — suppose you can only choose a number between 5 and 10.

THE "FIRST TO 100" GAME . . . PUPIL'S CHECKLIST

Try some simple cases	* Simplify the game in some way: e.g.:—play "First to 20" e.g.:—choose numbers from 1 to 5 e.g.:—just play the end of a game
Be systematic	* Don't just play randomly! * Are there good or bad choices? Why?
Spot patterns	* Are there *any* positions from which you can always win? * Are there other positions from which you can always reach these winning positions?
Find a rule	* Write down a description of "how to always win this game". Explain why you are sure it works. * Extend your rule so that it applies to the "First to 100" version.
Check your rule	* Try to beat somebody who is playing according to your rule. * Can you convince *them* that it always works?
Change the game in some way	* Can you adapt your rule for playing a new game where: — the first to 100 *loses*, (overshooting is not allowed) — you can only choose numbers between 5 and 10. — . . .

Support Materials

These Support Materials are divided into two parts—those that are included as part of the Module Book and those that accompany the video tape and microcomputer programs in the Resource Pack described below. Both are written under the same chapter headings:

1. Looking at lessons
2. Experiencing problem solving
3. How much support do children need?
4. How can the micro help?
5. Assessing problem solving

They will refer to each other and are usable together or independently.

This element in the Module aims to help in the following ways.

(a) It identifies those aspects of teaching style which are helping in developing children's problem solving skills.

(b) It offers a straightforward supportive way for teachers to explore these possibilities in their classroom and in discussion with their colleagues.

The aim is to go beyond the teaching suggestions made in the classroom materials and also to help teachers promote a certain amount of problem solving in their classrooms beyond the immediate context of this Module.

Each of the five chapters suggest activities—some only involve the teacher in looking at the material, some suggest trying something with another class, while others require a few teachers to get together for discussion.

A Resource Pack

The following material is available in the form of a Resource Pack (assembled in a presentation box).

(a) The Module Book *Problems with Patterns and Numbers*.

(b) A set of "master" sheets for photocopying for use by pupils and teachers.

(c) A video tape of 60 minutes' duration of extracts from problem-solving sessions in class using material from the Module with different teachers and groups of pupils, together with an explanatory booklet.

(d) A microcomputer disc containing five programs, together with explanatory booklets for the SNOOK and PIRATES programs included in the disc.

In addition, the Module Book *Problems with Patterns and Numbers* and the "master" sheets for photocopying are available as individual items.

A combined descriptive leaflet and order form is enclosed. Further copies are available from the Secretary.

References

Beeby, Terry, Hugh Burkhardt, and Rosemary Fraser. 1980. SCAN – Systematic Classroom Analysis Notation. Shell Centre.

Bell, A.W. 1982a. "Diagnosing Students' Misconceptions." Australian Mathematics Teacher 38 (1).

Bell, A.W. 1982b. "Treating Students' Misconceptions." Australian Mathematics Teacher 38 (2).

Bell, A.W. 1983. "Diagnostic Teaching." Zentralblatt der Didaktik der Mathematik 83 (2).

Bell, A.W. 1984. "Structures, Contexts and Learning." Journal of Structural Learning.

Bell, A.W., M. Swan, and G. Taylor. 1981. "Choice of Operation in Verbal Problems." Educational Studies in Mathematics.

Burkhardt, Hugh. 1981. "The Real World and Mathematics." Blackie.

Burkhardt, Hugh. 1982. "How Might We Move the Curriculum." 1982 BSPLM Oxford Conference. Shell Centre.

Burkhardt, Hugh. 1984. "How Can Micros Help in Schools: Research Evidence." Shell Centre.

Burkhardt, Hugh. 1985. "The Microcomputer: Miracle or Menace in Mathematical Education." In Proceedings of ICME 5, Marjorie Carss, ed. Birkhauser; revised version from Shell Centre.

Burkhardt, Hugh, Rosemary Fraser, et al. 1982. "Design and Development of Programs as Teaching Material." Council for Educational Technology.

Cockcroft Committee of Inquiry. 1982. Mathematics Counts. Report of the Cockcroft Committee of Inquiry. HMSO.

Fraser, Rosemary, Hugh Burkhardt, Jon Coupland, Richard Phillips, David Pimm, and Jim Ridgway. 1983. "Learning Activities and Classroom Roles." Shell Centre. To be published in Journal of Mathematical Behavior.

Hart, Kath (ed.). 1981. Children's Understanding of Mathematics, pp. 11-16. John Murray.

HMI Secondary Survey. 1977. "Aspects of Secondary Education." Report of the HMI Secondary Survey. HMSO.

ITMA. 1985. A catalogue of microcomputer resources by members of the ITMA Collaboration and associated groups and individuals. Shell Centre and College of St. Mark and St. John. Plymouth.

Shell Centre. 1984a. "Mathematical Education Review 1980-84." Shell Centre, Nottingham.

Shell Centre. 1984b. "Problems with Patterns and Numbers." Module in Testing Strategic Skills Program. Joint Matriculation Board, Manchester/Dale Seymour. Also available "The Language of Functions and Graphs."

Understanding and Teaching
the Nature of Mathematical Thinking[1]

Alan H. Schoenfeld
University of California at Berkeley

Overview

As suggested by its title, this paper focuses on the following two questions:

1. What are the most important dimensions of "thinking mathematically"? That is, can one identify fundamental aspects of mathematical knowledge and behavior present in those people who do well at mathematics (perhaps explaining a substantial proportion of their success) and often lacking in those who do not do well?

2. Having identified fundamental aspects of mathematical thinking, can one develop forms of mathematics instruction to teach such skills?

The bulk of my own work since the early 1970's, and a substantial amount of work in the mathematics education community (under the name of "problem-solving": see for example Krulik 1980 and Shumway 1980), has been devoted to these questions. This brief paper will summarize some of the results obtained over that period of time. A more extensive description of the research results and a framework for analyzing mathematical behavior may be found in my 1985 book Mathematical Problem Solving (hereafter called MPS); a more extensive discussion of practical suggestions for the classroom may be found in my 1983 book Problem Solving in the Mathematics Curriculum.

In what follows, it is assumed that the term "problems" refers to tasks that are truly problematic for the individuals who work them. That is, it is assumed that the individuals do not have at their disposal routine procedures for dealing with the tasks at hand. Where they do have such procedures we speak of exercises. Exercises are important, but they are not the focus of this discussion.

The most important idea underlying the work described here, the idea that distinguishes recent work on understanding and teaching thinking skills in complex domains from earlier work, is the idea of focusing on cognitive processes during problem-solving. A "product" approach to mathematical thinking concentrates on

362

identifying what the individual knows, generally by examining the final written solutions the individual produces to problems, with an emphasis on the correctness of the mathematical procedures employed. Obviously, the correctness of the approach an individual finally settles upon is vitally important. In process-oriented research, however, the scope of inquiry is far broader. The goal is to characterize as accurately as possible what an individual has done while trying to solve a particular problem. Typical research questions are the following. What kinds of knowledge were potentially accessible for that person to use? What knowledge did the individual choose to use and why? How effectively and efficiently was that knowledge used? What contributed to the ultimate success or failure of the problem-solving attempt?

This paper is organized along the lines suggested by Figure 1, which categorizes four dimensions of mathematical behavior: (1) cognitive resources and basic knowledge; (2) heuristics and other problem-solving strategies; (3) control or metacognitive behavior; and (4) belief systems. The next four sections of this paper are devoted to these four categories. Each section begins with a brief description of the nature of the category and the effects of behavior in that category on mathematical performance. It then turns to practical suggestions for dealing with that category in the classroom. A final section describes some of the ways that problem-solving instruction can affect students' mathematical behavior, and makes some summary recommendations regarding the nature of instruction designed to teach students how to think mathematically.

FOUR CATEGORIES OF KNOWLEDGE AND BEHAVIOR
NECESSARY FOR AN ADEQUATE CHARACTERIZATION
OF MATHEMATICAL PROBLEM-SOLVING PERFORMANCE

COGNITIVE RESOURCES: The mathematical knowledge possessed by the individual that can be brought to bear on the problem at hand.

— Intuitions and informal knowledge regarding the domain
— Facts
— Algorithmic procedures
— "Routine" non-algorithmic procedures
— Understandings ("propositional knowledge") about agreed-upon rules for working in the domain

HEURISTICS: Strategies and techniques for making progress on unfamiliar or non-standard problems; "rules of thumb" for effective problem-solving, including:

— Drawing figures; introducing suitable notation
— Exploiting related problems
— Reformulating problems; working backwards
— Testing and verification procedures

CONTROL: Global decisions regarding the selection and implementation of resources and strategies.

— Planning
— Monitoring and assessment
— Decisionmaking
— Conscious metacognitive acts

BELIEF SYSTEMS: One's "mathematical world view," the set of (not necessarily conscious) determinants of an individual's behavior.

— About oneself
— About the environment
— About the topic
— About mathematics

Figure 1
(Adapted from Mathematical Problem Solving)

Cognitive Resources and Basic Knowledge

The mathematics you know — the facts and procedures you have at your disposal when working problems — is the foundation upon which problem-solving performance is built. This section will focus on two findings about resources: the role of problem perception and problem representation in knowledge access, and the role of flawed or "buggy" performance.

Problem Perception and the Knowledge Base

The literature on "skill acquisition" makes it quite clear that highly competent performance in any complex domain — even domains in which "strategy" is considered quite important — is based on having at one's disposal a large number of pieces of almost immediately accessible knowledge. One sees a situation, recognizes it, and brings to bear the relevant information or tools for dealing with it. In reading, for example, we respond to entire words and phrases all at once without pausing to decipher individual letters. The words and phrases are called "chunks." In chess, a chunk may be a complicated configuration of pieces placed on the board. Chess masters recognize such configurations as single units and respond to them almost immediately — just as we respond automatically to the letters S, T, O, and P on a red octagonal sign that we perceive while driving (see, e.g., de Groot 1966). Simon (1980) estimates that experts in complex domains have "vocabularies" consisting of about 50,000 chunks of knowledge. Thus a large part of their expertise consists of routine, almost automatic responses to familiar situations.

This kind of perception also takes place in certain kinds of problem situations. Work by Hinsley, Hayes, and Simon (1977) indicate that after reading just the first few words of a problem statement, for example, "A river steamer...," subjects are already beginning to characterize the problem as a standard algebra problem that may be solved by the use of simultaneous linear equations. This initial categorization of the problem orients the individuals to the task at hand and brings to mind the relevant solution techniques. This is a general finding which has been replicated in other fields. In brief, stereotypical circumstances evoke stereotypical responses. Once they recognize the "generic" aspects of certain situations, competent problem solvers have almost automatic access to the appropriate procedures.

Research on problem perception in physics (Chi, Feltovich, and Glaser 1981) and in mathematics (MPS, Chapter 8) indicates the two-sided nature of this phenomenon. When problem perception takes place according to the appropriate criteria, problem-solving performance is enhanced. This is often the case with regard to "expert" solutions of problems. When problem perception takes place according to inappropriate criteria, severe difficulties can ensue. Misperceiving a problem may mislead the problem solver, resulting in the choice of an inappropriate solution approach or a wild goose chase. The research in both physics and mathematics suggests that students' problem perceptions are often based on superficial ("surface") characteristics of the problems, and that this frequently causes them difficulties in problem-solving.

The instructional "moral" drawn from this research is neither earthshaking nor new, but it is important. As Pólya (1945) admonishes, "First, you have to understand the problem." (p. xvii) The research indicates that experts spend more time doing so than novices, and that doing so pays off: Only when you know what the problem is really about should you begin working out its solution. The research in mathematics (MPS, Chapter 8) indicates that students can be successfully trained to analyze problems with care. When they do so, their problem perceptions, and their performance, become more expert-like.

The Role of Flawed or "Buggy" Performance

It has been known for some time that students tend to make certain kinds of mistakes with consistency. For example, Thorndike et al. (1926) document systematic error patterns in students' algebraic work. Recent work on error analysis has carried that kind of work forward in a number of mathematical domains.

The most detailed work has been done in elementary arithmetic, in particular the body of work represented by Brown and Burton's 1978 paper. The underlying notion is that many of the errors that students commit are not merely the result of a failure to comprehend — that is, slips of some sort — but rather the result of the consistent misapplication of a misunderstood rule. That is, the errors are like the "bugs" in computer programs, consistently producing (predictably) incorrect answers. Brown and Burton's analyses for subtraction resulted in the creation of a diagnostic test that allows one to predict, roughly half the time, not only which problems a particular student will solve incorrectly, but the precise digits of the incorrect answer the student will produce. Matz (1982) undertook a similar analysis for algebra, looking for consistent causes of incorrect answers in algebraic manipulations. Rosnick and Clement (1980) examine students' remarkably persistent difficulties with very simple word problems.

A very strong pedagogical message emerges from this cumulative body of studies. The prevailing, naive pedagogical model assumes the student is an "empty jug" waiting to be filled with knowledge, or a "blank slate" waiting to be written upon. According to this naive model, a student's failure to learn a particular technique is taken as a lack of mastery: the student hasn't "gotten it" yet. With this view of learning, the appropriate action to take is to show the student the procedure once again — carefully and slowly, until the student masters it. According to the view of the learning process derived from the studies mentioned above, the naive "show the students the procedure until they get it" approach is insufficient. The moral of those studies is that the students did get something — and that the something they got is consistent, but incorrect. For some reason they misinterpreted the procedure and learned an incorrect version of it — and now act quite consistently when asked to implement it. Simply showing the students the correct procedure once again may not do any good at all. Having already incorrectly interpreted the procedure, the student is not very likely to suddenly "see" the right way to do things. (By way of analogy, think of the difficulties music students have "unlearning" bad techniques on their instruments.) The research suggests that there should be a much greater focus in the classroom on elucidating students' interpretations of procedures — with an eye towards "debugging" incorrect ones — since the returns on repeated presentations of the "right" versions of the procedures are likely to diminish rapidly after a point.

Heuristics and Other Problem-Solving Strategies

In mathematics education the phrase "problem-solving strategies" is almost synonymous with "heuristics." First in How to Solve It and later in greater depth in Mathematical Discovery and Mathematics and Plausible Reasoning, George Pólya set out to describe the "mental operations typically useful for the solution of problems." Pólya's heuristics, rules of thumb for making progress when one is having difficulty with a problem, include the following: exploiting analogies, introducing and exploiting auxiliary elements in a problem solution, arguing by contradiction, working forwards, decomposing and recombining, examining special cases, exploiting related problems, drawing figures, and working backwards. There is some degree of consensus among mathematicians that these strategies are useful, and that mathematicians themselves tend to use them. Pólya's influence in mathematics education has been tremendous, and during the 1960's and 1970's there were a fair number of studies designed to test the hypothesis that teaching students to use problem-solving heuristics would improve their problem-solving performance. By and large, those studies produced disappointing results.

To sum things up briefly, one of the primary reasons for the lack of success in problem-solving instruction through heuristics is that the strategies themselves were very much underspecified. It is one thing to describe a strategy so that competent problem solvers can recognize it, and quite something else to describe a strategy in sufficient detail so that students can use it. It turns out that most of the "strategies" in How to Solve It are not really well-defined entities, but rather descriptive labels that subsume whole classes of more precisely defined strategies. For example, Chapter 3 of MPS provides a detailed examination of the following relatively straightforward heuristic strategy, "examining special cases."

S: To better understand an unfamiliar problem, you may wish to exemplify the problem by considering various special cases. This may suggest the direction of, or perhaps the plausibility of, a solution.

The use of this strategy is examined in the solution of five different problems, after which it becomes apparent that each problem is an archetype for a class of problems amenable to solution by a particular version of **S.** The description of **S** is merely a summary description of a number of closely related

368

strategies, each with its own particular characteristics. Five more precise (but still fairly broad) strategies of this type follow.

S_1: If there is an integer parameter n in a problem statement, it might be appropriate to calculate the special cases when n = 1, 2, 3, 4 (and maybe a few more). One may see a pattern that suggests an answer, which can be verified by induction. The calculations themselves may suggest the inductive mechanism.

S_2: One can gain insight into questions about the roots of complex algebraic expressions by choosing as special cases those expressions whose roots are easy to keep track of (e.g., easily factored polynomials with integer roots).

S_3: In iterated computations or recursions, substituting particular values such as 0 and/or 1 often allows one to see patterns (assuming the substitution causes no loss of generality). Such special cases allow one to observe regularities that might otherwise be obscured by a morass of symbols.

S_4: When dealing with geometric figures, one should first examine special cases with minimal complexity. Consider regular polygons, for example, or isosceles or right or equilateral rather than "general" triangles; or semi- or quarter-circles rather than arbitrary sectors, etc.

S_{5_a}: For geometric arguments, convenient values for computation can often be chosen "without loss of generality" (e.g., setting the radius of an arbitrary circle equal to 1). Such special cases make subsequent computations much easier.

S_{5_b}: Calculating (or when easier, approximating) values over a range of cases may suggest the nature of an extremum which, once "determined" in this way, may be justified in a variety of ways. Special cases of symmetric objects are often prime candidates for examination.

One cannot legitimately expect students to be able to implement strategies as nebulously defined as **S**. One needs to delineate the constituent substrategies and explain their use. That brings us to the second major issue regarding heuristic strategies — describing different ways of implementing the strategies in sufficient detail so that students can learn to implement them. The complexity of the implementation process has also been greatly underestimated. For example, the successful use of the strategy of "exploiting a simpler and related problem" depends on (a) thinking of using the strategy in the first place; (b) knowing which specific version of the strategy is appropriate for the current problem; (c) being able to generate <u>appropriate and potentially useful</u> simpler and related problems, using the chosen substrategy; (d) selecting a particular related and simpler problem to exploit, based on its potential utility; (e) solving that problem; and (f) being able to exploit its solution. This is decidedly nontrivial, and the fact that it was not taught explains the failure of most heuristic instruction.

A third issue concerns what might be called "strategy selection." If one has a small collection of powerful techniques that might be used on a new problem — say ten or so — then one may well find the right technique by trial and error or some sort of intuition. This would appear to be what happens when dealing with broad strategies such as the heuristics described in <u>How to Solve It.</u> With the detailed specification of those strategies as suggested above, however, we encounter a new difficulty. If each heuristic label represents a class of a dozen or so more finely defined strategies, then any collection of ten powerful techniques becomes a collection of over 100 specific strategies. This issue of knowing which strategies to use under what circumstances — and of when to give up on strategies that are not proving to be useful — becomes far more critical. This is part of the issue of "control," discussed in the next section.

The morals for pedagogy in the preceding are clear. To put things in simplest form: If you want students to be able to use a problem-solving strategy, you have to specify the strategy in sufficient detail so that they will be able to follow your instructions. (This specification includes training students to recognize when the particular strategy is likely to be useful.) Then you have to provide them with a substantial amount of practice in using the strategy — at first when they know it is the right strategy, so that they learn how to implement it — but later on in collections of assorted problems that are given to them out of context, so that they get practice in strategy selection as well as in strategy implementation. This is the "bad news": teaching heuristic strategies is far more difficult that it would at first appear. The "good news" is that teaching the strategies in this way (being mindful of the pedagogical lessons in the other sections of this paper) does indeed work. There is clear evidence that students can learn to use such strategies — not only on problems isomorphic to ones on which they have been trained, and not only on problems that are non-isomorphic but solvable by similar methods, but on problems that are quite dissimilar. (One part of my final exam in a problem-solving course consisted of problems I had selected from a problem competition according to the following criterion: A problem was placed on the exam only if after reading it I had no reasonable idea how it might be solved. The students did remarkably well, compared both to their pre-course performance and compared to a control group.)

Control and Metacognition

The previous two categories, cognitive resources and heuristics, characterize what might be considered the problem solver's "tool kit." Identifying the facts, procedures, and problem-solving strategies that are accessible to an individual provides an inventory of the knowledge the individual is capable of bringing to bear in a problem situation. However, what a person "knows" and how that person uses that knowledge can be two very different things. The ways that problem-solving approaches are chosen or ignored — and the paths students pursued or abandoned during the problem-solving process — are major determinants of success and failure.

"Control" and "metacognition" deal with what might be called resource allocation during problem-solving. In brief, how effectively does the individual use the knowledge potentially at his or her disposal? Examples of actions at the control level include (a) making certain you understand what a problem really calls for before jumping into a solution; (b) considering various ways to solve a problem and selecting a particular method based on an (obviously subjective) assessment of its potential usefulness; (c) monitoring progress "on line" and deciding to abandon a particular approach because it does not appear to be providing useful results; and (d) checking results after they have been obtained. Two major reviews by Ann Brown (1978; Brown et al. 1983) provide a general introduction to the psychological literature on the topic.

Colloquially speaking, one of the major difficulties students have during problem-solving is that they frequently go off on "wild goose chases." The following example describes rather typical behavior. Two students were asked to work as a team on a problem that asked them to characterize the largest triangle that could be inscribed in a given circle. Both students had done well in a multi-variate calculus course and were familiar with the analytical techniques that yield a solution to the problem. In fact, they had solved a comparably difficult problem on their final examination in the calculus course just a week before they were videotaped working this problem. The students read the problem, conjectured that the answer to the problem must be the equilateral triangle, and decided to calculate the area of the equilateral triangle. With no discussion of the merits of that approach, they began to work, and at once got bogged down in the calculations. On two or three occasions later on during the solution attempt, they slowed down and took a breather when they ran into difficulties, and then started up the calculations once again — never pausing to examine other possibilities, or even to question the utility of what they were doing. They wound up spending the total allotted time of twenty minutes engaged in calculating the area of the triangle; the videotape cassette clicked off while they were still doing so. At that point they were told the area of the equilateral triangle and asked in what way a knowledge of that number could have helped them arrived at a solution of the original problem — i.e., why they had spent 20 minutes seeking it. They were unable to say.

While this description may seem anecdotal, a rigorous research framework lies behind the analysis (see Chapters 4 and 9 of MPS for details). Moreover, this behavior is all too typical. Roughly 60% of the problem sessions that were videotaped and analyzed for that research demonstrated a similar pattern of behavior. Students who were familiar with the relevant mathematical techniques for solving a particular problem failed to use them because they chose to pursue inappropriate solution paths, did not curtail the inappropriate explorations, and did not exploit what they knew.

In general, the issue of overseeing one's problem-solving as one works on a problem — of checking to see that things are reasonable as one works along — goes unaddressed in mathematics instruction. Observations of competent problem solvers make it clear that it is possible to monitor and assess one's work "on line," and that doing so contributes substantially to problem-solving success. However, students are for the most part completely unaware of this aspect of problem-solving behavior. Asked what they do when they work problems, they will frequently say that they do "what comes to mind" — as though their minds were independent, autonomous entities over which they had no control. For that reason my courses in mathematical thinking focus extensively on decisionmaking processes during problem-solving. The idea is to make students aware of the importance of such decisionmaking and to train them to observe and intervene in their own problem-solving activities. The classroom sessions deal with this issue in four different ways.

Watching Videotapes of Problem Sessions

My research library includes a number of videotapes of students whose work clearly suffers because they make poor decisions at the control level. At the appropriate point in the problem-solving course, I show one or two of these videotapes to the students. Usually their reactions to the tapes are a combination of frustration ("Why are they doing that? It's so obviously wrong! Don't they realize they should stop that, and try something else?") and empathy ("I feel really sorry for them. That could be me."). That is exactly as it should be. In the discussions that follow, we discuss the fact that they do behave similarly — but that by standing back from their own work, as they did in observing the videotapes, they can see when their work is being unproductive. Once they are aware that their time is not being spent profitably, they can stop their wild goose chases and (try to) do something more useful.

Explicitly Modeling the Decisionmaking Process

It is difficult, but important, to resist the urge to present only "neat" solutions to problems. These tidy solutions present the product of our thinking, but they hide the processes that yielded them. Somewhat infrequently — but often enough so that the students get the point — I take students through a simulation of the whole problem-solving process, much like a blow-by-blow account — even though I know the answer to the problem, and could present the solution in a fraction of the time. In simulating a solution from scratch, I will: (a) make sure I know what the problem calls for, perhaps working a few cases to make certain I understand it; (b) raise a number of possible ways to solve the problem, and decide which to pursue; (c) review, after a few minutes have elapsed, what I have done so far and see whether that approach is still worth pursuing; and so on. Heuristic usage is overtly modeled in similar fashion. (Note: This approach can be tedious if used too often, but used sparingly it makes an important point.)

Serving as "Control" While the Class Works a Problem

Roughly one-third of the time in my problem-solving class is spent working problems as a group. When we do so I reserve for myself the role of "executive" whose job it is to check that things are going reasonably well. The sessions proceed as follows. I pose a problem and ask for suggestions. Frequently a first suggestion ("Why don't we do X?") is made within a few seconds. In that case — whether or not X seems to me to be a reasonable suggestion — I will ask if everyone understands the problem before we consider implementing X. If, as is frequently the case, the answer is no, we explore the problem for a while. When, as is also frequently the case, understanding the problem reveals X is not useful, this provides an opportunity for a brief lecture on the importance of making certain one understands a problem before jumping into its solution. Even if X turns out to be reasonable, I ask for other suggestions. If Y and Z are suggested, I ask the class which of the three approaches it would like to pursue and why. We discuss the relative merits of the suggestions and choose one. Whether or not I know that approach to be profitable we pursue it — but after five minutes have passed I ask the class to evaluate our progress and decide whether we should continue working on it or return to one of the other approaches. We proceed in this way. In other words, I raise the "control" issues but the class makes the decisions.

Intervening in the Class's Work

About one-third of the time, my problem-solving class is broken into small working groups of four or five students who work together on problems while I move through the class as a roving "consultant." During this phase of problem-solving I reserve the right to stop the students at any time and ask the following three questions:

What are you doing?

374

Why are you doing it?
How will you use the result when you are done?

Early in the class, the students are unable to answer these questions and are embarrassed when asked to do so. As a result, they start to defend themselves against the questions by preparing answers to them. By the end of the semester this behavior has become habitual, and the questions have served their purpose.

Belief Systems

The idea here is that students' understandings regarding the nature of mathematics establish the psychological context within which they do mathematics — and in consequence, these understandings shape the students' mathematical behavior. The results can often have strong negative effects on performance. For reasons of space limitation, the discussion here will be telegraphic. Two examples of beliefs and their impact will be discussed.

In a seminar I asked the fifteen very bright undergraduates who were present to use techniques of high school geometry to prove that in Figure 2a, the lengths of segments PV and QV are the same and that segment CV bisects angle PVQ. The students produced a correct proof in less than three minutes. I then asked the students to tell me how to contruct in Figure 2b the circle tangent to both lines with point P as the point of tangency to the top line. Students made the following conjectures, in order:

1. The center of the desired circle lies at the midpoint of PQ, where Q is the point on the bottom line that lies the same distance from V as does P (see Figure 2c).

2. The center of the desired circle lies at the midpoint of the segment of the arc drawn through P that lies between the two given lines (see Figure 2d).

3. The center of the desired circle lies at the midpoint of the segment of the perpendicular drawn through P and between the two given lines (see Figure 2e).

4. The center of the desired circle lies at the intersection of the perpendicular drawn through P and the bisector of vertex angle V (see Figure 2f).

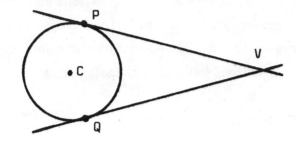

Fig. 2a

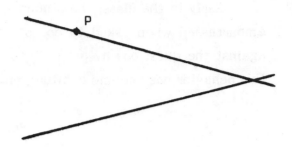

Fig. 2b

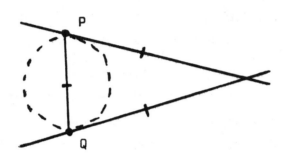

Fig. 2c

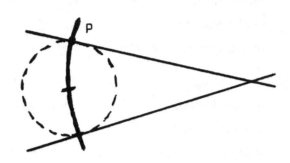

Fig. 2d

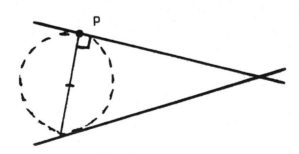

Fig. 2e

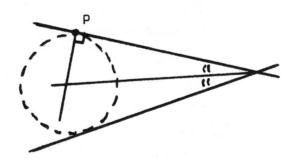

Fig. 2f

Asked which of the conjectures was correct, the students argued on purely empirical grounds (that is, which looked best) for about ten minutes — despite the fact that they had just produced the proof that resolved the issue and that the proof was still on the board. Why? It may appear to be an overstatement, but I will claim the students were blind to the results of the proof because the construction problem asked them to discover something — and part of the students' belief system is that proof has nothing to do with discovery. (For a more careful statement of this belief, and substantial documentation of it, see Chapters 5 and 10 of MPS.) A second belief commonly held by students, stated in equally brutal form, is the following: "All problems can either be solved in less than ten minutes, or they are impossible." This belief is an extrapolation of students' extensive experience with classroom and homework exercises, where the so-called "problems" are intended to be solved in a few minutes. (If you can't do so, it means you didn't understand the lesson.) As in the example above, there is an important behavioral corollary: Students will simply give up after having spent ten minutes trying to solve a problem.

How does one deal with such beliefs? In brief, by bringing them out into the open and challenging them. To deal with the belief that geometric proof is irrelevant to discovery and invention, I assigned students in the problem-solving class a series of problems that were "unlocked" by proof arguments, for example: "Why does the construction for an angle bisector work?" The class worked on the problem for about a half-hour. It concluded, without any prompting from me, that the sequence of compass markings created two congruent triangles — and therefore that the line between them, the line between corresponding (and therefore equal) angles, was guaranteed to be an angle bisector. After the solution was agreed upon, a student raised his hand and asked: "Are you trying to tell us that congruence is good for something?" This provided the opportunity for an open discussion of their conception of the role of proof and of how mathematical argumentation could actually be useful. Discussions that were similar in spirit took place over other problems I assigned, including problems that took the whole class a week or two to solve. (Moral: some problems can't be solved in ten minutes, but may well yield if you work on them for a week or two.)

Conclusion

A major thrust of this article has been that "thinking mathematically" means more than mastering mathematical facts and procedures. It includes having a repertoire of strategies for approaching new problems, being able to use one's mathematical knowledge effectively, and approaching mathematical situations with what might be called a "mathematical point of view." As indicated by the brief discussions above, these aspects of mathematical thinking can be addressed in the classroom — and if they are addressed explicitly, substantial changes in students' mathematical behavior can result. As a result of problem-solving instruction, students' perceptions of problem structure can become more expert-like. Students can learn to use heuristic strategies, even on problems unrelated to those studied in the class. They can become far more efficient problem solvers as a result of improved behavior at the control level. And perhaps their point of view becomes a bit more "mathematical." For more detail regarding these claims, see chapters 6–9 of MPS.

Note

1. The research described in this paper was supported by a grant from the Spencer Foundation, which the author gratefully acknowledges.

References

Brown, Ann L. 1978. "Knowing When, Where, and How to Remember: A Problem of Metacognition." In Advances in Instructional Psychology 1, R. Glaser, ed. Hillsdale, NJ: Erlbaum.

Brown, Ann L., John Bransford, Robert Ferrara, and Joseph Campione. 1983. "Learning, Remembering, and Understanding." In Handbook of Child Psychology, P.H. Mussen, ed., vol. III (J.H. Flavell and E.M. Markham, vol. eds.). New York: Wiley.

Brown, J.S., and R.R. Burton. 1978. "Diagnostic Models for Procedural Bugs in Basic Mathematical Skills." Cognitive Science 2: 155–192.

Chi, M., P. Feltovich, and R. Glaser. 1981. "Categorization and Representation of Physics Problems by Experts and Novices." Cognitive Science 5: 121–152.

de Groot, Adriaan D. 1966. "Perception and Memory Versus Thought: Some Old Ideas and Recent Findings." In Problem Solving: Research, Method, and Theory, Benjamin Kleinmuntz, ed. New York: Wiley.

Hinsley, Dan A., John R. Hayes, and Herbert A. Simon. 1977. "From Words to Equations: Meaning and Representation in Algebra Word Problems." In Cognitive Processes in Comprehension, P.A. Carpenter and M.A. Just, eds. Hillsdale, NJ: Erlbaum.

Krulik, S., ed. 1980. Problem Solving in School Mathematics (1980 Yearbook of the National Council of Teachers of Mathematics). Reston, VA: NCTM.

Matz, M. "Towards a Process Model for High School Algebra Errors." 1982. In Intelligent Tutoring Systems, D. Sleeman and J.S. Brown, eds. London: Academic Press.

Pólya, G. 1945. How to Solve It. Princeton: Princeton University Press.

Pólya, G. 1954. Mathematics and Plausible Reasoning. 2 vols. Princeton: Princeton University Press.

Pólya, G. 1962–65. Mathematical Discovery. 2 vols. New York: Wiley. Combined paperback edition, New York: Wiley, 1980.

Rosnick, Peter, and John Clement. 1980. "Learning Without Understanding: The Effect of Tutoring Strategies on Algebra Misconceptions." Journal of Mathematical Behavior 3, no. 1: 3–27.

Schoenfeld, Alan H. 1985. Mathematical Problem Solving. New York: Academic Press.

Schoenfeld, Alan H. 1983. Problem Solving in the Mathematics Curriculum: A Report, Recommendations, and an Annotated Bibliography. Washington, DC: Mathematical Association of America.

Shumway, Richard, ed. 1980. Research in Mathematics Education. Reston, VA: National Council of Teachers of Mathematics.

Simon, Herbert A. 1980. "Problem Solving and Education." In Problem Solving and Education: Issues in Teaching and Research, D. Tuma and F. Reif, eds. Hillsdale, NJ: Erlbaum.

Thorndike, E., M. Cobb, J. Orleans, P. Wald, and E. Woodyard. 1926. The Psychology of Algebra. New York: Macmillan.

Statistics and Probability in the
School Mathematics Curriculum:
A Review of the ASA/NCTM/NSF
Quantitative Literacy Program
Richard L. Scheaffer
University of Florida

Introduction

Our nation's public schools are in disrepute. By almost every measure, including the commitment and competency of teachers and the achievement of students, the performance of our schools falls far short of expected goals. As a result, our once unchallenged preeminence in commerce, industry, science, and technological innovation is being overtaken by competitors throughout the world. To sustain a complex and competitive economy, the schools must ensure the availability of large numbers of well-trained and capable persons. A critical prerequisite is proficiency in the basic skills of reading, writing, and calculating.

The decline of the quantitative skills of our youth has raised great concern, thus the need for statistics and probability to reinforce the four elementary operations of addition, subtraction, multiplication, and division (including fractions and percentages).

The need to collect, organize, display, and analyze data arises throughout our society, and is no longer the sole concern of a few specialists. Everyone is involved, whether as citizen or as worker. Unfortunately, the American education system was developed before this quantitative literacy became a major requirement.

Quantitative literacy is a complex of skills for and knowledge of the collection, display, and analysis of data. It must be an integral part of an academic program that includes modern methods of computing. Quantitative literacy consists in familiarity with very concrete subjects such as raw data and methods of collecting it, as well as the abstract ideas underlying the modeling of statistical issues by means of probability theory. In short, quantitative literacy is as complex and pervasive as verbal literacy. One cannot create, enjoy, or build without using skills of quantitative literacy as well as verbal literacy.

The American Statistical Association (ASA) recognizes that in today's society, all educated persons must possess quantitative literacy if they are to perform intelligently as informed citizens and hold jobs in increasingly technically oriented businesses and industries. Further, it recognizes that basic aspects of this quantitative literacy must be taught in pre-college (K-12) mathematics and in the natural and social science curricula at both the elementary and secondary levels. The Association supports the philosophy of active training programs in the area of statistics and probability for elementary and secondary mathematics, natural and social sciences school teachers.

Accordingly, ASA with the endorsement of the National Council of Teachers of Mathematics (NCTM) and with funding from the National Science Foundation is engaged in the development and delivery of a program of continuing education for mathematics teachers to prepare them for teaching modern statistical and probabilistic concepts effectively. An integral part of the project will be the development and production of teaching materials that, it is hoped, will lead to more effective teacher training and better tools for use in the classroom than are now available.

Statement of Objectives

The project entitled "A Program to Improve Quantitative Literacy in the Schools," is to be developed over a three-year period. The goal is to realize the following objectives:

1. Develop alternative guidelines for the teaching of statistics and probability as part of the pre-college (K-12) curriculum (scope of topics to be introduced, their rationale, the relevant applications, and indications as to where and how to teach the topics).

2. Develop a model in-service program for training teachers in modern statistical and probabilistic concepts and in methods of presenting these concepts effectively.

3. Produce a set of curriculum materials for teachers to assist them in teaching basic topics in statistics and probability in a variety of settings

and formats and encourage further development, testing, and evaluation of other programs in the natural and social sciences modeled on the experiences of this program.

4. Develop a mechanism to evaluate the teaching of statistics and probability in terms of effectiveness in teaching content in specific areas and aiding the teaching of other basic skills (reading and arithmetic).

Background Information and Issues
The General Need for Statistics and Probability

We live in an information society and must deal with quantitative information at all levels of endeavor. Raw data, graphs, figures, rates, percentages, probabilities, averages, forecasts, and trend lines are an inescapable part of our everyday lives, affecting decisions of health, citizenship, parenthood, employment, financial concerns, sports, and other matters. Today a person must have a facility in dealing with data and in making intelligent decisions based on quantitative arguments. Teachers, and then students, must acquire these facilities if our society is to grow and prosper.

To understand the latest news poll, the Consumer Price Index, or this week's rating of television, books, and recordings, students must possess the skill of reading tables, charts, and graphs. They must know how to interpret these displays correctly and be familiar with techniques of communicating quantitative material without distortion or omission of essentials. Students must develop habits that allow them to check data sources for internal consistency, and consistency with related known facts.

Few results in the natural or social sciences are known absolutely. Many are reported in terms of chances, or probabilities; e.g., the chance of rain tomorrow, the chance of living past the age of 60, the chance of inheriting a certain trait, the chance of getting a certain disease (or recovering from it). Students must obtain a feel for the concept of variability and be able to link it to the real problems they encounter every day.

The Needs of Industry and Government for Statistically Trained Personnel

Modern industry is becoming increasingly technical as society moves further and further into the electronic age. The world economic climate has forced industries into highly competitive modes of operation. The demand for instantaneous measurements of complicated processes and precise information for purposes of decision-making has made a significant impact on the kinds of skills industry now requires of its employees. Workers at every level have to be capable of interpreting charts, graphs, tables, and summary data. Statistical work is basic to any effort of improving the productivity and quality of most industries and permeates all areas of design, development, and production. By virtue of its broad applicability, statistics is interdisciplinary and inherent to all areas of science where quantitative information is required for decision-making.

Government agencies are now called upon to monitor many complex areas of society, particularly in the natural and social sciences, and to use extensive data sets for routine operational decisions. This work places heavy demands upon their employees' ability to reason quantitatively. Statistical skills are needed at almost all job levels throughout virtually all governmental agencies.

The proposed program has received letters of documented need and support from both industry and government with respect to requirements for upgrading statistics and probability in the schools. Companies and agencies represented include:

Bell Laboratories

E.I. duPont de Nemours

Ford Motor Corporation

General Motors Corporation

Eli Lilly & Company

NASHUA Corporation

Bureau of the Census

Bureau of Justice Statistics

National Bureau of Standards

National Center for Education
  Statistics

Criminal Justice Statistics
  Association

The World Bank

Current Views on the Importance of Statistics and Probability

In April 1983, the Madison Commission on Excellence in Education reported that we are a "nation at risk," in that we are settling for mediocrity in education and that our students are insufficiently trained in mathematics, science, and related areas. The report of this commission makes five major points about the teaching of mathematics, one of which is that high school students must be capable of understanding probability and statistics.

The National Council of Teachers of Mathematics (NCTM) urges (in An Agenda for Action: Recommendations for School Mathematics of the 1980's) that for students to be problem solvers, they should study "methods of gathering, organizing, and interpreting information, drawing and testing inferences from data, and communicating results."

In A Position Paper on Basic Mathematical Skills (1976), the National Council of Supervisors of Mathematics described as important basic skills (1) ability to read and make simple tables, charts, and graphs; and (2) knowing how mathematics is used to find the likelihood of future events. Building on this position paper, NCTM in its An Agenda for Action called for increased emphasis on (1) locating and processing quantitative information; (2) collecting data; (3) organizing and presenting data; (4) interpreting data; and (5) drawing measures and predicting from data.

In summarizing a September 1982 conference, the Conference Board of the Mathematical Sciences, underscoring what is fundamental in school mathematics, recommended for elementary and middle school mathematics that "direct experience with the collection and analysis of data be provided for in the curriculum to ensure that every student becomes familiar with these important processes." For secondary school this summary recommends that statistics and probability (among other topics) be regarded as fundamental and that appropriate topics and techniques from statistics and probability be introduced into the curriculum.

The PRISM (Priorities in School Mathematics) report (1981) sampled six categories of professional educators and three categories of lay persons. As goals for instruction in probability and statistics, over 80% of those sampled favored the use of data in subjects in which consumers in the real world must ordinarily process statistical information and organize data in easily interpretable forms. Strong support was expressed for teaching the following topics: (1) collection and organization of data; (2) reading and interpretation of statistical information; and (3) measures of central tendency (secondary school). To motivate the teaching of probability and statistics, those questioned strongly supported the use of real-world data, the performance of experiments, and recourse to problems from the sciences.

The American Association for the Advancement of Science (AAAS) has expressed strong interest in the work of the ASA-NCTM Joint Committee and its effort to influence the amount and type of statistics taught in the schools. The Joint Committee arranged a session on statistics at the Youth Symposium of the 1983 AAAS Annual Meeting, and presented a session on statistics in the pre-college mathematics curriculum for the 1984 AAAS Annual Meeting.

The Need for Teacher Training

A glaring problem in the teaching of statistics in elementary and secondary schools is that teachers of mathematics are often unprepared to teach courses in the subject. Many teachers of mathematics are trained to teach algebra and geometry to a broad spectrum of students but are not trained to teach statistical topics to students at any level. A teacher who is thoroughly grounded in a subject area and convinced of the worth of that subject will go to great lengths to introduce topics and prepare his/her own materials even if satisfactory texts are not available, and provide the motivation to make the subject alive and challenging. However, a teacher who has only minimal training in a subject will lack confidence in his/her ability to develop or present a course in the subject.

Statistics courses offered in colleges or universities to teachers or future teachers are usually one of two types: (1) formal courses in mathematical statistics that focus on probability distributions from a highly formal point of view, or (2) educational statistics courses, designed to help students deal with statistical problems arising in formal educational research. Neither of these approaches results in adequate preparation for teaching statistics and probability to elementary or secondary students.

For pre-college teachers to be successful with the materials this project intends to develop, thorough in-service training will be necessary. The goal is to introduce teachers to specific approaches and techniques for presenting new and unfamiliar topics and provide them with the opportunity to learn the material in much the same way they will have to present it later on to their students.

The Need for Materials in Statistics and Probability

A Joint Committee of the ASA and NCTM has compiled an extenisve bibliography of publications on statistics and probability appropriate for use in the school. However, most of these publications do not approach the topic from a point of view that is practical enough. Some examples follow:

- The Schools Council Project on Statistical Education, based in Sheffield, England, developed a fine series of booklets for students aged 11-16. This series, entitled, "Statistics in Your World," is extremely well done. The examples used are indigenous to British culture, however, and so would require adaptation prior to general use in the United States.

- Statistics and Information Organization is a loose-leaf kit of nearly 900 pages, developed under NSF funding as part of the Mathematics Resource Project at the University of Oregon. It is excellent supplemental material for the middle school teacher.

- John Camp, at Wayne State University, heads an NSF project developing short units relating statistics, probability, and micro-computer use to problem-solving. The four units developed are intended for upper elementary school and are undergoing field testing. They entail four to six weeks of instructional time, however, and only part of that is devoted to statistics and probability.

- The American Statistical Association's publications, Statistics by Example and Statistics: A Guide to the Unknown, have had wide impact. They were designed for other purposes than texts and achieved these purposes well. Statistics by Example contains examples appropriate for use from middle school through grade 12. Statistics: A Guide to the Unknown illustrated the wide range of applications of statistical procedures, but contains little statistical development. This publication, including the second edition and subsequent mini-book, has sold over 80,000 copies.

- _Teaching Statistics and Probability_, the 1981 NCTM Yearbook, serves as a useful reference. It contains classroom lessons, but too few at any level to serve as a text.

- The recent text, _Beginning Statistics with DATA Analysis_, by Mosteller, Fienberg, and Rourke, emphasizes subject matter. Its major drawback is that it requires mathematics (e.g., log transformations) which are beyond the experience of the general student.

Space does not permit discussion of other potentially useful materials. Those mentioned here include the most promising candidates for instructional use with students, and their limitations are characteristic of other materials.

Four booklets are under development by the Joint ASA/NCTM Committee expressly intended for teacher use and containing ready-to-copy pages of student exercises, suggested projects, and opportunities for "hands-on" involvement. They are arranged as teaching sequences and present topics in the spirit advocated by the joint committee. These booklets will be described in detail in a subsequent section.

The need for and the willingness to use such materials were expressed by the more than 550 teachers who attended mini-courses offered by the Joint ASA/NCTM Committee at NCTM meetings in the period 1974-76. Sample materials used with teachers at the ASA/NCTM Williamsburg-Leadership Conference (1981) were enthusiastically received. Presentations by members of the Joint Committee of the concepts and materials now under development at annual, regional, and state meetings of mathematics teachers have been met with great enthusiasm.

The Historical Involvement of ASA and NCTM in Teaching Statistics and Probability

As early as the 1920's, mathematics educators had seen the need to impart to students the ability to understand and interpret tables, graphs, and other forms of statistical data and had suggested that such training be made part of the mathematics curriculum. These thoughts were echoed throughout the 40's and 50's, but not much happened until 1967. In that year, Professor Frederick Mosteller of Harvard University made a presentation at the Annual Meeting of NCTM in which he called for the formation of a committee to examine how statistical education in the schools of this country might be improved. This speech was the impetus for the formation of the ASA/NCTM Joint Committee on the Curriculum in Statistics and Probability.

Initially, the Joint Committee focused on two major projects: (1) development of a collection of essays written by practicing statisticians to show the use of statistics in problems of national interest, starting with the publication, Statistics: A Guide to the Unknown; and (2) development of a collection of examples of applications of statistics for use in introducing statistical concepts into high school mathematics courses, leading to a series of four volumes entitled Statistics by Example.

After these publications, the Joint Committee turned its attention toward teacher training. Funded by royalties from earlier publications and by NSF, the committee organized a series of mini-conferences on statistics. These conferences were essentially short courses, or modules, on statistics, emphasizing important basic notions and sound examples. Seven such mini-conferences were held, reaching approximately 550 teachers throughout the country.

Since 1979, the Joint Committee has been gathering information and materials from a large group of mathematics teachers and statisticians with two goals: (1) develop curriculum guidelines on methods and content for teaching statistics and probability, and (2) develop materials that would allow teachers to introduce this content into the curriculum with flexibility. These curriculum ideas and materials were first presented at a conference held in Williamsburg, Virginia, in December 1981. Forty-five mathematics teachers and supervisors from across the country attended this conference. Their evaluations (which are available upon request) show almost uniform acceptance of the curriculum ideas and great enthusiasm for the materials. It is these materials that formed the basis of the four booklets now in preparation.

Several of the participants developed an informal publication entitled

"Network" as a communication medium and an information exchange for those interested in the teaching of statistics in primary and secondary schools. To date, seven issues have been distributed; copies are available free of charge. The "Network," disseminated throughout the United States and Canada, has over 1000 subscribers.

Paralleling the efforts of the Joint Committee, NCTM has been involved in activities on other fronts to encourage more statistics and probability in the curriculum. Most notable among these efforts is the 1981 Yearbook entitled Teaching Statistics and Probability. Members of the Joint Committee served as editor, associate editors, and contributors to the publication. This yearbook sold approximately 4,800 copies in 1981, a remarkable number by comparison to other yearbooks, and 1,100 copies in 1982, demonstrating a genuine continued interest in the subject. Also, the 1983 Yearbook, The Agenda in Action, contains two chapters dealing with probability and statistics.

In short, the two societies supporting this project, ASA and NCTM, have long and rich histories of involvement in the teaching of statistics and probability at the pre-college level. This project is a culmination of that involvement and experience, and can be expected to have dramatic effects on the teaching of these subjects in the future.

Project Description

Scope and Intent

The point of view underlying the proposed three-year project is that in today's complex society, it is essential for everyone (1) to be familiar with statistical and probabilistic reasoning, including the collection, summarization, display, and interpretation of data, and (2) to appreciate and accept the fact that all processes are subject to variability. In many situations, statistical reasoning does not require advanced or complicated mathematical formulation. However, appropriate ways to interpret data are not innate and must be learned. Therefore, it is proposed that statistical reasoning be taught primarily through the use of examples that are both interesting to students and related to current issues.

It is the intent of the project to realize the following:

1. Provide guidelines on the teaching of statistics and probability within the mathematics curriculum;

2. Develop a model in-service program for training teachers in modern statistical concepts and in methods of teaching these concepts;

3. Produce curriculum materials to assist teachers in the proper presentation of statistical and probabilistic concepts, and encourage further development in the natural and social sciences; and

4. Develop a mechanism to evaluate the effectiveness of materials and techniques for teaching statistics and other basic skills, such as reading and arithmetic.

Methodology

The project entitled, "A Program to Improve Quantitative Literacy in the Schools," has been designed by the American Statistical Association to respond to the critical need to modify the status of the mathematics and science curricula in the United States with particular emphasis on quantitative literacy.

During the course of the project, teaching guidelines will be developed after careful study of historical and present-day views of the needs for statistical training, and subsequent to evaluation of methods of delivering appropriate training to both teachers and students. The major focus of the project will be the development and delivery of a program of continuing education for elementary teachers and secondary teachers of mathematics and science that will prepare them to teach statistical and probabilistic skills and concepts effectively. Development and production of teaching materials, an integral part of the project, will result in more effective teacher training and better tools for use in the classroom than are now available. Development of both the in-service aspect and the teaching materials will be tied to past and current views of what is important in statistics, and will by influenced heavily by those topics that are likely to be needed by industry and government.

Since 1968, the Joint ASA/NCTM Committee has directed its efforts toward the development of innovative curriculum materials and methods to promote the teaching of statistical concepts. Recent work has focused on preparation of a series of four unified booklets to introduce basic skills of statistics and probability. These are designed for middle school (upper elementary and junior high school) teachers and students, although they will be valuable at the high school level as well. Topics covered include ways to collect, display, and interpret small sets of data; elementary probability; the use of simulation to understand topics in probability and to solve probability problems that are too hard to do conveniently with ordinary mathematics; and the use of the ideas of random sample and confidence interval to learn about a population.

The working titles of the booklets are: (1) "Exploring Data"; (2) "Probability"; (3) "Simulation"; and (4) "Information from Samples."

Drafts of these booklets currently exist and are being revised based on preliminary reviews by the authors (all of whom are members of the Joint Committee). A preliminary field test of these materials was conducted in the fall of 1983 without any additional funding from the National Science Foundation or other external source. About 25 mathematics teachers (including three elementary level teachers) were actually involved in this voluntary test. However, substantial resources are required to complete the full intent of the project, namely:

- Development and performance of in-service training for teachers in proper and effective procedures
- Completion of thorough field-testing and revision of teaching materials prior to final production
- Evaluation of the impact of the teacher training and materials on student performance.

Description of Teaching Materials

Taken together, the four booklets will provide a good introduction to ideas of statistics and probability which an educated citizenry should possess. They will help teach students how to deal with types of data and questions they will meet more and more frequently. The usefulness of statistics and probability in dealing with practical problems is shown through many examples. The emphasis is not on watering down advanced mathematical statistics and probability for younger students, but rather on experiencing real examples and hands-on applications.

All of the booklets follow the same basic style. For each of the several topics in a booklet, there is some brief explanatory material and questions for discussion at the start which the teacher can present to the class. Then there are worksheets (exercises) that can be duplicated for use by students. Students perform their own probability experiments and simulations, and analyze data presented to them. They also collect their own data and analyze the results. Throughout all the activities, they learn skills that have many applications and solutions to problems that are interesting and important.

"Exploring Data" uses various data sets (e.g., nutrition in fast foods, sales of the most popular children's books, accident rates for different automobiles, television ratings, baseball home run records, life expectancy values). The focus is on data sets that represent the kinds of issues citizens ordinarily deal with. Examples are presented and studied that are interesting to students. The statistical analyses concentrate on constructing and interpreting certain simple graphical displays. Numerical summaries are the median, maximum and minimum, quartiles, inter-quartile range, and mean. Graphs include the stem-and-leaf plot and box plot for displaying a single variable and scatter plot for displaying two variables. A simple method of fitting a straight line to a scatter plot is used, and scatter plots against time are smoothed and interpreted. Finally, simple graphical methods are used to extract, summarize, and display information in the data. The interdisciplinary nature of statistics allows illustrations from many areas of science and everyday activities.

"Probability" develops the basic notions through experiments that the students perform themselves involving simple spinners, dice, coins, cards, etc. The concepts of a pair of equally likely events, the probability of an event, and a pair of independent events are arrived at through many examples. Then tree diagrams are used to calculate — non-mathematically and without experimentation — probabilities of certain events. Neither "Probability" nor "Exploring Data" has the other as a prerequisite and the two of them could even be studied simultaneously.

The next booklet, "Simulation," builds on ideas introduced in both of the first two booklets. It shows how practical problems, some fairly complex, can be solved, at least approximately, by means of simple simulation techniques. Some of the data from the simulations are displayed using techniques from "Exploring Data." The examples also introduce and develop the idea of a mathematical model. Many of the problems can be solved by means of small manual simulations using coins, dice, random number tables, etc. For larger or more complicated problems, simulations are usually done using microcomputers, a tool that is now becoming widely available. Sample problems include finding the probability of exactly one girl in a three-child family; finding the probability of guessing five or more correct answers in a 25-question multiple-choice test; simulating the spread of an infectious disease; and the airline overbooking and no-show problem. The examples show that realistic and interesting problems can be stated and then solved using simulation ideas; however, the mathematics required for exact solutions is well beyond the level of the students. This booklet also takes advantage of the capabilities of microcomputers.

"Information from Samples" makes use of material developed in the earlier booklets to deal with basic problems of statistical inference, for example, if we have a sample proportion from a population, as in a common opinion poll, what can we conclude about the true (unknown) proportion in the entire population? Techniques from the second and third booklets are used to simulate this situation, and the results are displayed using plots presented in the first booklet. The notion of random sample from a population is important. From the simulations, the idea of a sampling distribution of the sample proportion is developed. By doing this for populations with different true proportions and comparing the results, the student is led to the idea of a confidence interval for the population proportion based on the proportion in the sample. Through experience with examples and simulations, the student learns the underlying ideas; there is no recourse to mathematical formulas beyond the student's understanding.

Guidelines for Teaching Statistics and Probability

An early effort of this project will be the development of guidelines for teaching statistics and probability in the pre-college curriculum. This document will designate the recommended components for teaching statistics and probability to this population and reflect those aspects that have been demonstrated as possible to teach effectively at these levels. It will also serve as a framework for developing other aspects of the project. Finally, the guidelines will be published separately and distributed so that others interested in the application of statistical education may use them.

The Role of the Computer

The microcomputer is becoming more readily available in elementary and secondary classrooms throughout the country. This is a major advantage in that large, real data sets can be managed efficiently only with the use of a computer. Likewise, simulations of probabilities of complex events can be completed satisfactorily only with such assistance. Alternative guidelines for computer use will be developed by this project with suggestions to be incorporated into the curriculum materials. Computer software will accompany most instructional units that will be developed, although the basic units will not require the use of a computer by either teachers or students.

Field Testing of Teaching Materials

A preliminary field test of the teaching materials described above was held in the fall of 1983. A more extensive field test begun in the fall of 1984 and now near completion was implemented shortly after the project began formally. The results of the preliminary field test were used to revise the booklets prior to the major test. Each of the teachers responsible for field testing the material will be visited by a member of the Task Force (project staff) who will monitor the process.

In-service Training for Teachers

In the spring of 1984, a model in-service training program to prepare 40 teachers in the use of the materials in the classroom was developed for presentation in December of that year. The model program was three days in duration consisting of two days of initial training and one day for an evaluation session during which teachers assisted with revisions for subsequent in-service programs and provided feedback on the training materials.

Beginning in the spring of 1985, teachers trained in the model in-service training program will begin to conduct workshops in their own areas, thus providing a multiplier effect for the project. Two continuing support groups (advisors) will be established: one for teachers using the materials in classrooms, the other for those replicating the in-service component in their local districts. Support will include "hot-lines" to specific task force members who will provide technical assistance for problems as they arise.

Also, in the spring of 1985, there will be four regional two-day in-service training sessions based on the model in-service program. These sessions will provide the basis for production of a videotaped version of the training sessions in the spring of 1985. The videotaped program will be disseminated to teachers unable to participate in any of the regional in-service sessions or replication workshops.

Videotapes will be produced under the direction of the Office of Continuing Education of the American Statistical Association. Since 1975, the Association has produced videotapes of numerous educational programs in the form of two-day short courses and special lectures that are distributed throughout the United States and abroad.

Evaluation and Dissemination

By the spring of 1985, an increasingly large number of teachers will have participated in the in-service aspect of the project, tested the teaching materials, and received training in the form of videotaped versions. Each aspect will be evaluated thoroughly prior to implementing procedures for widespread dissemination.

During the 1985-1986 school year the project's effectiveness will be evaluated. Evaluation mechanisms will be developed and studies conducted to assess how well students are learning skills and concepts in statistics and probability. Further, a mechanism will be established to permit the Joint ASA/NCTM Committee to continue monitoring the use of materials by teachers and revise them as necessary beyond the life of the project.

By the end of the project, four major evaluations will have taken place:

1. Materials. Assessment will be ongoing throughout the life of the project. The four booklets were revised prior to field-testing beginning in the fall

of 1984. At the model in-service program (December, 1984), time will be devoted to a thorough evaluation of the materials. This information will be used to revise the materials prior to the four regional in-service sessions (Spring 1985). Feedback from the 1984-1985 training sessions and from classroom field testing will be used as the basis for final production in the summer of 1985.

2. In-service. The final day of the December 1984 model in-service program will be devoted to evaluation. Feedback from participants on the effectiveness of the in-service session will be used to plan regional in-service sessions. Additional feedback useful for revising the in-service sessions will come from participants who will subsequently replicate in-service programs for teachers from their own school districts.

3. Videotapes. Videotapes will be evaluated from two perspectives: (1) technical quality of the videotapes; (2) (by the teachers) content and teaching techniques included in the videotapes.

A major question for which data will be gathered is whether the videotapes can be used effectively as part of a local in-service conducted by teachers who have not received training at a project in-service program.

A related question is whether the videotapes can be used as stand-alone training by individual teachers or by groups of teachers (e.g., the mathematics department of a high school) who have not attended a formal in-service program.

4. Student Performance. Data will be collected on student performance during the final year of the project. These data will be analyzed to learn: (1) how much of the new content of statistics and probability was absorbed; (2) the extent to which statistics and probability, as organized in the project materials, contribute to improved performance on basic mathematical skills; and (3) how much was attitude toward mathematics affected by the study of statistics and probability. Members of the Task Force will construct appropriate evaluation instruments and analyze the data collected.

Cognitively Guided Instruction: The Application of Cognitive and Instructional Science to Mathematics Curriculum Development[1]

Thomas P. Carpenter

Elizabeth Fennema

Penelope Peterson

University of Wisconsin-Madison

One of the fundamental lessons to be learned from the curriculum reforms of the 1950s and 1960s is that curriculum changes cannot focus exclusively on restructuring the selection and sequencing of content (National Advisory Committee on Mathematical Education 1975). Students' abilities to learn the content and the procedures teachers use to teach it must also be taken into account. One of the major recommendations to come from a recent conference on school mathematics was based on the premise that research on teaching and learning should play a central role in curriculum reform (Romberg 1984).

A viable body of research on children's learning of mathematics and on the teaching of mathematics is emerging that has clear implications for the mathematics curriculum. We are currently working on a project to investigate how this research can be applied to the teaching of addition and subtraction in the primary grades. Research in addition and subtraction provides one of the clearest analyses available of children's learning within a particular mathematics content domain. From an extensive body of research dating back over 50 years, a reasonably consistent and coherent picture has emerged of the development of addition and subtraction concepts and skills (Carpenter 1985a; Carpenter and Moser 1983; Riley, Greeno, and Heller 1983). This research specifies a taxonomy of problem types and strategies used by children and identifies major levels in the development of addition and subtraction concepts and skills. There now exist explicit models of children's cognitive processes which specify critical levels in the development of these concepts and skills. Specific hypotheses for structuring and sequencing content for instruction can be derived from this detailed analysis, which centers on problem types and children's solution processes. The analysis also provides a means of assessing children's knowledge and providing instruction for individual differences.

In the sections which follow, general principles of cognitive and instructional science that have clear implications for the teaching of mathematics are discussed. Specific examples are drawn from research on addition and subtraction, but the general principles apply to any content domain. Following the discussion of research, specific implications of this research for the design of instruction are identified. Examples from research on addition and subtraction again are used for illustrative purposes.

Cognitive Science Research
Learning as a Constructive Activity

One of the most fundamental contributions of current research is its conception of the learner. Most mathematics instruction has tacitly assumed that students learn what they are taught, or at least some subset of what they are taught. But current research indicates that students actively construct knowledge for themselves. Although instruction clearly affects what students learn, it does not determine it. Students are not passive recipients of knowledge; they interpret it, put structure on it, and assimilate it in light of their own mental framework. There is a growing body of research which suggests that children actually invent a great deal of their own mathematics (Resnick and Ford 1981) and that children enter school with highly developed informal systems of arithmetic (Fuson and Hall 1983; Gelman and Gallistel 1978; Ginsburg 1977).

Contrary to popular notions, young children are relatively succesful at analyzing and solving simple verbal problems. Before receiving formal instruction in addition and subtraction, most young children invent modeling and counting strategies which they use to successfully solve addition and subtraction problems (Carpenter, Hiebert, and Moser 1981; Carpenter and Moser 1983, 1984). The informal solution strategies invented have a clear relationship to the structure of the problems solved. Most young children directly model quantities described in a problem, apply actions to those models, and enumerate sets to determine an answer. Figure 1 includes a set of problems that provides different interpretations of addition and subtraction and consequently result in different solutions. To solve the second problem in Figure 1, the young children generally would represent the problem directly. They would construct a set of 5 objects, and then add more objects until there was a total of 13 objects and count the number of objects added. To solve the fourth problem, they would make a set of 13 objects and remove 5. The ninth problem would be solved by matching two sets and counting the unmatched elements.

The problem-solving analysis children naturally apply to simple word problems provides a much better model of problem-solving behavior than many of the superficial tricks for solving word problems children are often taught. Similarly, the strategies children at more advanced levels invent to solve addition and subtraction problems are more efficient and require a deeper understanding of the operations than the procedures they are asked to learn. Thus, children's informal knowledge of arithmetic constitutes a substantial basis for the development of number concepts and problem-solving skills. Most mathematics curricula fail to capitalize on this knowledge.

The Acquisition of Concepts

The curriculum programs developed in the 1950s and 1960s generally were based on the assumption that the instructional sequence of a topic should follow the logical and mathematical development of the content. Current research indicates that children do not necessarily acquire concepts by building up from the logical foundations of the concepts. Research is beginning to provide a picture of how concepts and skills actually develop in children. A reasonably clear map of the development of addition and subtraction concepts and skills has been constructed. This map provides a different picture of the development of addition and subtraction concepts than is reflected in most programs of instruction.

At the initial level of solving addition and subtraction problems, children are limited to solutions involving complete direct representations of the problem. They must use fingers or physical objects to represent each quantity in the problem, and they can only represent the specific action or relationship described in the problem. As a consequence, they cannot solve problems like the sixth problem in Figure 1, because the initial quantity is the unknown and therefore

Problem Type	Unknown		
Change	Result Unknown	Change Unknown	Start Unknown
Join	1. Connie had 5 marbles. Jim gave her 8 more marbles. How many marbles does Connie have altogether?	2. Connie has 5 marbles. How many more marbles does she need to have 13 marbles altogether?	3. Connie had some marbles. Jim gave her 5 more marbles. Now she has 13 marbles. How many marbles did Connie have to start with?
Separate	4. Connie had 13 marbles. She gave 5 marbles to Jim. How many marbles does she have left?	5. Connie had 13 marbles. She gave some to Jim. Now she has 5 marbles left. How many marbles did Connie give to Jim?	6. Connie had some marbles. She gave 5 to Jim. Now she has 8 marbles left. How many marbles did Connie have to start with?
Combine	7. Connie has 5 red marbles and 8 blue marbles. How many marbles does she have?		8. Connie has 13 marbles. Five are red and the rest are blue. How many blue marbles does Connie have?
Compare	9. Connie has 13 marbles. Jim has 5 marbles. How many more marbles does Connie have than Jim?	10. Jim has 5 marbles. Connie has 8 more than Jim. How many marbles does Connie have?	11. Connie has 13 marbles. She has 5 more marbles than Jim. How many marbles does Jim have?

Figure 1

Classification of Word Problems

400

cannot be represented directly with objects.

Children's problem-solving strategies develop concurrently along two dimensions — abstractness and flexibility. There is a shift to more abstract counting strategies like "counting on" or "counting back." For example, to solve the second problem in Figure 1, the child would recognize that it was unnecessary to construct the set of 5 objects, and instead would simply count from 5 to 13, keeping track of the number of counts. As their strategies become more abstract, children also begin to be more flexible in their choices of strategy and are able to use strategies that do not directly model the action or the relations described in the problem. For example, the counting strategy used for problem 2 might also be used for problem 4 or problem 9.

The transition to more abstract and flexible strategies takes place over an extended period of time. There is a period during which children use both physical modeling and counting strategies and are able to transform certain types of problems but not others. However, before they master recall of number facts, almost all children become skilled in using counting strategies, and during this period they develop the ability to solve a wider range of problems. At this point, children begin to integrate the different interpretations of addition and subtraction represented by their modeling and counting strategies, and can solve problems like problem 6 that cannot be modeled directly. These counting and modeling strategies play a critical role in children's problem solutions for some time.

Structure and Understanding

Prevailing theories of instruction in mathematics have often been based on assumptions about whether it is more important to develop understanding or teach skills. Current research is beginning to provide some perspective on the intricate relationship between understanding and skill development (Resnick and Ford 1981). It is also beginning to sort out exactly what constitutes understanding. In the curriculum of the 1960s, the understanding issue was addressed by the use of precise language, the specification of basic principles like commutativity and associativity, the reliance on formal mathematical justification or proof, and an emphasis on mathematical structure. The research on children's learning of mathematics indicates that any theory of understanding must consider how the learner thinks about a problem or concept. The role of understanding is to fit information into the learner's existing cognitive framework. Thus, we have to take into account the knowledge of the mathematics under consideration the learner brings to the situation, discover how conceptual knowledge is connected to procedural skills, and encourage the integration of related concepts.

Research also is beginning to document the critical role that structure plays in understanding and problem-solving (Chi, Glaser, and Rees 1982; Greeno 1978). Cognitive science research suggests that structures imposed in information in long-term memory and networks of relationships between related concepts facilitate recall and make it more likely that the information will be used in problem-solving (Carpenter, 1985b; Chi et al. 1982). In general, it appears that it is important to stress relationships between concepts, especially higher-order relationships that are related to ways concepts may be used to solve problems. The analysis of conceptual networks constructed by experts or capable students provides a framework for organizing instruction to emphasize important correspondences between related concepts.

Structure and understanding play a key role even in something as routine as learning number facts at recall level. Before all addition facts are completely mastered, some children use a small set of memorized facts to derive solutions for addition and subtraction problems involving other number combinations. These solutions usually are based on doubles or numbers whose sum is 10. For example, to find $6 + 8 = ?$, a child might recognize that $6 + 6 = 12$ and then recall that $6 + 8$ is just 2 more than 12. Although derived facts seem to require a great deal of insight about numbers, it appears that they are used not only by a handful of bright students. In a three-year longitudinal study, over 80% of the children used derived facts at least once, and 36% used derived facts consistently at some period during the study (Carpenter and Moser 1984). These results suggest that derived facts play an important role for many children in the learning of number facts. Most instruction fails to capitalize on the network of relationships between number facts or on children's inclination to build upon those relationships.

Instructional Science Research

In the 1960s and 1970s, most research on teaching followed a dominant process-product paradigm (Peterson and Walberg 1979). In process-product research (also called "teaching effectiveness research"), the researcher designs and uses a low-inference classroom observation instrument to observe the classroom process, both teacher behavior and student behavior. The "product" is student achievement, usually measured by student scores on standardized achievement tests (cf. Dunkin and Biddle 1974). The frequencies of observed "process" variables are then correlated with these scores to identify teacher behaviors related in a significant way to student achievement. Once promising teaching behavior variables have been identified, teachers are trained to implement the behaviors in their classrooms in an experimental study. Growth in student performance in the experimental classrooms is compared to growth in student performance in control (no treatment) classrooms to see if training in the selected behaviors has had a positive impact on student performance.

In tse mid-1970s, some researchers began arguing that process-product research was necessarily limited by its restricted focus on only observable phenomena (such as teacher and student behavior). Excluded from concern were teachers' cognitive processes during classroom instruction. Shulman and Elstein (1975) suggested that researchers view the teacher not only as one who engages in certain classroom behaviors, but also as an active processor of information before, during, and after classroom instruction.

The rationale for this perspective of the teachers was presented most clearly in a report produced by Panel 6 as part of the National Conference on Studies in Teaching convened by the National Institute of Education in June 1974. The panelists argued that:

> It is obvious that what teachers do is directed in no small measure by what they think. Moreover, it will be necessary for any innovation in the context, practices, and technology of teaching to be mediated through the minds and motives of teachers. To the extent that observed or intended teachers behavior is "thoughtless," it makes no use of the human teacher's most unique attributes. In so doing, it becomes mechanical and might well be done by a machine. If, however, teaching is done, and in all likelihood, will continue to be done by human teachers, the question of the relationships between thought and action becomes crucial (National Institute of Education 1975, 1).

Research on teachers' thought processes and decisions has burgeoned in the decade since the publication of the Panel 6 report. Comprehensive reviews of this research have been done by Shavelson and Stern (1981) and more recently by Clark and Peterson (in press). Rather than attempt to provide an exhaustive review of this research here, we will briefly summarize Clark's and Peterson's conclusions.

Clark and Peterson start by noting that the research shows that thinking plays an important part in teaching and conclude that the image of a teacher as a reflective professional as proposed originally by Panel 6 (National Institute of Education 1975) is not far-fetched. As thoughtful professionals, teachers have more in common with physicians and lawyers than with technicians. It also appears that teachers plan for instruction in a wide variety of ways and that their plans have real consequences in the classroom. Third, during interactive teaching, teachers are continually thinking, and from their reports, make decisions frequently — once every two minutes. Finally, teachers have theories and belief systems that influence their perceptions, plans, and actions in the classroom.

Concurrently with the concern about teachers' cognitive processes, researchers began pointing out the importance of students' cognitive processes during classroom instruction as mediators between teacher behavior and student achievement. Doyle (1975, 1978) argued that a major weakness of the dominant process-product research paradigm was its focus on student behavior alone; students' covert responses during instruction were simply not taken into account. He described a mediating-processes paradigm for classroom research and indicated that, although overt student behavior may be a useful proxy of what goes on inside a student's head, it is a relatively gross measure of cognitive processes. Doyle suggested that the covert cognitive processes that operate during teaching and learning have to be defined more precisely.

Much research on students' thought processes has been conducted during the past decade. Research has focused on what students perceive to be the teacher's behavior (Winne and Marx 1982), students' perceptions of teachers' behavior (Weinstein, Marshall, Brattesani, and Middlestadt 1982), and students' reports of their own understanding of the mathematics content as well as the kind of cognitive processes and strategies they engaged in during class (Peterson, Swing,

Braverman, and Buss 1982; Peterson and Swing 1982; Peterson, Swing, Start, and Waas 1984). (See Weinstein 1982 and Wittrock 1986 for complete reviews of this research.)

In their studies of fifth-grade students, Peterson et al. (1982) and Peterson, Swing, Stark, and Waas (1984) found that student ability and mathematics achievement were significantly related to reports students gave of their thoughts during classroom instruction in mathematics. Thoughts such as attending to the lesson, understanding the lesson, and either engaging in a variety of specific cognitive processes or engaging in them more frequently appeared important. Peterson et al. (1982) also found that student engagement in mathematics as assessed by classroom observers was unrelated to student achievement. Thus, students' reports of their own understanding of the mathematics lesson and their cognitive processes during instruction may be more reliable and more valid indicators of classroom learning than observers' judgment of students' attention. In other words, this research suggests that the quality of time students spend attending to the mathematics task — the actual cognitive processes involved in processing the mathematics information presented in class — may be equally important, or possibly even more important, than that quantity of time spent engaged in the mathematics task.

Implications for Instruction

While there is no established paradigm for the development of materials for classroom instruction that applies results of cognitive and instructional science, principles have emerged to guide curriculum development. The following guide our current work.

Content Selection and Sequence

The first principle is that the instructional sequence should be consistent with the sequence in which concepts and skills develop naturally in children. Case (1985) has proposed that instruction should actually recapitulate the natural developmental sequence. In other words, instruction should be sequenced to reflect major levels in the acquisition of a concept or skill and be designed to guide children in the use of some of the more productive strategies employed by the most capable students. There are four basic implications of cognitive science research pertaining specifically to the selection and sequencing of instruction in addition and subtraction in the early primary grades.

Early and Continuing Emphasis on Word Problems

In the traditional curriculum, computational skills have to be learned before children can be taught to solve even simple word problems. Word problems have often appeared only at the end of a unit and have emphasized the computational skills taught in the unit. Whatever value such exercises may have for practicing the computational skill, they do very little to teach problem-solving. Because students are often taught that they will be subtracting the smaller number from the larger in each problem, the exercises encourage the learning of computational skills by rote and discourage children from reading or thinking about problems. These exercises may actually contribute to the deterioration of children's problem-solving skills. Furthermore, the exercises often provide only a limited exposure to different types of addition and subtraction problems. The research summarized above shows clearly that even before they receive formal instruction in addition and subtraction, young children are able to solve a wide variety of addition and subtraction problems by modeling the problems with manipulatives or by means of different counting strategies. Based on this research, it appears that word problems are appropriate for introducing addition and subtraction. In fact word problems should not just be integrated into the mathematics curriculum (Carpenter et al. 1981); they should form the basis of the curriculum.

Instruction in Counting Strategies

Recent studies indicate that the progressive use of more sophisticated counting strategies such as counting on[2] is important in the acquisition of addition and subtraction concepts and skills (Carpenter and Moser 1984; Groen and Parkman 1972). Although most of the research has suggested that children discover counting on for themselves without formal instruction (Carpenter and Moser 1982; Groen and Resnick 1977), there is evidence that counting on can be taught (Secada, Fuson, and Hall 1983). In a clinical study based on an analysis of the component subskills underlying counting on (Fuson 1982), first-grade children were successfully taught how to count on by having them first learn each of the requisite subskills (Secada et al. 1983). This suggests designing instruction to facilitate the development of the critical stage in which counting strategies are paramount.

Learning Number Facts

Although young children are able to solve addition and subtraction problems by counting, this is an inefficient and cumbersome process when it is a matter of computing or estimating large numbers. Even though there have been recommendations to decrease emphasis on computational skills, it is nevertheless generally agreed that learning number facts at the recall level continues to be an important objective of the mathematics curriculum (Conference Board of the Mathematical Sciences 1982).

Instruction in number facts, however, often fails to capitalize on the rich network of relations between number facts. There is mounting evidence that instruction which teaches children strategies for relating number facts is effective (Steinberg, in press; Thornton 1978; Thornton, Jones, and Toohey 1983). These studies suggest that instruction in specific strategies greatly increases the rate at which they are applied in deriving unknown facts and leads to significant gains in arithmetic achievement.

Representing Word Problems

Although young children develop reasonably sophisticated processes for analyzing and solving addition and subtraction word problems, they have difficulty relating this knowledge to the formal mathematical procedures they learn in school. Children learn quite easily how to write number sentences for simple addition and subtraction problems (e.g., the first and fourth problems in Figure 1), but often have difficulty writing number sentences for other problem types, even when they can solve them by modeling and counting (Carpenter et al. 1983; DeCorte and Verschaffel 1983). The difficulties children experience in representing word problems occur because the canonical representations they have available ($a + b = ?$ and $a - b = ?$) do not always correspond to their informal solutions or to the semantic structure of the problem. For children "$a - b$"

represents a separating action that corresponds to their solution to problem 4 in Figure 1. As a consequence, they have no difficulty representing or solving this type of problem. Problem 2 (Figure 1), however, is solved by adding elements to another set and keeping track of the number of elements added. This problem corresponds more closely to the noncanonical number sentence $5 + ? = 13$. There is evidence that children readily represent certain problems like the second, third, fifth, and sixth problems in Figure 1 using noncanonical sentences ($a + ? = b$, $? + a = b$, $a - ? = b$, $? - a = b$), because these representations correspond to the way they think about the problem (Moser and Carpenter 1982; Carpenter, Moser, and Bebout, in press). These results suggest that instruction which includes noncanonical as well as canonical number sentences would correspond more nearly to children's informal knowledge of mathematics than would instruction which attempts to immediately represent all problem situations in canonical form.

Adaptation of Instruction

The second principle is that instruction should be appropriate to each child's level of knowledge and should facilitate growth to successive levels. To achieve these goals, teachers have to assess each child's knowledge and skills at regular intervals, and also plan instruction to take into account the wide range of knowledge and skills that exist in any classroom at any given time.

If it were assumed that children bring little background to instruction and simply learn what is taught, there would be little need to worry about assessing what students know or about individualizing instruction. However, children develop informal systems of mathematics outside of the classroom, rather than simply absorbing what they are taught. They structure and interpret the mathematics presented to them in light of their existing knowledge. At any point in time in any classroom, different children will be operating at different levels in terms of ability to solve different kinds of problems. And because new learning is based on previous knowledge, children benefit from individually adapted instruction.

Research also indicates that children who are able to solve the same problem often use different solution strategies, a fact with serious implications for instruction. Just because a child can readily write the answer to addition problems does not mean the child has committed certain addition facts to memory. The instructional needs of a child who solves such problems by counting is different from the needs of a child who has memorized the relevant number facts. Thus, in planning instruction, it is important that teachers attend to the strategies children use to solve problems, not just whether the problems are correct or incorrect. This requires new methods of assessing students' knowledge.

Research on addition and subtraction has provided techniques for use in assessing students' knowledge and solution strategies by observing how they apply these strategies to a relatively limited set of problems. Such techniques have been refined through clinical interviews and adapted in a natural way to classroom use (cf. Steinberg, in press). This research also provides a reasonably clear map of children's levels of strategy, a map which can be used to assess each child's abilities. Problems and solutions are well-scaled. If a child is able to solve a specific problem by applying a certain strategy, this says a great deal about how that child will solve a number of other problems, and consequently what instruction is appropriate. For example, if a child cannot solve the first or fourth problem in Figure 1, that child probably cannot solve any of the other types of problems, and there is little value in assessing the child's ability to solve more difficult problems. If a child can solve the sixth problem, the child can probably solve any of the first five as well. Similarly, if a child cannot count on, instruction that emphasizes derived number facts is probably beyond the child's ability.

Once a child's level of knowledge has been ascertained, appropriate instruction must be chosen. Although there are some examples of successful individualized programs, providing different activities for different children has often proved to be difficult. One of the outcomes of research on addition and subtraction is a clearer picture of the rich diversity in solution strategies for problems that on the surface appear to be similar. The same problem can be solved by means of different strategies that represent different levels of development. Within the context of basic addition and subtraction with the same range of numbers, different problems are appropriate for children at different levels of knowledge. As a consequence, the same basic activity can be adapted to each individual child's needs by providing appropriate problem sets for different children or the means for children to solve problems using procedures appropriate for their level of knowledge.

Instruction in addition and subtraction may be individualized along several dimensions. Different types of word problems are clearly scaled by difficulty, and appropriate selection of problem types can provide for a wide range of individual differences. Strategy choice can also provide a dimension for individualization. Children can solve the same types of problems by modeling them with physical objects or using counting strategies or number facts that are either recalled or derived. By structuring problems in appropriate ways, instruction can facilitate the development or use of particular strategies when appropriate for a particular child. Distinctions between concept and skill development should make it possible to individualize instruction to reflect different needs of children at a given time with respect to developing a concept or practicing a skill.

Students as Active Cognitive Processors

Students should be considered active processors of information. Students actively construct knowledge about mathematics concepts and skills and actively construe the teaching and learning situation in the classroom. To be effective, mathematics instruction must be built on these principles. The basic structure of the typical mathematics classroom might have to be changed to promote students' active cognitive processing during mathematics instruction.

The picture of the typical classroom currently is one of extensive teacher-directed explicating and questioning in the context of whole-group instruction followed by students working on paper-and-pencil assignments at their seats. In a recent study of 36 fourth-grade mathematics classrooms in Wisconsin, Peterson and Fennema (1985) found that 43% of the classtime in mathematics was spent in whole-group instruction and 47% with students doing seatwork.

Recent research by Webb (1982) and Peterson and her colleagues (Peterson and Janicki 1979; Peterson, Janicki, and Swing 1981; Peterson, Wilkinson, Swing, and Spinnelli 1984) has suggested that an effective alternative or adjunct to whole-class instruction is to have students work togther in small cooperative groups. For example, Peterson adapted small-group cooperative learning techniques to the format of mathematics instruction typically used in elementary classrooms. In this approach, the classroom teacher teaches the day's mathematics lesson for approximately 20 minutes, and then students work together on their mathematics seatwork assignments in small mixed-ability groups of four students. It was found that positive effects on mathematics achievement apparently are achieved as a result of task-related interaction in the group. Students work on their own mathematics seatwork, but when a student has a problem, another student helps out. Research indicates that students learn by explaining why the answer is incorrect and helping the students come to see the correct answer. In addition, the receiver of the explanation may benefit from receiving an explanation that describes the kinds of strategies and processes a student should use to solve the problem. (See, for example, Webb 1982; Peterson, Wilkinson, Swing, and Spinelli 1984). One can also argue that students learn effectively in small cooperative groups because they become active information processors rather than passive recipients of information that has been presented to them.

Moreover, research shows that "less direct methods of instruction seem to be better for promoting student independence, creativity, and learning of higher level skills" (Peterson 1979). As Fennema and Peterson (1985) suggest, learning of higher level skills in mathematics requires the development of independence and autonomy in the learner. Direct instruction, by definition, places control of instruction and learning squarely in the hands of the teacher and offers little opportunity for student independence and autonomy in learning. This suggests that direct instruction would not be the most effective method of promoting learning of higher level skills in mathematics.

Teachers as Active Cognitive Processors

Research on teachers' thought processes to date supports the thesis that the teacher is a reflective, thoughtful individiual, and, moreover, that teaching is a complex and cognitively demanding process. Teachers' beliefs, knowledge, judgments, thoughts, and decisions have a profound effect on the way they teach as well as on the way students learn in their classrooms. These factors also affect how teachers perceive and think about the training and new curricula they receive and the extent to which they implement the training and curricula as intended by

411

the developers.

From this research it is clear that any reform in mathematics teaching and education in the 1990s must actively involve teachers in the very process of reform for it to be effective. Reform efforts that view the teacher as a passive recipient, such as attempts to develop "teacher-proof" curricula, are likely to fail. In contrast, reforms that take into account teachers' beliefs, perspectives, and knowledge, and involve teachers in planning and decision-making and treat them as professionals are more likely to succeed. Instructional design should not be prescriptive; instead, it should provide a framework, a knowledge base, alternative materials for instruction, and a variety of appropriate assessment techniques to enable teachers to make informed decisions.

Conclusion

Most cognitive and instructional science research does not provide radical new conceptions of how children acquire mathematical concepts and skills or revolutionary methods of instruction. Many of the findings are consistent with the intuition and observations of experienced teachers and curriculum writers. But current research provides a level of precision and rigor that offers some hope of moving things forward in the development of the mathematics curriculum rather than simply riding another swing of the pendulum. For example, mathematics programs developed in the 1960s included noncanonical number sentences. For the most part, noncanonical number sentences were dropped from the curriculum because teachers found it difficult to teach the approach that was used at that time. Noncanonical sentences were solved by transforming them to correspond canonical sentences. Current work suggests that such transformations require relatively advanced concepts (Briars and Larkin 1984; Riley et al. 1983). However, children can solve noncanonical sentences directly using the same processes they use to solve corresponding word problems (Blume 1981; Groen and Poll 1973). Thus, the difficulty was not with noncanonical number sentences; it was with the approach used to teach them. Results from recent studies of addition and subtraction provide a new perspective on noncanonical number sentences and how they should be taught. Whether instruction involving noncanonical number sentences is appropriate is thus once again open to debate.

There are no simple formulas for applying cognitive and instructional science to instruction. Prescriptions for instruction do not follow immediately from this research, and additional research is needed to determine the most effective way to make the connection. There is no single ideal program of instruction that will come out of this effort. However, development of new curricula should be consistent with what we know about the processes of learning and instruction. If we are going to make real progress in curriculum reform, we need to build upon this knowledge.

Notes

1. The research reported in this paper was supported in part by a grant from the National Science Foundation (Grant No. MDR-8550236). The opinions expressed in this paper do not necessarily reflect the position, policy, or endorsement of the National Science Foundation.

2. To solve a problem like 8 + 5 by counting on, a child starts counting at 8 and counts on 5 more counts. This requires an understanding of several basic counting principles and the acquisition of specific counting skills (Fuson 1982).

References

Blume, G. 1981. "Kindergarten and First-Grade Children's Strategies for Solving Addition and Subtraction, and Missing Addend Problems in Symbolic and Verbal Problem Contexts." Unpublished doctoral dissertation, University of Wisconsin-Madison.

Briars, D.J. and J.H. Larkin. 1984. "An Integrated Model of Skill in Solving Elementary Word Problems." Cognition and Instruction 1(3): 245-296.

Carpenter, T.P. 1985a. "Learning to Add: An Exercise in Problem Solving." In Teaching and Learning Mathematical Problem Solving: Multiple Research Perspectives, E. Silver, ed. Hillsdale, NJ: Lawrence Erlbaum.

Carpenter, T.P. 1985b. "Research on the Role of Structure in Thinking." Arithmetic Teacher 32: 58-60.

Carpenter, T.P., J. Hiebert, and J.M. Moser. 1981. "Problem Structure and First-Grade Children's Initial Solution Processes for Simple Addition and Subtraction Processes." Journal for Research in Mathematics Education 12: 27-39.

Carpenter, T.P., J. Hiebert, and J.M. Moser. 1983. "The Effect of Instruction on Children's Solutions of Addition and Subtraction Word Problems." Educational Studies in Mathematics 14: 55-72.

Carpenter, T.P. and J.M. Moser. 1983. "The Acquisition of Addition and Subtraction Concepts." In The Acquisition of Mathematical Concepts and Processes, R. Lesh and M. Landau, eds. New York: Academic Press.

Carpenter, T.P., and J.M. Moser. 1984. "The Acquisition of Addition and Subtraction Concepts in Grades One Through Three." Journal for Research in Mathematics Education 15: 179-202.

Carpenter, T.P., J.M. Moser, and H.C. Bebout. In press. "The Representation of Basic Addition and Subtraction Word Problems." Journal for Research in Mathematics Education.

Case, R. 1985. Intellectual Development: Birth to Adulthood. New York: Academic Press.

Chi, M.T.H., R. Glaser, and E. Rees. 1982. "Expertise in Problem Solving." In Advances in the Psychology of Human Intelligence, R. Sternberg, ed. Hillsdale, NJ: Lawrence Erlbaum.

Clark, C.M. and P.L. Peterson. 1986. "Teachers' Thought Processes." In Third Handbook of Research on Teaching, M.C. Wittrock, ed. New York: Macmillan.

414

Conference Board of the Mathematical Sciences. 1982. "The Mathematical Sciences Curriculum K-12: What Is Still Fundamental and What Is Not." Washington, D.C.: National Science Foundation.

DeCorte, E. and L. Verschaffel. 1983. "Beginning First-Graders' Initial Representation of Arithmetic Word Problems." Paper presented at the annual meeting of the American Educational Research Association, Montreal.

Doyle, W. 1975, April. "Paradigms for Research on Teacher Effectiveness." Paper presented at the annual meeting of the American Educational Research Association, Washington, D.C.

Doyle, W. 1978. "Paradigms for Research on Teacher Effectiveness." In Review of Research in Education 5, L.S. Shulman, ed. Itasca, IL: F.E. Peacock.

Dunkin, M.J. and B.J. Biddle. 1974. The Study of Teaching. New York: Holt-Rinehart and Winston.

Fuson, K. 1982. "An Analysis of the Counting-On Solution Procedure in Addition." In Addition and Subtraction: A Cognitive Perspective, T.P. Carpenter, J.M. Moser, and T.A. Romberg, eds. Hillsdale, NJ: Lawrence Erlbaum.

Fuson, K.C. and J.W. Hall. 1983. "The Acquisition of Early Number Word Meanings: A Conceptual Analysis and Review." In The Development of Mathematical Thinking, H. Ginsburg, ed. New York: Academic Press.

Gelman, R. and C.R. Gallistel. 1978. The Child's Understanding of Number. Cambridge, MA: Harvard University Press.

Ginsburg, H. 1977. Children's Arithmetic: The Learning Process. New York: Van Nostrand.

Greeno, J.G. 1978. "Understanding and Procedural Knowledge in Mathematics Education." Educational Psychologist 12(3): 262-283.

Groen, G.J. and J.M. Parkman. 1972. "A Chronometric Analysis of Simple Addition." Psychological Review 79: 329-343.

Groen, G.J. and M. Poll. 1973. "Subtraction and the Solution of Open Sentence Problems." Journal of Experimental Child Psychology 16: 292-302.

Groen, G.J. and L.B. Resnick. 1977. "Can Preschool Children Invent Addition Algorithms?" Journal of Educational Psychology 69: 645-652.

Moser, J.M. and Carpenter, T.P. 1982. "Using the Microcomputer to Teach Problem-Solving Skills: Program Development and Initial Pilot Study." (Working Paper 328.) Madison: Wisconsin Center for Education Research.

National Advisory Committee on Mathematical Education. 1975. "Overview and Analysis of School Mathematics, Grades K-12." Washington, D.C.: Conference Board of the Mathematical Sciences.

National Institute of Education. 1975. "Teaching as Clinical Information Processing: Report of Panel 6, National Conference on Studies in Teaching." Washington, D.C.: National Institute of Education.

Peterson, P.L. 1979. "Direct Instruction Reconsidered." In Research on Teaching: Concepts, Findings, and Implications: 57-69, P.L. Peterson and H.J. Walberg, eds. Berkeley, CA: McCutchan.

Peterson, P.L. and E. Fennema. 1985. "Effective Teaching, Student Engagement in Classroom Activities, and Sex-Related Differences in Learning Mathematics." American Educational Research Journal 22(3): 309-335.

Peterson, P.L. and T.C. Janicki. 1979. "Individual Characteristics and Children's Learning in Large-Group and Small-Group Approaches." Journal of Educational Psychology 71(5): 677-687.

Peterson, P.L., T.C. Janicki, and S.R. Swing. 1981. "Ability and Treatment Interaction Effects on Children's Learning in Large-Group and Small-Group Approaches." American Educational Research Journal 18(4): 453-473.

Peterson, P.L. and S.R. Swing. 1982. "Beyond Time on Task: Students' Reports of Their Thought Processes During Classroom Instruction. Elementary School Journal 82(5): 481-491.

Peterson, P.L., S.R. Swing, M.T. Braverman, and R. Buss. 1982. "Students' Aptitudes and Their Reports of Cognitive Processes During Direct Instruction." Journal of Educational Psychology 74(4): 535-547.

Peterson, P.L., S.R. Swing, K.D. Stark, and G.A. Waas. 1984. "Students' Cognitions and Time on Task During Mathematics Instruction." American Educational Research Journal 21: 487-515.

Peterson, P.L. and H.J. Walberg, eds. 1979. Research on Teaching: Concepts, Findings, and Implications. Berkeley, CA: McCutchan.

Peterson, P.L., L.C. Wilkinson, S.R. Swing, and F. Spinelli. 1984. "Merging the Process-Product and Sociolinguistic Paradigms: Research on Small-Group Processes." In The Social Context of Instruction: Group Organization and Group Processes: 125-152, P.L. Peterson, L.C. Wilkinson, and M. Hallinan, eds. Orlando, FL: Academic Press.

Resnick, L.B. and W.W. Ford. 1981. The Psychology of Mathematics for Instruction. Hillsdale, NJ: Lawrence Erlbaum.

Riley, M.S., J.G. Greeno, and J.I. Heller. 1983. "Development of Children's Problem-Solving Ability in Arithmetic." In The Development of

Mathematical Thinking, H. Ginsburg, ed. New York: Academic Press.

Romberg, T.A. 1984. Chair. "School Mathematics: Options for the 1990s." Report of Conference held December 5-8. Madison, WI: University of Wisconsin.

Secada, W.G., K.C. Fuson, and J.W. Hall. 1983. "The Transition from Counting-All to Counting-On in Addition." *Journal for Research in Mathematics Education* 14: 47-57.

Shavelson, R.J. and P. Stern. 1981. "Research on Teachers' Pedagogical Thoughts, Judgments, Decisions, and Behavior." *Review of Educational Research* 51: 455-498.

Shulman, L.S. and A.S. Elstein. 1975. "Studies of Problem Solving, Judgment, and Decision Making: Implications for Educational Research." In *Review of Research in Education* 3, F.N. Kerlinger, ed. Itasca, IL: F.E. Peacock.

Steinberg, R. 1985. "Instruction on Derived Facts Strategies in Addition and Subtraction." *Journal for Research in Mathematics Education* 16(5): 337-355.

Thornton, C.A. 1978. "Emphasizing Thinking Strategies in Basic Fact Instruction." *Journal for Research in Mathematics Education* 9: 214-227.

Thornton, C.A., G.A. Jones, and M.A. Toohey. 1983. "A Multidimensional Approach to Thinking Strategies for Remedial Instruction in Basic Addition Facts." *Journal for Research in Mathematics Education* 14: 198-203.

Webb, N.M. 1982. "Student Interaction and Learning in Small-Groups." *Review of Educational Research* 52: 421-445.

Weinstein, R.S. 1982. "Students in Classrooms." *Elementary School Journal* 82: 397-540.

Weinstein, R.S., H.H. Marshall, K.A. Brattesani, and S.E. Middlestadt. 1982. "Student Perceptions of Differential Teacher Treatment in Open and Traditional Classrooms." *Journal of Educational Psychology* 74: 678-692.

Winne, P.H. and R.W. Marx. 1982. "Students' and Teachers' Views of Thinking Processes Involved in Classroom Learning." *Elementary School Journal* 82: 493-518.

Wittrock, M.C. 1986. "Students' Thought Processes." In *Handbook of Research on Teaching*, 3rd ed., M.C. Wittrock, ed. New York: Macmillan.

Lessons Learned from the First Eighteen Months of the
Secondary Component of UCSMP

Zalman Usiskin

University of Chicago

The title of this paper comes from the title of a similar paper by Ed Begle, written after 14 years of work on SMSG. Begle's paper contains what I believe to be superb advice. We have followed much of his advice and can confirm more. He wrote:

> ...the textbook has a powerful influence on what students learn. If a mathematical topic is in the text, then students do learn it. If the topic is not in the text, then, on average, students do not learn it. ...This is an important finding, since the content of the text is a variable that we can manipulate. In fact, it seems at present to be the only variable that on the one hand we can manipulate and on the other hand does affect student learning.

Today we know that manipulating tests can also affect student learning, but, in theory, tests should follow rather than lead the curriculum. So we have taken Begle's advice and have concentrated on textbook production rather than on teacher training, which some would advocate. Yet Begle's statement also has implications for teacher training, for the textbook used by teachers has a powerful influence on what teachers feel they need to know. If a mathematical topic is in the text, then teachers feel they should know what it is about. If not, then teachers feel they can ignore it. Thus only after the creation of text materials are we planning to work on teacher training.

We have also concentrated on changing the content we want taught rather than the approach we wish teachers to take in teaching. Again we are following Begle dicta: "(T)he successful exploitation for instructional purposes of an interaction between a student's aptitude and a pedagogical treatment has not been demonstrated" and "(T)he best predictors of performance at the higher cognitive levels of understanding, application, and analysis seldom include computational skills."

Begle drew a number of conclusions concerning individual differences among learners. In what seems to me an obvious attack on Piagetians, he said "...we see now that the location of specific topics in the curriculum should be based not on the age of the student but rather on the overall structure of mathematics."

Begle was emphatic in believing that mathematics can be learned by all. "...the difference in ability between the low achievers and the above-average students is quantitative but not qualitative."

We can give preliminary confirmation to these statements. Reactions to our Transition Mathematics materials are almost identical among 7th, 8th, and 9th grade students and teachers, despite the fact that the 7th graders are students who score in the top half of students nationally and the 9th grade students are in the bottom half. Things that are easy for one grade level are easy at all grades; things difficult for one are difficult for all; specific problems that are nice at one level tend to be nice at all levels; and so on.

There is a pervasive view in our country that paper and pencil skill is easy, while theory and application are hard. Begle wrote

> ...a decade ago...the opinion was frequently expressed that modern mathematics was suitable only for students with high IQs. However, all our data analyses point in the opposite direction. No such cutoff point appears. There is no reason why any student should be deprived of the advantages of modern mathematics.

Replace the word "modern" with "applications of" and you will find a preliminary lesson learned from our project. We are not hearing any complaints from teachers regarding the amount of applications in Transition Mathematics. Quite the contrary: the applications seem to bring an excitement and an interest to the classroom that was not there before.

A Multidimensional View of Understanding

The result of all this is that we believe that the understanding of mathematics is a multidimensional entity, in the sense that there are independent components that constitute what might be called "full" understanding. There are, I believe, at least five dimensions to this understanding. Four are given strong play in UCSMP materials, in what is called the SPUR approach: Skills, Properties, Uses, and Representations.

The acronym SPUR oversimplifies and distorts what each dimension entails. But before detailing these dimensions, let me indicate the kind of thinking that does not fit this conception of understanding. The presumption that a student has to know how to do multiplication before he can know how to use that multiplication is in effect saying that these two things, doing and using, are in the same dimension of understanding, with using further out on the line than doing. Ordering ideas or concepts in terms of difficulty is only appropriate if these items are in the same dimension.

This multidimensional approach to understanding conflicts somewhat with the organization of knowledge found in Bloom's Taxonomy of Educational Objectives or with those organizations found in NLSMA (Begle and Wilson 1970), NAEP, or IEA studies. In each of these there is an ordering of "knowledge" or "skill," "comprehension" and "application," with the "skill" considered easy or prerequisite for "comprehension," and both the first and second considered easier than and prerequisite to "application." However, within each dimension of SPUR these are suitable categories. The notion of several independent dimensions found in SPUR is closer to the "computation," "concept," and "application" independent sections found in some standardized tests than to the organizations of test items found in research studies.

A first dimension of understanding is the skill-algorithm dimension. It is a full dimension in the sense that it extends from rote memorization through very difficult analyses of procedures. To get a sense of this and other dimensions, consider the skill of multiplication of fractions. A typical question is to multiply 2/3 x 4/5; and it takes little skill to do this. But not all problems are similar. The expert uses a shortcut to do 2/3 x 3/5, a different shortcut for 2/3 x 6/5, and still another for 2/3 x 60. If there are more than two fractions to be multiplied, other shortcuts may be used. The expert knows how to adapt these strategies for mixed numbers, what to do if the numerators or denominators are decimals, and so on. Thus, one of the easiest skills in all of arithmetic still has intricacies. With more complex skills, we might compare methods and invent new algorithms. Those are the highest attainments in this dimension.

A second dimension of understanding concerns mathematical underpinnings. Understanding of multiplication of fractions in this dimension means knowing why the order of fractions can be switched (a low-level attainment) or how the algorithm may be derived from the field properties (a high-level attainment). We know that this dimension is independent of the first because not only do many people have a skill without knowing anything about the general properties underlying the skill, but the reverse is often true as well, though not necessarily for the multiplication of fractions. You and I are probably rather knowledgeable regarding the mathematical theory behind integration, but that has little relation to our ability to calculate integrals of particular functions. The highest attainments in this dimension are the proofs of new theorems. We may view the new math movements of the 50's and 60's as an attempt to incorporate this dimension of understanding into the curriculum.

A third dimension of understanding is the use-application dimension, as measured by the ability to apply a concept. Obviously, one may be able to multiply fractions without having the foggiest notion of when or how this idea might be applied. Yet it is easy to teach students that, given two independent events, the probability of their both happening is the product of the probabilities of each, and that determining this probability often involves multiplying fractions. That is a low-level attainment in this dimension. A higher-level attainment might be the realization that a situation can be treated in this way that does not look that way, as for example, realizing that winning two games in a row might involve the multiplication of fractions. In this dimension, the highest attainment is the discovery of new applications. Current movements to raise the importance of applications in the curriculum have increased the visibility of this dimension.

These three dimensions are preeminent as goals of schooling in mathematics, but the order of their importance depends upon who is asked. The typical adult views understanding as synonymous with the ability to operate in the skill dimension. Ask an adult if he or she understands multiplication of fractions, and a typical response may be "I think (or I'm not sure) I remember." Many mathematicians, on the other hand, do not view skill in any degree as an aspect of understanding. If a college professor says that most of his prospective elementary school teachers do not really understand multiplication of fractions, we know exactly what he means — they don't know the mathematics behind the algorithm.

Most engineers do not consider either skill or ability to undertake the mathematical derivation as true understanding of the mathematics. The ability to use the mathematics is the ultimate test.

These are not the only ultimate tests. To psychologists, the ability to represent mathematics, either concretely or by pictures or by metaphors, constitutes actual understanding. This is the representation-metaphor dimension. Is the student capable of representing multiplication of fractions by area-by-array models, or perhaps in the language of stretchers and shrinkers? Is the student capable of giving a situation — real or fanciful — depicting multiplication of fractions? (Sometimes there is overlap with the application or other dimensions.) We know this dimension to be independent of the others as well: There are students who can demonstrate subtraction with borrowing using bundles of sticks but who cannot do the same subtraction with the symbolic representation; and there are students who can manipulate symbols but who know no concrete embodiments and have no picture in their minds of what they are doing.

In Transition Mathematics, we try to work in all four of these dimensions simultaneously, because all of them are important and because we feel that students have tendencies to favor one or more of the dimensions in attaining knowledge about a concept.

There is at least one more dimension, the historical-cultural dimension. Who first multiplied fractions? This is a low-level question, even if we do not know the answer. How did fractions arise? — a higher-level question. We have all had the experience of a concept being widened, even changed, when we find out the origins of the idea. Often what happens is that we gain an appreciation of the elegance of modern notation, theory, and usage. In general, discussing a different dimension than those one is familiar with — such as giving an application of complex numbers, or doing the mathematics behind fractals, or showing how to calculate cube roots with a four-function calculator — always widens one's view of a concept. Reversing this idea, it might be reasonable to define a concept as an idea that is multidimensional.

In the SPUR framework, if problem-solving applies to the confrontation of situations with which the solver is unfamiliar, then problem-solving may appear in

422

any dimension.

One of the lessons learned from the first eighteen months of work with this conceptualization is that it provides a convenient way of explaining what we are about and of ensuring that students are exposed to a variety of viewpoints for each topic. It is not a perfect sorter of knowledge; like geometric figures, many ideas are more than one-dimensional. But this conceptualization enables us to begin with application, or with theory, or with skill, or with a representation, without feeling that we are compromising the student's ultimate understanding of the concept.

Scientific Calculators

We have learned that there is a significant difference between the use of calculators and the use of scientific calculators. The first draft of Transition Mathematics was written assuming calculators. Some students had scientific calculators but most had four-function calculators. For the revision, we made a decision to require scientific calculators because at an advisory board meeting, the question was raised and we could not think of sufficient cons to counter the pros: scientific calculators enable us to enter very large and very small numbers, e.g. the population of the world; four-function calculators give error messages or cannot do operations that can be performed simply because of their insufficiency and thus lead students to a narrow view of number operations; special keys such as the π key and ! keys can be utilized to advantage; and a scientific calculator is the one to buy if a student wants one calculator to be adequate for many years of schooling.

The curriculum implications are considerable. Traditionally, powers and exponents, particularly negative exponents, come late in most books at this level. With scientific calculators, the sooner the better. Scientific notation is a necessity, not a frill. Square roots are a cinch. A wealth of applications involving larger and smaller numbers are accessible. Thus both scope and sequence of content are affected.

The practical considerations involved in requiring calculators are not as difficult as most mathematics educators believe. High schools using our materials require students to buy the calculators; junior high schools tend to buy classroom sets. Only solar-powered calculators are considered, so there is no problem with batteries. (When asked, we recommend the TI-30 SLR because it has the needed functions and is easy to use.) Schools can buy three classroom sets of scientific calculators for the cost of one computer. But that need not be the way they are bought. One school district in our area has the perfect solution. They considered a calculator as a textbook. They formed a "calculator adoption committee" and distributed calculators in September along with other books. If a student loses his calculator, the school follows its policy on lost textbooks. The calculator costs only a little more than other textbooks and has about the same life span. This school has taught us a lesson that we are in turn sharing with other school districts.

The teachers unanimously report that students enjoy working with the calculators and love them. Most teachers report that it enables them to go through many more problems than they could before. The only problem is that at times they would prefer students not use them, so they don't always allow them. We have not heard of a single parental complaint.

Grouping of Students

Above are some of the positive lessons we have learned. There are negative ones as well, and some concern grouping. Last year I wrote Transition Mathematics as I was teaching a class of ninth graders who did not get into algebra. In the particular school where I was working — Glenbrook South High School, situated in Glenview, an affluent suburb of Chicago — these students were in the 10th-25th local percentile. As is true in many high schools, the teachers prefer not to teach students of this relative ability. The classes are supposed to be tough, but mine was the easiest class I had taught in years.

Four students in my class were classified as LD, learning disabled. I would never have known this from their performance. Were they cured? If so, why were they still so classified? In fact, we know that students' initial classifications as LD are usually a result of reading or language problems. But for years they have been treated differently in all subjects, and it is considered a surprise when they can do anything well. But these kids are not dumb! All of them could read

424

without hesitation, and many could read quite well. Nor are they disabled in mathematics, except as a result of an impoverished background.

They are not alone. Most students in this class came from low-level classes in 7th or 8th grade in junior high school or an even lower-level class in this high school into which they had been placed because they were poor at paper-and-pencil skills. I earlier indicated my belief that such skills are not necessarily low-level and that they are not a necessary prerequisite for all sorts of understandings in mathematics. But "good" elementary schools do not operate with this belief. Arithmetic skills are treated as if they are easier than anything else and the prerequisite for studying everything else. When students are found to be deficient in these skills, the teacher gives them additional work. In essence, <u>we discover what a person is bad at, and then give him or her nothing else.</u> While the other students are learning geometry, probability, and having experiences with all sorts of problems, these students had some paper-and-pencil skills, e.g. they could multiply decimals. But they could not order decimals. Of what use is it if you can multiply but not order? The extent of their conceptual impoverishment can be seen from the following question, hard even after a few days of instruction: "Jarmila Kratochvilova recently set a world record for women in the 400-meter run. Her time was 47.99. If she ran this distance one tenth of a second faster, what would the new record be?" Only three students in a class of 27 answered this question correctly.

The scores of students from Japan and Hong Kong demonstrate that average students can learn mathematics at a level we think requires giftedness and acceleration. I can only conclude that our practice of denying groups of students the opportunity to learn mathematics for the ostensible purpose of remediation is effectively <u>decelerating</u> large numbers of students. School districts that teach algebra in the eighth grade may view this practice as acceleration, and it may be wise politically for them to take this attitude so that parents will not feel bad if their child doesn't make it, but in UCSMP we have come to view the teaching of algebra in the eighth grade as necessary to avoid deceleration.

Furthermore, the more we delay algebra for these students (and for average students), the harder we may be making the course. For if algebra is indeed a language to be mastered, we know from experience with foreign languages that the older you are, the more difficult the learning becomes.

Standards

The problem of standards is difficult and multifaceted. One of the Transition Mathematics teachers asked, "I've never had a class that learned so much, but my grades are lower than usual. What should I do?" A second noted that she had some students who were failing her class, but she didn't want to transfer them out, because they would get so much more out of this class.

The biggest difference we've seen in our 7th and 8th grade classes compared to our 9th grade classes — in our area, the former are in K–8 or junior high schools while the latter are in senior high schools — is that many junior high school teachers use the same percentage scale regardless of content. Most commonly, 90% is an A, 80% a B, 70% a C, and 60% passing. It makes no difference whether the content is multiplication of whole numbers or graphing of lines or applying percent.

We thought we knew about this problem. In the first pages of the teacher's edition read by these teachers, we stated that a fixed curve should not be used. We elaborated on this point. But not until we further elaborated to these teachers that normally much of what they have usually taught is review and that little involves any understanding other than memorized skills; that we have to have less review of old things and ought to cover many more aspects of understanding; and that we have recourse to continual review only because we do not expect immediate mastery, were we able to get the point across. It took a full half-year.

The problem seems to be even greater in earlier grades in the elementary school, where customary levels of mastery are even higher. Fixed grading scales are a barrier to the asking of anything but routine questions.

A second question regarding standards is "What constitutes sufficient performance for entrance into algebra?" Here standard practice defies logic. Consider a student who scores at the 80th percentile nationally on a test designed for entrance into algebra. What are the chances that the student will be placed into algebra? I purposely have not mentioned the grade level of the student

426

because the grade level seems to be at least as important as the score in determining whether the student gets into algebra. Our experience this year is with students taking such tests in grades 6 through 9. Here is what I can report.

If the student is a 6th grader, he or she will not be placed in algebra in the 7th grade because it is not available. The score means next to nothing. Even the 90th percentile is not enough in most districts unless the parents complain and/or the student is a troublemaker. There is one school district in which Transition Mathematics is being used at the 6th grade level with the best students. These are students equal to the best anywhere in the country. Although we are testing these students, the use is unofficial in the sense that the 6th graders do not constitute a target population and cannot be used to test the materials. (Note added later: Our testing shows them to be the students who score highest after using the materials.) It took special efforts to convince even this very forward-looking school district to offer an algebra course for these students in the 7th grade. In fact, the first reaction of the school district was to impose higher standards for admission into 7th grade algebra than they had for admission into 8th grade algebra.

Now back to our hypothetical student. If the student taking the entrance test is a 7th grader, he or she will not be placed in algebra in the 8th grade because you are supposed to be in the top 5% or 10% to get into that course. Scoring at the 80th percentile is not high enough.

If the student scoring at the 80th percentile is an 8th grader, he or she will be placed into algebra the next year, no questions asked.

If the student is a 9th grader or above and scores at the 80th percentile, of course he or she will get into algebra. In fact, such a student gets into algebra not by a score on any test, but by passing a prerequisite class, even with a C. Standards are lowered for older students so that no standardized test results are needed.

The general pattern thus emerges, the older you are, the easier it is for you to get into algebra. Yet just the opposite should be the way schools operate. It should be easier for a younger student to get into algebra if the student has demonstrated the same amount of knowledge. After all, if a student has learned

427

things faster, it would follow that any shortcomings would be more quickly erased. But it is a turf battle, not subject to logic. Senior high schools tend to believe that only their teachers can teach algebra. They allow feeder schools to teach algebra only to those students who are deemed so bright that they cannot be destroyed, to avoid parental complaints, and because they need those students to populate their Advanced Placement classes. They justify their practice of carefully controlling who takes algebra not by extolling the virtues of all those who have performed well, but by pointing to those few students who really didn't do so well in the course after all.

There is an analogy with the consideration of calculus in the high school by many college professors of mathematics.

The implication for UCSMP from their lesson is that we will have to determine our own standards for admission into our algebra. The problem for us is that we will not be able to know what these standards should be until after we have tested our materials.

A disturbing number of high school students today fail algebra. I do not know the exact percent of students who fail or withdraw before failing, but it seems to be between 15% and 25%. I have heard of city schools in which the failure rate is 50%. In a county outside Washington D.C., the mathematics supervisor reported the following: There are 23 high schools in the county, and they can be ranked by the mean scores on 8th grade tests of the students they accept. The school that has the highest-scoring students also has the highest percentage of failures in algebra. It seems clear that there exist today no consistent standards for passing algebra.

In our curriculum, we are planning to teach algebra one year earlier than normal. We are planning to teach it to many students who in the past would not be allowed to take algebra so early. (Like the case of teaching a foreign language, this earlier teaching should result in better performance.) And we are planning to introduce considerable differences in our materials from the existing first-year algebra texts. It is clear that the lack of standards is going to present problems in implementation.

There seems to be a dearth of literature regarding the problems of standards that I've presented here. We are going to have to suggest reasonable standards for our course. How should we do this? We welcome advice.

Let me summarize what we have learned from the first 18 months of the project. We have confirmed many of the lessons learned in earlier projects and benefited from them. We have developed a theory of understanding that helps us in conceptualizing, writing, testing, and discussing our curriculum. We know how to implement calculators and we strongly recommend that secondary schools utilize scientific calculators. We have seen first-hand the detrimental effects of segregating students into activities or classes whose purpose is solely remediation, and we have come to view such segregation as effectively decelerating large numbers of students. Finally, we have identified a variety of problems caused by non-uniform, non-existent, or illogical standards of performance. We have learned that we know almost nothing about how to deal with these. We might take solace from the saying of Lao-tzu, "To know that you do not know is the best," but Lao-tzu also said: "He who knows does not speak." How will we find out the answers?

References

Begle, Edward G. 1973. "Some Lessons Learned by SMSG." The Mathematics Teacher 66 (March): 207-214.

Begle, Edward G. and James W. Wilson. 1970. "Evaluation of Mathematics Programs." In Mathematics Education, the Sixty-ninth Yearbook of the National Society for the Study of Education. Chicago: NSSE.

Bloom, Benjamin et al. 1956. Taxonomy of Educational Objectives: Cognitive Domain. London, Longmans.

National Assessment of Educational Progress. 1981. Mathematics Objectives. 1981-82 Assessment. Report No. 13-MA-10. Denver: Education Commission of the States.

Usiskin, Zalman. 1979. Algebra Through Applications. First-Year Algebra via Applications Development Project, University of Chicago, 1976. Reston, VA: National Council of Teachers of Mathematics.

Usiskin, Zalman et al. 1985. Transition Mathematics. Chicago, IL: University of Chicago School Mathematics Project, Department of Education, University of Chicago.

Teaching Addition, Subtraction, and Place-Value Concepts

Karen C. Fuson

Northwestern University

This paper discusses current progress in the use of research findings (concerning how children think about mathematical concepts) in developing crucial units in the elementary school arithmetic curriculum. Each unit is designed to teach representations of concepts that are cognitively and mathematically more advanced than those children know at the beginning of the unit. Research findings on developmental sequences in children's thinking about these concepts and the cognitive and practical difficulties they experience in using them serve as the basis for the design of the units. The goal is to develop units that can be taught by regular classroom teachers with only a small amount of in-service training. Skills and concepts prerequisite to each unit are to be specified, and the units themselves will be designed to fit into a variety of curricular approaches and classroom organizations. Suggestions for appropriate time-of-year placement within grade levels for above-average, average, and below-average students will be made, based on classroom trials.

There is considerable concern at present about the mathematics achievement of American children as compared to that of children in other countries (e.g., Japan, Soviet Union). In general, topics are presented later, and in some cases considerably later, in American curricula (Fuson, Stigler & Bartsch 1986). Thus, an important goal is to demonstrate that certain crucial computational concepts can be taught much earlier than is now the case in American classrooms. Further goals are to teach these concepts in ways that will prevent misunderstandings and confusion from arising, help children apply the concepts more readily, and use no more class time than is now the case so as to make room for teaching certain mathematical concepts now often largely ignored in the elementary school curriculum (primarily geometry, measure and problem-solving).

The development of each unit starts with a study of the research literature and an examination of the usual textbook approaches to teaching the concepts in the unit; proceeds to an identification of cognitively and mathematically more mature representations the unit will attempt to provide; and finally specifies means of supplying such representations. A preliminary version of the unit is developed, after which it is taught by the project staff to a group or classroom of children, or project staff work with a teacher responsible for this initial teaching. The progress of individual children is followed quite closely during this period to facilitate making changes in the unit as needed if the concepts are not being learned. The unit then follows a second progression, starting from instruction by project staff to instruction by teachers with no direct contact with project staff. Here project staff may co-teach a unit with a teacher while development and modification of the unit continues. Later on teachers may teach the unit independently based on lesson plans developed by the project and possibly a short in-service meeting with project staff. During this stage teaching is monitored quite closely by project staff to determine precisely how teaching of the unit is diverging from the desired path, and adjustments are made in the lesson plans to minimize such divergence. At later points in this progression teachers who have already taught a particular unit become in-service advisers for new teachers. Other means of indirect in-service training for new teachers of the unit may also be employed. We view this second progression as a vital but often ignored aspect of the development of new curricula whose omission can interfere with the implementation of new approaches no matter how meritorious they may seem.

In the 20 months of the project to date, two units (addition and subtraction of single-digit numbers with sums between 8 and 18) have passed through the second progression with new teachers teaching the units who had been taught by teachers who had taught them once before. Three units (addition and subtraction of multi-digit numbers with between two and ten places, and application of place-value concepts) have reached the stage where they are taught directly from lesson plans. Three other units (addition and subtraction word problems) have moved from the assessment of research literature and ordinary teaching approaches through development to initial staff teaching of trial units. Considerable project effort has been expanded in assessing the effectiveness of the developed units. This paper will summarize results to date concerning the teaching of the first five units.[1]

Teaching Counting-On for Addition and Counting-Up for Subtraction

Research indicates that children in many countries follow a developmental progression in the counting procedures they spontaneously invent to solve addition and subtraction problems (Carpenter and Moser 1984; Davydov and Andronov 1981; DeCorte and Verschaffel 1984; Fuson 1982; Fuson 1984; Svenson and Broquist 1975; Steffe, von Glasersfeld, Richards, & Cobb 1983). For addition problems, children move from object counting all (e.g., for 8 + 6, "1,2,3,4,5,6,7,8,9,10,11,12, 13,14" where 8 objects and 6 objects are counted all together) to either object counting on (for 8 + 6, "8,9,10,11,12,13,14," with six objects available to keep track of when to stop counting) or verbal sequence counting all (for 8 + 6, counting "1,2,3,4,5,6,7,8,9,10,11,12,13,14," keeping track only of the words beyond 8 by using successively extended fingers or some other means), and finally verbal sequence counting on ("8,9,10,11,12,13,14," keeping track of the six words counted past 8 by successively extending fingers or by some other keeping-track method). This development may extend over a two- or three-year period and varies somewhat with the size of the numbers involved (more advanced performance occurs when one or both numbers are very small). Most children complete this progression sometime in the first three grades of school. Children also move beyond this progression of counting solution procedures to direct memory of addition facts and to the use of certain addition facts to produce other facts. At any given moment most children use a mixture of various solution procedures, with the procedure used depending on the size of the numbers involved and on individual preferences.

The developmental progression for solving symbolic subtraction problems (e.g., 14 - 8 = ?) by counting is not so clear. Almost all American textbooks introduce subtraction as object take-away (14 objects then take-away 8 objects; how many objects are left?) Take-away is thus the common solution procedure used by children when objects are available. This take-away conceptualization seems to lead children most naturally to verbal sequence counting down ("14,13,12,11,10,9,8,7,6" 14 - 8 is 6, where the eight words counted down from the 14 are kept track of by successively extending fingers or by some other method). For some problems, children also use verbal sequence counting down to (for 14-11, "14,13,12,11" while extending three fingers) or verbal sequence counting up to (for

14 - 11, "11,12,13,14" while extending three fingers). The counting–down solutions are much more difficult for children than is the counting-up-to procedure because counting backwards is not nearly as well automatized as is counting forwards (Fuson, Richards, and Briars 1982; Bell and Burns 1981); children show other difficulties with counting down as well (Baroody 1984; Fuson 1984; Steinberg 1984). For verbal subtraction problems, younger children (e.g., first graders) typically model the verbal problem directly with either an object or verbal sequence solution procedure, and then later (most children by third grade) use the subtraction solution procedures more flexibly, with many choosing the simpler counting-up-to procedure even for take-away problems (Carpenter and Moser 1984).

In the first year of the project, we developed a teaching unit to move children from the first step in the addition progression (object counting all) through a middle step (object counting on) to the final step in the counting solution procedures (verbal sequence counting on). The first part of this unit was based on teaching skills identified as necessary for counting on (Secada, Fuson, & Hall 1983). The second part addressed a practical problem we had observed in those children who use their fingers to keep track of numbers they are counting on. Whenever the number to be counted on was greater than 5, children would put down their pencils to use the fingers on both hands for keeping track; some also put down their pencils for every problem to keep track with their preferred (writing) hand. This putting down and picking up of the pencil consumed a great deal of time. Thus a method of keeping track of up to nine words counted on through the use of the fingers of only one hand (the nonwriting hand) was desirable. The Chisanbop finger patterns for the numbers 1 through 9 were chosen. However, in our application, finger patterns are successively matched with the counting on from the first addend; thus, the fingers show the second addend as they do when children spontaneously use their fingers for keeping track. Contrast this with the way fingers are usually used in Chisanbop: there the fingers count up the final sum count and words are said aloud for the second addend "1,2,3,4,5,6". Our finger pattern method (see Figure 1) was taught in the first year quite successfully by five regular classroom teachers to average and above-average first graders and to below-average second graders (Fuson and

Secada, in press).

In the first year of the project a teaching unit was also developed for teaching children how to solve symbolic subtraction problems such as 13 - 8 by counting up from 8 to 13. This unit began by presenting take-away, comparison, and equalizing story problems, followed by a discussion of how each of these situations involves subtraction. These stories then provided the settings for teaching subtraction as counting up to, with the number of words counted being kept track of by means of finger patterns. Note that with finger patterns it is actually easier to subtract than add since counting in subtraction stops when the big number (e.g., 13) is reached; the finger pattern is then just read off as the answer (e.g., 5). For addition problems, the finger patterns need to be overlearned so that children can recognize a given finger pattern (i.e., the one for the second addend) and stop counting on when it is reached. The same five teachers taught the subtraction unit (Fuson and Hall 1984; Fuson 1986). Children had no trouble learning to count up for subtraction. Their subtraction performance on the more difficult single-digit problems (one addend between 5 and 9 and the other between 3 and 9) was equivalent to their addition performance in terms of both speed and accuracy. Interviews with children taught to subtract by counting up also indicated good and equivalent performance on take-away, comparison, and equalizing subtraction story problems. By comparison, typical American performance at the first grade shows take-away performance above the other two.

In the second year work on these units addressed a number of concerns. To start with, the five first-year teachers had had three 30-minute in-service sessions with project personnel prior to teaching; two came before or while teaching the addition unit and one came before teaching the subtraction unit. To test one method of extending the use of the units without direct project help in one school, in the second year no in-service sessions were given by project staff, and in the other only a single session was given. Instead, the first-year teachers in the two project schools demonstrated how to teach the finger pattern units to their colleagues.

Second, in the first year skills tests were given regularly throughout both units to track progress and determine where to modify them. Project personnel graded all skills tests overnight and gave a list of low-scoring children to the teachers the next day. Clearly, the units had to be "teachable" without this extra help from the project, so in year two the staff did no grading of skills-progress tests, though they did grade pretests and posttests for research purposes.

Third, teaching the units was extended to include below-average first graders and average second graders; the latter group was included so that they would be able to find the single-digit answers needed in learning the multi-digit addition and subtraction and place-value units.

Fourth, in the first year all teachers, whether teaching average first graders or below-average second graders, taught the units at the same time: in December for the addition unit and in March or April for the subtraction unit. It seemed to make sense to fit these units into a progression of teaching addition and subtraction skills and concepts, varying the time of year for teaching them by the age and mathematics ability of the children. Therefore, in the second year teachers were given the option of deciding when to teach the unit, after which project staff discussed with the teachers a sensible teaching progression and placement of units within the year by age and ability for addition and subtraction concepts and skills.

Fifth, advancing children to the use of more sophisticated and efficient counting procedures is not the end of single-digit addition and subtraction teaching. Many children can move on to even more rapid and low-effort (automatic) solutions such as heuristic strategies (Carpenter and Moser 1984) and memorized facts. Two steps were taken within the project this year to facilitate this further step. Within both the addition and subtraction units it was stressed that some facts were already known and that finger patterns were only to be used for unknown facts. Each worksheet was changed so that it contained two small number problems which the children knew (e.g., 3 + 2), so the children would not get into the habit of routinely using finger patterns for every problem no matter how simple. Second, the performance data from both years indicated that the idea of counting on with finger patterns was quite firmly embedded in most children by the second month after completion of the unit. Therefore, the second-month addition review was followed by a short teaching overview of various heuristic addition strategies (e.g., doubles: 6 + 7 = 13 because 6 + 6 = 12). In the second grade, these overviews were given after the second-month addition finger pattern posttest or during the place-value multi-digit addition unit, whichever came first. It was our intention that such overviews be undertaken periodically throughout the year. However, the number of other activities undertaken in all classes resulted in rather haphazard efforts at helping children move beyond the use of counting strategies.

Sixth, a serendipitous finding from the first year of the project was that a short (twenty-minute) review of counting on with finger patterns two months after completion of the addition unit produced a significant rise in performance. Therefore in the second year we examined the effect of such spaced review and practice (on the test) for the finger-pattern units. Monthly posttests were given at the start of the lesson to assess addition or subtraction performance, finger-pattern counting on or counting up was reviewed, and post-review tests given during the same class. During later months, heuristic strategies rather than counting strategies were the focus of the reviews.

Finally, there still remained some concern about the possible risk of teaching counting on for addition before children were ready for it, and about teaching subtraction as counting up rather than as counting down. For subtraction the main concern was that the take-away approach was already known and therefore was "more natural" to children, so that counting up might have an adverse effect on this conceptual basis for subtraction. An interview study was undertaken with first graders who had been taught subtraction by counting up in the second year and with first graders in other schools taught subtraction by take-away.

Second-Year Results

The time for implementation of the teaching units and the amount of related instruction which had to precede the units varied considerably between classes, mainly as a function of ability level. Eight of the ten classes required between 3 and 5 weeks to complete the addition units. In the remaining two classes (the low-ability Grade 1 classes), these units were preceded by similar units in which

436

smaller sums (between 8 and 12) were involved, and so a total of about 8 weeks was spent for addition in those classes. The subtraction units required less than 3 weeks in all the classes.

The performance data (left-hand columns in Tables 1 and 2) indicate that both the addition and subtraction teaching were successful at every ability level in both grades. Children in every classroom improved significantly on the counting-on test, the addition test, and the subtraction test. Subtraction posttest scores did not differ significantly from the addition posttest scores. As in the earlier study, once children knew how to count on, teaching subtraction as counting up was a relatively simple matter. This seems to be true partly because counting up is easier to execute than is counting on: the feedback loop for stopping the counting when counting up is visual and does not depend upon overlearning the finger patterns.

The decline in some addition scores from the initial posttest to the first one-month test may be due to the teaching of subtraction during that period. Interview results in the first year indicated that some children had forgotten how to add using the finger patterns once they had learned to subtract (Fuson 1986). However, there were several classes where no such decline occurred even though subtraction had been taught in that period. Perhaps initial learning was not so strong in the former classes or perhaps the teachers in the latter classes took steps to reduce interference. Note that in two of the classes where there was a decline in addition performance, a brief review was sufficient to restore it to the posttest level. It would seem worthwhile to study such interference effects to learn how they might be minimized (Hall 1984).

Some combination of the initial test and brief review appear to be beneficial, in that performance often improved from the first to the second test on those occasions. However, the separate contributions of the initial tests and reviews within these sessions could not be determined from our data. In general, there seems to have been improvement in performance over the several-month period following the posttest. However, in some cases it was not possible to track continuously any improvement in the course of the year because of ceiling effects that necessitated a shift to more difficult tests. One purpose of the monthly sessions was to begin development of a program of spaced brief review to retard the process of forgetting and encourage a level of relatively effortless performance. Although the advantages of spaced review have been well known for many years, schools in the United States typically do not implement such procedures in any systematic way.

Teachers used the skills progress tests themselves to varying degrees. Most teachers used the tests and student worksheets to monitor the progress of individual students and then helped children with individual difficulties. In both the addition and subtraction units, there was a longish plateau during which many children were having trouble with some aspect of the new procedure. During this period, it was helpful for the teacher to review important aspects of the procedure at the beginning of each period and only then to help individual students. For some reason, two teachers made little use of the skills progress tests and, in general, did not work at teaching the units. They did not check to see how individual children did on the skills progress tests, and consequently did not know who was experiencing difficulty. Nor did these teachers carry out adequate reviews at the beginning of the class period or work sufficiently with individual children. One of these teachers had taught the units in the first year of the project. Her children were more successful in that year than in the second year when she no longer gathered information on her own on how each student was doing and only rarely even bothered to walk around during class to see how the children were doing. Rather, much of the math period was used by the teacher to work on some other project (such as making bulletin boards) while the children worked on the worksheets. These two teachers happened to be the teachers of the two average first-grade classes. The two teachers of the below-average first grade classes, on the other hand, spent a great deal of time reviewing and helping individual children. The fact that the below-average first graders scored almost as well as the average first graders (see Tables 1 and 2) is probably largely due to these teacher attributes. What is clear from the second year is that the addition and subtraction units must be taught if they are to be successful. A teacher cannot simply make an initial presentation or two and then become involved in unrelated activity while children work on the worksheets alone.

The second year results indicate that teaching of both units can extend up

438

through average second-grade classes and down to below-average first graders, although in the latter case there may be a need for a first counting-on unit with smaller sums (between 8 and 12). Thus even many first graders who are below average in mathematics background and ability seem capable of functioning above grade level in addition and subtraction if taught the counting-on and counting-up units.

Individual trials this year by first-grade teachers have indicated more clearly a reasonable progression of addition and subtraction concepts and skills within which our units fit. Skills which are prerequisite for the units include the ability to read and write numerals up to 20, produce an accurate number word sequence up to 20, count objects accurately up to 20, and count up in the number word sequence starting from an arbitrary number (e.g., "8,9,10,11,12,etc."). In the first year, the units were taught in December, after all the children had been exposed to a conceptual approach to addition with objects for small numbers and knew the sums to 8 as immediate facts. In the second year, the two above-average and the two average first-grade classes began immediately with the addition finger pattern unit. It became clear to both teachers and the project observer that the unit was not moving as it had in the first year (three of these four teachers had taught the unit in the first year). All the teachers then did conceptual work with small number addition. This went quite quickly (a few days) for the above-average first graders but took more time for the average first graders. Work on these conceptual prerequisites and on the skill prerequisites for the below-average first graders took several months.

These trial-and error efforts have produced a sequence for the addition conceptual work on small numbers (sums to 8) which seems to function quite well. This sequence was largely worked out by two of the project teachers from other materials and has now been used by other teachers. The first step here is to encourage children to use objects on part-part-whole sheets to add small numbers: The top half of such sheets consists of one section called the WHOLE while the bottom half is divided into two PARTS. Children put objects for each addend on the part sections of the sheet and push all the objects together onto the whole part of the sheet to find the answer. Children then move on to worksheets that start with the use of dots and symbols for both addends, progress to worksheets

439

where only the first addend is expressed in symbols (children can then count on the dots for the second addend), and finally use worksheets where both addends are expressed only in symbols. Similar conceptual work for small number subtraction is then done, with subtraction presented as a problem in which the whole and a part is known. Practice with fact-family flashcards (triangles with a number at each corner) facilitates memorization of these small number facts. The second-year work to date indicates that proper time placement of finger pattern units in first-grade classes would seem to be in the second and third months for above-average children, in the fourth and fifth months for average children, and sometime in the sixth through eighth month for below-average children.

The addition finger-pattern unit was taught in the second year with insufficient conceptual preparation for average first graders. However, observations of some children indicate that such teaching is not necessarily harmful; rather children assimilate the finger patterns to the more primitive solution procedures they have already constructed. One child was observed to use finger patterns in an object counting-all solution procedure: for 8 + 6 she counted out loud producing the successive finger patterns for 1 through 8 on one hand, after which she counted from 1 to 6 while producing the finger patterns for 1 through 6 on the other hand. Then she counted all the fingers, touching again the finger patterns for 1 through 8 on the first hand, continuing to count "9,10,11, 12,13,14" while touching the successive finger patterns for 1 through 6 on the other hand. The finger patterns enabled her to represent a number above 5 on each hand, and she could therefore represent all 14 objects on her hands. The finger patterns thus avoid a common problem confronting children who still need to use an object counting-all solution procedure: they run out of fingers at ten and cannot represent sums above ten just by using their fingers. The second observation was of a child who used a verbal counting-all solution procedure. She said out loud "1,2,3,4,5, 6,7,8" while doing nothing with her fingers and then made the finger patterns for 1 through 6 while continuing to count out loud "9,10,11,12,13,14." The finger patterns therefore seem to be helpful for all four common addition counting solution procedures. Teaching them to children who are not ready to advance to counting on does not seem to be detrimental (because children seem to adapt them to their own conceptual state) and can even be advantageous (because 18 objects can be represented). However, children definitely do need prior work with objects and with small sums and differences before moving on to the finger-pattern units.

The analysis of the interview data comparing children taught subtraction as counting up and those taught only take-away is still underway.[2] However, one finding bears on the issue of how "natural" counting up for subtraction is for first graders. Randomly chosen children were interviewed after they had been taught counting on for addition but before subtraction had been taught. Thirteen of the 15 above-average first graders who had been taught counting on for addition spontaneously invented counting up to with finger patterns to solve at least one verbal subtraction story (take-away, compare, and equalize stories were given). That is, they figured out how to use finger patterns to keep track as they counted up from one known number to a larger known number in order to find the difference. Therefore counting up for subtraction seems to be rather a natural solution procedure for children who have been taught counting on with finger patterns for addition.

The data presently available from the second year of the project seem quite positive. Teachers are able to teach other teachers the finger pattern units, the units seem to fit nicely into an identifiable progression of addition and subtraction teaching, monthly reviews seem to be a powerful means of maintaining and improving performance, and children taught to count on prematurely seem to assimilate the finger patterns into their more primitive conceptual state. The units allow children to learn how to solve the more difficult single-digit addition and subtraction problems during the first year of school. These results will, of course, be modified as necessary by further analysis of data collected this year.

Teaching Multi-Digit Addition and Subtraction

In the United States, the usual approach to the teaching of algorithms for multi-digit addition and subtraction has at least three major attributes. The first is to start by teaching place-value ideas (by chip trading, place-value charts, or other means), after which the addition and subtraction algorithms are taught under the assumption that children will understand the algorithms because they have learned the prerequisite place-value notions. The second attribute is to parcel out place-value notions and the addition and subtraction algorithms in little pieces over a period of several years: children first learn to add two-place numbers (tens and ones), then a year later how to add three-place numbers, and another year later learn four-place problems, etc. The third attribute is the assumption that children first have to learn to add and subtract using problems that do not require any renaming (trading, carrying) before learning to add with renaming; the latter problems occur in textbooks much later (Fuson, Stigler, & Bartsch 1986). Thus, children's introduction to these algorithms typically extends over three years at least, and there is little opportunity to learn the algorithm as an algorithm, i.e., a repetitive procedure that works the same way in each column regardless of place value. It is therefore difficult for children to learn the real power of these algorithms or to feel powerful themselves as they work large addition and subtraction problems. Place-value concepts also prove quite difficult for young children to learn, so it seems a questionable strategy at best to base the teaching of algorithmic procedures on a good understanding of these concepts.

Our approach is to reverse the usual order of teaching. We teach the addition and subtraction algorithms concurrent with teaching various place-value concepts and begin this combined teaching with four places (the ones through thousands place). Children are introduced to the algorithms by means of a set of physical materials that embody base-ten ideas but also have pieces that are physically larger in the higher places (see Figure 2). When children understand an algorithmic procedure well enough to do it symbolically without recourse to base-ten manipulatives, they move on to symbolic problems involving five to ten places. With this approach, children learn place-value concepts as they learn the addition and subtraction algorithms, but these aspects of place-value are meaningful from the beginning because children see them in use in the algorithm. Trading for both addition and subtraction is fairly easy to understand with the base-ten manipulatives, so we teach problems with trading right away. Thus these problems seem natural rather than special "hard" problems.

We also tried a modification of the usual subtraction algorithm. Usually, one regroups (trades, borrows) in a column if the top number has to be larger and then subtracts in that column, moves to the left and regroups if necessary and then subtracts <u>in that column</u>, etc. Thus, children must alternate between two very different procedures. It is much easier for young children to do <u>all</u> the necessary trading (regrouping, borrowing) first and then do <u>all</u> the subtracting. This method also emphasizes that the <u>top number</u> in the problem is the <u>larger whole</u> which children can then see separated into two parts — the part that is the bottom number and the part that is the answer. In the problem below, 8625 was rewritten so that <u>every</u> top number was bigger than every number in the second row and then <u>all</u> the subtractions were done at once. The number 8625 was separated by the subtraction into two parts: a 3907 part and a 4718 part. This trade-first procedure is similar to but not identical with the low-stress algorithm discussed by Hutchings (1975).

Usual Algorithm

```
              1  15
    8    6    2    5
-   3    9    0    7
    ─────────────────
              1    8
```

trade, then subtract
trade, then subtract, etc.

Simpler Subtraction Algorithm

```
    7   16    1   15
    8    6    2    5
-   3    9    0    7
    ─────────────────
```

first step
do all trades

```
    7   16    1   15
    8    6    2    5
-   3    9    0    7
    ─────────────────
    4    7    1    8
```

second step
subtract each column

The base-ten embodiment we use was chosen partly to give children an intuitive base for metric concepts. Thus the pieces for the ones place are white Cuisenaire rods (1 cm x 1 cm x 1 cm), the tens pieces are orange rods (10 cm x 1 cm x 1 cm), the hundreds pieces are orange flats (10 cm x 10 cm x 1 cm), and the thousands pieces are orange large cubes (10 cm x 10 cm x 10 cm). We made most of the large cubes out of posterboard and had many flats cut cheaply at a local lumberyard. A large cardboard calculating sheet with columns and rows was used to help the children keep the organization of the wood neat. To add, one number is made with base-ten pieces in the top row and the other number is made in the second row. The pieces in the ones column are combined (added) and a trade is made into the next column if there are too many pieces (i.e., 10 or more) to record in the ones column (see Figure 2). Children <u>record</u> the results of the actions in this column <u>before</u> doing anything with the pieces in the next column. The function of using the base-ten pieces is to provide children with a representation of the addition procedure; it is <u>not</u> just to get answers. The links between the operation with the wood and the operation with the symbols must be very close, if the wood is to provide meaning to the symbols (Bell, Fuson & Lesh 1976).

To subtract with base-ten pieces, the larger number is made in the top row with the wood and the smaller number in the second row (see Figure 3). (Subtraction can actually be done in several ways; it is fun for faster children to try to figure out some other ways of doing it with the base-ten blocks.) The top number in every column must be the whole for that column (the parts are the numbers in the second and bottom rows). If the whole (top number) in a given column is not bigger than the part (the number in the second row), more objects can be obtained in that row by trading a piece from the next larger column of the top number. After each column is traded, the trade is recorded on the symbolic problem (see Figure 3). Again, the reason for using the base-ten pieces is to help make sense out of what the children do with the symbols. The link between operations with the wood and operations with the symbols needs to be a very close one. All trades are then made (each one is recorded on the problem just after it is made), and then all the subtractions are done (each one is recorded as it is made). The subtraction can be done by counting up from the part to the whole (by asking "How many more do we need to make the whole?"), by using "over tens" (adding from the bottom number up to the traded ten and then adding the top number to that number), or by any other method.

The multi-digit addition and subtraction units were taught to children after they had learned to count on with finger patterns for addition and count up with finger patterns for subtraction. At that point they knew how to find the answers in any given column.

We developed lesson plans and children's worksheets for both the addition and subtraction multi-digit algorithms and began testing our approach by teaching both algorithms ourselves to an above-average second-grade class. Because this went well, we began to work with two teachers of average second graders (age 7), doing the teaching ourselves for the first two to four days. Once the teacher understood the approach, the teacher took over the teaching. Two teachers of below-average second graders and two teachers of above-average first graders then also taught the units to their classes.

First- and second-grade topics related to place-value concepts were also identified. Six tests of these concepts were constructed and given before the addition instruction. Instruction on these concepts followed the addition testing. Because classes began the multi-digit addition instruction at various times in the school year, they did not all have an opportunity to complete the instruction and testing on all the place-value concepts and on subtraction.

The performance data in Tables 3, 4, and 5 indicate that the addition and subtraction instructional units were successful at each of the ability levels and grades taught. Mean scores for every classroom improved significantly and considerably from the pretest to the posttest on both timed and untimed tests. For addition, five of the classes had means of 88% or above on the test of 2- and 3-digit sums, while the other two classes had a mean above 75% correct. The means on the more difficult 4- and 6-digit problems ranged from 45% to 85% correct performance. Scores dropped on all addition tests after the teaching of multi-digit subtraction. Some children seemed to experience interference between the two multi-digit operations. However, scores generally rose to posttest levels or above on the test given in the next month.

In those classes where there was time for follow-up at least two months after instruction, after two or three months children had gained considerable speed at multi-digit addition; scores on the timed test rose between 27% and 66% above the posttest score. Children were fairly accurate on the timed test (between 67% and 90% correct of the problems tried), though the accuracy data did not show any uniform pattern over time. The high initial scores on the easy (2- and 3-digit) untimed test left little room for improvement. On the more difficult (4- and 6-digit) test, only the class with five months of test data showed a significant improvement.

A single ten-digit plus ten-digit addition problem was given to test whether the children could generalize the multi-digit addition procedure. To test nonrote application of the trading procedure, three nonadjacent columns of the problem required no trading (carrying, regrouping); one point was given for each correct digit in the answer. Children loved doing such large problems. The posttest means here ranged from 8.4 to 9.4 out of a possible 10, indicating highly accurate trading and single-digit calculation when children are highly motivated.

In the case of subtraction, all three classes for whom testing was completed were near ceiling on the test of 2- and 3- digit differences and had means of 50% to 80% on a test of 4- and 6-digit differences. In the two classes for which results at least two months after instruction are available, children got considerably (about 50%) faster at subtracting multi-digit numbers. All three classes showed competence in dealing with problems having zeros in the minuend; the class with the most months of practice reached ceiling on this task. The posttest means on the ten-digit subtraction problem were 8.5 for all classes, again indicating highly accurate trading (borrowing, regrouping) and single-digit calculation. Subtraction instruction was begun but not finished in the other four classes. The crowded end-of-year schedule did not permit even partial testing of subtraction results except in the high first-grade class in school A. A single ten-digit subtraction problem was given to this class. The mean class score on this test was 9.7 out of 10; 18 of the 21 children tested had perfect papers. Thus the subtraction instruction was successful in this class.

The errors children made on the addition and subtraction pretests, immediate posttests, one-month posttests, and the final posttests were examined. For both addition and subtraction, the teaching was responsible for a massive drop in the number of errors and a shift in the kinds of errors made. The errors which reflected little or no comprehension of place value showed a huge decrease. There were some new errors due to mistakes in the trading procedure and others due to interference between trading in addition and trading in subtraction, but these occurred only infrequently (see Fuson, in press, for details).

The results found for the place-value concepts tests are in Table 6. Because most of the posttests were given all at one time rather than immediately after instruction, they functioned as delayed rather than immediate posttests. In the future they might be given both immediately following instruction and at some standard delay period. Performance improved significantly on every test in every class, and in almost every class the improvement was considerable. However, further instruction and testing would seem to be warranted in every area except the "words to symbols" test, where posttest scores were very high. Because specific teaching instructions were not given to teachers about these concepts, except to use the wood in doing the teaching, efforts to ascertain especially effective types of instruction for these concepts might be helpful. For example, the method employed by the grade two average-class school B teacher to teach the topic on word-symbol with trades rather than the method used by the school A teacher would seem preferable. However, in all these tests the wood does seem to be quite an effective support for teaching these concepts. That the children were able to do so well on so many place-value concepts would seem to indicate that their multi-digit addition and subtraction procedures could not simply be rote procedures without an understanding of the underlying bases for the procedure.

Interviews were held with children whose test scores were low on either the addition or subtraction tests to ascertain the nature of their difficulties and the way that using, talking about, or thinking about the procedure with the wood could correct the symbolic procedure (see Fuson, in press, for more details). Ten of the 33 interviewed children did all problems correctly; six more did all of the trading correctly but made an error in the single-digit addition or subtraction. Ten children were able to self-correct their procedure when they were told to think about the wood. Five children had to use the wood. After using the wood for 1-4 problems, these children were able to do problems symbolically without the wood and to think about the wood out loud to describe what they were doing. Two children self-corrected their subtraction procedure with zeros in the top number when asked to explain how they had done their (incorrect) problem.

The results of these interviews underscore an important aspect of the instructional use of materials that physically embody mathematical attributes. In our earlier work with teachers and in this present work with children, we stressed that one must relate the physical material (here, the wood) very closely to the mathematical symbols. If the meanings in the physical material and in any procedure with the physical material are to be transferred to the symbols, each step in the physical procedure must be related to the corresponding step in the symbolic procedure. Furthermore, we stressed the need for children and teachers to develop a vocabulary in which the physical materials and the symbols could be discussed. Both of these important steps are taken at the beginning of instruction, when the children are moving from the physical materials to the symbols. The

447

interviews point out an important, but often neglected, third step that occurs later on in instruction: children and teachers can move in the other direction, from the symbols to the physical materials, to check or correct their symbolic procedure. Furthermore, this move can frequently stop short of the use of the actual physical materials, for many children will have formed a mental representation of the physical materials sufficient for use in checking or self-correcting. That is, they can re-present the physical materials to themselves mentally. The words for the materials and for the symbols can be used at this point by a child to "talk through" the use of these mental representations or by both teacher and children to communicate their mental representations. The interview data indicate that many children can produce and use such mental representations. The instructional goal is to teach the child how to do such talking through spontaneously whenever a problem is encountered.

These results indicate that multi-digit addition and subtraction can be successfully taught much earlier than is now the case in American schools and that the almost universal practice of doling out one more place in each successive grade (two-digits in second grade, three-digits in third grade, etc.) is unnecessary and may even be counterproductive. Use of physical materials that represent the mathematical symbols and procedure is quite effective for learning both multi-digit addition and subtraction procedures and place-value concepts.

Conclusion

The success of units on multi-digit addition and subtraction and single-digit addition and subtraction indicate that children in American schools can learn the computational aspects of whole number addition and subtraction by the end of second grade (first grade for above-average children). Children evidently also can learn place-value concepts along with and after learning the multi-digit algorithms. The results of units on multi-digit addition and subtraction also indicate that for most children the counting-on and counting-up procedures for single-digit addition and subtraction are sufficiently automatic to allow children to concentrate on new aspects of adding and subtracting (trading) in the multi-digit procedures. Clearly, children do not need to learn all their addition and subtraction facts before they can learn to add and subtract large numbers.

Table 1

Counting On and Addition Test Means by Class

Grade Ability School	Counting On Pre	Post	Addition Pre	Post	1 Month BR^a	AR^b	2 Month BR	AR	3 Month BR	AR	4 Month BR	AR	5 Month BR	AR
1 Low A	3	< 11	1	< 9	9	11 s	9	11						
1 Low B	3	< 11	2	< 10 s	$^-$6	< 10	9	11						
1 Average A	3	< 14	2	< 10	$^+$12	12$^+$	X	X						
1 Average B	3	< 12	2	< 11 s	$^-$5	6$^-$	X	X	$^-$8	< 13	10	< 13$^+$		
1 High A	5	< 12	3	< 13 s	13	< 16$^+$	X	X	14	16$^+$	48	< 51	55	< 59
1 High B	6	< 13	7	< 13 s	$^-$11	< 15	X	X	$^+$16	< 19	32	< 38	52	< 66
2 Low A	10	< 14	11	< 14 s	14	< 17$^+$	X	X	16	< 17$^+$	47	52	61	59
2 Low B	8	< 14	7	< 14 s	$^-$11	11$^-$	X	X	15	< 17$^+$	45	44	47	< 51
2 Average A	10	< 15	11	< 17 s	18	18	X	X	18	< 18$^+$	65	65	66	> 60
2 Average B	12	< 15	13	< 17 s	16	17	X	X	$^-$16	15$^-$	65	66	60	< 68

Note. The Counting On scores are the mean number of items correct on a 2-minute 20-item test of addition problems with dots and numerals given for each addend; scores rise considerably when children change from counting all the dots to just counting on the dots for the second addend. Addition scores are the mean number of items correct on a 2-minute 20-item test of the more difficult single-digit sums--those with sums between 11 and 18, excluding doubles $(a + a)$. The addition scores for 4M and 5M of the bottom 6 rows in the table are the mean number of items correct on a 3½ minute test of all 100 addition facts. The ability groups are math ability (assigned before any of our instruction) based on tests and teacher judgment.

< indicates a t-test significant at the .05 level.

- indicates a score significantly lower than the posttest $(p < .05)$.

+ indicates a score significantly higher than the posttest $(p < .05)$; the scores on the 100 item test cannot be tested with the posttest.

s denotes the timing of the subtraction teaching. ^{a}Before Review ^{b}After Review

X denotes no test given.

Table 2

Subtraction Test Means by Class

Grade Ability School	Pre	Post	Subtraction 1 Month BR[a]	1 Month AR[b]	2 Month BR	2 Month AR	3 Month BR	3 Month AR	4 Month BR	4 Month AR	5 Month BR	5 Month AR
1 Low A	1	< 11										
1 Low B	2	< 8	9	9	8	11						
1 Average A	2	< 9										
1 Average B	0	< 11	9	9	10 <	13						
1 High A	1	< 16	16 <	17[+]	16 <	18[+]	56 <	60	52	55		
1 High B	2	< 14	[+]16 <	17[+]	[+]18	18[+]	25 <	37	X	59	61	65
2 Low A	3	< 13	12	16	12 <	15[+]	43 <	55	45	49		
2 Low B	2	< 16	15	16	[-]13 <	17	39	45	41	44		
2 Average A	10	< 17	[+]18	18[+]	[+]18	18[+]	27 <	30	72 <	77		
2 Average B	7	< 15	15	16	[+]18	18[+]	66 <	75	64 <	72		

Note. Subtraction test scores are the mean number of items correct on a 2-minute 20-item test of the more difficult single-digit differences--those with sums between 11 and 18, excluding doubles (a + a). The scores for 3M, 4M, 5M of the bottom 6 rows in the table are the mean number of items correct on a 6-minute test of all 100 subtraction facts. The ability groups are math ability (assigned before any of our instruction) based on tests and teacher judgment.

< indicates a t-test significant at the .05 level.

- indicates a score significantly lower than the posttest ($p < .05$).

+ indicates a score significantly higher than the posttest ($p < .05$); the scores on the 100-item test cannot be tested with the posttest.

[a]Before Review   [b]After Review

450

Table 3

Mean Number Correct and Mean Percent Correct on Timed Addition Test of Two-Digit to Six-Digit Sums

Grade Ability School	n	Pre	Post		1 Month BR[a]	1 Month AR[b]		2 Month BR	2 Month AR	3 Month BR	3 Month AR	4 Month BR	4 Month AR	5 Month BR	5 Month AR
1 High A	23	0.4 <	6.5	s	6.4	6.0									
1 High B	22	1.5 <	6.9												
2 Low A	14	1.8 <	5.7	s	4.9	6.1	s								
2 Low B	17	1.9 <	5.2	s											
2 Average A	25	1.3 <	6.3		+8.3 < 9.1+			6.1	6.3	+7.8 < 9.1+					
2 Average B	21	2.7 <	5.9	s	5.2	4.9		6.2 <	7.5+						
2 High A	24	X	5.9	s	5.0	5.6		6.6	6.9+	+8.4	8.5+	+8.0 <	9.5+	+8.7 <	9.8+
1 High A	23	15% <	74%	s	84%	86%									
1 High B	22	19% <	86%												
2 Low A	14	27% <	74%	s	68%	77%	s								
2 Low B	17	19% <	67%	s											
2 Average A	25	15% <	80%	s	87%	87%		75% > 65% s		90%	86%				
2 Average B	21	52% <	81%	s	73%	66%		75%	81%	88%	84%				
2 High A	24	X	79%	s	75%	78%		84%	89%	88%	84%	76%	86%	85%	89%

Note. Each score in the top 7 rows is the mean number of correct answers on an 11-item 3-minute test of 2, 1, 4, 1, 3 addition problems of 2, 3, 4, 5, and 6-digit numbers, respectively. One 2-digit problem had no carries (trades), and 2, 3, 3, 1, and 1 problems had 1, 2, 3, 5, 6 carries (trades), respectively. Each score in the bottom 7 rows is the percent of problems correct out of the problems a child tried within the 3-minute time limit. The ability groups are math ability (assigned before any of our instruction) based on tests and teacher judgment.

< and > indicate a t-test significant at the .05 level.
+ indicates a score significantly higher than the posttest (p < .05).
s denotes the timing of the subtraction teaching.
x denotes no test given.

[a] Before Review   [b] After Review

451

Table 4

Mean Number Correct on Untimed Addition Tests of 2- and 3-Digits and of 4- and 6-Digits

Grade Ability School	Pre		Post		1 Month BR[a]	1 Month AR[b]		2 Month BR	2 Month AR	3 Month BR	3 Month AR	4 Month BR	4 Month AR	5 Month BR	5 Month AR
1 High A	0.1	<	3.8	s	3.5	3.8									
1 High B	0.5	<	3.6												
2 Low A	1.0	<	2.6	s	2.8	3.1	s								
2 Low B	0.9	<	3.1	s											
2 Average A	0.4	<	3.5	s	3.8	3.5	s	⁻2.4	2.6⁻	3.4	3.9				
2 Average B	2.3	<	3.7	s	⁻2.4	2.4⁻		3.4	3.6						
2 High A	X		3.5	s	3.2	3.4		X	X	3.7	3.7	3.6	3.8	3.6	3.7
1 High A	0.0	<	3.4	s	⁻2.5	2.7⁻									
1 High B	0.4	<	2.9												
2 Low A	0.9	<	2.6	s	2.5	2.5	s								
2 Low B	0.2	<	1.8	s											
2 Average A	0.2	<	3.2	s	3.3	3.0	s	2.1	1.9⁻	2.8	3.2				
2 Average B	1.0	<	2.6	s	1.9	2.4		2.6	2.8						
2 High A	X		2.8	s	2.0	2.1		X	X	3.2	3.0	⁺3.3	3.3⁺	⁺3.1	3.2⁺

Note. Each score in the top 7 rows is the mean number of correct answers on a 4-item untimed test of 2 2-digit and 1 3-digit addition problems with one carry (trade) each and 1 3-digit problem with 2 adjacent carries (trades). Each score in the bottom 7 rows is the mean number of correct answers on a 4-item untimed test of 2 4-digit and 2 6-digit addition problems with 2 non-adjacent, 2 adjacent, 5 adjacent, and 6 adjacent carries (trades) respectively.

< indicates a t-test significant at the .05 level.

− indicates a score significantly lower than the posttest (p < .05).

+ indicates a score significantly higher than the posttest (p < .05).

s denotes the timing of the subtraction teaching.

X denotes no test given.

[a]Before Review   [b]After Review

452

Table 5

Mean Number Correct and Mean Percent Correct on Timed Subtraction Test and

Mean Number Correct on Untimed Tests of 2- and 3- Digits, of 4- and 6-Digits, and of Minuends with Zeros

Grade Ability School	Pre	Post	1 Month BR[a]	1 Month AR[b]	2 Month BR	2 Month AR	3 Month BR	3 Month AR	4 Month BR	4 Month AR	Possible Score
2 Average A	1.0	< 7.5	⁻6.4	7.0	8.3	8.9⁺					(14)
2 Average B	1.0	< 5.1	5.7	6.3⁺	⁺6.7	7.4⁺					
2 High A	1.0	< 6.4	6.7	x	⁺7.6	7.8⁺	⁺8.3 < 9.3⁺		⁺9.0	9.6⁺	
2 Average A	7%	< 76%	67%	72%	77%	80%					(100%)
2 Average B	11%	< 71%	68%	71%	69%	76%					
2 High A	9%	< 80%	82%	80%	84%	84%	82%	86%	83%	81%	
2 Average A	0.1	< 3.2	3.1	3.4	3.2	3.4					(4)
2 Average B	0.2	< 3.1	3.0	3.1	3.3	3.6					
2 High A	0.3	< 3.5	x	3.2	x	x	3.6	3.5	3.6	3.4	
2 Average A	0.1	< 2.6	2.3	2.2	2.2	2.6					(4)
2 Average B	0.0	< 2.0	2.2	2.3	2.5	2.6					
2 High A	0.1	< 2.6	x	2.9	x	x	3.0	3.2	⁺3.2	3.1⁺	
2 Average A	.0	< 4.7	4.5	4.2	4.9	5.1					(6)
2 Average B	.0	< 3.6	3.9	4.2	3.8	4.6					
2 High A	.0	< 3.9	x	5.0⁺	x	x	⁺5.6	5.1	⁺5.2	5.5⁺	

Note. The addition tests in Tables 1 and 2 were used to construct the subtraction tests for the top 12 rows above. Each subtraction item is an inverse problem of the corresponding addition item. The timed subtraction test also had 3 problems at the end with 0's in the minuend; six minutes were given for this test. Each score in the bottom 3 rows is the mean number of correct answers on a 6-item untimed test of 2, 2, and 2 subtraction problems of 2-, 3-, and 4-digits, respectively, with 3 problems with one 0 in the minuend, 2 problems with 2 0's, and 1 problem with 3 0's in the minuend.

✔ indicates a t-test significant at the .05 level.

− indicates a score significantly lower than the posttest (p < .05).

+ indicates a score significantly higher than the posttest (p < .05).

x denotes no test given.

[a]Before Review   [b]After Review

453

Table 6

Mean Percent Scores on Place-Value Tests

	Add Three Numbers		Write Uneven Addition		Circle the Bigger		>,< Signs		Words to Symbols		Word-Symbol with Trades	
	Pre	Post	Pre	Post	Pre	Post	Pre	Post	Pre	Post	Pre	Post
1 High A	0 < 55		0 < 75		50 < 80		0 < 70		2 < 97		5 < 63	
1 High B									5 < 92			
2 Low A	5 < 60		0 < 65						3 < 97			
2 Average A	0 < 55				28 < 79		X	55	50 < 83		3 < 14	
2 Average B	10 < 65				61 < 81		X	73	57 < 87		4 < 63	

Note. < indicates a t-test significant at the .01 level.

454

The finger patterns for 1 through 9 are made by touching certain fingers and/or the thumb to some surface such as a table. Thus there is kinaesthetic as well as visual feedback for the finger patterns. The finger patterns use a subbase of 5. The thumb is 5, and 6 is the thumb plus the 1 finger (6 = 5 + 1), 7 is the thumb plus the two fingers (7 = 5 + 2), etc. The motion from 4 to 5 is a very strong and definite motion--the fingers all go up and the thumb goes down, all in one sharp motion with the wrist twisting.

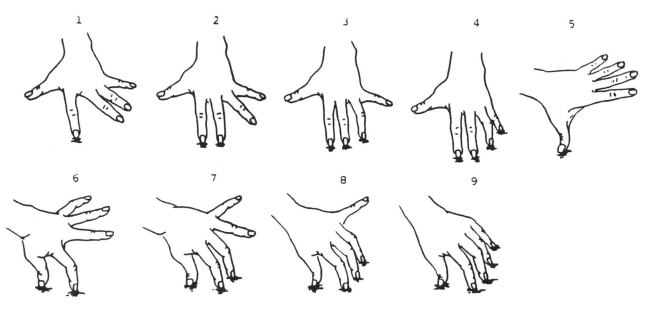

The finger patterns above are the patterns used in Chisanbop. We use the finger patterns differently from the way in which they are used in Chisanbop. We use them in the way that children spontaneously use fingers on both hands to keep track--the fingers just match the counting words which are counting on through the second addend (or counting up to the sum in subtraction).

Addition	Subtraction
8 + 5 = ?	13 - 8 = ?

The counting-on procedure:
1. Count on 5 more words from 8.
2. Stop when finger pattern for 5 is made.
3. Answer is last <u>word</u> said (13).

The counting-up procedure:
1. Count up from 8 to 13.
2. Stop when say 13.
3. Answer is what the <u>hand</u> says--the finger pattern for 5.

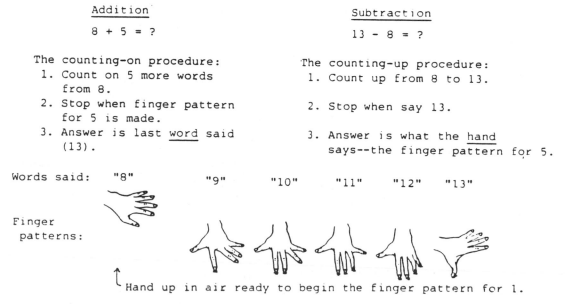

Words said: "8"   "9"   "10"   "11"   "12"   "13"

Finger patterns:

↑ Hand up in air ready to begin the finger pattern for 1.

Figure 1: Counting On and Counting Up To with Finger Patterns

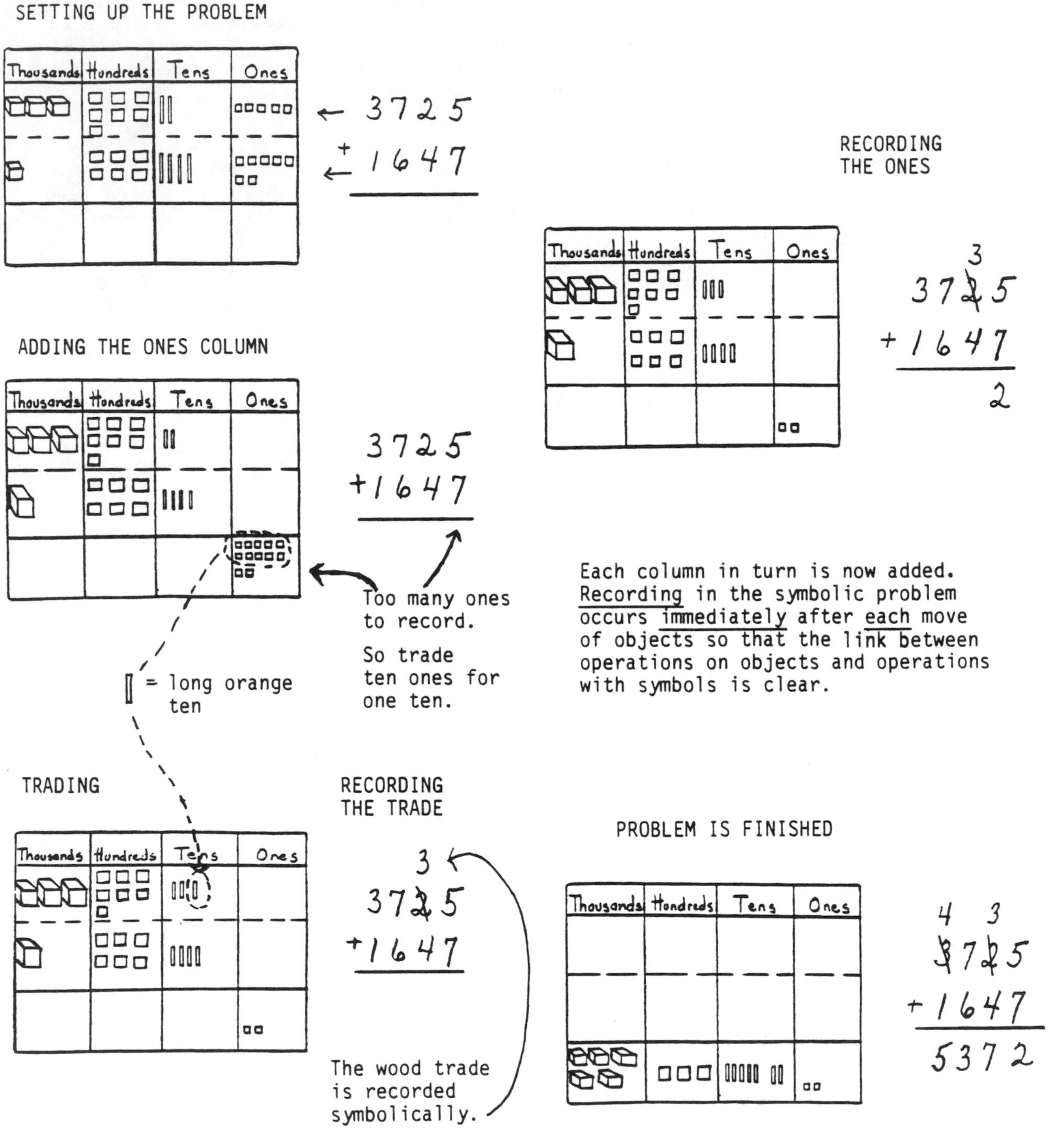

SETTING UP THE PROBLEM

$$\begin{array}{r} 3725 \\ +1647 \\ \hline \end{array}$$

ADDING THE ONES COLUMN

$$\begin{array}{r} 3725 \\ +1647 \\ \hline \end{array}$$

Too many ones
to record.

So trade
ten ones for
one ten.

☐ = long orange
ten

RECORDING
THE ONES

$$\begin{array}{r} {}^{3} \\ 3725 \\ +1647 \\ \hline 2 \end{array}$$

Each column in turn is now added.
Recording in the symbolic problem
occurs immediately after each move
of objects so that the link between
operations on objects and operations
with symbols is clear.

TRADING

RECORDING
THE TRADE

$$\begin{array}{r} {}^{3} \\ 3725 \\ +1647 \\ \hline \end{array}$$

The wood trade
is recorded
symbolically.

PROBLEM IS FINISHED

$$\begin{array}{r} {}^{4\ 3} \\ 3725 \\ +1647 \\ \hline 5372 \end{array}$$

Adding and Recording with Our Base Ten Wood
Figure 2

SETTING UP THE PROBLEM

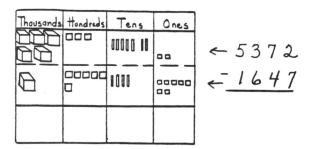

FIRST: Do all trading

Need more ones in order to subtract. Trade one ten for ten ones.

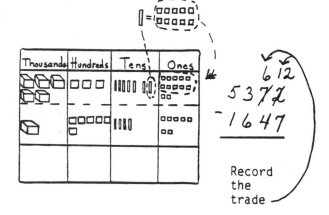

Record the trade

Now make each necessary trade, recording each trade on the symbolic problem immediately after trade is made.

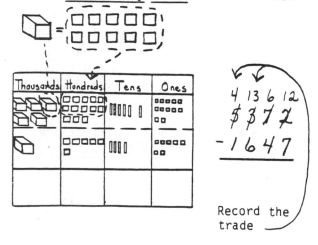

Record the trade

SECOND: Do all subtracting

RECORD COLUMN BY COLUMN

SUBTRACT THE ONES   RECORD

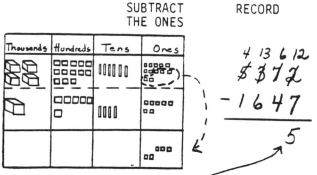

To subtract 12-7, a child can count up from 7 to 12, know the fact, go over 10 (7 + 3 + 2), or any other method.

Each column in turn is now subtracted. Recording in the symbolic problem occurs immediately after each move of objects. Each column is recorded before the next column is subtracted with the wood.

PROBLEM IS FINISHED

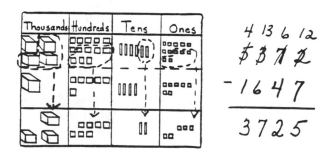

Subtracting and Recording with Our Base Ten Wood
Figure 3

457

Notes

1. The references were updated as this paper went to press, so some papers may report work that followed the writing of this paper.

2. Some of the analyzed data appear in Fuson and Willis 1986.

References

Baroody, A.J. 1984. "Children's Difficulties in Subtraction: Some Causes and Questions." Journal for Research in Mathematics Education 15:203-213.

Bell, M. and J. Burns. April 1981. "Counting, Numeration, and Arithmetic Capabilities of Primary School Children." Proposal submitted to the National Science Foundation.

Bell, M.S., K.C. Fuson, and R.A. Lesh. 1976. Algebraic and Arithmetic Structures. New York: The Free Press.

Carpenter, T.P. and J.M. Moser. 1984. "The Acquisition of Addition and Subtraction Concepts in Grades One Through Three." Journal for Research in Mathematics Education 15:179-202.

Davydov, V.V. and V.P Andronov. 1981. "Psychological Conditions of the Origination of Ideal Actions." (Project Paper 81-2) Wisconsin Research and Development Center for Individualized Schooling, The University of Wisconsin.

DeCorte, E., and L. Verschaffel. 1984. "First Graders' Solution Strategies of Addition and Subtraction Word Problems." In Proceedings of the Sixth Annual Meeting of the North American Chapter of the International Group for the Psychology of Mathematics Education, J.M. Moser, ed., 15-29. Madison, WI: University of Wisconsin, Madison.

Fuson, K.C. In press. "Roles of Representation and Verbalization in the Teaching of Multi-Digit Addition and Subtraction." European Journal of Psychology of Education.

Fuson, K.C. 1986. "Teaching Children to Subtract by Counting Up." Journal for Research in Mathematics Education 17:172-189.

Fuson, K.C. 1984. "More Complexities in Subtraction." Journal for Research in Mathematics Education 15:214-225.

Fuson, K.C. 1982. "An Analysis of the Counting-on Solution Procedure in Addition." In Addition and Subtraction: A Cognitive Perspective, T.P. Carpenter, J.M. Moser, and T.A. Romberg, eds., 67-81. Hillsdale, NJ: Lawrence Erlbaum Associates, Inc.

Fuson, K.C., and J.W. Hall. 1984. "Introducing Subtraction as Counting Up." In Proceedings of the Sixth Annual Meeting of the North American Chapter of the International Group for the Psychology of Mathematics Education, J.M. Moser, ed., 9-14. Madison, WI: University of Wisconsin, Madison.

Fuson, K.C., J. Richards, and D.J. Briars. 1982. "The Acquisition and Elaboration of the Number Word Sequence." In Progress in Cognitive Development, Children's Logical and Mathematical Cognition 1, C. Brainerd, ed. New York: Springer-Verlag.

Fuson, K.C. and W.G. Secada. In press. "Teaching Children to Add by Counting On With Finger Patterns." Cognition and Instruction.

Fuson, K.C., J.G. Stigler, and K. Bartsch. 1986. "Grade Placement of Addition and Subtraction Topics in China, Japan, the Soviet Union, Taiwan, and the United States." Manuscript submitted for publication.

Fuson, K.C. and G. Willis. 1986. "Subtracting by Counting Up: More Evidence." Manuscript submitted for publication.

Hall, J.W. 1984. "Mathematics Instruction. Some Notes on Practice." Unpublished manuscript, Northwestern University.

Hutchings, B. 1975. "Low-Stress Subtraction." The Arithmetic Teacher, 22:226-232.

Secada, W.G., K.C. Fuson, and J.W. Hall. 1983. "The Transition From Counting-all to Countingon in Addition." Journal for Research in Mathematics Education 14, 47-57.

Steffe, L.P., E. von Glasserfield, J. Richards, and P. Cobb. 1983. Children's Counting Types: Philosophy, Theory, and Application. New York: Praeger.

Steinberg, R. 1984. A Teaching Experiment of the Learning of Addition and Subtraction Facts. Doctoral dissertation, University of Wisconsin - Madison, 1983. Available from Dissertation Abstracts International, 44, 3313A.

Svenson, O. and S. Broquist. 1975. "Strategies for Solving Simple Addition Problems: A Comparison of Normal and Submormal Children." Scandinavian Journal of Psychology 16, 143-151.

Mathematical Preparation for College

Jeremy Kilpatrick
University of Georgia

At a time when many people have virtually given up attempting to change the school mathematics curriculum in any but cosmetic ways, it is heartening to see the efforts being made by the University of Chicago School Mathematics Project to develop new instructional strategies and an applications-oriented mathematics curriculum for schools in the Chicago area. I hope I can contribute to these efforts in a small way today by discussing some work that has been done along the same lines by the Educational EQuality Project of the College Board.

I shall begin by outlining the proposals for the college preparatory curriculum in mathematics that have been made by the Educational EQuality Project. Then I shall consider how these proposals might be implemented, especially as they relate to the development of students' competencies in speaking, listening, and writing, as well as the more general competencies that relate to monitoring and directing one's thinking. My general thesis is that an applications-oriented curriculum is needed, not simply to provide greater motivation for the study of mathematics, but also to support the development of higher-level thinking skills.

The Educational EQuality Project

The Educational EQuality Project began in 1980 as a 10-year effort by the College Board to strengthen the academic quality of secondary education and to ensure equality of opportunity for postsecondary education for all students. Both the "E" and the "Q" are capitalized in the name of the project to indicate its combined emphasis on quality and equality, on strengthening the academic preparation of all potential college entrants.

In 1983, the College Board published a little booklet entitled Academic Preparation for College: What Students Need to Know and Be Able to Do. The booklet presented a set of Basic Academic Competencies—that is, the broad intellectual skills that are needed in and developed by all fields of study—and a set

460

of Basic Academic Subjects—that is, the specific fields in which more extensive preparation is needed for college work.

The six Basic Academic Competencies are reading, writing, speaking and listening, mathematics, reasoning, and studying. Computer competency is identified as "an emerging need," and competency in observing is identified as "a competency to consider." The six Basic Academic Subjects are English, the arts, mathematics, science, social studies, and foreign language. You will notice that mathematics is unique in appearing in both lists. This dual role reflects the dual position that mathematics has historically occupied in the school curriculum. It is, on the one hand, a "tool" subject that provides students with the skills needed to deal with quantitative problems in any field, and, on the other hand, one of the liberal arts that educated people are supposed to have mastered. Students need to be exposed to both faces of mathematics.

Mathematics as a Basic Academic Competency

The study of mathematics in school builds upon and develops the broad intellectual skills in mathematics that students need in all fields of college study. Here is the list of Basic Academic Competencies in mathematics given in Academic Preparation for College:

- The ability to perform, with reasonable accuracy, the computations of addition, subtraction, multiplication, and division using natural numbers, fractions, decimals, and integers
- The ability to make and use measurements in both traditional and metric units
- The ability to use effectively the mathematics of:
  — integers, fractions, and decimals
  — ratios, proportions, and percentages
  — roots and powers
  — algebra
  — geometry
- The ability to make estimates and approximations, and to judge the reasonableness of a result
- The ability to formulate and solve a problem in mathematical terms

461

- The ability to select and use appropriate approaches and tools in solving problems (mental computation, trial and error, paper-and-pencil techniques, calculator, and computer)
- The ability to use elementary concepts of probability and statistics.

Mathematics as a Basic Academic Subject

The above competencies in mathematics can be developed and should be used in all school subjects. Academic Preparation for College identifies five broad categories of skills that define the additional preparation in mathematics needed by all college entrants. These skills should be the outcomes of the study of pre-college mathematics:

- The ability to apply mathematical techniques in the solution of real-life problems and to recognize when to apply those techniques
- Familiarity with the language, notation, and deductive nature of mathematics and the ability to express quantitative ideas with precision
- The ability to use computers and calculators
- Familiarity with the basic concepts of statistics and statistical reasoning
- Knowledge in considerable depth and detail of algebra, geometry, and functions.

The first two outcomes should result from all mathematics instruction. High school students need to see mathematics applied to problems in the real world and how people in various career fields use mathematics. No one can anticipate the particular applications that a student will use as an adult, but every student can be given practice in applying a variety of mathematical ideas and methods. The effective application of mathematics requires skill in formulating and solving problems. That skill, in turn, requires a mastery of basic mathematical knowledge and processes, explicit instruction in various techniques, and extensive practice.

All high school students preparing for college should learn to use the language and notation of mathematics effectively. Students should have frequent opportunities to see mathematical rules develop in a logical sequence. Although Euclidean geometry can provide a fertile ground for gaining an appreciation of the axiomatic aspect of mathematics, it should not be the exclusive source of experiences with deductive reasoning. Experiences in pattern recognition and inductive reasoning should be organized so as to lead to the development of mathematical laws and concepts.

The more specific preparation needed for achieving the last three outcomes in the list is given below.

Computing. Every college entrant will need:

- familiarity with computer programming and the use of prepared computer programs in mathematics
- the ability to use mental computation and estimation to evaluate calculator and computer results
- familiarity with the methods used to solve mathematical problems when calculators or computers are the tools.

Those students who are expecting to major in science, engineering, or business in college or who expect to take advanced courses in mathematics or computer science will need additional preparation in computing. They should be able to do more than simply use programs and have a general familiarity with how computer methods work. They will need more extensive mathematical proficiency, namely,

- the ability to write computer programs to solve a variety of mathematical problems
- familiarity with the methodology of developing computer programs and with the considerations of design, structure, and style that are an important part of this methodology.

Statistics. Every college entrant will need

- the ability to gather and interpret data and to represent them graphically
- the ability to apply techniques for summarizing data using such statistical concepts as average, median, and mode
- familiarity with techniques of statistical reasoning and common misuses of statistics.

If students intend to pursue majors or careers requiring advanced mathematics courses in college, they will need more preparation in statistics, covering:

- understanding of simulation techniques used to model experimental situations
- knowledge of elementary concepts of probability needed in the study and understanding of statistics.

Algebra. Every college entrant will need:
- skill in solving equations and inequalities
- skill in operations with real numbers
- skill in simplifying algebraic expressions, including simple rational and radical expressions
- familiarity with permutations, combinations, simple counting problems, and the binomial theorem.

Those students who will take advanced mathematics in college will need additional preparation in high school, including:
- skill in solving trigonometric, exponential, and logarithmic equations
- skill in operations with complex numbers
- familiarity with arithmetic and geometric series and with proofs by mathematical induction
- familiarity with simple matrix operations and their relation to systems of linear equations.

Geometry. Every college entrant will need:
- knowledge of two- and three-dimensional figures and their properties
- the ability to think of two- and three-dimensional figures in terms of symmetry, congruence, and similarity
- the ability to use the Pythagorean theorem and special right triangle relationships
- the ability to draw geometrical figures and use geometrical modes of thinking in the solving of problems.

Those students who plan to enter fields that will require them to take advanced mathematics courses in college will also need at least:
- application of the role of proofs and axiomatic structure in mathematics and the ability to write proofs
- knowledge of analytic geometry in the plane

- knowledge of the conic sections
- familiarity with vectors and with the use of polar coordinates.

Functions. Every college entrant will need:
- knowledge of relations, functions, and inverses
- the ability to graph linear and quadratic functions and to use them in the interpretation and solution of problems.

Students who will study advanced mathematics in college will need:
- knowledge of various types of functions, including polynomial, exponential, logarithmic, and circular functions
- the ability to graph such functions and to use them in the solutions of problems.

The Proposed Curriculum

In an attempt to suggest how the competencies and outcomes above might be achieved, the Mathematical Sciences Advisory Committee of the College Board has prepared a booklet for mathematics teachers (College Board 1985). One chapter in the booklet describes a curriculum for secondary schools designed to implement the proposals outlined in Academic Preparation for College. The curriculum may not be as strongly applications-oriented as some might prefer, but it does stress applications.

Although the curriculum is aimed at preparing students for college, it seeks to be as inclusive as possible. The committee has taken the view that no student should be denied preparation for college in mathematics solely because of decisions made in the first two years of high school. Provisions need to be maintained for equitable and flexible access to college preparatory courses, and in particular, opportunities should be available for students who may have been shunted into courses such as general mathematics, consumer mathematics, or business arithmetic. The committee attempted to bridge the gulf between college preparatory courses and dead-end courses such as general mathematics by formulating a new course, Computational Problem Solving. This course would be expressly designed to prepare students who are not ready to begin the regular college preparatory curriculum.

In formulating a proposed curriculum, the committee had two guiding principles. The first was that all students who are capable of entering college should be provided as much preparation in mathematics as possible before leaving high school. It follows, therefore, that in their academic preparation for college, students should have a continuous experience in mathematics. The committee has proposed a structured set of courses for college-capable students in which mathematics is taken each year. "College-capable" students means those students who are capable of entering a postsecondary educational institution. In many high schools, they might be from one half to two thirds of the student body. Ordinarily, they are drawn from the top three quarters of the ability distribution. The proposed courses attempt to alleviate some of the problems that are caused when students enter college without having studied mathematics in the last year or two of high school.

The second guiding principle was that as much as possible, students should have a common mathematics curriculum in high school. The committee rejected the idea that students should be sorted into categories at the eighth or ninth grade, with one group receiving a mathematics curriculum that prepares them for college and the other dead-end courses that lead nowhere. Students who are slow to mature in mathematics or who make a late decision to attend college should not be handicapped by the mathematics courses they have taken in high school. Instead, they should be given courses that will prepare them to enter the college-preparatory curriculum we have outlined. They should not be shunted off into so-called remedial courses that only exacerbate the difficulties they are designed to alleviate.

Depending on the composition of its student body and faculty, and on the needs, wishes, and support of the community, a high school might offer all or part of the proposed curriculum. At the heart of the curriculum is a 3-year sequence of courses entitled Mathematical Topics 1-2-3. This sequence includes the basic topics in computing, statistics, algebra, geometry, and functions that are outlined above. Entry into the sequence is possible at any time during the student's high school career.

Students who are prepared to begin Mathematical Topics 1-2-3 at the end of

Grades 7 or 8 will enter the sequence in either Grade 8 or 9. If they begin the sequence in Grade 8, they will have the opportunity to take at least one Advanced Placement course in Grade 12. Students needing additional preparation or additional academic maturity at the end of Grade 8 can take a Computational Problem Solving (CPS) course in Grade 9 and enter the sequence in Grade 10. Students who begin the sequence later than Grade 10 will fall short of meeting all the outcomes specified above. Colleges might require such students to make up the deficiency before they are admitted or might provide noncredit courses that would complete the preparation.

Computational Problem Solving Course

The CPS course is intended for students who lack either the preparation or the maturity to begin Mathematical Topics 1-2-3. Such students are usually identified because they are deficient in skills of computational arithmetic; greater deficiencies, however, are created by an insufficient understanding of mathematical concepts and processes. The purpose of the course is to improve the students' skill and understanding so as to build a foundation for the study of algebra, geometry, functions, statistics, and computing. The course should expose students to various facets of mathematics and give them some opportunities to experience success with mathematics.

The CPS course would be designed to provide interesting and challenging applications of arithmetic, statistics, and computing to real-world problems. Students would use calculators and — where possible — computers to solve problems. The course should not subject students to lengthy and repeated paper-and-pencil computations. Heavy emphasis would be placed on developing an awareness of the history, scope, and nature of mathematics. The course would be organized around a half dozen or so modules that would be chosen and organized in light of the interests and abilities of both students and teacher.

It is important to stress that the proposed CPS course is not meant to be simply a retreaded course in general mathematics. The course is intended to prepare students for the study of college preparatory mathematics. It is not an arithmetic review course. It is designed to help students who, for whatever

467

reason, are not prepared to go on to algebra and other topics. The course should be forwardlooking — that is, it should attempt to anticipate key concepts, such as functions and variables, and to provide students with the background to understand such concepts through methods that will be largely numerical and computational. The level of abstraction should purposely be kept low, and most of the problem material should be drawn from real situations which are of interest to students. At Ohio State University Joan Leitzel and Alan Osborne are conducting a project to construct four units at Grade 7 and four at Grade 8 that exemplify the modules one might find in a CPS course.

Mathematical Topics 1-2-3

In proposing Mathematical Topics 1-2-3, which would ordinarily span three successive years, the Mathematical Sciences Advisory Committee encourages schools to organize the topics in computing, statistics, algebra, geometry, and functions in whatever manner they see fit. They can preserve the current algebra/geometry/advanced algebra structure, or they can construct an integrated mathematics curriculum in which topics in each of the major areas are studied each year and the relationships among topics are stressed.

As the name "Mathematical Topics 1-2-3" might suggest, the committee favors the integration of topics. A more efficient and flexible organization of the content is possible if one is not bound by year-long courses devoted to a single theme. A truly integrated high school curriculum in mathematics would allow computing and statistics to be used as a context for learning algebra; would draw upon geometric concepts to illustrate ideas from algebra, computing, statistics, and functions; and would make use of functions as a means of unifying much of the other content. Rather than being organized according to concepts and techniques, the curriculum could be organized around problems, with concepts and techniques from various parts of mathematics brought in as they contribute to understanding and solving a problem. The committee has taken the view that an integrated curriculum can more easily incorporate applications of mathematics and mathematical modeling than can a curriculum divided along content lines.

The committee recognizes that strong forces of tradition, textbooks, tests, and college requirements — not to mention the powerful force of inertia — may hamper schools in the task of reorganizing the mathematics curriculum for the college-bound. That such a reorganization is not impossible, though, can be seen in the recent experience of New York State in devising a unified mathematics curriculum for its schools. The Unified Mathematics books by Howard Fehr and his colleagues (Fehr et al. 1972, 1974) and the materials produced by the School Mathematics Project in Great Britain (Thwaites 1972) also illustrate an integrated approach. Nonetheless, a reorganized curriculum is not easily attained. The proposed topics can be incorporated into the existing algebra/geometry/advanced algebra course structure if necessary.

Upper Level Courses

Many students will take the Mathematical Topics 1-2-3 sequence in Grades 9 to 11, leaving them a year for additional preparation in mathematics. Depending on its staff, resources, and student body, a high school might offer a selection of several Grade 12 courses for these students.

Students not planning to pursue careers requiring advanced mathematics might profit from the additional study of topics such as logic, combinatorics, and probability and statistics. They could also study programming methods.

Students planning for college study in fields that require advanced preparation in mathematics or computer science might want to take one or more upper-level courses in Grade 12. The content of these courses would include topics such as trigonometry, elementary functions, statistics and probability, discrete methods, and programming methods. The courses should be designed to prepare students both for the study of calculus and for the study of computer science. Students who begin the Mathematical Topics 1-2-3 sequence in Grade 8 could, with an appropriate selection of courses in Grade 11, take an Advanced Placement course in Calculus of Computer Science in Grade 12.

Speaking, Listening, and Writing

There is much more that could be said about the proposed curriculum and how it relates to the Basic Academic Competencies. I want to concentrate in these remarks on the twin competencies of speaking and listening and the competency of writing, partly because I think they have been given insufficient attention in most mathematics classes and partly because they seem especially important for any application-oriented curriculum.

At the Open University in England, a course entitled "Developing Mathematical Thinking" (Floyd 1981) has been offered for several years. One of the approaches advocated in that course is captured in the phrase "do, talk, and record": "Children do all manner of activities, talk about them with each other and their teachers, and record some of what happens" (1981:3). The do-talk-and-record approach allows teachers to develop students' skills in speaking and listening — as well as writing and reasoning — in the context of mathematical investigations.

For many schoolchildren, mathematics seems like an endless chain of paper-and-pencil exercises to be worked out. In the words of Joe Lipson (1974), the children have heard the notes of mathematics, but they have not detected the melody. When children work on mathematical activities together, they come to see how various ideas are related, and they develop a vocabulary for talking about the ideas. Of course, it is possible to learn mathematics by oneself, studying textbooks and writing out the solutions to problems. But by speaking with others and listening to their responses, one gets a different perspective on one's mathematical ideas.

It is possible that with the decline of the daily recitation — in which students were called to the front of the mathematics classroom to present their solution to an assigned problem — the skills of speaking and listening began to decline. Few people today would advocate returning to the formal recitation, but many teachers seem to have gone to the other extreme. Their students are never called upon to express their ideas orally to their classmates and the teacher. In such classes, mathematics has become a bookkeeping exercise in which one assignment succeeds another in a seemingly endless stream, and nobody except the teacher ever "talks mathematics." The teacher who gets students talking about mathematics with each other finds that the study of mathematics can help improve speaking and listening skills and that speaking and listening can help students improve their understanding of mathematics.

It is important, too, for students to write in the mathematics class — that is, to write more than simply the answers to exercises. Through the regular use of writing activities, teachers can demonstrate to students how writing interacts with understanding. We sometimes think we have to understand something before we can write about it, but writing is more often an aid to understanding. We may not understand something very well until we have written (or spoken) about it. When students keep a journal in a mathematics course, they can see for themselves how their understanding of the major ideas in the course changes over the term. When they document a computer program they have written, they learn to anticipate the questions that a reader might raise about how the program operates. When they write brief papers to report on their investigation of a problem, they have the opportunity to reflect on the processes they have used in solving the problem.

Teachers should expect students to provide more than simply the answers to assigned exercises and problems. Students should be encouraged and assisted to write up their solutions so that someone else can look at the solution and follow the student's line of reasoning. Many teachers have students keep a notebook containing their written assignments rather than having them hand in each day's work. The notebook can be checked at convenient intervals, and it can be used to demonstrate to students the value of recording their work in a form that will later be intelligible to them.

Students need to learn to write for themselves and for their peers. They need opportunities to engage in "pre-writing" discussions that will help them shape the message they want to convey. They need opportunities to present their tentative products to other students for advice and helpful criticism. They need not submit to the teacher everything they write, and the teacher need not feel compelled to grade everything they submit. The teacher needs to play the role of editor rather than judge, helping students, too, to be editors for each other's work.

Monitoring and Directing One's Thinking

Students also need opportunities to reflect on their own thinking. Metacognition is the currently popular term to describe the knowledge you have about your own thinking in general, the knowledge you have about how you are thinking right now, and the control you have over your thoughts. Good students usually develop their own metacognitive techniques for monitoring and controlling their thinking processes, but most students seem to need some instruction and practice.

Some recent work by Linda Anderson and her colleagues (Anderson, Brubaker, Alleman-Brooks, and Duffy 1984) suggests how school practice may inhibit the development of the skills of reflection and self-awareness in some students and encourage it in others. Consider what happens when a teacher gives a student some mathematical task that the student finds too difficult or confusing and insists that the student stay busy to complete the work. If the student has been successful in learning mathematics up to this point, such assignments will have been relatively rare. Consequently, the student is likely to seek help from the teacher or some other source. The student develops some skills in learning how to learn — checking progress, finding what to do when you do not understand — even though the formal classroom instruction has not explicitly dealt with such skills. In contrast, if the student has not been a successful learner of mathematics, then difficult or confusing tasks are a matter of course. Under pressure to stay busy, the student who is accustomed to doing meaningless mathematical tasks is not likely to try to learn how to guide his or her own learning. In other words, the cognitively rich get metacognitively richer, while the cognitively poor get metacognitively poorer.

Bob Davis has noted that good students, after solving or working hard on a difficult mathematics problem, often seem to be deep in thought, turning the problem over in their minds in an "after-the-fact analysis" (Davis and McKnight 1979: 101). When questioned, the students can replay the problem solution, much as an expert game player can replay a game from memory. Davis conjectures that after-the-fact analysis may be a time when students consolidate their knowledge and develop metacognitive knowledge about their procedures.

If Davis is correct, mathematics teachers might well make more provision for students to make after-the-fact analyses of their work on a problem. By keeping a record of their work, students can be encouraged to compare the relative effectiveness of various procedures they are using. Teachers report that getting a student to look back over a problem he or she has solved is one of the most difficult teaching tasks they face. In part, that difficulty may arise because of the emphasis in so many assignments on solving as many problems as possible in as short a time as possible. When a teacher shifts to assigning activities that require more extended investigation than the typical textbook word problem demands, students may see more benefit in looking back at their work.

The development of students' metacognitive knowledge should be given a high priority in mathematics instruction. Students who have trouble monitoring their progress in mathematical problem solving, for example, can be given practice in paraphrasing the problem, summarizing what they have done with the problem thus far, predicting what difficulties the problem is likely to present, and discussing with other students various approaches to solving the problem. They can learn to ask themselves questions during problem solving that will help them keep track of their progress. John Mason and his colleagues (Mason et al. 1982) have developed some useful techniques that help students become their own questioners and that give them suggestions of what to do when they are stuck. Their book, Thinking Mathematically, contains a wealth of interesting problems and suggestions for the teacher who would like to develop reasoning skills in the mathematics classroom.

Some Final Observations

Applications of mathematics should be given a central place in the mathematics curriculum and in mathematics instruction. The characteristics of applications that make them so difficult to fit into the traditional curriculum — their demand for knowledge from other fields, their open-endedness, their complications — are just the characteristics needed to promote the skills of speaking, listening, writing, reasoning, and metacognition that I have been stressing.

Because applications depend so heavily on teachers who are comfortable exploring new ideas and departing from the routines of "answer-giving" mathematics teaching, a more applications-oriented curriculum will require that teachers be active and central participants in the process of curriculum change. The last wave of curriculum reform, in the 1960s, came down from above; teachers were not, for the most part, key actors in the drama. No matter who writes the textbook or the standardized test, however, the curriculum that ultimately reaches the student is fashioned by the teacher. No one expects the teacher to construct a curriculum single-handedly, but if college preparatory mathematics is to be reshaped so that more students have more opportunities to see applications of mathematics at work in their courses, teachers will need to join together in partnerships to bring that about. The Mathematical Sciences Advisory Committee, like the University of Chicago Mathematics Project, both exemplifies and encourages such partnerships.

References

Anderson, L.M., N.L. Brubaker, J. Alleman-Brooks, and G.G. Duffy. 1984. Making Seatwork Work (Research Series No. 142). East Lansing: Michigan State University, Institute for Research on Teaching.

College Board. 1983. Academic Preparation for College: What Students Need to Know and Be Able to Do. New York: College Board.

College Board. 1985. Academic Preparation in Mathematics: Teaching for Transition from High School to College. New York: Author.

Davis, R.B., and C.C. McKnight. 1979. "Modeling the Processes of Mathematical Thinking." Journal of Children's Mathematical Behavior 2(2): 91–113.

Fehr, H.F., J.T. Fey, and T.J. Hill. 1972. Unified Mathematics: Courses I - III. Reading, MA: Addison-Wesley.

Fehr, H.F., J.T. Fey, T.J. Hill, and J.S. Camp. 1974. Unified Mathematics: Course IV. Reading, MA: Addison-Wesley.

Floyd, A., ed. 1981. Developing Mathematical Thinking. London: Addison-Wesley.

Lipson, J.I. 1974. "IPI Math—An Example of What's Right and Wrong with Individualized Modular Programs." Learning 2(8): 60–61.

Mason, J., L. Burton, and K. Stacey. 1982. Thinking Mathematically. London: Addison-Wesley.

Thwaites, B. 1972. The School Mathematics Project: The First Ten Years. Cambridge: Cambridge University Press.

On the Value of Examinations

H. Halberstam and A. Peressini

University of Illinois at Champaign-Urbana

I am pleased to have this opportunity to reflect briefly on the value of examinations in mathematics before such an expert audience. My presentation has two objectives:

1. To argue in favor of examinations in mathematics as a means of
    a. defining objectives
    b. securing adequate levels of class preparation on the part of teachers, and commitment to study on the part of students
    c. achieving acceptable levels of enduring mastery
2. To describe a voluntary testing program of college preparatory mathematics in its third year of trial in Illinois which is about to reach significantly large numbers of students.

General Observations on Examinations

I had better begin by saying that I use the word "examination" to distinguish what I have in mind from the plethora of mushy quizzes and soft tests that lie on the face of American education like a rash of measles, forever measuring no more than promise and forever silent on questions of achievement. Nor do I want the word "examination" to stand for a means of checking minimal competency, either of students or of teachers, nor yet of schools, school districts, or even states, and I envision any serious examination system as providing safeguards against such misuse.

Given these disclaimers, let me attempt to describe what I do mean. I shall use the term "examination" (for the present purpose, always in the context of mathematics) for any process that tests, by written means, the possession of knowledge, understanding, and skills at levels that not only cannot be achieved solely on the basis of a last-minute effort, but on the contrary require for their attainment careful and structured preparation over a significant period of time. Each examination should serve the teacher as a concrete expression of what some portion of the curriculum aims at, and provide the student with a challenge — a

476

challenge to develop habits of work, the capacity to reason and the acquisition of mastery — as well as a career pointer. An examination should not force class work to focus too narrowly on examination success; but this tendency has to be acknowledged as inevitable and therefore precautions have to be taken. First, the examination should be of a quality that, notwithstanding the said tendency, is such that a creditable performance can be interpreted as a meaningful achievement. Second, the examination might well require evidence of project work or experimentation in the course of the semester. This would allow for the promotion of originality and tenacity, and would be very much in line with current emphasis in the discipline of mathematics on the importance of experiment, modeling, and heuristics. (Such activity could parallel the recording of experiments in science and the promotion of verbal facility in modern languages.) Third, the style and format of examining should be under constant review, stable enough to provide some security of expectation but not so stable as to become predictable.

An examination should be externally set and graded, but some teachers should at all times be involved in the examining process; also all participating teachers should be encouraged to comment on it. At the end of the round of examining, feedback should be provided, certainly to the teachers and, if appropriate, to the candidates, so as to secure the utmost benefit from the process.

The type of examination I have in mind is not a competition. Its dominant objective is to elicit what students know and can do and are likely to be able to do in the future, after many years of schooling. Its makeup has to be sufficiently varied to admit application to a large proportion of the school population. Given that over 60% of all Americans participate in some form of tertiary education, one might have in mind this proportion as the target group. The task of mounting examinations of this breadth is certainly not an easy one, but it is well worth attempting, and models do exist.

It is important, of course, that students be examined at levels of difficulty appropriate to their ability, attainment, and stage of development. There is a begging-the-question element to this assertion. It presupposes that these levels can be determined with some degree of accuracy. I have two comments to offer: my experience is that in broad terms those levels are not so difficult to set, and, on the other hand, the less we expect of children, the less by and large do they achieve. (When I first taught in the U.S. in 1955, at a selective private university, I was astonished to see, when supervising tests, that in the face of what I regarded as straightforward questions, students displayed the same sighs of anguish, distress, and incredulity at the sheer injustice of it all as do British students faced with more demanding exam papers. There must be a law governing human response to challenge; without preparation we react to challenge with a degree of anxiety entirely independent of the magnitude of the challenge!)

The 1982 Cockcroft Report, which is full of good sense on most matters relating to mathematics education, stresses as a fundamental principle that "examinations should not undermine the confidence of those who attempt them." This is an admirable sentiment (shared by all but the most savage examiners), but adherence to it without question, in light of the problems, both real and imagined, facing the disadvantaged, the less able, and so on, will emasculate any examination system, and beyond that any disciplined training a society devises for the preparation of its youth. In mathematics above all other subjects, at whatever level of study, the road to success is paved with discouraging experiences. Good teaching in mathematics does not avoid such experiences, but prepares us to transcend them, at least some of the time. The doctrine of easy success in the exercise of thought is pernicious; it even runs counter to the instinct we all possess to operate at the point where our mind begins to be stretched (as witness crossword puzzles and brain teasers, even in the popular press). The experience of wrestling with the angel before we ascend the ladder is something we all need!

(I'm sure I'm not the only professional mathematician to fall prey to anxiety when confronted with a problem; if I'm told to relax because the problem is very easy, I become even more anxious! A distinguished mathematician once told me that every time he finished a paper he felt overwhelmed by a feeling of certainty that never again would he have an idea for another one — and so on.)

In a recent article Hassler Whitney speaks of a loss of faith in our children's ability, which has led us into doing things for them that they could better do for themselves and thereby deprives them of the exercise of responsibility. I think (though he might not agree with me in this interpretation) that making it easy for children to acquire phenomenal grade point averages on the basis of thin curricula or easy tests — or, all too frequently, on the basis of both — is a good illustration of Whitney's remark. (The other day one of our students burst into tears when given 67% on a test — up to that point she'd only received A's!) In this context, let me remark that no serious examination can have outcomes in which an A corresponds to success with at least 90% of the questions. A good examination should offer a choice of questions and a mixture of levels of difficulty. (We might recall Bertrand Russell's dictum that even the ten commandments should be prefaced with the instruction not to obey more than five!)

In concluding this portion of my remarks, let me say that I do not ascribe any absolute virtue to examinations as a means of pacing the learning process. Perhaps in that perfect society towards which we all strive, the process of concept formation, the development of originality, and the enjoyment of intellectual effort will all be so well understood as to be expertly fostered through absorbing classroom activity directed by highly professional teachers whose worth in society is amply recognized. All I claim is that in an imperfect world one has to try harder, not less, and with whatever means come to hand, to achieve even quite modest objectives.

The Illinois Universities Test of College Preparatory Mathematics: Preliminary Report for 1984–85

Description of the Program

The Illinois Universities College Preparatory Mathematics Testing Program is designed to provide college-bound high school juniors with a measure of the adequacy of their preparation for the study of college-level mathematics, and to encourage them to take any additional mathematics courses they may need during their senior year in high school. A similar testing program that has been in operation in Ohio for several years has resulted in a 40% increase in senior year mathematics enrollments among college-bound juniors, as well as improved performance on the mathematics placement test at Ohio State by students from participating high schools.

Development of this testing program began in 1982 as a cooperative project of the mathematics departments of all the state universities in Illinois. Initial pilot testing of the program was conducted during the spring of 1983 with approximately 2,000 students at 20 selected high schools. The program was modified and pilot tested again during the spring of 1984 with approximately 2,400 students from 22 selected high schools. The Illinois Board of Education assisted with the sample selection and test distribution for the second pilot test. After further revisions, the first large-scale implementation of the testing program took place in the northeast sector of Illinois during the 1984-85 school year.

The Illinois Universities Test of College Preparatory Mathematics (I.U. Test) consists of the following three parts:

Part 1: Elementary Algebra

Part 2: Geometry and First Semester Advanced Algebra

Part 3: Second Semester Advanced Algebra and Trigonometry

Parts 1 and 2 are multiple-choice tests, each with a 45-minute time limit. These two parts are administered between January 15 and February 15. The answer sheets are machine scored and the school and student reports prepared by the UIUC Measurement and Research Division for distribution through the test coordinators at the participating schools.

Part 3 of the test, which also has a 45-minute time limit, consists of a combination of multiple choice and open-response items. It is administered after the students have completed advanced algebra and trigonometry, usually near the end of the junior year or at the end of the first semester of the senior year. This part of the test is scored at the participating high school. An answer key, a partial-credit guide, and standard cutoff scores are provided as part of the test package to assist the local test coordinator with scoring the test and reporting the test results to the students.

The I.U. Test is significantly different from the quantitative segments of the ACT or SAT tests. The latter tests are primarily tests of quantitative aptitude, while the former is a test of mastery of some of the basic concepts and skills required for college-level mathematics courses. In fact, the items on the I.U. Test are identical or very similar to items that have been used on the mathematics placement tests of the participating universities.

The test report package for Parts 1 and 2 that is distributed to the test coordinators for each participating high school includes:

1. Two copies of each student report, one for the student and one for the school files

2. A roster of participating students at that school with total score and subscores for each student

3. A school report that lists the percentage of students in that school that responded correctly to each item, the percentages of students at various course levels in that school who passed Part 1 or Part 2 or both parts, and summary demographic information on the students in that school.

In addition to listing subscores in elementary algebra, geometry, and advanced algebra, each student report includes a recommendation based on these subscores, advice about high school mathematics preparation for college study, and information about the mathematics requirements for various college majors.

No statewide norms for test performance are provided, and the school report is designed to discourage use of the test results to rank students or their teachers. The sole purpose of the test is to provide students with a measure of their preparation for college-level mathematics.

High schools wishing to participate in the testing program are required to specify one of their mathematics teachers as a local test coordinator and to provide the administrative support necessary to conduct the test under the conditions specified in the letter inviting schools to participate. The participating high schools are also required to make arrangements for substitute teachers and travel so that their test coordinator can attend a pretest training meeting and post-test reporting and evaluation meeting.

The 1984–85 Testing Program

Although the Illinois Universities Committee on College Preparatory Mathematics Tests[1] believed that progress with developmental work and the results of the two pilot tests justified implementing the testing program on a statewide basis during the 1984–85 school year, adequate funds were not available for this purpose. The Department of Mathematics at UIUC made $25,000 available to the program from state funds that were appropriated for school assistance programs. These funds, together with faculty time and travel funds contributed by the other mathematics departments, made it possible to offer the test program to approximately 10,000 students during this school year. The Committee decided to proceed with implementation of the program at this level during the 1984–85 school year and at the same time to make every effort to obtain the state funding necessary for full statewide implementation during the 1985–86 school year.

Because of these budgetary constraints, the 1984–85 testing program was offered only to those high schools in the northeast sector of Illinois (roughly north of I–74 and east of Route 51). In mid-September, a letter was sent to all high school principals in this region explaining the purpose of the testing program and the fact that it would be necessary to limit the total test population to 10,000 students in the current year. These principals were invited to apply for participation and to make the required school commitment with the understanding that if more than 10,000 tests were required, a selection of applicants would be made that would reflect the school demographics throughout the state (that is, approximately 83% public high schools; 30% in Chicago, 25% Chicago suburban, 45% downstate, and so on).

Applications were received from 170 high schools in the region; the total testing population was approximately 27,000 students. The Committee selected 74 schools for full participation in the program, and offered the other applicant schools the opportunity to use the test if they had the facilities and staff to print, score, and report the results to their students and to the Committee. Thirty-five of these schools, with a total testing population of 4,000 students, were willing and able to participate in the program on this limited basis. Thus, the 1984–85 testing program served a total of 14,000 students in 109 high schools in the region.

Two pretest training sessions were conducted for the test coordinators, one in Champaign-Urbana on November 17 and the other in Chicago on December 1. The test coordinators were allowed to choose the session they wished to attend but no school was allowed to participate in the testing program unless its test coordinator attended one of these training sessions. The objectives of these training sessions were to explain the purpose of the testing program, to describe alternative procedures for administering the test, and to determine how each school intended to implement the program. The sessions, which lasted approximately four hours, turned out to be extremely valuable because they clarified a number of misunderstandings about the program; they prompted the test coordinators to plan their own school's test administration carefully; they helped the Committee to understand some of the difficulties faced by the schools in administering such a program; and, last but by no means least, they fostered a spirit of cooperation between the high schools and the universities that helped to resolve problems that did arise.

Test packets for Parts 1 and 2 were mailed to test coordinators at the participating high schools in mid-December. For those schools selected for full participation, these packets included:

1. an adequate number of test booklets and answer sheets for the students taking the test at that school

2. a complete set of instructions for administering the test and returning the answer sheets

3. a test administration form to be completed by the test coordinator that would provide detailed information on actual administration of the test and feedback on any problems encountered

4. a return postcard to indicate that the test packet had been received.

Schools participating in the program on a limited basis were sent one copy of Parts 1 and 2 of the test, a complete set of instructions for administering the test, a test administration form, a reminder that the test results were to be reported to the Committee, an answer key, and a return postcard to indicate that the test packet had been received.

Printing of Part 3 of the test package was delayed until after the pretest training sessions for test coordinators so that an accurate estimate of the required number of tests would be available. Test packets for Part 3 were mailed to the test coordinators in early January with contents similar to those included in the test packets for Parts 1 and 2.

The administration of Parts 1 and 2 in the schools during the prescribed testing period of January 15 to February 15 proceeded very smoothly. Some delays in testing were encountered because of school closings due to weather conditions; however, nearly all of the answer sheets were received by February 20. Machine scoring of the answer sheets and preparation of the school and student reports were completed well ahead of schedule. It was possible to return the test results to the test coordinators on February 27 at post-test meetings conducted simultaneously in Chicago and Champaign-Urbana. These post-test meetings were used to review the school and student reports and to discuss the test results in qualitative terms. In keeping with program policy, no information was provided that would permit easy performance comparisons between schools. A final program evaluation form was distributed to each test coordinator at these meetings to survey how the test results are utilized by the participating schools. These forms are due for return to the Committee by April 1, and the results of this survey will be included in the final report for this program.

Preliminary Report of 1984–85 Test Results on Parts 1 and 2

As a matter of policy, the Illinois Universities Committee on College Preparatory Mathematics will not publish statewide norms for the test parts or other information that would allow specific performance comparisons between schools. Instead, each student report for Parts 1 and 2 of the test includes a specific recommendation based on that student's performance on each part. Appendix A lists the recommendations used on the student reports that correspond to the various combinations of performance on Parts 1 and 2. This recommendation matrix has been distributed to the test coordinators to assist them in advising students concerning their test results. Thus, each participating student is told whether or not his or her performance is up to standard on Parts 1 and 2. Moreover, the test coordinators may use the student answer sheets to determine specific areas of weakness in a given student's performance.

In general, the performance of students on Parts 1 and 2 was consistent with current mathematics placement testing results at the various state universities.

At this stage, Part 3 of the test has been administered in only a few high schools. Most participating high schools plan to administer that part of the test near the end of the current semester in their trigonometry classes and others will not administer that part until late next fall. The final report for the 1984-85 test program will include some description of the test results available for Part 3 by the end of the current semester.

A preliminary item analysis was conducted after approximately 40% of the answer sheets for Parts 1 and 2 had been scored. This analysis reveals a number of interesting and informative points concerning student performance on these parts. The results on selected items illustrate these points without revealing overall norms (see Appendix B).

Changes Under Consideration

The Illinois Universities Committee is proceeding with plans for statewide implementation of the testing program for the 1985-86 school year on the assumption that the required state funding will materialize. On the basis of the response to the program offering this year in the northeast sector of the state, a conservative estimate of the 1985-86 test population would be 40,000 students. No problems were encountered in this year's administration, scoring, and reporting schemes that would suggest that the system will not function well at the anticipated 1985-86 level. If the funding level for the program proposed in the UIUC FY 86 budget is maintained through the budgetary process, adequate funds will be available to operate the program with the anticipated fourfold increase in the testing population.

The preliminary item analysis of approximately 40% of this year's test population for Parts 1 and 2 indicates that these two test parts are very good measurement instruments. Only two items had poor discrimination qualities and the reliability of both parts exceeded .8. The Standard Error of Measurement for the test indicates that the results would vary less than 2 points overall if readministered to the same population.

The test administration reports submitted by the test coordinators indicate that four or five items can be added to Part 1 and two or three items to Part 2 without lengthening the time limit for the tests. The Committee plans to replace some items and add new items to achieve better content coverage and item discrimination.

The Committee also recognized the need for follow-up programs to assist the participating schools and students. A summer self-study review program in algebra and trigonometry for entering UIUC students has been in place and working rather effectively for several years. The Committee may decide to make this program available to high school students participating in the testing program. The Committee is also planning workshops and other forms of consultative services for high school mathematics teachers, counselors, and administrators interested in improving their college preparatory mathematics programs.

The final evaluations of the testing program by the test coordinators are scheduled to reach the Committee by April 1. The Committee will review these evaluations and the final item analysis to decide on other changes that might be required to improve the effectiveness of the 1985-86 program.

It is clear from the foregoing account that this testing program falls short of the kind of examination described in the first part of the paper. Our hope is, given time and adequate resources, there will be a greater use of questions that are both more problem-oriented and more skill-based. We have no doubt that such a development will, at long last, infuse with real meaning those sadly enigmatic statements by students of having had two, or three, or four years of exposure to high school mathematics. We have no doubt too that Commerce, Industry, and Higher Education will all benefit. We must take urgent note of the fact that high tech industry is beginning to create its own educational programs out of sheer frustration, and that even highly selective universities have increasingly been forced to devote portions of the first two undergraduate years to imparting a high school education.

We are pleased to report that the State of Illinois has provided sufficient recurrent funding to operate the testing program on a statewide basis at the level anticipated earlier in this paper for the 1985-86 school year. Over 40,000 students from 385 high schools participated in the 1985-86 program.

During the summer of 1986, the Illinois Board of Higher Education provided funding to pilot test two new high school courses for college bound students with background deficiencies in college preparatory mathematics. One of the courses, Precollege Mathematics, is patterned after an intensive summer course that has been offered for several years to students entering the University of Illinois at Urbana-Champaign with low mathematics placement scores. The other course, Pre-Advanced Algebra, which was developed at Ohio State University in conjunction with their high school testing program, is designed for high school students who have completed Elementary Algebra and Geometry but do not have the command of these subjects that is necessary to perform well in Advanced Algebra. Instructional materials and teacher workshops for both courses will be offered to high schools participating in the 1986-87 testing program.

Appendix A

Student Report Recommendations for Parts 1 and 2

Part 2 / Part 1	Part 2 Score (Max = 20)	
	Low < 14	High ≥ 15
P A R T LOW (< 12)	You need a substantially better understanding of elementary algebra in order to avoid being placed in a remedial algebra course in college.	Your elementary algebra skills appear to be weak. You need to improve these skills to succeed in college level mathematics courses. Your performance in the combined areas of geometry and advanced algebra (Part II) appears to be adequate.
1 S C O R E MARGINAL ≥ 12 and < 15	Your elementary algebra skills appear to be marginal. You need to continue study of algebra during your senior year in order to enroll in a college level mathematics course.	Your elementary algebra skills appear to be marginal. You need to improve these skills to succeed in college level mathematics courses. Your performance in the combined areas of geometry and advanced algebra (Part II) appears to be adequate.
HIGH ≥ 16 (Max=20)	Your understanding of elementary algebra (Part I) is commendable. However, your performance in the combined areas of geometry and advanced algebra (Part II) is below the level required for the study of college level mathematics courses.	You are proceeding nicely. Keep up the good work! However, in order to study calculus you need to complete advanced algebra and a year of precalculus, including trigonometry.

Appendix B

All of the following are multiple-choice items with five possible responses.

Item 16. $x - x^{-1} = ?$

1) 23% gave the correct response overall
2) 30% answered $\frac{x-1}{x}$ overall

   24% answered -1 overall
3) 70% of the top 20% answered this item correctly
4) Of the bottom 40%, 4 times more responses were either

   $\frac{x-1}{x}$, -1, $2x$

   than the correct response (Only 10% of this group responded correctly.)

Item 13. If $C = \frac{5}{9}(F - 32)$, then $F = ?$

1) 48% gave the correct response overall
2) 20% responded $\frac{9}{5}(C + 32)$ instead of $\frac{9}{5}C + 32$
3) 90% of the top 20% answered correctly
4) Of the bottom 40%, the $\frac{9}{5}(C + 32)$ responses outnumbered the correct responses 3 to 2

Item 10. $x^4 (x^3)^2 = ?$

1) 59% responded correctly overall
2) 21% overall just multiplied all of the exponents (x^{24})
3) Of the lowest 20%, there were over twice as many that multiplied the exponents to get x^{24} as those who selected the correct response and just as many added the exponents to get x^9 as those who selected the correct response.

489

Item 19. The equation of the line

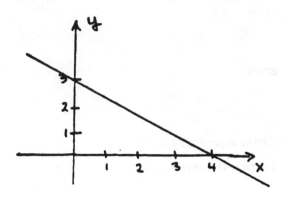

1)   48% responded correctly overall

2)   80% of the top 20% responded correctly

3)   Only 20% of the lowest 40% responded correctly

4)   21% overall took the slope as $\frac{3}{4}$ while 40% of the lowest 40% used this as the slope

Item 23.

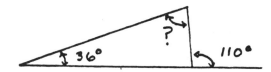

1) 56% correct responses overall

2) 93% of the top 20% answered correctly

3) Only 21% of the bottom 40% answered correctly.

4) Twice as many responded with the complementary angle 54 than the correct response in the lowest 40%

Item 27.

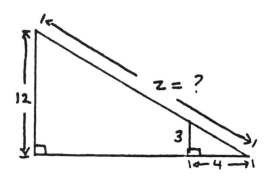

1) 44% responded correctly overall

2) Only 15% of the lowest 40% responded correctly

Item 32. Solve for x: $2x^2$ $9x + 5$

1) 32% responded correctly overall
2) Only 65% of the top 20% responded correctly
3) Only 11% of the lowest 40% responded correctly

_____

Item 37. Points of intersection of $x + y = 2$ and $y^2 = x$

1) 45% correct responses overall
2) 20% overall said that these curves do not intersect
3) No intersection outnumbered the correct response 4 to 3 among lowest 40%

_____

Item 39. (word problem setup)

Given: TV service contract = $40/yr
          TV set cost = $500
          Cost of TV + service contract for x years = ?

1) 73% correct response overall
2) 20% overall selected 540x
3) In the lowest 40%, half as many responded 540x as the correct response

Notes

1. The 1985 membership was:

    Ray Boehmer (Test Coordinator)

    Dale Brandenburg (UIUC-Testing Office)

    Mervin Brennan (IBOE, Consultant)

    Robert Kuller (NIU, Math)

    Wendell Meeks (IBOE, Consultant)

    Roy Meyerholtz (EIU, Math)

    Frank Pedersen (SIU-C, Math)

    Faustine Perham (UNI, Math)

    Tony Peressini (SIU-E, Math)

    Paul Phillips (SIU-E, Math)

    Lawrence Spence (ISU, Math)

Organizing for the Classroom Ideas Drawn from the History of Mathematics*

R.J.K. Stowasser

Technische Universität, Berlin, West Germany

Overview

The problem of "math dropouts," always a serious one, has increased now that the slogan "mathematics for everyone" has come into fashion. The subject matter of mathematics is normally presented as a meaningless mass of knowledge, unrelated to the experience of the students and totally uninteresting. It is presented with sterile "logical connections" which the students seldom understand or appreciate, and as a result they are prone to error. This has given rise to a flourishing enterprise — empirical research — which studies and characterizes the symptoms without producing a cure.

We need a new approach to the teaching of mathematics, but there is little hope that it will emanate from this psychological perspective. Epistemological perspectives and historical sources offer a much greater hope.

The history of mathematics abounds in outstanding examples of simple but powerful ideas for organizing the content of the curriculum in a meaningful way. Traditionally, the same idea is treated in several disguised forms within the so-called "spiral curriculum." Thus, for example, similarity governs the equality of fractions, y = cx, the classification of triangles, and so forth. Do students see how these seemingly disparate ideas can be related in one unifying idea, or do they simply memorize a collection of facts (e.g., 20 theorems in geometry)? Similarly, the Pythagorean theorem is closely related to the equation of the circle and to simple properties of trigonometric functions, yet they are separated in the curriculum by years.

In contrast, I first suggest the liberal use of Ockham's razor, then the connecting of simple but powerful ideas one to another within a simpler network. In this way we are led to a position from which Pascal's "esprit de finesse" (from

* I would like to express my appreciation to Dr. Georg Booker (Brisbane, Australia), Benno Mohry (Brussels, Belgium), and Prof. Avrum I. Weinzweig (Chicago, U.S.A.) for their great assistance in preparing this paper.

a few central points, short paths can be laid to many stations) will more broadly govern the process of thinking, and by extension, the process of learning.

Some impressive examples will show that the less able students will especially profit from such an organization of mathematical knowledge.

Following some general comments on "organizing ideas," I present two paradigms:

- A problem-solving approach to one of Pascal's ideas (for organizing the divisibility rules found in Arabic writing), suitable for 11-year-olds
- A proposal of Leibniz to base geometry on distance and symmetry, adapted to the classroom.

More examples of this kind — primarily produced at the Technische Universität, West Berlin — are given in the references, especially on real numbers,[1] similarity,[2] and recursion.[3]

Organizing Ideas

Insight, understanding, and comprehension are all metaphors reflecting the perception of the primary relationship of the various parts to the whole.

Our first view of a person or of a landscape from a mountain top gives us a "total meaning" which is simpler and more easily retained than the multitude of individual "meanings" of the constituent parts. The total meaning (the idea or the plan) provides an ongoing clue to the parts, and how the parts connect in an integrated framework.

Mathematical "objects" (such as numbers, similarity, functions, and fields) are infinitely simpler than people or landscapes, and are not "learned" in the same way. Moreover, they are not seen with real eyes, but with Platonic eyes. There are some mathematical concepts for which it is appropriate first to learn the "parts." However, the usual practice is always to direct teaching to these individual parts without any real view of the "whole" — the networking of these parts. This "atomized" mathematics, much easier for unimaginative teachers and perhaps more suited to "learning for the test," is seemingly justified by referring to psychological principles such as the principle of "isolation of difficulties." The problem is that this approach reduces the subject to a collection of isolated facts, leaves the student completely unmotivated, and leads to error-producing behaviors. This no doubt accounts for the lack of success in the teaching of mathematics to all students.

Attempts to correct the problem after the fact such as error correcting, remediation, and overcoming "math anxiety," provide great opportunities for psychological "research" but do little to repair the damage.

There is a great deal of lip service and well-intentioned advice given regarding the use of history in the teaching of mathematics, but very few of the examples of the kind I would like to discuss have been worked through.

In the history of mathematics I have sought out <u>ideas</u> that were
- influential in the development of mathematics
- simple, useful and powerful

and which at the same time could act as
- "centers of gravity" within the curriculum
- nodes in a cognitive network.

In this sense I call them <u>organizing ideas</u>.

Often, in the course of history, new central ideas were developed out of a reorganization of the old stock of knowledge, which led to a better general map, a "higher point of view."

"La vue synoptique" brings to light associations hitherto hidden. This approach does not operate by "longues chaînes de raison" (Descartes); nor does Euclid's "l'esprit de géométrie" govern the process of cognition and, by analogy, the process of learning. Rather, the approach is governed by a mode of thinking related to Pascal's "esprit de finesse": <u>to lay out short paths from a few central points to many stations.</u>[4]

Pascal himself provides an excellent example. At the age of sixteen, he reorganized the knowledge of conics passed down by Desargues, revealing the "mysterium hexagrammicum" (since called Pascal's theorem): the high point surrounded by a number of close corollaries.

Having drawn organizing ideas out of their historical context, one is left with the difficult task of processing them into more or less comprehensive teaching modules (or even textbook chapters). The result can be a problem field in which, instead of dozens of theorems, a few organizing ideas operate as means for problem solving.

Concentrating on a few simple and at the same time powerful ideas which organize their environments and which are connected one to another within a simple network aids less-able students as well. Their lack of success derives to a large extent from the fact that they are unable to organize their thoughts with respect to a complex field in which the connections are presented in the usual logical-systematic way, and where teaching is used to administer only spoonfuls of the subject matter, disconnected and without depth.

Combing through history, pasting together ideas, looking for interesting problems of very different provenance, even some routine exercises might not always please "real" historians. Junior high school teachers, however, should find some building blocks for interesting problem-oriented teaching modules in the following examples.

A Textbook Chapter Based on an Idea of Pascal

Suppose you are required to teach some divisibility rules. (The West German Lehrplan (syllabus) requires teaching the rules for the divisors 2, 4, 8, 5, and 9.) Pascal, when he was living in a monastery near Paris, was confronted with the same pedagogical task. He thought about the problem, and wrote a little-known paper on a general "digit-sum rule."

Generalizing Pascal's approach, we will develop a quick algorithm for remainders. The organizing idea behind it provides far more than a mere understanding of all the special cases (divisibility by 2, 4, 8, 5, and 9).

How can students discover this organizing idea? Pascal does not answer this pedagogical question. He wrote the paper as a mathematician, explaining the algorithm by some examples (9-, 7- rule) and then proving it by induction. His approach is certainly not appropriate for an interesting lesson on divisibility rules for 11-year-olds.

In my approach, with the very familiar clock face, Pascal's idea, and the more general idea of congruence, is revealed in a very simple way.

I quote from Stowasser (1982b).

In the 11-year-olds' daily life experience the clock is just the right thing to start with. I have a big cardboard clock without a minute hand. The hour hand is on the 12.

A pupil is called to the blackboard. He is told to write down the number of hours the hour hand should move. He writes down an unpronounceable number of hours which goes from left to right across the blackboard. Three seconds after the last written figure I put the hour hand on the right hour. For example, the pupil writes

20450102230531234567890246813579025414O3.

I put the hour hand on the 7. The pupils check by dividing through by 12. One page is filled. Plenty of mistakes. In the right hand corner is written the only interesting thing: the remainder.

The pupils know that I am not a magician, especially that I am not good at mental arithmetic. Of course I do not reveal the trick. The pupils will work for it, discover it. A prepared work sheet asking "what time is it", that means asking for the remainders regarding 12, shows a pattern: the remainders of 12 (Zwölferreste) divided into the powers of 10 (Zehnerpotenzen) from 10^2 on, are all equal, thank God.

$R_{12} (10^n)$ is constant for $n > 2$.

Every 10th power pushes the hour hand 4 hours ahead.

I assume, otherwise it has to be dealt with, the pupils really know what the abacus is, that they can see a decimal number consisting of powers of 10.

Now my mystery trick in arithmetic is solved. No matter how many digits there are in front, the hour hand simply jumps to and fro among three positions (beyond the tens).

I prefer to do the rotation of the hand mentally instead of on the actual clock. In the example R_{12} (2106437822) my mental arithmetic looks like this:

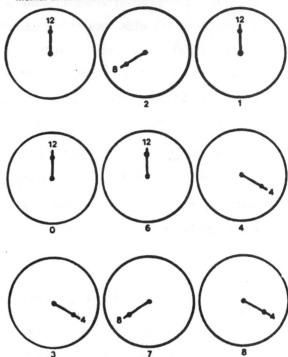

The 22 hours at the end, being out of the routine, put the hour hand in the final position:

The 11-year-olds even understand my enquiry whether the calculations on the working sheet confirm that $R(10000000000) = 4$. The reason why – hidden reasoning by induction because of the recursively defined powers of 10 – can be found by 11-year-olds with a little help.

4 hours remain from 10^2 hours after taking away the half days. From $10^3 = 10 \cdot 10^2$ hours remain $10 \cdot 4$ hours. From 10^3 hours remain therefore again 4 hours after taking away the half days. We proceed in the same way for $10^4 = 10 \cdot 10^3 \ldots$

498

No doubt, some pupils saw only the repetition in the division scheme:

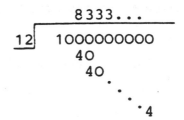

Whatever the ability of 11-year-olds to deal with loops in programming their microcomputers, this experiment disproves the old didactic prejudice that even much older students cannot understand recursion.

Do you think that our 11-year-olds can now handle a non-Babylonian clock? Of course, and not only an 11-clock, but a 37-clock, if necessary. The 2-, 4-, 8-, 5-, and 9-clocks are really trivial cases.

What of the divisibility rule for 11 and the rules required by the syllabus? Our pupils understand the reduction principle:

$$R\ (2352843) \underset{11}{\equiv} R\ (2-3+5-2+8-4+3)$$

$$\underset{11}{\equiv} R\ (9) \underset{11}{\equiv} 9$$

Therefore 2,352,843 is not a multiple of 11.

The pupils have grasped the idea of <u>congruence</u> which (together with the idea of recursion implemented in the decimal system) completely organizes the mathematics lesson around the divisibility rules (and much more, as a look into the history of number theory shows (see Stowasser 1985a)).

I take this opportunity to attack another psychological principle. Contrary to some popular prejudices, in teaching a mathematical idea it is not necessary to burden the students with a number of examples from which the idea is to be abstracted. It is far more effective for the teacher to expose the idea by an appropriate paradigm such as, in our case, the Babylonian clock. Having grasped the idea from that paradigm, the student must then apply (transfer) this idea to new situations. Aristotle has already pointed out the difference between research (many examples) and teaching (the best example, or paradigm).

Leibniz's Proposal for Basing Geometry on Distance and Symmetry, Adapted for 11–13–Year–Olds[5]

The following sketch of materials prepared for the classroom was inspired by a letter of Leibniz to Huygens dated September 8, 1679 (von Engelhardt 1955). In this letter, Leibniz was motivated by a desire to find an algebra (his Characteristica Universalis) by which the truth of philosophical arguments could be determined. Contrary to Descartes, who translated geometry into an algebra of numbers, Leibniz proposed an algebra in which geometric objects and relations would be represented by symbols and manipulated according to certain rules. The proof of a theorem would be reduced to a calculation with these symbols.

Leibniz complained that Euclid's foundation of geometry was too cumbersome. For his purpose, Leibniz chose <u>distance</u> as a fundamental concept, to which he related <u>symmetry</u>. All that was needed to specify elementary geometric entities in space were points and the distance between them. Thus a <u>plane</u> was defined to be a set of points equidistant from two fixed points. A <u>circle</u> is the set of points at a fixed distance from two points.

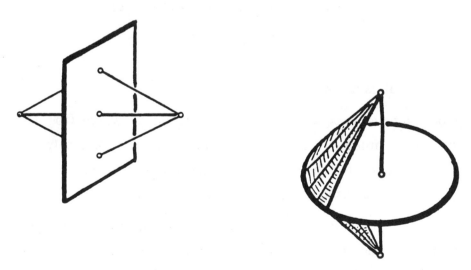

A <u>line</u> is a set of points equidistant from three nonlinear points.

Not only is this approach to geometry fundamentally different from Euclid's foundations of geometry, which dominated the first systematic treatment of geometry in the schools, but the order of development is essentially reversed. Yet it possesses logical rigor, is straightforward and intuitive, and shows the deductivist approach based on Euclid to be unnecessarily complicated and obscure. As such, Leibniz's suggestions to base geometry on the concepts of "distance" and "symmetry" are in accord with contemporary viewpoints on school geometry and hold promise for introducing such elementary geometric ideas to children.

Introducing the Concept to Children

As a starting point, it is only necessary to take from this proposal of Leibniz the idea that distance is the fundamental notion of geometry. A problem which can set the scene for such a development could be one of determining a system of "school zones" so that children attend the school nearest to their home (Booker, Mohry and Stowasser 1984; Mohry, Otte and Stowasser 1981). From a city map like that below, perhaps their own city, children should first identify the schools, marking them in some distinctive manner.

A first-level problem would then be to have the students locate certain addresses (by reference to coordinates, by locating their own home, and so on) and determine which school people living at such an address should attend. This could be done by using a ruler, a piece of string or even a compass, and the children could be led to see that a set of circles centered on the home could be used to determine the closest school.

At the next level, the problem becomes one of determining which areas of the city would be assigned to a particular school, so that the focus shifts to circles centered on the schools. The general question, "How can the map be divided up to show the zones for each school?" can now be posed.

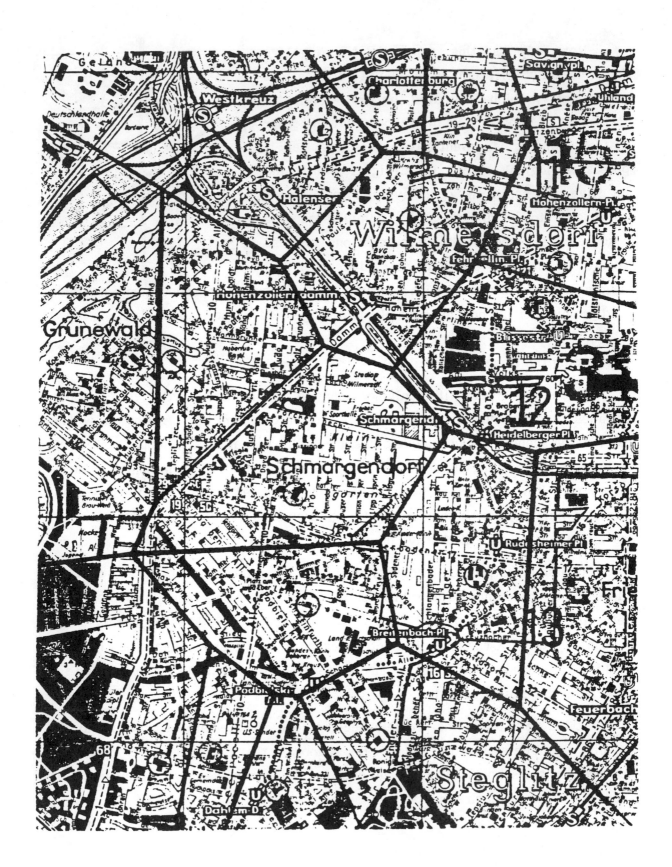

As a first step, the problem must be <u>simplified</u> to enable the fundamental nature of the problem to be grasped, so only two schools should be considered. Circles may be constructed around one of the schools (s_1), and questions asked to elucidate from the children that beyond some radius, points are closer to one school than the other. Drawing circles of the same radius around the other school as well shows how the zones are formed, and this may be made clearer by using successively finer sets of concentric circles.

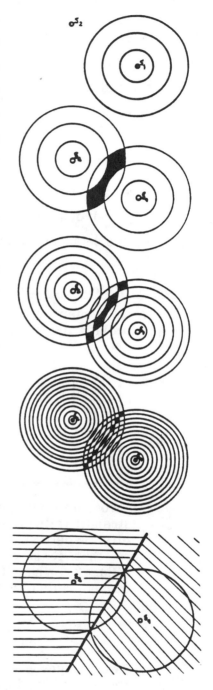

This in turn leads to the idea that the dividing line between the two schools must be a straight line formed by the points of intersection of these circles; that is the perpendicular bisector of the line joining the two schools.

This point can be verified by a paper-folding procedure. Thus two distinct but related methods for determining the dividing lines have been established: by <u>symmetry</u>, folding one point onto the other, or by <u>distance</u>, using circles of the same radius.

An understanding of how the dividing lines are determined can now be applied to all the schools. It is clear by inspection that the zones must be convex, and mathematical arguments can be used to justify this conclusion at a level appropriate to the level of the children. Consider just one school. Each dividing line cuts the whole region into two parts, the nearer of which contains those parts in the school zone. The part further away is in another zone. If you shade or even cut off, one at a time, each of the parts not in the zone, a convex region remains.

The problem of the school zones has been solved very simply by an application and interpretation of the idea of a dividing line. At the same time, this application has built up an understanding of the essential nature of symmetry and distance; it has demonstrated the essence of geometry which Leibniz spoke about.

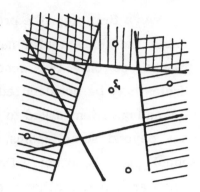

Generalization of Method

Once the fundamental idea of the dividing line and its relationship to symmetry and distance have been grasped, extensions of this approach can be made using a sequence of problems (see below).

Problems Which Are Direct Applications of the Method

1. On a map or diagram with several points but only some dividing lines, provide the remaining dividing lines and locate the regions.

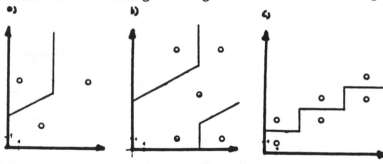

2. Two small villages are adjacent to a straight railway line to the nearest big city. Where should a station be placed so that it is equidistant from each village? What if the railway line is curved, rather than straight?

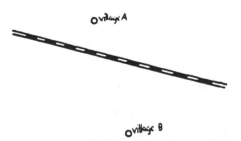

3. Two villages are 5 km apart. A new railway line is to be constructed, but there is no land available closer than 3 km to either village. The station must be equidistant from both and more than 5 km from either village. Can such a position be found?

Problems which require the process to be reversed

1. On a map or diagram with all of the dividing lines but only 1 point, locate the other points.

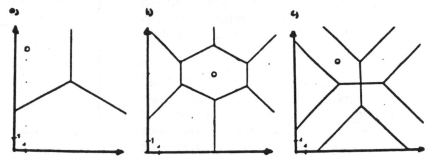

2. Given one point and some of the boundary lines, locate the fewest number of points and complete the regions.

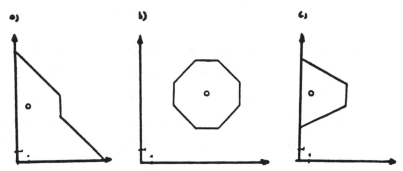

3. Provide copies of uniformly distributed or symmetric patterns of points on which the children could sketch the dividing lines free hand, and then color the resulting patterns.

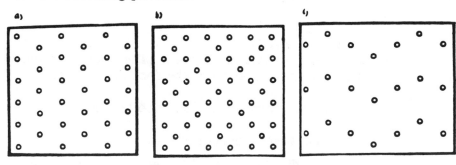

Generalization of Method

1. Consider distributions where the distances from one point to the other are proportional rather than equal. (This gives rise to conic sections as dividing lines rather than straight lines.)

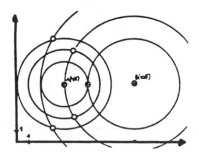

2. In a more realistic situation for school zones, the distances might be measured along roadways rather than simply in a straight line. A hypothetical city in which streets cut at right angles would not be too difficult, but could the original problem also be solved this way?

3. Electoral boundaries for municipal elections are drawn up so that approximately the same number of electors are in each electorate. Could this method be extended to draw up such "zones"?

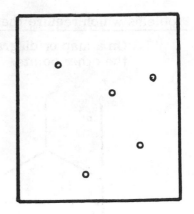

Old Problems Made Easy

This method adapted from Leibniz can also be used to solve more readily problems that in the past have relied on formal and mechanical applications of the Pythagorean theorem. A similar approach is expected in schools today, so that such problems are usually left until the children have become familiar with Pythagoras' result, and occur only as routine applications of it. In "Chin Chang Suan Sin" (九 章 算術 — The Nine Books of Arithmetic), published in China some 2,000 years ago (Vogel 1968), an interesting collection of such problems was posed and solved in the following manner:

- Problem 13 from Book IX.
  A bamboo stick has a height of 10 feet. The upper portion is bent and falls to touch the ground 3 feet from the foot of the stick.

  Question: at what height is the bamboo broken?

The solution was given thus: Take the distance to the foot; multiply it by itself; divide this product by the length. Subtract this from the height of the bamboo; halve the difference. This gives the height of the break.

3
3 x 3
(3x3)÷10
10−(3x3÷10)
[10−(3x3÷10)]÷2
$4\frac{11}{20}$

The solution expected in schools today is similarly routine, although symbolic notation is used in the application of the Pythagorean theorem.

506

On the other hand, once children are familiar with the properties and power of the dividing line process used in the earlier problems, only a small step is needed to quickly and simply solve the new problems. The method is intuitively obvious and does not require the Pythagorean theorem at all. However, it is not necessarily easy for the child, who must be guided into its use, nor is it so easy for

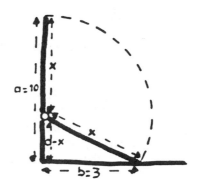

the teacher, who is asked to forego a reliable but mechanical tool. Indeed, it is unusual for a teacher even to consider or encourage such an intuitive approach.

The means to achieving such a transfer has been stressed by Polya (1948). It is to "work backwards," to begin by assuming the result and then seeking a step or series of steps that will provide the solution. Such an approach, of course, is simply the analytical method known and used by the Greeks. Returning to the original problem, the following thinking provides the solution (Mohry, Otte, and Stowasser 1981):

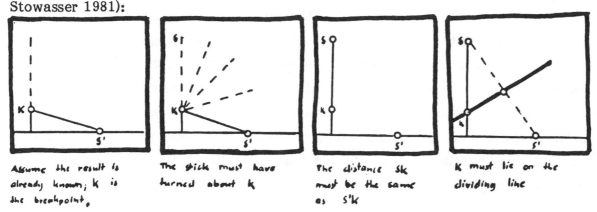

Merely using the notion of <u>symmetry</u> to find the dividing line fixes the point K, and the height of the break can be measured.

The other problems from this source, and from similar collections from Babylonian tablets of earlier times, can also be solved far more simply using the notion of symmetry than by using the solutions that accompany them.

- Problem 6 from Book IX.
  An aquarium has a square base of side 10 feet. A reed grows at the center of the tank, extending 1 foot above the surface. If the reed is pulled to one side, it just reaches the edge of the aquarium at the water

507

line.

Question: What is the depth of the water and what is the length of the reed?

- Problem 8 from Book IX.
A wall is 10 feet high. A plank leans against the wall so that its height is the same as that of the wall. If the plank is moved so that the end resting on the ground is 1 foot further away from the wall, the plank is just able to rest along the ground.

Question: What is the length of the plank?

- Problem 10 from Book IX.
A double door is ajar, with both doors equally open. The distance from the opening to the doors is 1 foot, and the two doors are 2 inches apart.

Question: What is the width of the double door?

Conclusion

A way to present important ideas and applications of elementary geometry has been drawn from the history of mathematics, not for anecdotal use, nor in its original form, but as a basis for developing a meaningful and motivating approach. Such an approach clearly has potential for contributing to a rebirth of geometry in the school: to rescue it from the constraints of the Euclidean perspective and to bridge the gap which has opened between the "old" ways of Euclid and the "new" transformational geometry. It has also brought to the fore a type of thinking and use of intuitive ideas that lie at the heart of successful problem solving.

Notes

1. See Stowasser (1983b). Lagrange in his "Leçons élémentaires sur les mathématiques donées à l'école normale en 1795", teaching about 1300 prospective teachers from all over France to form the staff of local teacher colleges, presents a very simple but universal method to solve geometrical construction problems. The infinite halving procedure, similar to the metric system, used already by Stevin to solve engineering problems, became later with Bolzano a very simple proof method. One and the same proof scheme works and can be successfully applied by even high school students to all the fundamental theorems about real sequences and continuous or differentiable functions (theorems about cluster points, intermediate, mean, and extremal values, and so on). For most people these facts lie in the deep sea. But actually they float on the surface, held together and ruled by the simple idea of repeated halving. The insight that measuring and proving are like the head and tail of a coin should soon have important consequences for high school teaching.

2. See Leibniz (1981a). This is an undated manuscript (1679?) on analysis situs (analysis of position), in which Leibniz criticizes Euclid's cumbersome handling of the simple idea of similarity and presents his own powerful approach. He looks at invariants when changing units by linear transformations. A very modern idea, which could unify several chapters of school mathematics. Some lessons in the spirit of Leibniz's idea. See also Stowasser (1981); von den Steinen (1981).

3. See Stowasser and Mohry (1983), which is based on a dissertation of Jacob Steiner, "Einige Gesetze über die Teilung der Ebene und des Raumes" ("Some Theorems About the Division of the Plane and the Space" (1826)); Stowasser (1985b), which is addressed to teachers to create interesting teaching units from the collected basic material; and Stowasser (1978). See also Stowasser and Mohry (1978), which concentrates on the fundamental idea of recursion as a tool for modeling, especially combinatorial situations, rather than for proofs. Difference equations treated before differential equations, linear methods used as heuristic means to find, for example, Binet's formula (1843) for the Fibonacci sequence (1202), polynomial interpolation, and so on offer a good opportunity for general practice and preparation for the harder material to follow in the analysis, linear algebra, and stochastic courses. The booklet explains in a manner suitable for 16-17 year olds a philosophy of mathematics teaching which combines Polya's problem-solving approach and Wagenschein's (Wittenberg's) paradigmatic teaching. Although the materials appear as a rule in a somewhat historical setting, we do not follow Toeplitz's unsuccessful genetic approach. Irreverently "ransacking history for teaching purposes" may offer better opportunities to reach the classroom.

   The result is a collection of 14 problem areas, including simple patterns; Lucas' problem; coloring, handshaking; Bachet de Meziriac's weighing; Jacob Bernoulli's figurate numbers; Jacob Steiner's intersection problems; from Leonardo of Pisa to Binet; Pascal numbers; Montmort-

Bernoulli letters; and population problems.

More about the authors' philosophy and the solutions of the problems can be found in a booklet (Stowasser and Mohry 1980).

4. ". . . il faut tout d'un coup voir la chose d'un seul regard, et non par progrès du raisonnement, au moins jusqu'à un certain degré . . ." (Pascal, Pensées, ed. Lafuma 512: the difference between l'esprit de géométrie and l'esprit de finesse.)

5. This is the main part of a jointly written paper (Booker, Mohry, and Stowasser 1984). For a more comprehensive version, see Mohry, Otte and Stowasser (1981).

References

van der Blij, F., K. Kiesswetter and R.J.K. Stowasser. 1984. "Forschungsauf-
gaben" für Lehrer und Schüler ("Research Problems" for Teachers and
Students). Reprint No. 2. West Berlin: Technische Universität Mathematics
Department.
  A collection of 6 pieces under this heading in the journal Mathematiklehrer ("Mathematics Teacher"), 1981–83. The intention was to stimulate
teachers themselves to learn mathematics by doing. If they are led to certain
interesting points in history, they may explore the offered problem areas and
finally be tempted to create an attractive "family" around a central problem
for the classroom. Topics include (1) recursively generated number arrays
with Leibniz learned from Huygens and Pell as starting points; (2) the problem
field of "geometric constructions" (with various tools), opened by the research
program of Jacob Steiner; (3) the problem field of "plane linkages" introduced
by David Hilbert; (4) tangency problems solved by geometric non–calculus
techniques (anticipating the calculus), motivated by a memoir of Denis
Diderot; and (5) fragments from Euler about the Bernoulli approximation for
the roots of algebraic equations.

Booker, G., B. Mohry and R.J.K. Stowasser. 1984. Geometry Reborn: A Unifying
Idea for Teaching Geometry Drawn from the History of Mathematics.
Reprint No. 1. West Berlin: Technische Universität Mathematics
Department.
  A proposal of G.W. Leibniz to found geometry on distance (and symmetry)
from a letter to Huygens (August 9, 1679) prepared for pupils age 11-13. An
attractive "opening problem" with a nice family. An interesting use of
symmetry to solve problems that have relied on applications of the
Pythagorean theorem.

von Engelhardt, W. 1955. Gottfried Wilhelm Leibniz: Schöpferische Vernunft.
Münster, Germany: Böhlan-Verlag.

Keitel-Kreidt, Ch. and E. Papamastorakis. 1984. Mathematiklernen aus der
Mathematikgeschichte: auswahl aus Descartes' "Geometrie" ("Learning
Mathematics from the History of Mathematics: A Selection from Descartes'
"Geometry"). In Beiträge zum Mathematikunterricht.
  A prospectus of a booklet in the making, centered around geometrical
construction problems analyzed by algebraic means.

Leibniz, G.W. 1981a. "Im Umkreis einer fundamentalen Idee: Ähnlichkeit." In
Mathematiklehrer 1: 11-12.

Leibniz, G.W. 1981b. "Trennlinien und Trennflächen zur Grundlegung der
Geometrie." In Mathematiklehrer 1: 28-29.

Mohry, B., M. Otte, and R.J.K. Stowasser. 1981. "Trennlinien. Problemsequenzen
zu einer integrativen Unterrichtsidee." Parts 1, 2. In Mathematiklehrer (1):
21-27; (2): 25-27.
  This is a more comprehensive version of Booker, Mohry, and

511

Stowasser (1984).

Polya, G. 1948. <u>How To Solve It</u>. Princeton, NJ: Princeton University Press.

von den Steinen, J. 1981. "Problemsequenzen zum Thema y = c · x in der Geometrie" ("Problem Sequences Concerning y = c · x in Geometry"). Parts 1, 2.
In <u>Mathematiklehrer</u> (1): 13-17; (2): 28-31.
    A reaction to Leibniz's idea of similarity.

Stowasser, R.J.K., ed. 1977-78. "Historische Aspekte für Mathematikdidaktik und
Unterricht." In <u>Zentralblatt für Didaktik der Mathematik</u> (4): 185-213; (2):
57-80.

Stowasser, R.J.K. 1978. "Ransacking History for Teaching." In <u>Zentralblatt für</u>
<u>Didaktik der Mathematik</u> (2): 78-80.

_____. 1981. "Eine Anwendung von Leibnizens Methode der Massstabs-
änderung: Das Pendelgesetz" (An Application of Leibniz's Method of Changing
Units by Linear Transformations: The Law of the Pendulum). In <u>Mathematik-</u>
<u>lehrer</u> (1): 12-13.

_____. 1982a. <u>Geometry and Problem Solving</u>. Proceedings of the
International Colloquium on Geometry Teaching, organized by the Belgian
Subcommission of ICMI at Mons, 1982 (G. Nöel, ed.): 207-226.
    Message: The renewal of geometry teaching means above all its
emancipation from more elaborate knowledge and from the logical-
systematic Euclidean organization. Geometry should not be taught en bloc
and self-reliant, but in a network of overlapping problem families around a
few simple fundamental ideas linking together geometry, algebra, number
theory, and analysis. Such an approach can hinder the normal passivity in
learning, and support problem-solving activities, leading to more intelligent
behavior in new situations. The paper shows by example how the history of
mathematics is helpful.

_____. 1982b. "A Textbook Chapter from an Idea of Pascal." In <u>For the</u>
<u>Learning of Mathematics</u> 3(2), David Wheeler, ed.
    First published in German in <u>Mathematiklehrer</u> 1982 (2), R.J.K. Stowasser
and Roland Fischer, eds. A French translation is also available: West Berlin:
Preprint No. 4, Technische Universität Mathematics Department, 1984.

_____. 1983a. <u>An Excerpt of Dürer's Treatise on Mensuration with the</u>
<u>Compass and Ruler in Lines, Planes, and Whole Bodies to Be Used for</u>
<u>Teaching</u>. Preprint No. 3. West Berlin: Technische Universität Mathematics
Department.
    Dürer's constructions of regular polygons of n sides for all n < 17 serve as
a starting point to reflect on algebraic equations and to get the idea of
continuous fractions.

_____. 1983b. <u>Stevin and Bolzano: Practice and Theory of Real Numbers</u>.
Preprint No. 4. West Berlin: Technische Universität Mathematics

Department.

_____. 1985a. <u>Alte und Junge Anwendungen der Kongruenzrechnung in der Schule: Pascal, Fermat, Gauss, Rivest</u> ("Old and New Applications of the Idea of Congruence in the Schools: Pascal, Fermat, Gauss, Rivest"). Preprint No. 1. West Berlin: Technische Universität Mathematics Department.

_____. 1985b. <u>A Collection of Problems From History Attractive for Teaching Recurrence Methods.</u> Preprint No. 2. West Berlin: Technische Universität Mathematics Department.

Stowasser, R.J.K. and T. Breiteig. 1984. "An Idea from Jacob Bernoulli for the Teaching of Algebra: A Challenge for the Interested Pupil." In <u>For the Learning of Mathematics</u> 4 (3): 30-38.
    Jacob Bernoulli's <u>Ars conjectandi</u> (1713) offers a readable discussion of the binomial numbers and derives from them formulas for evaluating the power sums $(1^k + 2^k + 3^k + \ldots n^k)$ — the Bernoulli polynomials. "The wonderful properties of the figurate numbers" traced by Bernoulli where "eminent secrets from the whole of mathematics are hidden..." (Bernoulli) can be obtained by interested pupils. But there is enough material for the normal classroom too: a "family" of problems for the general practice of basic algebraic techniques.

Stowasser, R.J.K. and B. Mohry. 1978. <u>Rekursive Verfahren. Ein problemorientierter Eingangskurs</u> ("Recursive Procedures: A Problem-Oriented Preparatory Course"). Hannover, West Germany: Schroedel Verlag.

_____. 1980. <u>Didaktik und Lösungen zum Kursbuch Rekursive Verfahren.</u> Hannover, West Germany: Schroedel.

_____. 1983. "Die Idee der Rekursion und der Isomorphie im Umkreis von Steiners Raumteilung und Eulers vertauschten Briefen: für talentierte Schüler verfasst" ("The Idea of Recursion and of Isomorphism: Problem Sequences Prepared for Talented Students"). In <u>Mathematiklehrer</u> (2): 2-10.

Vogel, K. 1968. <u>Neun Bücher Arithmetischer Technik.</u> Braunschweig: Friedrich Vieweg & Sohn.

Coded Graphical Representations:
A Valuable but Neglected Means of Communicating
Spatial Information in Geometry

Claude Gaulin

Université Laval

Ewa Puchalska

Université de Montréal

Introduction

In an earlier presentation, Max Bell gave an overview of the Elementary Component of the University of Chicago School Mathematics Project. We were particularly pleased to find the following listed among the objectives:

Visual displays — representing non-numerical information

Diagrams such as blueprints, circuits, machinery parts, etc. Graphs showing relationships . . .

Geometric relations

Projections as applied to perspective drawing, contour maps, or representing the world on flat maps.

In this paper, we wish to stress the fundamental importance of such means of communicating spatial information, and to put a particular emphasis on one type of graphical representation which remains generally neglected in school mathematics curricula.

Coded Graphical Representations

In order to communicate information about objects, locations, directions, relations and transformations in three-dimensional space, various graphical means are commonly used: drawings, sketches, schemas, diagrams, maps, plans, graphs, charts, and so on. Such illustrations often contain information conveyed by means of a code, i.e. a structured system of signs and/or words complying with some conventions and rules of arrangement (Laborde 1982). The code may include: letters; numerals; marks like dots, asterisks, crosses, triangles or other simple geometrical shapes; pictograms; arrows; contour lines; colors or shades; hachures; some continuous or dotted lines; and so on. Of course, the meaning of all the signs used must be clear, either explicitly (by means of a legend or of verbal explanations in the illustration) or implicitly (because the signs are used in a well known, conventional way). In this paper, we are going to call a <u>coded graphical representation</u> (CGR) any coded illustration in which the meaning of at least one of the signs used is given explicitly. Here are a few examples of CGRs.

<u>Example 1</u> (Diagrammatic map of lines of the London Underground. The seven lines are shown in distinctive colors.)

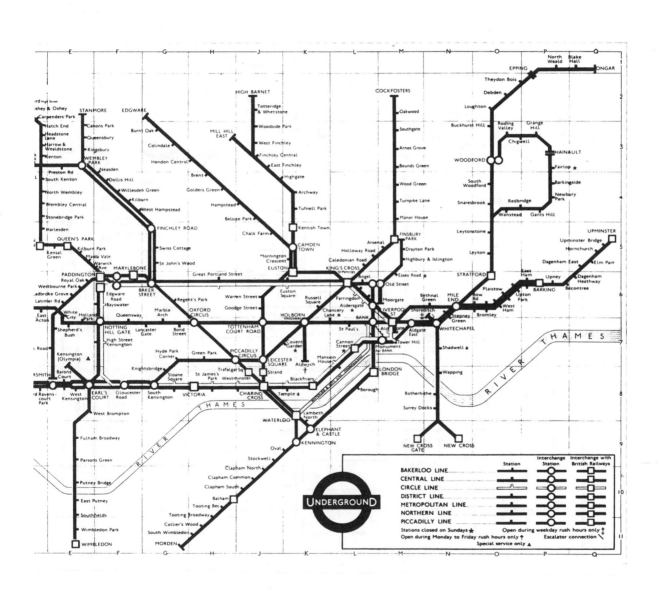

Example 2 (Map of the area near the University of Chicago campus)

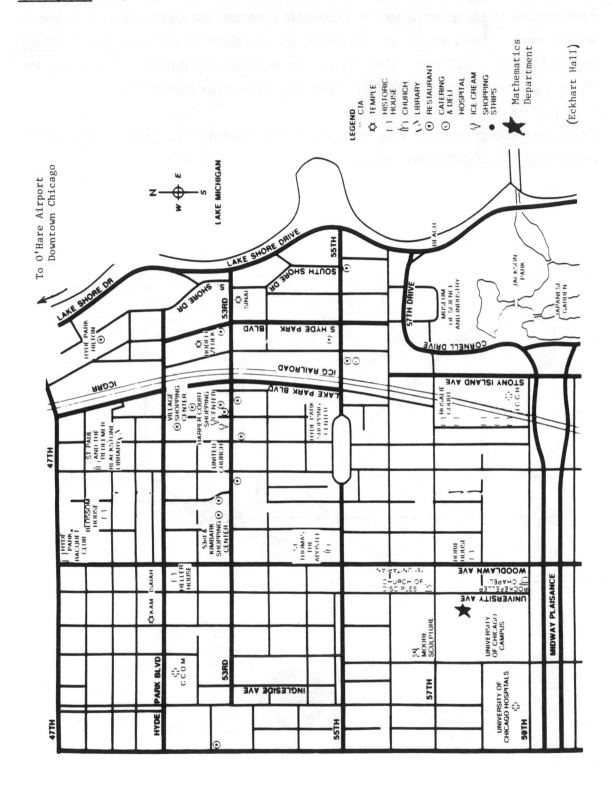

(Eckhart Hall)

516

Example 3 (Topographic map of a cross-country skiing center near Quebec City)

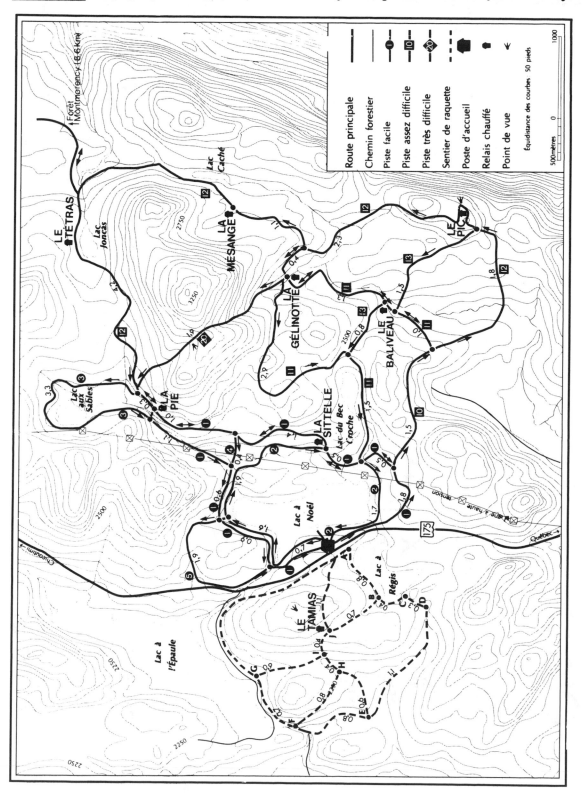

Example 4 (Part of the instructions for assembling a piece of furniture)

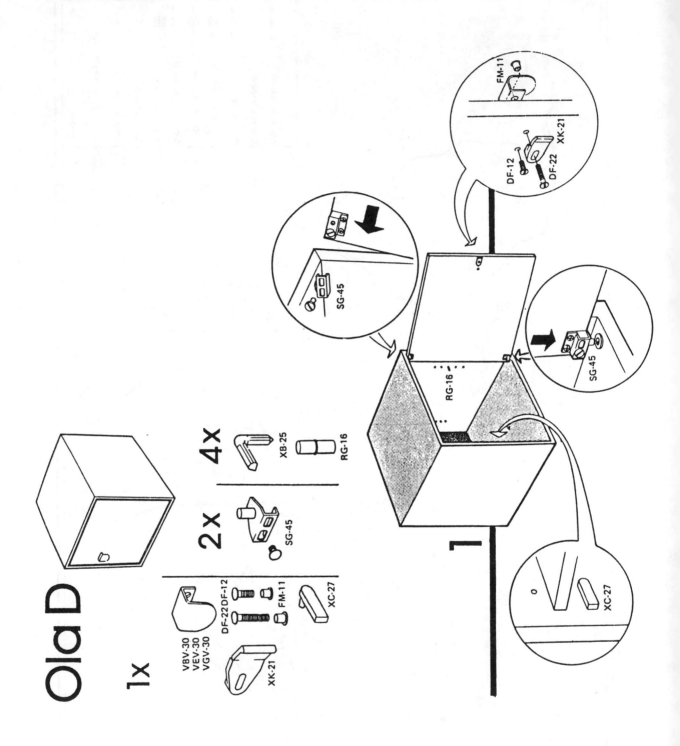

Example 5 (Anatomical diagrams)

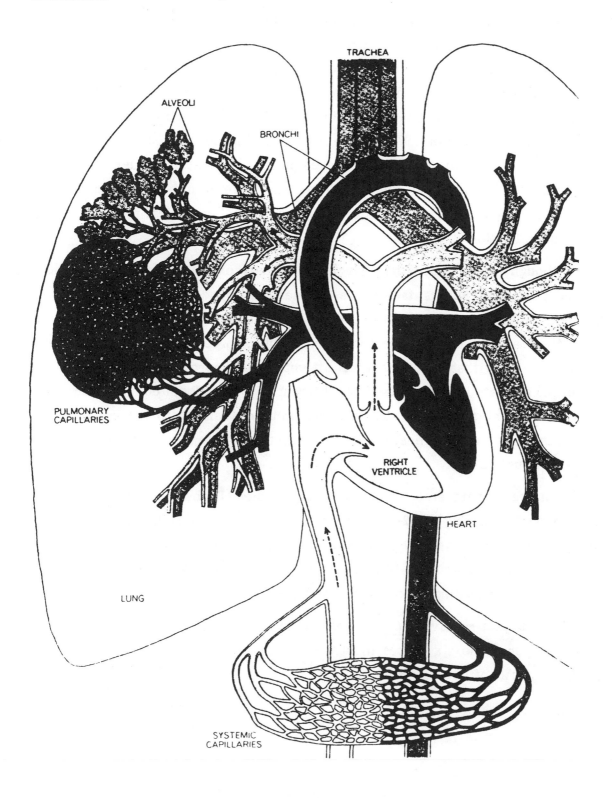

The Origin of Our Interest in CGRs

From 1978 to 1980, materials were developed at Laval University (Gaulin et al. 1980) to serve as a basis for two courses called "Explorations Géométriques" which have been offered since as part of an important long-distance in-service teacher education program in mathematics (PPMM) for elementary teachers of the Greater Quebec City area. Following guidelines which one of the authors had suggested earlier (Gaulin 1974), these materials placed particular emphasis on activities intended to foster the development of spatial visualization and geometric intuition in upper elementary grades. Several units were also devoted to the exploration of various types of (two-dimensional) graphical representations of polycubical solids.

Although such activities involved actual experiments with physical materials, many teachers taking the courses experienced great difficulties in making and interpreting such graphical representations, which of course tended to hinder their ability to visualize the corresponding three-dimensional shapes and relations. Notwithstanding several attempts to improve those units, such difficulties persisted. This aroused the interest of a few PPMM collaborators, who decided to investigate more closely some problems related to the use of plane representations to communicate spatial information and to foster the development of spatial visualization.

As a preliminary step, two exploratory studies (Puchalska and Gaulin 1983) were conducted during 1982-83, with approximately five hundred pupils about equally distributed among 4th grade (10-11 years old), 6th grade (12-13 years old), 7th grade (13-14 years old), 9th grade (15-16 years old) and 11th grade (17-18 years old). The following task was administered in each of the 21 classes.

Each subject received <u>one</u> geometrical solid made of congruent plastic cubes glued together. The six shapes illustrated below were distributed about equally among the pupils of each class.

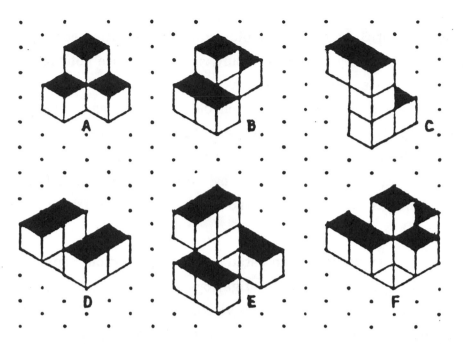

In addition to one plastic shape, each subject was given a sheet of paper with a brief instruction and plenty of space to answer. Half of the pupils got the following ("algorithmic") instruction:

- Imagine that one of your friends lives in France and that he has a whole box of plastic cubes, all of the same size. Now you would like your friend to build a shape like the one you have in front of you. Prepare a message you could send him so that he can build it. You may give your explanations using words or drawings, as you wish.

while the other half got the following ("descriptive") instruction:

- Imagine that one of your friends lives in Ottawa. Near his home, there is a store selling all kinds of shapes made of plastic cubes. Now you would like your friend to go to that store and buy you one shape like the one you have in front of you. Prepare a message you could send him so that he can recognize that shape. You may give your explanations using words or drawings, as you wish.

The type of communication task used (preparation of a "message") was inspired by the work of Guy Brousseau in France.

One of the objectives of those pilot studies was to observe the types of responses (verbal or graphical or mixed) spontaneously given by the subjects. Two major observations were made which are relevant here. The first one was the great variety of types of productions obtained among all age groups. The second was the predominance of "Coded Orthogonal Views" (COVs) among all age groups. (By definition, any COV of a solid S consists of one particular orthographic view of S, along with supplementary information conveyed by means of a code.) This was an unexpected result, since COVs are not taught in Quebec schools (except for a few students taking a specialized course in technical drawing) and since the instructions given did not make any suggestion to that effect. (During the completion of our studies, we were informed that a communication task of a similar type had been used, although with a different methodology, in experiments conducted by Bessot and Eberhard (1982a, 1982b). These researchers, as well as Franchi and de Azevedo (1983) later in Brazil, found results like ours.)

We were so intrigued and interested by the results of our exploratory studies that we decided to conduct two investigations concurrently as a follow-up. The first aims at studying the nature and development of the ability to communicate spatial information by means of Coded Orthogonal Views, as well as the influence of social interaction on that ability (Gaulin, Noelting, and Puchalska 1984; Noelting, Puchalska and Gaulin 1985). The second one more generally aims at studying coded graphical representations (of which COVs obviously are one particular case) and their didactical aspects in relation to the teaching of geometry.

The following pages give a brief overview of the main types of students' productions we have obtained in the pilot studies:

- verbal descriptions
- descriptions by means of many side views (explicitly mentioned)
- descriptions layer by layer
- descriptions by means of one "coded orthogonal view"
- attempts at perspective drawing
- other types

Verbal Descriptions (Translated from French)

● The shape I am going to describe has 4 cubes, 24 faces, 36 edges, 21 corners and it is red.

> (10-year-old; Shape A)

● Put one cube on the right. Put another cube on the left and another cube at the center. Put another one above the center.

> (10-year-old; Shape B)

● You must have a cross, with one block on top in the middle. There is also one block on the other face of the cross, but it is above.

> (9-year-old; Shape F)

Descriptions by Means of Many Side Views (Explicitly Mentioned)

from above

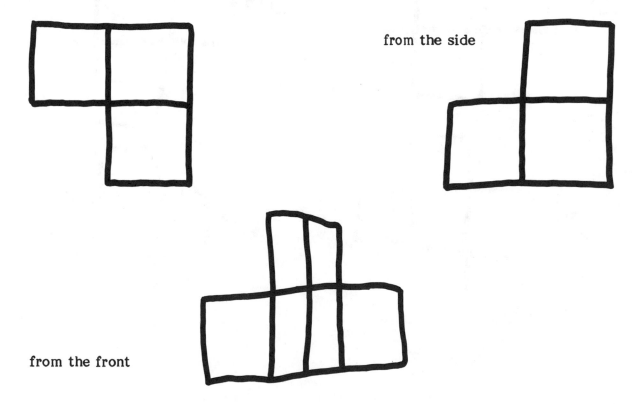

from the side

from the front

> (9-year-old; Shape A)

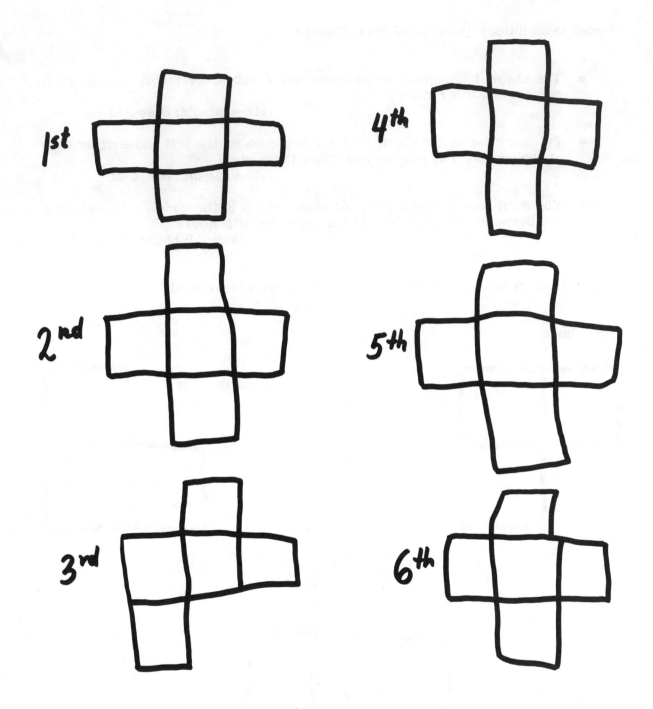

1st

4th

2nd

5th

3rd

6th

(9-year-old; Shape F)

524

Descriptions Layer by Layer

bottom   top

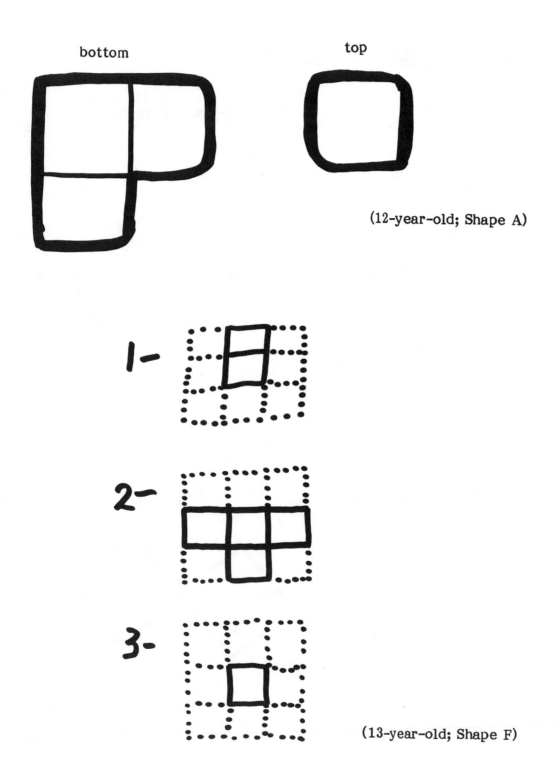

(12-year-old; Shape A)

1-

2-

3-

(13-year-old; Shape F)

Descriptions by Means of One "Coded Orthogonal View"

There is a cube above this one.

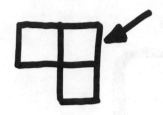

(12-year-old; Shape A)

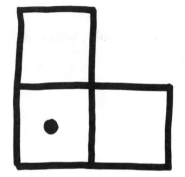

Place your fourth cube above the one marked with a ●.

(14-year-old; Shape A)

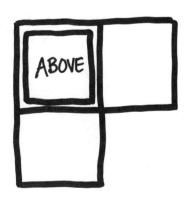

(13-year-old; Shape A)

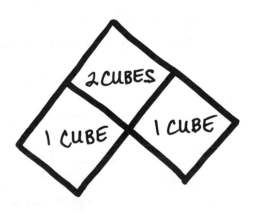

(13-year-old; Shape A)

from the side

no hachure	☐	means that it projects from the others
small numeral	(1)	gives the step for constructing the shape
large numeral	3	gives the number of cubes stuck together
hachure	▨	means that it is recessed from the others

(16-year-old; Shape F)

Attempts at Perspective Drawing

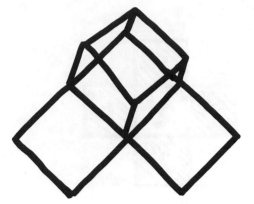

(9-year-old; Shape A)

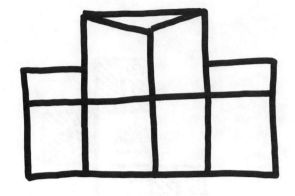

(13-year-old; Shape A)

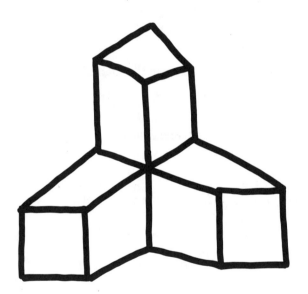

(13-year-old; Shape A)

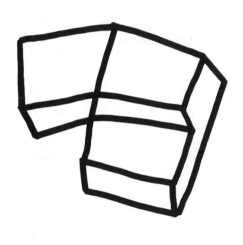

(13-year-old; Shape A)

528

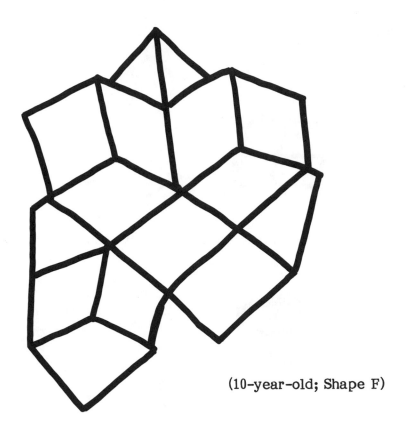

(10-year-old; Shape F)

 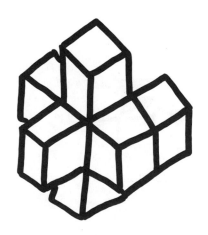 

(14-year-old; Shape F)

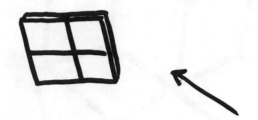

a) You take 4 cubes and make a platform.

b) You take one of the corners

and place it here slantwise.

(14-year-old; Shape A)

Reasons for Emphasizing CGRs

Graphicacy as a Basic Educational Objective

The ability to use various types of graphical representation is one aspect, among others, of what is sometimes called "graphicacy" in England, i.e. "the communication of spatial information that cannot be conveyed adequately by verbal or numerical means" (Balchin 1972).

> Balchin and Coleman describe <u>literacy</u>, <u>numeracy</u>, <u>articulacy</u> (subsequently superseded by <u>oracy</u>) and <u>graphicacy</u> as the four "acies", or "aces" for short . . . Neither words nor numbers are superior or inferior as modes of communication. They are only more suitable or less suitable for particular purposes, and each ranges from the very simple to the extremely complex. They complement each other and achieve their highest level of communication when properly integrated . . . Balchin argued that the discussion ought to be more concerned with modes of communication and that the three Rs should be replaced by the "four aces". He also noted that in France a similar discussion was taking place around the concept of "four languages" in communication skills and the need to teach them to all pupils. These four languages corresponded exactly with the four modes of communication distinguished in Britain.
>
> Meanwhile it has been argued by Boardman (1976) that geography teachers share with their colleagues in other subjects, especially English and Mathematics, responsibility for ensuring that graphicacy is developed by all pupils before they leave school . . .
>
> (Boardman 1983, preface)

As a geography educator, Boardman speaks of graphicacy, insisting particularly on maps of all kinds and photographs. However, we feel that his arguments apply equally, perhaps with even greater strength and more relevance, when "graphicacy" is defined in a more general way so as (1) to include the communication of nonfigural information by means of diagrams, schemas and graphs having a low degree of "iconicity," i.e. representing abstract relationships (see Bertin 1967, 1977; Herdeg 1981); and (2) to take also into account the non-social instrumental role of graphical representations for concept formation, problem solving and more generally organizing thought (see Van Sommers 1984, Biehler 1982, Chernoff 1978).

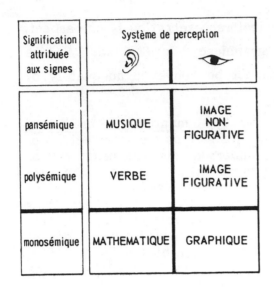 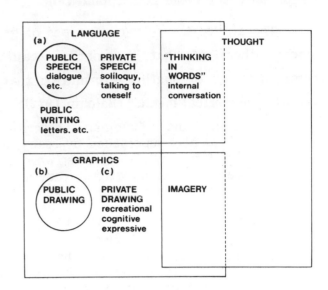

Bertin (1967, 1977)   Van Sommers (1984)

Developing graphicacy as a basic educational objective is clearly a multidisciplinary enterprise in which mathematics education has an important contribution to make, just like geography education, art education, language education, and so on. Providing all pupils with opportunities to explore a variety of types of graphical representation — including CGRs — appears to be one of the features of such a contribution.

The Ever-Increasing Practical Importance of Such Representations

Coded graphical representations of various types are commonly used in a great number of practical situations and disciplines for conveying spatial information. Here are a few examples:

- maps, plans and sketches: topographical, geographical, geological, meteorological, architectural, for finding directions (e.g. in a shopping center), for trains / buses / subway / airlines networks, and so on.
- diagrams and flowcharts giving instructions: for assembling (e.g. a piece of furniture or parts of a construction kit), for operating a

machine, for sewing or knitting or crocheting, and so on.

- scientific or technical descriptive drawings: in anatomy, botany, mechanics, engineering; models of atoms and molecules, and so on.

Taking into account the pervasiveness of computerized graphical displays in today's society, the ability to communicate with (or at least to interpret) coded graphical representations is likely to be more and more needed in the near future.

The Need for Reestablishing the Development of Spatial Intuition as One Major Goal for Teaching Geometry

For years mathematics educators like Bishop, Clements, Mitchelmore, Tahta and others have claimed that one major goal for teaching geometry that has been overlooked during the "new math" wave and ought to be reestablished is the development of students' spatial intuition, including their ability to visualize and to communicate spatial information by various means. Taking into account the existing confusion in the definition of terms like "spatial ability," "spatial intuition," and "visualization," we prefer to think of such an objective with reference to the two types of ability constructs that have been proposed by Alan Bishop (1980) in order "to help mathematics educators focus on relevant training and teaching research":

> 1. The ability for interpreting figural information (IFI). This ability involves understanding the visual representations and spatial vocabulary used in geometry work, graphs, charts, and diagrams of all types. Mathematics abounds with such forms and IFI concerns the reading, understanding, and interpreting of such information. . . .

> 2. The ability for visual processing (VP). This ability involves visualization and the translation of abstract relationships and nonfigural information into visual terms. It also includes the manipulation and transformation of visual representations and visual imagery.

> (Bishop 1983: 184)

Some knowledge of graphical representations and their tacit conventions is a necessary condition for the development of IFI and probably also, in our opinion, for the development of some aspects of VP. Accordingly, some experience with various types of graphical representations, and with CGRs in particular, is certainly needed for developing the students' spatial intuition and their ability to visualize shapes and relations in three dimensions. We wish to stress the point that we are not talking here of any systematic instruction, but rather of familiarization through occasional exploratory activities.

The Need for Diversity of Such Representations

With all due respect to many art educators and to psychologists who stick to the Piagetan tradition for explaining the genesis of the representation of space by individuals, we wish to support strongly the point argued by Josiane Caron-Pargue (1979, ch. 1) that perspective drawing is just one among many modes of graphical representation, each one having its characteristics and its merits. The results obtained by L. Páez Sanchez (1980) with adults having little formal schooling and the observations made by Caron-Pargue let us hypothesize the development of "simultaneous, but inequally accessible" abilities corresponding to different modes of graphical representation: perspective or isometric drawing, coded orthogonal views (as in cartography or in technical drawing), representations by means of layers or sections, and so on. Obviously such a hypothesis entails the importance in schools of exploratory activities of communication of spatial information by means of various types of representations, including CGRs.

The Specific Advantages of CGRs

Because they make use of codes, CGRs constitute a particularly powerful and efficient means for communicating spatial (as well as non-spatial) information. As a matter of fact, codes offer some specific advantages:

- High condensation power. Any symbol has the potential to convey very complex information; its multiple use in one CGR makes the representation still more economical.
- Simplicity of use. No drawing techniques are required; signs can be used easily and quickly.
- Potential for representing non-visible information in complex situations, e.g., hidden parts which cannot be drawn in perspective.

The following example gives a good illustration. It shows the 29 essentially different pentacubes (solids obtained by juxtaposition of 5 congruent cubes, no

534

partial superposition of faces allowed).

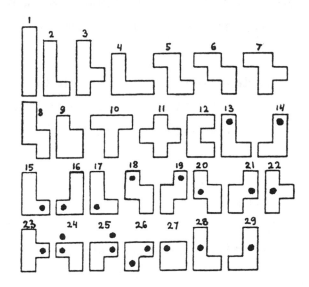

(J. Carstensen, "Legespiele." <u>Der Mathematik Unterricht</u>, 2/80)

Suggestions for a Few General Guidelines for Emphasizing CGRs in School Curricula

1. It goes without saying that students need to be involved in a sufficiently great number of <u>preliminary exploratory activities involving real problematical situations and manipulative materials</u> before any attempt is made to focus on the use of graphical means — and CGRs in particular — to represent three-dimensional shapes and relations.

2. The objective is <u>not</u> to teach students ready-made conventional codes (like those used in cartography or in technical drawing), but rather to give them opportunities to actively explore a variety of situations where codes are used — or must be invented — to communicate spatial information. What is most important here are the <u>processes</u> of decoding and encoding, not the product.

3. The exploration of coded (as well as non-coded) graphical representations should preferably form a recurrent theme over many years in the curriculum, rather than an isolated teaching unit or chapter.

4. Such exploratory activities appear particularly well suited for the upper elementary and the junior high school levels and ideally they should be used, in a complementary way, in the teaching of geometry, geography and perhaps other disciplines.

5. One must keep in mind that an important objective of such activities is to support the development of students' spatial intuition and visualization ability, besides acquainting them with several means of graphical representation.

References

Association of Teachers of Mathematics. 1982. Geometric Images. Derby, England: ATM.

Balchin, W.G.V. 1972. "Graphicacy." Geography 57 (3).

Bertin, J. 1967. Sémiologie graphique — Les diagrammes, les réseaux, les cartes. Paris: Gauthiers Villars.

Bertin, J. 1977. La Graphique et le Traitement Graphique de l'information. Paris: Flammarion. English version (1981): Graphics and Graphic Information-Processing. New York: W. de Gruyter.

Bessot, A. and Eberhard, M. 1982a. "Représentations d'assemblages de cubes au C.M." Grand N, March 1982. IREM, Université de Grenoble.

Bessot, A. and Eberhard, M. 1982b. "Représentations d'assemblages de cubes au C.E." Grand N, December 1982. IREM, Université de Grenoble.

Biehler, R. 1982. Explorative Dateanalyse — eine Untersuchung aus der Perspektive einer deskriptiv-empirischen Wissenschaftstheorie. Institut für Didaktik der Mathematik, Universität Bielefeld.

Bishop, A. 1974. "Visual Mathematics." In Proceedings of the ICMI-IDM Regional Conference on the Teaching of Geometry. Institut für Didaktik der Mathematik, Universität Bielefeld.

Bishop, A. 1980. "Spatial and Mathematical Abilities: A Reconciliation." Paper presented at the conference on Mathematical Abilities at the University of Georgia at Athens, U.S.A.

Bishop, A. 1982. "Towards Relevance in the Teaching of Geometry." In Proceedings of the International Colloquium on Geometry Teaching, G. Noël, ed. Université de l'Etat à Mons, Belgium.

Bishop, A. 1983. "Space and Geometry." In <u>Acquisition of Mathematics Concepts and Processes</u>. Lesh, R. and Landau, M., eds. New York: Academic Press.

Boardman, D. 1976. "Graphicacy in the Curriculum." <u>Educational Review</u> 28 (2).

Boardman, D. 1983. <u>Graphicacy and Geography Teaching</u>. London: Croom Helm.

Boero, P. 1981. <u>La geometria per la conoscenza del territorio</u>. Istituto di Matematica, Università di Genova.

Bowman, W.J. 1968. <u>Graphic Communication</u>. New York: Wiley.

Brousseau, G. 1972. "Processus de mathématisation." Bulletin de l'Association des Professeurs de Mathématiques de l'Enseignement Public (France), February 1972.

Caron-Pargue, J. 1979. "Etude sur les représentations du cube chez des enfants de 3 à 11 ans — La représentation et le codage de propriétés spatiales." Thèse de doctorat de 3e cycle en psychologie. Université de Paris.

Caron-Pargue, J. 1985. <u>Le dessin du cube chez l'enfant — Organisations et réorganisations de codes graphiques</u>. Berne: Peter Lang.

Chernoff, H. 1978. "Graphical Representation as a Discipline." In <u>Graphical representation of multivariate data</u>, Wang, P.P.C., ed. New York: Academy Press.

Commission Internationale pour l'Etude et l'Amélioration de l'Enseignement des Mathématiques (CIEAEM). 1981. <u>Actes de la 33e rencontre internationale organisée à Pallanza, Italie sur le thème "Processus de géométrisation et de visualisation"</u>, edited by M. Pellerey.

De Lange, J. 1984. "Geometry For All: No Geometry at All?" <u>Zentralblatt für Didaktik der Mathematik</u> 84 (3).

Dreyfus, T. and Eisenberg, T. "On Spatial Reasoning: Implications for Teacher Education." Unpublished paper. Holon and Beer Sheva, Israel.

Franchi, A. and de Azevedo, M.V.R. 1983. "Relatorio de uma experiência." Unpublished paper. São Paulo: Póntifica Universidade Católica.

Freeman, N.H. 1980. <u>Strategies of Representation in Young Children</u>. New York: Academic Press.

Gaulin, C. 1974. "Genuine Geometrical Activities for Elementary Schools." In <u>Proceedings of the ICME-JSME Regional Conference on Curriculum and Teacher Training for Mathematical Education</u>. Tokyo: Japanese Society for Mathematical Education.

Gaulin, C. et al. 1980. <u>Explorations Géométriques</u>, Tomes I et II. Québec: P.P.M.M., département de didactique, Université Laval.

Gaulin, C. Noelting, G. and Puchalska, E. 1984. "The Communication of Spatial Information by Means of Coded Orthogonal Views." In Proceedings of the 6th Annual Meeting of PME-North American Chapter, Carpenter, T. ed. University of Madison, Wisconsin.

Giles, G. 1979. 3-D Sketching Series. DIME Projects. University of Stirling / Oliver and Boyd, Scotland.

Herdeg, W. 1981. Graphis Diagrams — The Graphic Visualization of Abstract Data. (4th ed.) Zurich: Graphis Press Corp.

Institut voor de Ontwikkeling van Het Wiskunde Onderwijs. 1976. "Five Years IOWO — IOWO Snapshots." Educational Studies in Mathematics 7 (3).

Keates, J.S. 1982. Understanding Maps. London: Longman.

Laborde, C. 1982. "Langue naturelle et écriture symbolique: deux codes en interaction dans l'enseignement mathématique." Thèse de doctorat en didactique des mathématiques, Université de Grenoble.

Lesh, R. and Mierkiewicz, D., eds. 1978. Recent Research Concerning the Development of Spatial and Geometric Concepts. Columbus, Ohio: Eric/Smeac.

Liben, L.S., Patterson, A.H. and Newcombe, N., eds. 1981. Spatial Representation and Behavior Across the Life Span. New York: Academic Press.

Martin, J.L., ed. 1976. Space and Geometry. Columbus, Ohio: Eric/Smeac.

Middle Grades Mathematics Project (W. Fitzgerald et al.). 1986. Spatial Visualization. Menlo Park: Addison-Wesley.

Mitchelmore, M.C. 1980. "Three-Dimensional Geometrical Drawings in Three Cultures." Educational Studies in Mathematics 11, 205-216.

Morris, C. 1971. "Writings on the General Theory of Signs." The Hague: Mouton.

Müller, P. 1981. "Raumgeometrie in der Schule und für die Schule." Mathematica didaktica 4, 155-168.

Noelting, G., Puchalska, E. and Gaulin, C. 1985. "Levels in the Ability to Communicate Spatial Information by Means of Coded Orthogonal Views." Paper presented at the AERA Annual Meeting, Chicago, March 31, 1985.

Páez, Sanchez, L. 1980. "La représentation graphique de l'espace chez l'enfant et chez l'adulte peu scolarisé." Thèse de doctorat de 3e cycle en didactique des mathématiques. Université Paris VII.

Piaget, J. and Inhelder, B. 1948a. La représentation de l'espace chez l'enfant.

Paris: Presses Universitaires de France. English translation (1956): The Child's Conception of Space. London: Routledge and Kegan Paul.

Piaget, J., Inhelder, B., and Szeminska, A. 1948b. La géometrie spontanée de l'enfant. Paris: Presses Universitaires de France. English translation (1960): The Child's Conception of Geometry. London: Routledge and Kegan Paul.

Puchalska, E. and Gaulin, C. 1983. "Representations on Paper of Three-Dimensional Shapes." In Proceedings of the 5th Annual Meeting of PME-North American Chapter, Volume 1, Bergeron J. and Herscovics, N. eds. Université de Montréal.

Schmid, C.F. and Schmid, S.E. 1979. Handbook of Graphic Presentation. (2nd ed.) New York: Wiley.

School Mathematics Project 1985. SMP 11-16 (series of booklets). Cambridge: Cambridge University Press.

Southworth, M. and S. 1982. Maps — A Visual Survey and Design Guide. Boston: Little, Brown and Co.

Van Sommers, P. 1984. Drawing and Cognition. Cambridge: Cambridge University Press.

Visual Education Curriculum Project 1981. Do You See What I Mean? Learning Through Charts, Graphs, Maps and Diagrams. Canberra, Australia: The Curriculum Development Centre.

Instructional Processes in the
Teaching of Mathematics

Alphonse Buccino and James W. Wilson
University of Georgia

Introduction

Improving the teaching and learning of mathematics is of utmost importance to the education of our students and to the society of the future. We must have a much larger pool of scientific talent, more people with expertise in technologies, and a better informed citizenry. The quality of the lives our citizens lead in the future will depend much more than it ever has on whether or not they are equipped with mathematical tools for thinking about the problems that confront them individually and collectively. We must have a renewed commitment to the learning of mathematics — and therefore to its teaching.

This paper will focus on instructional processes in the teaching of mathematics, i.e., one area of mathematics education. The focal point could have been mathematics curriculum, teacher education in mathematics, or some other area. Indeed, these things are interrelated, part of a system, and concentrating on improving instructional processes in mathematics class will involve some awareness, work, and consequences in other aspects of mathematics education. The action, however, is inside classrooms. Thus, work that improves the quality of the instruction that goes on in classrooms has immediate and widespread applicability. There is sufficient relevance and leverage implicit in the study of instructional processes in the teaching of mathematics to influence mathematics instruction at any level.

Recent Developments

The modern era in mathematics and mathematics education began roughly with the post-World War II years, the late forties and early fifties. The importance of science and mathematics, and of education in these subjects, received increased emphasis in the late fifties in response to the launching of

Sputnik. During most of this period, the emphasis was clearly on the substantive content of mathematics. Under the auspices of the National Science Foundation, a dual strategy was employed: curriculum development and teacher education. The content emphasis affected both. Teaching methods and pedagogy had lower priority for study (National Science Foundation 1959).

With curriculum development, the concern was to bring materials up to date and assure their accuracy from a scientific and mathematical point of view. With teacher education, it was generally assumed that improved content preparation of teachers would lead to improved instruction. Whatever the priorities were, we must recognize that there was no background of research and knowledge of what actually takes place in schools to guide what was done.

Good, Grouws, and Ebmeier (1983) point out that until the 1970's few studies presented data based on classroom observation, and many researchers and critics of education argued or assumed that little variability in what students learn could be attributed to differences in teacher behavior or classroom environment. Indeed, Gage (1960) and Dunkin and Biddle (1974) each concluded that the thousands of studies of teacher effectiveness provided very little insight into teacher behavior and student achievement. Much of this research examined teacher characteristics rather than classroom processes to explain student achievement.

Now, some thirty to forty years later, the trends and the attitudes toward education generally, and toward teacher education in particular, appear to be pretty much the same. States are calling for increased content preparation in teacher education programs. There is little explicit attention to pedagogy. A somewhat new wrinkle is an increased emphasis on field experiences, student teaching, or internships as important factors in the preparation of teachers (Berliner 1986).

The lack of emphasis on pedagogy is unfortunate. First, between 1955 and 1975, disciplined inquiry into the teaching and learning of mathematics began to emerge as a legitimate field of interdisciplinary study. The times saw a rediscovery of the importance of meaning in the teaching and learning of mathematics and a retrospective look at the work of Brownell (1951) and others

541

who had taken the trouble to look into mathematics classrooms as a part of their research.

Studies of classroom processes in the teaching of mathematics were done by a few mathematics educators during this period. Fey (1970) analyzed the verbal activity of teachers and students in mathematics classes. His work was important as a pioneering effort, suggesting attention be given to several conceptual and methodological issues in this type of research. One important result has come to be regarded as an essential assumption of experts in and out of mathematics education. That is, the structure and content of mathematics exert a definite influence on the pattern of discourse in mathematics classes. Stated more strongly, there is something different about mathematics classes. If we wish to improve instructional processes in mathematics classes, we should observe and study mathematics classes. In a different context, this point has been elaborated upon by Resnick and Ford (1981) in arguing for a psychology of mathematics education.

Another development of the study of instructional processes in mathematics classes during the period 1955-1975 occurred at the University of Illinois (Cooney, Henderson, and Davis 1972). Henderson and his students pursued several studies of concept moves in mathematics teaching. The methods book was in large part a synthesis of a long-term research project.

The two handbooks of research on teaching that emerged during this period were of enormous significance. Each has become a classic reference in the field. Particularly noteworthy in the second handbook is the article by Rosenshine and Furst (1973) on the use of direct observation to study teaching.

The timing of the Rosenshine and Furst paper coincided with a rapid acceptance within mathematics education of naturalistic studies and the emergence of qualitative analysis as a research tool. Also, 14 volumes of Soviet Studies in the Psychology of Learning and Teaching Mathematics were translated and published between 1968 and 1973 (Kilpatrick, Wirszup, Begle, and Wilson 1968-1975). The teaching experiment and the ascertaining study (Kantowski 1979; Rachlin 1983) were new research tools of particular interest to mathematics educators. These developments led to attempts to address much more complex

542

and realistic research problems in the teaching and learning of mathematics.

Important work in the study of instructional processes in mathematics was conducted during the 1970's by interdisciplinary teams at the University of Texas (Brophy and Evertson 1976), the Far West Regional Laboratory (Berliner and Tikunoff 1976), and the University of Missouri (Good, Grouws, and Ebmeier 1983), among others. This work tended to extend the traditions of teacher effectiveness studies, with attention to the direct observation of mathematics classes. These were significant improvements over previous studies, but all were limited by the conceptualization of mathematics achievement and the implicit goals of mathematics as being defined by standardized tests. All concentrated on elementary school or junior high school mathematics.

Another important development in mathematics education during the 1970's was that many investigators began to examine a variety of issues in children's early number learning. Significant contributors included Gelman and Gallistel (1978), Ginsburg (1977), Romberg, Carpenter, and Moser (1981), and Steffe, von Glasersfeld, Richards, and Cobb (1983). These investigators drew from a variety of traditions, including the emphasis on meaning in learning mathematics, the developmental psychology of Piaget (and others), and a constructivist epistemology. The constructivist orientation has led to particular attention and awareness of children's algorithms (as contrasted with adult-dictated algorithms) in learning mathematics.

Cooney (1980) presented an overview of the teacher effectiveness research and identified clarity, variability, and time on task as particular variables to explore in depth in mathematics classes. In his research now in progress, however, Cooney is studying how the teacher's system of beliefs about mathematics determines instructional processes and, in turn, how the teacher's belief system is modified by teaching experience. A teacher whose mathematical belief system is centered on problem solving, reasoning, and application would be expected to make quite different instructional decisions from the teacher whose mathematical belief system placed primary emphasis on algorithmic skills.

The second problem with the downgrading of pedagogy that we see today is misunderstanding of the role of practice in the preparation of teachers. If thirty years ago we believed that improved content preparation of mathematics teachers would be sufficient for improved instruction, today we seem to believe that a modicum of practice added to the content preparation will do the job. In America, the notion of learning through practice is quite deep-seated.

Yet, reliance on subject matter competency and practice teaching may not be enough to insure adequate instruction. John Dewey (1904) pointed out that learning to teach is not just a matter of practice but also of analysis. The role of practice or internship is not to help one learn to teach by teaching, but to provide material for the analysis of what teaching is and ought to be. We must increase our efforts along these lines.

Current Status

In 1980, the National Council of Teachers of Mathematics published a statement on the mathematics curriculum for the 1980's. This report, Agenda for Action, has already had widespread influence. The most significant emphasis is on problem solving and applications of mathematics as the central theme for all curricula. Additional recommendations included a call for redefining basic skills in mathematics to be more inclusive than arithmetic algorithms (for example, measuring, estimating, data organization) and to make extensive and effective use of calculators, computers, and other technologies.

These recommendations raise serious issues about instructional processes in mathematics. For example, there are many pressures on teachers to emphasize only algorithmic skills. During the 1970's the heavy emphasis on behavioral objectives, the back to basics movement, the growth of testing and assessment, and the trivialization of material in textbooks all led to an emphasis on skills, often without attention to meaning. Further, the curriculum is full and teachers tend not to have training or authority to rearrange, modify, or delete. How then can a teacher find time for emphasizing problem solving? In fact, a major thrust in research, development, and dissemination must come in the area of teaching strategies to incorporate problem solving at all levels of the curriculum, with students of any ability level. Similar statements can be made for the efficient and effective use of computers.

The several lines of inquiry that may be productive in the study and improvement of instructional processes in mathematics seem to include the efficient and effective use of student learning time, response to a changing curriculum (e.g., problem solving and computers now, something else later...), the role of meaning, the role of reasoning processes, and the real and perceived demands of the system (e.g., tests to some extent dictate the curriculum).

Kilpatrick and Wilson (1983) raised concern over the failure of various segments of society and the educational community to take seriously the teaching of mathematics. There has been a severe "devaluation" of mathematics teaching. Too many people have little regard for either the profession or the activity of mathematics teaching. Kilpatrick and Wilson argue that:

> The image of mathematics teaching must be changed, and the only honest way to do that is to change the substance as well ... the professional life of mathematics teachers needs to become more rewarding and ... that might be done by expecting teachers to do more.

Kilpatrick and Wilson then develop a view of professional teaching as less paperwork but more decision making. In their vision, the professional teacher has multiple roles: the teacher as mathematician, the teacher as a developer (or modifier) of curriculum, and the teacher as researcher. The implementation of career ladder concepts could provide a vehicle for these roles to be realized. In this context instructional processes include more than the delivery of instruction.

Probable Directions

Research on teacher effectiveness in mathematics classes has to be more attuned to the variety of expected outcomes in student achievement and the evolution of our abrupt changes in curriculum. For example, research that demonstrates the importance of a highly structured presentation where students are learning to perform an algorithm may not apply to a lesson where students are learning to write a computer program. Effectiveness variables demonstrated for a trigonometry class may not apply to a remedial mathematics class (caution should also be exercised in drawing conclusions as to why they might not apply). Thus, extensions of the research program of Active Mathematics Teaching (Good, Grouws, and Ebmeier 1983) should include not only different levels of classes but also different student outcome measures.

The Agenda for Action identifies several areas in which research and development is needed in instructional processes. Here is the beginning of a list of examples:

1. How to conduct group problem solving
2. How to incorporate a computer in a demonstration lesson
3. How to include student programming as a problem solving tool
4. How to teach estimation skills
5. How to include problem solving and applications in a remedial mathematics class

At the early elementary level there appears to be a real need to understand algorithms, concepts, and processes from the child's view. Children are constantly inventing algorithms — often wrong — and teachers need instructional strategies to identify these algorithms and to help children construct their mathematics concepts and understanding.

Attention to meaning in the context of learning mathematical skills will also receive emphasis in the next few years. Much of the poor performance of students that labels them as remedial can be traced to the rote learning of skills. Meaningful learning to get something right the first time is the appropriate long-range solution. Rachlin (1983 and current projects) has used learning experiments to examine the development of meaning during the learning of skills.

Finally, research in the career development of teachers, as opposed to early training, will explore teachers' involvement in curriculum development or modification, development and evaluation of instructional strategies, the effect on beliefs about mathematics and mathematics teaching, and its effect on students.

References

Begle, E.G. 1979. Critical Variables in Mathematics Education. Washington, D.C.: Mathematical Association of America.

Berliner, D. 1986. "Laboratory Settings and the Study of Teacher Education." Journal of Teacher Education, 36: 2-8.

Berliner, D. and W. Tikunoff. 1976. "The California Beginning Teacher Evaluation Study: Overview of the Ethnographic Study." Journal of Teacher Education, 27: 24-30.

Brophy, J. and C. Evertson. 1976. Learning from Teaching: A Developmental Perspective. Boston: Allyn and Bacon.

Brownell, W.A. 1951. "Psychological Considerations in the Learning and Teaching of Arithmetic." In Teaching of Arithmetic. Tenth Yearbook of the National Council of Teachers of Mathematics. New York: Teachers College Press.

Cooney, T.J. 1980. "Research on Teaching and Teacher Education." In Research in Mathematics Education, R.J. Shumway, ed. Reston, VA: National Council of Teachers of Mathematics.

Cooney, T.J., K. Henderson, and E.J. Davis. 1972. Dynamics of Teaching Secondary School Mathematics. Boston: Houghton-Mifflin.

Dewey, J. 1904. "The Relation of Theory to Practice in Education." In The Relation of Theory to Practice in the Education of Teachers, National Society for the Study of Education. Third yearbook, Part I. Bloomington, IL: Public School Publishing Co.

Dunkin, M. and B. Biddle. 1974. The Study of Teaching. New York: Holt, Rinehart and Winston.

Fey, J.T. 1970. Patterns of Verbal Communication in Mathematics Classes. New York: Teachers College Press.

Gage, N.L. 1960. Research Resume, 16. Burlingame, CA: California Teachers Association.

Gage, N.L., ed. 1963. Handbook of Research on Teaching. Chicago: Rand McNally Co.

Gelman, R. and C.R. Gallistel. 1978. The Child's Understanding of Number. Cambridge: Harvard University Press.

Ginsburg, H. 1977. Children's Arithmetic: The Learning Process. New York: Van Nostrand.

Good, T.L., D.A. Grouws, and H. Ebmeier. 1983. Active Mathematics Teaching. New York: Longman.

Kantowski, M.G. 1979. "Another View of the Value of Studying Mathematics Education Research and Development in the Soviet Union." In Mathematics Education in the Soviet Union, A.B. Davis, T.A. Romberg, S.L. Rachlin, and M.G. Kantowski, eds. Columbus, OH: ERIC.

Kilpatrick, J. and J.W. Wilson. 1983. "Taking Mathematics Teaching Seriously: Reflections on a Teacher Shortage." Paper presented to the National Institute of Education Conference on Teacher Shortages in Science and Mathematics. Washington, D.C.

Kilpatrick, J., I. Wirszup, E.G. Begle, and J.W. Wilson, eds. 1968-1975. Soviet Studies in the Psychology of Learning and Teaching Mathematics Volumes 1-14. Reston, VA: National Council of Teachers of Mathematics.

National Science Foundation. 1959. Ninth Annual Report. Washington, D.C.: National Science Foundation.

Osborne, A., ed. 1977. An In-Service Handbook for Mathematics Education. Reston, VA: National Council of Teachers of Mathematics.

Rachlin, S.L. 1983. "A Description of the Clinical Methodology Used in Two Studies of Learning Difficulties in Elementary Algebra." Paper presented to the Annual Meeting of the Research Council for Diagnostic and Prescriptive Mathematics. Bowling Green, OH.

Resnick, L.B. and W.W. Ford. 1981. The Psychology of Mathematics for Instruction. Hillsdale, NJ: Lawrence Erlbaum.

Romberg, T.A., T.P. Carpenter, and J. Moser., eds. 1981. Addition and Subtraction: A Developmental Perspective. Hillsdale, NJ: Lawrence Erlbaum.

Rosenshine, B. and N. Furst. 1973. "The Use of Direct Observation to Study Teaching." In The Second Handbook of Research on Teaching, R.M. Travers, ed. Chicago: Rand McNally Co.

Steen, L.A. and D.J. Albers, eds. 1981. Teaching Teachers, Teaching Students: Reflections on Mathematical Education. Boston: Birkhauser.

Steffe, L.P., E. von Glasersfeld, J. Richards, and P. Cobb. 1983. Children's Counting Types: Philosophy, Theory, and Application. New York: Praeger.

Part 3: Technology-Supported Learning

The Impact of Computer Science on Pre-College Mathematics:
A Research Program

Anthony Ralston

State University of New York at Buffalo

It is unusual for developments in science or technology to lead to immediate and significant changes in the subject matter or approach to teaching at the pre-college level, or even in the first two years of college. Neither of the two recently attempted "revolutions" in American education — those in reading ("Why Johnny Can't Read") and mathematics (the "New Math") — was the direct result of developments in the discipline, although the New Math was motivated in part by Sputnik. In the case of reading, the revolution resulted from ideas conceived by educators in response to perceived problems in the educational system; in the case of mathematics, change grew out of insights related to the broad sweep of intellectual development in the subject (Kline 1974). The thesis of this paper is that the development of computer science over the past two decades requires unprecedented steps in the United States. Significant changes in the school mathematics curriculum must be considered, and changes in the approach to much of pre-college mathematics teaching must be contemplated as well.

That computer technology has an important and expanding role to play in education, particularly in mathematics education, is virtually a truism. Indeed, almost all of the activity that relates computing to education focuses on the technology of computers — hardware and software — rather than on the science of computers. It may not be unfair to say that the educational establishment, like Americans more generally, is so mesmerized by the wonders of this technology that it has not seen beyond the forest to the equally important issues raised by the development of computer science. No matter how wonderful the hardware is and no matter how flexible and powerful it may be as a classroom assistant, the revolution — if there is to be a revolution — depends as much on the infusion of the ideas, values and paradigms of computer science into pre-college mathematics education as on the introduction of microcomputers and other computer technology into the classroom.

Three current aspects of the way in which computer science influences mathematics education make this a propitious and even necessary time to do a thorough study of the pre-college mathematics curriculum. These are:

1. The rapidly increasing importance of discrete mathematics in the college mathematics curriculum, particularly in the first two years, not as a threat to the current calculus sequence, but rather as a coequal partner with calculus (Ralston and Young 1983). This development is partly a result of the mathematical needs of computer science undergraduates, though its main intellectual stimulus is the effect of computer science on the very nature of applied mathematics and research in applied mathematics.

2. The use of algorithms in secondary school mathematics. Although this idea has been in the air for some years now, there seems to be little realization that teaching students how to use algorithms means more than specifying a rule for doing this or that task. True algorithmic teaching requires a notation for the expression of algorithms; a focus on the development of algorithms rather than just their application; and an emphasis on the analysis of the properties of algorithms (Maurer 1984; Maurer 1985). Of course, in the elementary school grades the notation must be as simple in its context as LOGO is for its purpose. Similarly, in the elementary grades any analysis of algorithms would have to be very rudimentary. The value of an algorithmic approach, however, is that it enables students to develop a "top-down" mode of thinking, encouraging them to solve problems in a systematic, step-by-step fashion. The algorithmic approach also makes it possible for even average students to discover mathematical ideas and facts on their own.

3. The development of symbolic mathematical systems (such as Macsyma and muMath). This development has the same potential impact on the secondary school mathematics curriculum, and indeed on the first two years of the college curriculum (Heid 1983; Wilf 1982), that hand calculators may yet have on the elementary school curriculum. The development of these systems will affect both the content of the secondary school curriculum (because various topics now taught will no

longer be relevant when such systems are widely available) and the **approach** to teaching high school mathematics (because it will no longer be possible to justify a need for good manipulative skills in certain areas).

All three of these developments result from the growth of computer science as a discipline. (The last, of course, is also a result of specific computer science research and the development of associated software technology.) Each by itself suggests some changes that might be made in the pre-college mathematics curriculum. Together they raise the possibility of a substantially altered, more effective pre-college mathematics curriculum. Of course, to have any chance of being widely implemented, proposed changes must be very carefully studied, developed, and sold to pre-college mathematics educators. Here I outline a research program to develop the framework, with specific examples, of such an altered curriculum.

Outline of the Proposed Research Program

1. It is necessary to look both macroscopically and microscopically at the content of, and motivations for, the current secondary school mathematics curriculum to ascertain

    a. which areas or topics in the curriculum may no longer be justified by intellectual development or manipulative skill development

    b. which topics can and should be approached from different perspectives due to items (2) and (3) in the previous section.

2. The results of such a review should then be used

    a. to suggest general changes in the secondary and elementary school mathematics curricula, in the sense that the recommendations should focus on subject matter broadly (e.g., what is the role of plane geometry in the high school curriculum?) and on the goals of the mathematics curriculum (e.g., what knowledge and skills should one expect a child to have at the completion of elementary school (sixth grade)?); and

    b. to give many concrete and detailed examples of

i.   how current topics could be better approached using modern paradigms

ii. mathematical topics not commonly taught to pre-college students at present which ought to be considered for inclusion in pre-college mathematics

iii. topics usually in the current curriculum whose inclusion can no longer be justified on mathematical or broader intellectual grounds

iv. how traditional topics might be taught in different orders than they are now in order to take advantage of computer technology and changed needs to prepare students for later topics.

3. The poor performance of American children vis a vis their counterparts in other countries has been documented in a variety of recent studies. A particularly noteworthy case in point is the study comparing performance in the United States, Japan, and Taiwan (Stevenson et al. 1986). Thus, all of the above needs to be done in a framework in which current American practice and trends are compared with what is happening in other developed countries.

A More Detailed Exposition of the Proposed Research Program

I begin with examples of the questions that need to be asked about the pre-college mathematics curriculum in light of the developments discussed at the beginning of this proposal. It should be noted that the way the questions are asked reflects a point of view developed by the author through considerable recent study of the college mathematics curriculum. But I do not wish to suggest that these are the only questions that might be asked or even necessarily the best ones. Rather, I present them to orient the reader toward the general philosophy that motivates this paper. Some of the questions are intentionally "far out"; no matter how unlikely it is that radical changes in the pre-college mathematical curriculum might actually take place, it may be worthwhile asking — and studying — questions that entail radical change if only to challenge long-entrenched assumptions and promote a dialogue on the subject as a whole.

Question 1

How much of the subject matter of high school plane geometry is justifiable today on either mathematical or intellectual grounds? (Discussion questions: Can the goal of teaching patterns of logical thinking, which has always been one of the important motivations for the teaching of plane geometry, perhaps be equally well achieved using an algorithmic approach to the teaching of other kinds of mathematics at this level? What is the best balance between rigor and informality in the presentation of proofs in plane geometry? How can the notion of proof in geometry be blended effectively with the notion of proof in algebra?)

Question 2

Are the following goals for a kindergarten through sixth-grade mathematics curriculum feasible: (a) an understanding of the processes of arithmetic; (b) manipulative skill only with one digit number facts (Anderson 1983); and (c) the development of estimation ability for arithmetic computations in general? (Discussion questions: If calculators are rapidly reducing the societal value of good arithmetic manipulative skills, why not concentrate on the skills that would be needed in a world where calculators are ubiquitous? How might we introduce a curriculum that deemphasized computational skills in the context of standardized tests that emphasize these skills? How would fractions and decimals be handled in such a curriculum? How much correlation is there between drill and practice and an understanding (however interpreted) of the mathematics being practiced (Usiskin 1983)?)

Question 3

If symbolic mathematical systems will soon (say, in a decade) be as readily available to children as hand calculators are now, what portions of the high school curriculum will have to change in adapting to this new environment? (Discussion questions: Is facility in polynomial algebra in any way more important than facility in long division (which the Cockcroft report (Cockcroft 1982) recommended abolishing as a subject of instruction in schools in England and Wales)? Is comprehension of symbolic processes possible in the absence of considerable skill at performing such processes? Does the availability of symbol manipulating systems suggest the introduction into the secondary school curriculum of topics that would be otherwise impractical?)

Question 4

If the development of arithmetic manipulative skills is becoming less important, can real arithmetic instruction start in, say, the third grade on the basis of Piaget-type studies (Piaget 1953) that show when children are capable of "understanding" certain concepts? (Discussion questions: What kind of arithmetic familiarization would then be appropriate before the third grade? If the amount of arithmetic instruction before the third grade were to be sharply decreased, what should take its place?)

Question 5

What would be the impact of an algorithmic approach to the teaching of the bulk of pre-college mathematics? (Discussion questions: Can very young students be taught to think algorithmically to the extent of approaching arithmetic and algebra generally from this perspective? If so, what kind of algorithmic language or notation would be appropriate for such students? If an algorithmic approach becomes fairly pervasive in pre-college mathematics, how can algorithms be used to enhance the presentation of various topics in pre-college mathematics (e.g., the solution of simultaneous linear equations)?)

Many more questions like these could be posed. But these five should be enough to get across the main point: What is needed is a thorough study of the current pre-college curriculum, with the goal of developing significant recommendations concerning the entire curriculum.

As to the current curriculum, the following issues are appropriate for study:
1. Consider broadly the goals of each of the major portions of the secondary and elementary school curricula (e.g., arithmetic, algebra, geometry), and determine what relevance they have in light of the impact of computer science on mathematics. The merits of a unified (integrated)

secondary school curriculum should also be considered.

2. Although the precollege mathematics curriculum in this country is by no means a monolith, there is nevertheless a considerable amount of commonality between the topics taught under each of the general rubrics referred to in (1). Topics that are commonly or universally part of the pre-college mathematics curriculum should be studied and their relevance assessed; the usual approaches to these topics might be more effective if influenced by the values or paradigms of computer science. In particular, I believe it should be possible to show how an algorithmic approach to mathematics (in the sense described earlier, not in the sense of rote learning) can be an effective and unifying theme for the entire curriculum.

3. Discrete mathematics is rapidly becoming more important in the college mathematics curriculum (Ralston and Young 1983). Inevitably this will have an effect on the high school curriculum. Which topics in discrete mathematics are appropriate for the secondary school curriculum and when and how can they be introduced in the most effective manner?

4. Symbolic mathematical systems can perform much of the algebraic and trigonometric manipulations that are staples of high school mathematics. Deciding whether the development of the corresponding manipulative skills can still be justified depends mainly on the degree of correlation between manipulative skill and understanding. The existing experimental evidence on this question needs to be examined and experiments defined whose results might shed further light on this important question. (There is surprisingly little literature on this subject; much of the work most often referred to was done by Brownell and is almost 50 years old (e.g., Brownell 1938).) I note here the importance of the premise that secondary school mathematics must concentrate on the development of deployable skills (Hilton 1984) rather than on just those skills that are traditional (Saxon 1982).

5. The elementary school curriculum poses a different set of problems. One issue which should be studied is how early and in what manner an algorithmic approach can be used in the elementary grades. (Elementary

557

school teachers sometimes claim they already use an algorithmic approach in teaching arithmetic. But an algorithmic approach in this context does not mean the rote learning of computational rules (Maurer 1984); rather, as noted earlier, it means encouragement of the design, understanding, and discovery of algorithms together with the use of a suitable notation in which to express them.) In much the same sense that LOGO provides a language in which young children can communicate with a computer (Papert 1980), an algorithmic notation (not a programming language) needs to be developed which young children could use to express algorithmic ideas.

6. How might the overall shape of the elementary school curriculum be affected by the computer science developments discussed above? If the development of manipulative skills is not to be the major goal of the elementary school mathematics curriculum, we need to consider what arithmetic and related subject matter is appropriate and when it should be introduced. The possibility of radical changes in both subject matter and timing of mathematics instruction need to be considered. One option is to design a radically new curriculum — not with the expectation of using it in the foreseeable future, but rather to provide a model against which to gauge more practical changes.

7. If the subject matter of pre-college mathematics as well as the approach to teaching the curriculum is to be significantly changed, there are important implications for teacher training, both pre-service and in-service. How should the curriculum of pre-service programs to train mathematics teachers be changed? What kind of in-service training would be appropriate for teachers who might be introducing the new content and using the new methods suggested by this research? This point is particularly important because the current pre-college teachers of mathematics will continue to be the dominant cadre for the next 15 to 20 years.

8. There is a pervasive belief in the United States that other countries are rapidly outdistancing us in mathematics education. Insofar as this belief is correct, it is obviously a serious threat to the status of American

science and technology. But differences in the structures of educational systems and the goals of elementary and secondary education in other countries may mean that corresponding test scores and similar data are misleading. In any case, it would be worth devoting considerable effort to studying current trends in mathematics education in Western Europe, Japan, and the Soviet Union in order to compare other approaches to the challenges to mathematics education discussed heretofore.

Research and Curriculum Change

A research program of this nature can only be of value insofar as it influences the course of mathematics education in the United States. Therefore, it can only be considered the first step in a much broader program involving many people and aimed at the actual implementation of new curriculum ideas and approaches to teaching.

Indeed, before the results of any research program in mathematics education can be implemented in the classroom, there are a number of necessary intermediate steps. These include careful translation of research results into curriculum materials, pilot testing and revision of these materials, training of teachers who will present the new materials, development of appropriate testing instruments for the new materials and, for major changes, the dissemination of appropriate public and parental information programs. Unless careful attention is paid to all these steps, even the best planned research and the most innovative new ideas in mathematics education will not find their way into the curriculum. Since the need for significant change in the American pre-college mathematics curriculum is great, so also is the need to get on with the research which will be needed to inform these necessary changes.

References

Anderson, Richard D. 1983. "Arithmetic in the Computer/Calculator Age." In The Future of College Mathematics, A. Ralston and G.S. Young, eds. New York: Springer-Verlag.

Brownell, William. 1938. "Two Kinds of Learning in Arithmetic." Journal of Educational Research 31: 656-664.

Cockcroft, W.H. 1982. Mathematics Counts. Report of the Committee of Inquiry into the Teaching of Mathematics in Schools. London: Her Majesty's Stationery Office.

Conference Board of the Mathematical Sciences. 1982. The Mathematical Sciences Curriculum K-12: What Is Still Fundamental and What Is Not. Washington, DC: National Science Foundation.

Fey, James T., ed. 1984. Computing and Mathematics: The Impact on Secondary School Curricula. Reston, VA: National Council of Teachers of Mathematics.

Heid, M. Kathleen. 1983. "Calculus with muMath: Implications for Curriculum Reform." The Computer Teacher (November): 46-49.

Hilton, Peter. 1984. "Current Trends in Mathematics and Future Trends in Mathematics Education." For the Learning of Mathematics 4: 2-8.

Kline, Morris. 1974. Why Johnny Can't Add — The Failure of the New Mathematics. New York: St. Martin's Press.

Maurer, Stephen B. 1984. "Two Meanings of Algorithmic Mathematics." The Mathematics Teacher 77: 430-435.

Maurer, Stephen B. 1985. "The Algorithmic Way of Life Is Best." College Mathematics Journal 16: 2-5.

National Science Board Commission on Precollege Education in Mathematics, Science and Technology. 1983. Educating America for the 21st Century.

Papert, Seymour. 1980. Mindstorms — Children, Computers and Powerful Ideas. New York: Basic Books.

Piaget, Jean. 1953. "How Children Form Mathematical Concepts." Scientific American (November): 74-79.

Ralston, Anthony. 1985. "The Really New College Mathematics and How It Is Going to Change the High School Curriculum." 1985 Yearbook of the National Council of Teachers of Mathematics.

Ralston, Anthony. In press. "The Effect of a New College Mathematics

Curriculum on the Secondary School Curriculum." <u>Proceedings of a Symposium on Mathematics Education at the International Congress of Mathematicians, 1983.</u>

Ralston, Anthony and Young, Gail S., eds. 1983. <u>The Future of College Mathematics: Proceedings of a Conference/Workshop on the Future of the First Two Years of College Mathematics.</u> New York: Springer-Verlag.

Saxon, John. 1982. "Incremental Development: A Breakthrough in Mathematics." <u>Phi Delta Kappan</u>: 482-484.

Stevenson, H.W., S.-Y. Lee, and J.W. Stegler. 1986. "Mathematics Achievement of Chinese, Japanese and American Children." <u>Science</u> 231: 693-699.

Usiskin, Zalman. 1983. "A Proposal for Reforming the Secondary School Mathematics Curriculum." Paper read to the 61st Annual Meeting of the National Council of Teachers of Mathematics.

Wilf, Herbert S. 1982. "The Disc with the College Education." <u>American Mathematical Monthly</u> 89: 4-8.

The WICAT Computer-Based Elementary Mathematics Project:
A Curriculum in Development

Warren D. Crown

Rutgers University and WICAT Systems

Introduction

During the current decade the use of computers in elementary mathematics education has increased dramatically, and we expect it will continue to increase. But there has also been an increase in the frustration felt by teachers and other professionals in the field at the poor quality of the commercial software that has appeared in the marketplace. The lack of quality is understandable when one considers the backgrounds and resources of the creators of the materials. Most are educators with little programming or technical experience and their materials are shallow, unprofessional-looking, and not very responsive to individual users' needs. Other software producers are small entrepreneurial groups. Some of them do have a good hardware and software orientation, but lack any overall picture of school curriculum, children's learning styles, and appropriate instructional strategies.

There are, of course, notable exceptions to this general description. Pioneering work with special-purpose programs like LOGO, SemCalc, and Geometry Supposer deserves all the praise it has received. Other programs that treat single curriculum topics are frequently very well done. Examples of this type of program are Green Globs and Rocky's Boots.

What has been missing from the field, however, is a well-rounded, comprehensive, integrated software curriculum. For the past two years, a development team at WICAT Systems has been working to develop such a curriculum. Although the development is not complete, we have reached the point of marketing a first version of the program. This paper describes what we perceive to be the state of elementary mathematics curriculum in this country at the time of development and the background and philosophy of the resultant program.

562

Elementary Mathematics in the U.S.

Characterization

The major publishers of elementary mathematics curricula in this country follow almost identical scope and sequence documents. They cover arithmetic of whole numbers, decimals, and fractions; measurement; geometry; descriptive statistics; and applications of those topics. Occasionally, a topic which is presented at a given grade level in one series may appear at the following grade level in another series, but the difference is never greater than a single level. The placement of topics is affected as much by state guidelines and by "what the competition is doing" as by thoughtful analyses of the average skills of a particular age group or by reviews of the research findings in the area. So, in effect, we can be said to have a reasonably well-defined national elementary mathematics curriculum.

It should also be said that the textbooks define the actual as well as the official curriculum. More so than in any other area, teachers of elementary mathematics see their textbooks as their programs. They deviate from them rarely and feel ill-prepared to change their focus. Time and again surveys of elementary teachers show that they feel mathematics is the subject they are least competent to teach and the one they least enjoy teaching. As a result, most teachers spend little time or effort designing creative and challenging mathematics instruction outside of the suggestions of their textbooks. The typical elementary teacher works through the text with his or her students page by page, planning to reach the end of the book at the end of the year. (Most books have roughly 360 student activity pages — 2 pages per day for 180 days.) Even so-called "individualized" programs frequently provide nothing more than the opportunity for the student to go through the book at his or her own pace.

This standard curriculum is embodied in a series of nine single-grade-level textbooks that cover kindergarten through eighth grade. The overwhelming emphasis in these books is on computation, with the dominant instructional sequence being a focus on algorithmic skill followed by plenty of practice. In a recent popular third-grade textbook, for example, 51% of the pages are devoted to the teaching and practice of computational algorithms. Another 20% of the pages focus on problem-solving and are comprised predominantly of one-step story problems giving rise to more whole number computation. The remaining 29% of the text covers numeration (12%), measurement (10%), geometry (5%), and probability, descriptive statistics, and graphing (2%).

The attention given to geometry, measurement, and descriptive statistics in the elementary grades is one of the few vestiges of the "new math" curriculum movement. The treatment of these topics in most available texts, however, has been significantly watered down in recent years and has ended up as a scattered and repetitive focus on definitions, recognition of basic forms, and manipulation of formulas.

The situation with respect to problem-solving is more complex. The National Council of Teachers of Mathematics proposed in 1980 that problem-solving become the focus of school mathematics in this decade. The proposal met with wide-ranging acceptance, and practically all education publishers have produced a "problem-solving" elementary mathematics series since then. The new emphasis, however, seems for the most part to have resulted in just a greater amount of what had already been present in textbooks — pages of one-step story problems easily solved by a student who understands the content in the most recently completed chapter. Some series deal more effectively with this issue than others, attempting to include mixed sets of interesting, real-world applications of mathematical concepts and skills, but those texts are in the minority and much more needs to be done. As a result, many children progress through school mathematics with very little awareness of the usefulness of the skills they have learned.

There is also very little attempt in the standard curriculum to teach students how to deal with problems whose solutions may not be immediately obvious. Children frequently fail to make even the simplest connections between the mathematical operations they know how to perform and the relationships among the elements in a problem situation.

Effects

The effects of traditional elementary mathematics instruction have been well documented and publicized over the past few years. Probably the most reliable source of data on this issue is the National Assessment of Educational Progress. Recent NAEP documents have shown that whole number computation performance among the elementary school population is relatively good. When given arithmetic problems involving the four basic operations, most children can calculate correct answers at, or only slightly later than, the age at which such skills are taught in the standard curriculum. Similarly, when given simple, one-step word problems, the children's performance is also satisfactory. About 80% of the nine-year-olds in the most recent sample solved such an addition story problem and 87% of the thirteen-year-olds correctly solved addition and subtraction problems. The percentages drop slightly for problems involving the other operations.

Results for other types of problem-solving skills are not so positive, however. When given multi-step problems or non-standard problems (problems that require some original thinking or creative solution strategy), the children's performance drops precipitously. Percentage scores for these types of problems show that only 10-30% of the nine- and thirteen-year-olds successfully deal with them. Apparently, the reasoning and analysis skills necessary are neither taught nor practiced in the classroom.

Results from the geometry, measurement, and other non-arithmetic sections of the assessment follow the same general trend. Children are able to identify common geometric figures, measure line segments, and interpret simple graphs, but when asked to apply these skills in a real-world setting or in any other than standard form, performance drops significantly.

These children don't know how to apply the mathematical skills they've mastered. The skills and concepts they have learned are only classroom skills. The children are able to apply computational algorithms only when presented with problems which are in exactly the same form as the ones they practice with every day. Their ability to see other uses for, and extensions of, their skills is severely limited.

Hope

It seems that there is growing public concern about these issues. Certainly, comparisons between the quality of mathematics education received by American children versus Russian children, the comments about mathematics and science education in the recent report by the President's Commission on Education, the increasing publicity given to the National Assessment and SAT results, and the reports of widespread teacher shortages in mathematics and science are all indications of increasing public awareness. The ultimate outcome of this rumbling may be a demand for more rigorous, challenging, and real-world-oriented mathematics experiences for young children. Professionals in the field have been arguing the need for such curriculum changes for more than ten years.

There are also abundant data available to show that children are capable of doing more with mathematics than typically demanded of them in the traditional curriculum. Special experimental curricula designed during the past two decades nearly always produced increased student achievement, improved student affect, and gave students a better sense of the richness of mathematics. Programs such as the Madison Project, the Miquon School project, the Comprehensive School Mathematics Program, and the Unified Science and Mathematics in the Elementary School program offered students a rich variety of mathematical experiences to which students responded in a positive way. The children in those programs frequently dealt very effectively with topics the standard curriculum reserves for much older children. Unfortunately, these programs never seem to be implemented widely because they are seen as being commercially risky. Also, the reactions of teachers to the approaches are unpredictable. In the hands of teachers who are not specially trained, there is no guarantee that the strategies will be successful. And without the immense marketing effort that supports the sales of the best-selling texts, the experimental programs seem to fade away. Most of the aforementioned programs can now be found only on the bookshelves of old mathematics educators.

There is yet another source of data supporting the thesis that children can handle more mathematics than they are usually asked to deal with; namely, research on young children's abilities. This research suggests that children frequently develop real-world number skills such as notions of measurement, fraction vocabulary, and counting abilities several years before they encounter those topics in the school curriculum. There is little relation between the standard sequence and the out-of-school lives of children. In designing the WICAT curriculum, we proceeded by assuming that parents and teachers were looking for a series of mathematical experiences that will broaden the scope of the traditional curriculum. The WICAT program attempts to instill in children an appreciation of the usefulness of numbers and mathematics.

The WICAT Curriculum

The following three excerpts from the documents of well-respected organizations characterize one of the significant trends in mathematics education over the past five years.

> Any list of basic skills must include computation. However, the role of computational skills in mathematics must be seen in the light of the contributions they make to one's ability to use mathematics in everyday living. In isolation, computational skills contribute little to one's ability to participate in mainstream society. Combined effectively with the other skill areas, they provide the learner with the basic mathematical ability needed by adults.
> Position Statement on Basic Skills. National Council of Supervisors of Mathematics, 1978.

> There must be an acceptance of the full spectrum of basic skills and recognition that there is a wide variety of such skills beyond the mere computational if we are to design a basic skills component of the curriculum that enhances rather than undermines education.
> An Agenda for Action, Recommendations for School Mathematics of the 1980's. National Council of Teachers of Mathematics, 1980.

> Looking at these (National Assessment) results across a wide range of tasks at all grade levels, it appears that American schools have been reasonably successful in teaching students to perform routine computational and measurement skills, and to answer questions assessing superficial knowledge about numbers and geometry. It is encouraging to note positive change on items assessing knowledge and skills not only in numerical computation, but also in geometry and measurement. On the other hand, it appears from the low percentages of success on some items that schools have thus far taught only a small percentage of students how to analyze mathematical problems or apply mathematics to nonroutine situations.

The findings of this (1982) assessment suggest that the mathematics curriculum needs reexamination in light of currently available computational technology. The hand-held calculator is nearly ubiquitous, and students are much more likely to be using computers today than they were even a few years ago. If computations can be done by machine, students need a much better understanding of the relationship between problem situations and the operations necessary for finding solutions. Moreover, a technologically literate population needs to be skilled in analyzing and managing quantitative data, estimating answers to calculations and judging the reasonableness of results.

These new skills should not simply be appended to the existing curriculum. It is likely that some of the current curriculum could be revised or even eliminated. For example, the computational algorithms are often difficult to relate to associated mathematical concepts. Given current computational technologies, it is time to reconsider the utility of such algorithms and examine alternate, easier-to-understand procedures.

The results of this assessment also suggest that underlying concepts and skills in geometry and measurement have not been mastered by enough students. This suggests a need for a carefully developed curriculum sequence of concepts and skills in the areas of measurement and geometry just as we have for whole numbers, fractions and other areas of mathematics.
The Third National Mathematics Assessment: Results, Trends, and Issues. National Assessment of Educational Progress, 1983.

The recent emphasis in curriculum development has been on problem-solving. Future emphases will probably focus on the mathematical needs of a technologically advanced citizenry. In light of these trends, and in keeping with our own philosophies, we have designed a WICAT Mathematics Curriculum that in its final form will include essentially all the concepts and skills found in the standard curriculum, but with emphases on different learner outcomes and mathematical content.

Rather than emphasize rote pencil-and-paper computational skills applicable almost exclusively in a symbolic setting, the WICAT Mathematics Curriculum underscores the usefulness of mathematics in everyday life. In support of this primary emphasis the WICAT program emphasizes conceptual understanding and flexibility of mental imagery. Insight into the mathematics involved is clearly essential for any application to the real world. The curriculum in final form will emphasize mental arithmetic: the development of alternate, more practical computational strategies, estimation, approximation, and testing for reasonableness of computed results.

The curricular approach to achieving these learner outcomes includes:

- introducing concepts and operations in a real-world context
- providing a variety of applications and interpretations for each concept
- using well-established modeling techniques for basic understanding
- providing for active learner involvement
- providing traditional algorithms only after conceptual understanding and problem-solving skills have been established
- greater emphasis on geometry than in the traditional curriculum.

The WICAT Mathematics Program is organized horizontally in five strands: (1) Numeration, (2) Operations, (3) Geometry, (4) Applications, and (5) Drill and Practice Computation. There is also a vertical organization by grade level pairs: pre-kindergarten-kindergarten, first-second, third-fourth, and so on. Each of the strands except Drill and Practice Computation spans the entire grade level sequence from pre-kindergarten through eighth grade. Computation will begin at the first-second grade level.

Numeration

The Numeration strand is comprised of games and activities focusing on the meaning of number. It begins with early experiences with whole numbers and escalates to treatments of fractions, decimals, and integers.

Number Models

Throughout the Numeration strand, models of numbers are built as instructional tools to give the student concrete referents for the abstract concepts. At the outset, these models appear on the screen simply as illustrations of the meaning of a given type of number. For example, 243 is shown by means of base-ten blocks.

The second step allows the student direct control over the graphics generation routines to enable him or her to build pictorial models of other numbers suggested by the computer. The program then checks the resultant model and gives appropriate feedback.

Counting

Another set of recurring activities in the Numeration strand will be counting tasks, only some of which have been implemented so far. Counting skills will be stressed as an important introduction to computation. Exercises on naming the number which is 1, 10, or 100 more than some other number based on intelligent application of place-value concepts is a crucial pre-addition task. Skip counting by 2's or 3's or 10's helps prepare the child to deal with multiplication. Harder tasks such as counting back and counting on from a given arbitrary starting number are similarly useful in developing good number sense. Much of this counting activity will result in the student inputting numbers through his keyboard, but some will necessarily involve oral responses that can be monitored by a parent or teacher.

Order, Comparison, and Equivalence

The last major component of the Numeration strand are activities that stress order, comparison, and equivalence relations. These are mostly game formats in which the student identifies numbers or number expressions greater than, less than, or equal to some target number chosen by the computer. Or, the student may be trying to guess a hidden number, in which case the computer returns feedback as to the accuracy of the guesses.

Operations

The Operations strand concentrates on the meanings and real-world uses of the basic arithmetic operations. Its purpose is to introduce the student to the operations through motivational applications, show the student how to use the number models created in the Numeration strand to model the processes and help him or her develop the rudiments of meaningful algorithms with manageable numbers, and finally challenge the student to develop useful mental arithmetic strategies.

Meanings of Operations

There are several real-world meanings for each of the four basic arithmetic operations. One of the major failures of the traditional curriculum is its disinclination to present them explicitly or frequently enough to students. Subtraction, for instance, is almost exclusively presented as "take-away" (a proper and appropriate meaning for some of subtraction), but there are many subtraction situations which are not take-away situations. Comparison situations involve two quantities whose difference is of interest (I have $13 and you have $16. How much more do you have than I?) and shift situations involve some phenomenon which has started at a given level and dropped to a lower level on some sort of scale. (The temperature last evening was 75 degrees but it dropped 23 degrees overnight. What was the temperature in the morning?). The result of "hiding" most of the uses of the operations from children is that when they are finally encountered, they are not recognized and classified as mathematical processes the children are familiar with. It is because the only situations recognizable as subtraction are those in which some objects are "taken-away" that simple solutions of many real-world problem situations in which subtraction is the helpful operation are overlooked.

It is important for the child to see the operations in as many real-world contexts as possible. Also, that exposure must include as many conceptually different uses of the operation as possible and each of the models of the operation presented must be consistent with the uses. We avoid, for instance, using a take-away model to represent a comparison situation. Although there is some practice included in this strand, its purpose is not to offer drill and practice with computation. The exercises are designed to enable the student to achieve a firm grasp of the meaning and uses of the operations rather than to become algorithmically proficient.

Algorithms and Mental Arithmetic

An outgrowth of the modeling of operations is the use of some models to show more complex computation. For instance, an area model of one-digit multiplication (How many square-foot tiles would cover the floor of a nine foot by six foot bathroom?) grows very nicely into a model of two-digit multiplication, showing each of the two partial products in the traditional algorithm distinctly.

Through interaction with the screen models, the student should develop nice intuitive notions of how these more complex operations work.

The algorithms taught at this point are those based on the models presented. A useful two–digit multiplication algorithm accessible to young children is one in which each of the two partial products is computed and their sum then found. This type of computation is not only conceptually simple when presented correctly, but, with increased understanding, leads to greatly increased ability to do mental computation and estimation. These two skills are major objectives of this program strand.

Children who have learned strategies like these for mental computation have a much greater appreciation of, and facility with, numbers. Moreover, this type of instruction is precisely the kind which is addressed in the document excerpts presented at the start of this section of the paper. Involvement with mathematics at this level gives children a sense of understanding and meaning that can be directly translated into more intelligent use of operations and concepts in problem–solving settings.

Geometry

With so little geometry in most pre–secondary school work, recent research shows that the majority of the 50% of students in high school who elect geometry have little background for that course. One result is that only about one-third of those who do elect geometry succeed in learning how to compose nontrivial proofs. The geometry competence of those who do not elect high school geometry is no doubt worse. Yet geometry is potentially a highly practical subject for nearly everyone, essential for continued work in mathematics and useful in a wide variety of trades and other occupations.

The persistent failure of significant geometry in the actual mainstream curriculum is accounted for in part by the poor mathematics preparation of teachers, but it also reflects pedagogic difficulties for which the microcomputer offers significantly new solutions. For example, geometry is a highly visual subject but appropriately rich and precise graphic presentations are difficult with only chalkboard and cheaply duplicated materials. Even the best chalkboard or printed presentations (or even overhead projector presentations) are necessarily static, while the most suitable presentations very often are dynamic, showing a number of examples, special cases for generalization, and so on. Three-dimensional concepts are especially difficult to convey merely with chalkboard or print. Transformations are very difficult to convey with the primitive graphics available to most teachers. Interactive possibilities are obviously severely limited in conventional classrooms. One of our main hopes is that the WICAT course will help overcome these pervasive pedagogic difficulties with graphics and interaction.

The WICAT K-8 geometry activities aim for excellent skills and intuition in geometry, including coordinate geometry and transformations. Taken as a whole, they teach certain specific prerequisites for success in standard high school geometry and also give a solid start on the coordinate geometry essential in most mathematics courses beyond elementary school.

The geometry activities are built on top of a sophisticated graphics tool developed at WICAT called GeoDraw. GeoDraw gives the user considerable control over the construction and transformation of geometric figures. Other rich graphics programs for microcomputers exist, but GeoDraw may be unique in the extent to which its V-draw, Construct, and Transform Subsystems, with Measure and Axes, explicitly model standard contemporary approaches to geometry.

In the lessons themselves, the user is prompted to use the GeoDraw functions to their fullest potential to create geometric figures and procedures that are then labeled to teach basic concepts. In the free explorations that follow, the student can extend his or her understanding by using the tool in any manner desired.

Applications

The Applications strand of the WICAT curriculum has presented more difficulties than any of the other strands in our effort to implement our philosophies within the constraints of a microcomputer environment. It was our original goal to offer the student many interesting and entertaining mathematical problem situations and then provide a variety of problem specific "helps" for students who needed them. We are part-way to achieving that goal, but the constraints of a small internal memory, relatively little available disk space, and a finite amount of room on the display screen have been more troublesome here than anywhere else.

In its current form our one problem-solving interaction presents the student with mathematical problem situations and offers helps which are selectable from the menu that accompanies each problem presentation. These include hints (the last in a sequence of which is a direct solution of the problem in question); a picture, diagram, or chart; additional information not included in the problem statement; a restatement of the same problem with more manageable numbers; and a similar, but easier, problem. Any given problem statement is accompanied by a menu containing some of these options. They are rarely all present for any single problem.

Computation

The Computation strand of the WICAT curriculum focuses on the pencil-and-paper algorithms that are traditionally the major part of elementary mathematics instruction. It covers all four basic operations with whole numbers, decimals, and fractions in a straightforward practice mode. The algorithmic helps offered are limited to the placement of the next digit in the computation and regrouping aids when necessary. There is no attempt to build a lot of meaning into these procedures nor is there any attempt to relate the procedures to real-life situations or to the number models created earlier in the Numeration or Operations strands. The helps are useful to the student who needs to see complex symbolic manipulations broken down into short, relatively simple steps.

Conclusion

Many pieces of high-quality educational mathematics software are available to deal with single curriculum topics or broadbased thinking skills. The WICAT Curriculum Project is, though, an attempt to produce a prototype comprehensive, software-based elementary math program. The components of the curriculum produced so far exemplify the project's commitment to dealing with many traditional topics at a conceptual level which forces the student to come to grips with the mathematical meanings involved. The degree of interaction required from the student, the on-screen, manipulative, numeration models, the attention to uses of operations, and the de-emphasis of rote computational procedures are all aspects of the program that we hope will provide students with a qualitatively different kind of educational experience in mathematics from that usually encountered.

Learning Mathematics Through Modern Media and the
THE System, a Prototype Videodisc Learning System

Fumiyuki Terada
Waseda University

Introduction and Overview

A case study (Fujita and Terada 1983) has shown us that in Japan, as in many other countries, mathematics education in a society with modern technology faces a certain dilemma. (1) The level of competence in mathematics required for citizens to live and work in such a society is much higher than in traditional societies. If financial and other conditions permit, this requirement leads to an extremely high rate of advancement to senior secondary education; for instance, in recent years the percentage of Japanese students going on to secondary school has been as high as 94% of the age group. This means that the average aptitude of high school students is not as high as it was formerly, and that the diversity in aptitude among students in a single classroom is remarkable. (2) On the other hand, in order to maintain and develop the society's high level of science (natural and social) and technology, it is necessary to have a number of students of higher aptitude (moderately gifted students) study mathematics through some advanced level in high school so that they will be well-prepared for further study and for their future professions. In addition, particularly in Japan, there are strong social forces which reject explicit differences or differentiation in education, and therefore a single nationwide high school curriculum is set by the government.

In this setting, as described in detail in Fujita and Terada (1984), Fujita, Terada and Shimada (1984), and Fujita (these Proceedings), we propose to adopt as the goal of mathematics education in Japan the cultivation of mathematical intelligence, realized through fostering mathematical literacy for all students and maximizing the mathematical strength and potential of the brighter students. With regard to the mathematics curriculum itself, we propose organizing a flexible "core and option modules" type program.

We would like to emphasize in this paper that in order to achieve our goal it is absolutely necessary to create an educational environment which facilitates individualized instruction, meaning that each student can choose the level and the pace of his study. This is feasible only if suitable information technologies are employed, since the number of qualified mathematics teachers and the number of man-hours are too limited to take care of large numbers of students at various levels. With this in mind, we have tried to develop a technological system which can provide individualized instruction of a highly interactive type and supplement classroom teaching. Our goal has been realized as the THE (Terada-Hirose Educational) system, the objectives, functions, and features of which are described below. It is our belief that the THE system can be effectively applied in various academic contexts for both mathematics and other subjects, and that it exemplifies the promising and far-reaching role of information technology in teaching and learning through the new media, both in school and at home.

Individualized Instruction and Desiderata for a System of Information Technology for Teaching and Learning

As in many other countries, a very common form of education in Japan is classroom instruction in which a teacher works with a group of 40-50 students. Needless to say, there are various merits to this form of group instruction: for instance, the teacher can easily identify gifted students, some students may work harder because they aspire to be the best in the class, and students can learn the importance of discussions and academic cooperation. In fact, if the teacher is good and the students are well-prepared, classroom instruction can be stimulating and inspiring. However, particularly in mathematics, individualized teaching and/or learning should supplement classroom instruction if the students in the class are at diverse levels. An average student can easily be discouraged by the difficulties he encounters in mathematics, for instance, slowness in computation, an abstract notion that he cannot get a realistic feel for, or fast-paced instruction on a topic that he cannot follow. Such feelings can occur even when the student is capable, properly guided and assisted by an individualized instruction program. Diversity in the level of the students and flexibility in the curriculum increase the need for individualized instruction.

As a matter of fact, individualized instruction in the form of tutorial instruction for high school students is rather popular in big cities in Japan. Tutors in these cases are usually university students who do tutoring as a part-time job in the evening. Obviously, the quality of this type of individualized instruction depends on the tutor and on his knowledge, experience and personality, and not all families can afford the fee. The situation reaffirms that in order to offer individualized instruction of reliable quality at a reasonable cost to all the students who need it, we must somehow make use of information technology and invent a new system which can be supported by excellent courseware.

More concretely, a desirable teaching system using modern media which can be expected to offer a good approximation of tutoring by an experienced and able teacher should have the following characteristics.

a) The system should have a rich variety of teaching materials at its disposal, from which the items needed at any given time can be easily extracted.

b) The system should be capable of offering instruction through both voice and image, so that the same educational effects can be obtained as with live teaching. Moreover, the system should be capable of offering appropriate lectures which match the level of the individual learner.

c) The system should be able to diagnose the learner's level of achievement, ability, and weak points and, on the basis of the results, judge whether the learner should move ahead or repeat certain topics, i.e., the system should be able to select the most appropriate path for each learner.

d) The system should be able to understand and respond to questions from the learner, i.e., to offer high-quality interactive instruction.

If we review traditional systems or media for education with characteristics a–d in mind, there are none which satisfy all the requirements. Books, radio, television, cassette tape recorders, videotape recorders, and micro-computers fulfill only some of these conditions and consequently cannot be used for supplementary individualized instruction that is good enough to provide substantial assistance to students and teachers.

Optical videodiscs seem to be the most convenient medium for constructing an electronic system to meet conditions a–d, since they have two important features:

1) Videodiscs can store a considerable amount of teaching material. One

side of a single disc can store more than 50,000 still images.

2) The image is very clear and sharp and can be held for any length of time.

If we combine videodiscs with a computer, we can construct a system with properties e-h below, which we have actually done for the THE system.

e) A desired item can be searched and promptly called up from an extensive bank of topics with the aid of the computer.

f) Making use of screen images accompanied by a voice (oral explanation and advice), the system can offer extremely good instruction which the learner feels comfortable with. A synchronized presentation of still images and narration is quite effective in this connection.

g) By giving the computer a standard for diagnosing the level of the learner's knowledge at the outset, we can make the system automatically judge the learner's achievement and ability, provide him with appropriate topics, and guide him along the most suitable path.

h) With the aid of the computer, the system can grasp the learner's question and promote interaction between the instructor (= the system) and the learner.

Structure and Functions of the THE System

The THE macrosystem is composed of three main parts: the courseware system, the software system, and the hardware system.

(1) The courseware system is divided into five components.

a) The display component comprises a database of still and/or motion pictures with an accompanying sound track and provides the required screen images.

b) The problem component stores and supplies the required problems, which are classified according to their level and other characteristics, and the answers to them.

c) The evaluation component tests and evaluates the learner's understanding and/or progress on the basis of both his own scores and the overall pattern of scores for the sample population.

d) The selection component chooses problems for the learner to try and material to study currently, using the information supplied by the evaluation

579

component.

e) The control component manages the other components and records the test result data (i.e., scores and overall tendencies) for each learner as supplied by the evaluation component.

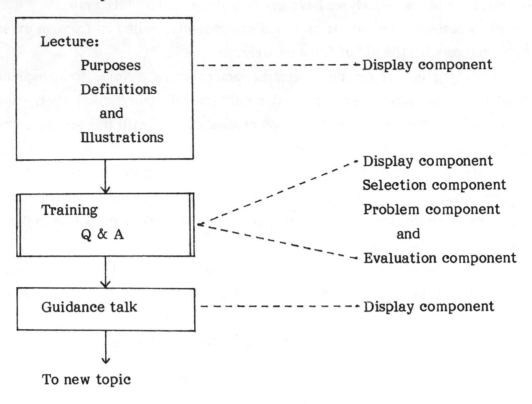

Diagram 1

The courseware paradigm supported by these components is illustrated in Diagram 1. The learner's practical experience with the THE system takes the following form. At the outset the learner is guided by motion pictures, animation, and narration. He is asked about his background in the subject he is going to study and about his preference regarding the starting level. The system recognizes his responses and then gives a test for an initial evaluation of the learner's level and ability. Using the result of this initial test, the system selects an appropriate path for the learner and leads him to a specific topic he should study first. This topic is taught by means of a lecture and an interactive lesson consisting of a graded series of "questions and answers." If the learner does well, he can exit to another

580

Details of "Training Q & A"

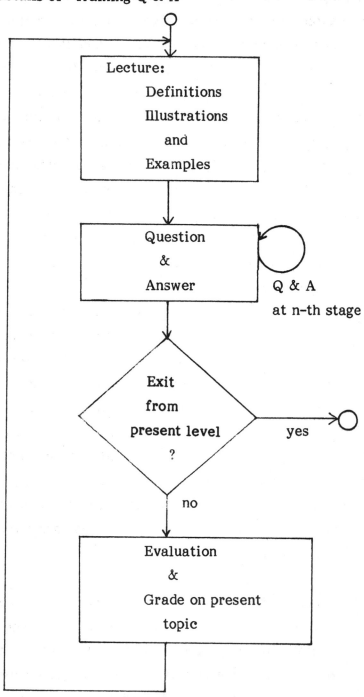

Problems are chosen
according to answers at
(n-1)th stage and
what they show.

Next lectures are chosen
according to the student's
grade and needs.

Diagram 2

topic after completing a few steps. Otherwise, he is asked to complete more steps until he acquires some mastery of the topic. Meanwhile, the problems (questions) that the system provides at each stage are chosen on the basis of the results of all preceding tests. If the learner feels it necessary, he can listen to the lecture again or can take time to think it over as much as he likes. This kind of interactive session usually continues either for a predetermined length of time or until some preset condition is fulfilled. To conclude each training session, the instructor (= the system) evaluates the learner's achievement and makes suggestions for the learner's subsequent study. The history of all interactions between each learner and the system is recorded, together with the learner's individual data, including the results of all tests. These data are analyzed and stored for future reference. A flow chart of the "Training Q & A" stage is shown in Diagram 2.

(2)   The courseware is realized through the software system, which runs the various electronic devices and coordinates the above-mentioned five components. Visual instruction is given via the display component in the form of still pictures, motion pictures or animation supplied from videodiscs. Practice problems are stored on videodisc as well as on floppy discs. Interactive use of these problems is managed by the selection system. Decision trees to determine each learner's path of study are traced (branches are chosen), taking into account the individual learner's responses to questions, overall achievement, and the level of difficulty of the topics according to standard criteria stored in the database; these judgments are made by the evaluation component. Diagram 3 shows how related courseware components are linked by the software system.

(3)   The hardware system is a typical off-the-shelf personal computer connected to an optical laser videodisc player. At present, the connection is made through a videodisc interface. The videodisc which we have adopted is the "UPC (Universal Pioneer Corporation) laser videodisc system" for educational and industrial use. The disc is 30 cm in diameter and 2.5 mm thick and weighs 180 grams on average. Audio and visual information is recorded on the discs from master videotapes or films. Each side of the disc has 54,000 uniquely-numbered frames, any one of which can be retrieved independently. A TV frame (NTSC) is generated by each

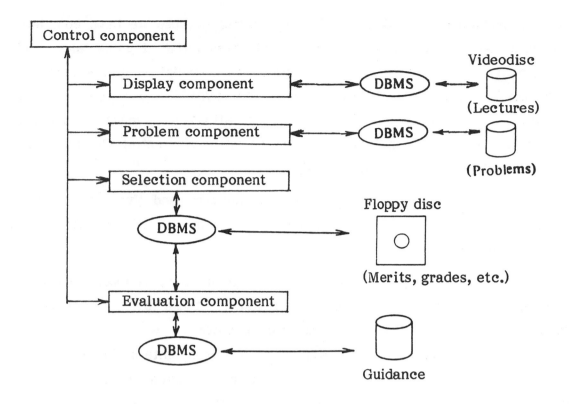

Diagram 3

rotation of the disc, i.e., by the information recorded on a single band of the disc. This means that 30 minutes' worth of motion pictures can be recorded on each side. The hardware system of the THE system permits a number of convenient functions, such as dual audio channels, stop motion, frame by frame review, slow motion, auto stop, rapid scan, and, particularly, direct and almost immediate access to any desired frame. Also, the high color quality and resolution of the image were appreciated by the students who had the experience of learning to use the THE system.

Some Advantages of the THE System and Courseware Available for It

Although instructional use of the THE system is in principle not limited to mathematics, our efforts have been concentrated so far on producing courseware for mathematics at the senior high school level. The THE system can be used not only for standard instruction starting from the beginning or from elementary knowledge of the subject, but also for intensive training of mathematical ability, especially problemsolving ability. For the former purpose we have completed courseware on "Trigonometric Ratios" and "Geometric Figures and Their Measurement"; for the latter, we have produced material on "Combinatorics," "Probability," "Vectors," and "Sequences." Moreover, we are now compiling courseware on "Linear Algebra (Linear Transformations in a Plane)" belonging to the first category, and on "Analytic Geometry," "Differential Calculus," and "Integral Calculus" in the second category.

From our experience in producing this courseware, and given the nature of the THE system, we feel that the following topics in high school mathematics would be suited to the THE system.

1) Topics involving geometric figures and geometric relations. Liberal use of visual material assists the learner in understanding these topics. Examples are trigonometric ratios, vectors, coordinate geometry, etc.

2) Topics involving calculations and manipulations. Interactive instructions promotes the learner's ability in these areas. Examples are algebraic manipulations, differential calculus, integral calculus, etc.

3) Topics involving abstract concepts. The system can provide many concrete examples through which the learner can form a realistic idea of abstract concepts. Examples are permutations, combinations and discrete probability.

4) Topics involving the generalization and refinement of notions. The learner can form ideas of stepwise extended notions by stepwise learning through the THE system. Examples are notions such as real numbers, mappings, matrices, etc.

To be fair, we should at this point mention some drawbacks of the THE system as it presently exists.

1) The format of the learner's answer is limited. Most questions are answered simply with numerical values. This may cause a bias in the assessment data.

2) The learner can input only final answers, and the system cannot trace or evaluate the learner's intermediate efforts.

3) There are difficulties with giving problems which take many steps for their solution or require string manipulations.

It might be interesting to note that these defects of the THE system are similar to those found in the so-called "Common Primary Entrance Examinations for Universities" in Japan, in which examination papers (in the form of mark sheets) of 400,000 applicants are graded by computers. Perhaps it would not be too optimistic to expect that these drawbacks or deficiencies in the operation of the THE system will be resolved soon, given the rapid progress of information technology (certainly with fifth-generation computers).

Finally, we would like to state our belief that the extensive use of applications of information technology like the THE system can be a valuable supplement to traditional classroom teaching, and can in addition create better educational environments both in the school and at home; it can release teachers from routine burdens and can offer finely-tuned individualized instruction to low achievers as well as gifted students. Moreover, the extensive and detailed data concerning the learner's responses which the THE system can accumulate should provide a reliable basis for improving both curriculum and teaching methods.

References

Fujita, H. and F. Terada. 1983. "Mathematics Education in a Society with High Technology." Proceedings of ICMI–JSME Regional Conference on Mathematical Education. Tokyo.

Fujita, H. and F. Terada. 1984a. "Mathematical Education in a Society with Modern Technology: A Case Study in Japan." Project Presentation, ICME 5, Adelaide.

Fujita, H. and F. Terada. 1984b. "The Present State and Current Problems of Mathematics Education in Senior High Schools of Japan." AG4, ICME 5. Adelaide.

Fujita, H., F. Terada, and S. Shimada. 1984. "Toward a Mathematics Curriculum in the Form of the Core and Option Modules." AG4, ICME 5. Adelaide.

Terada, F. and H. Fujita. 1981. "The Present State and Current Problems in General Mathematics Education at Japanese Universities." Proceedings of the Seventh International Conference on Improving University Teaching, 327. Tsukuba.

Terada, F., et al. 1982. "THE System – Instruction Using Videodisc and a Microcomputer." Educ. Technol. Res., 6, 29–36.

Using the Computer To Teach Geometry

Mary Grace Kantowski

University of Florida

Introduction

Geometry is perhaps the most controversial area of the secondary school mathematics curriculum in the country. The content of a high school geometry course, how it should be taught and, in fact, if geometry should be included in the curriculum at all have been topics of numerous publications in recent years (Brown 1982; Grunbaum 1981; Hoffer 1981; Sherard 1981; Usiskin 1981). More and more high school students do not elect to take geometry, and for others geometry is the terminal secondary mathematics course (National Advisory Committee (NACOME) Report, Table 1, p. 7). Whether this phenomenon occurs because the mathematics requirements have been fulfilled or because difficulties often encountered in geometry discourage students from selecting mathematics courses subsequently is a question open to debate. The fact remains that many students never go beyond the study of geometry, with the result that potential scientists, engineers, and mathematics and science teachers are lost to those professions because they lack the "critical mass" of mathematics content necessary for more advanced study.

Some geometry should be included in the mathematical experiences of all students. We live in a geometric world — the ordinary problem-solving tasks that occur on a typical day are often geometric in essentials. However, the geometry course as it is taught does not reflect the dynamics of the real world. In most schools, geometry is taught as a static subject with emphasis on its logico-deductive aspect. Undefined terms, definitions, and axioms are presented to students as constituting a conceptual framework they must work within, even though this framework does not exist in their cognitive structures. One of the leading Soviet researchers in visualization, Chetverukhin, notes (1971) the two aspects of Euclidean geometry — the spatial-visual and logico-deductive. He observes that in the Soviet Union, considerable emphasis is placed on the logico-deductive aspect of geometry in the secondary school, as in this country, and suggests a need for more emphasis on the spatial-visual aspect. This emphasis on the two-fold nature of geometry has led to a curriculum based on the work of Van

Hiele (Wirszup 1974) and to research on the development of spatial abilities (Vladimirskii 1971; Yakimanskaya 1971). In our country, geometry in the secondary school has become synonymous with classical Euclidean geometry approached from a purely logico-deductive point of view with little or no emphasis on the spatial-visual or intuitive aspects of the subject. Most geometry courses consist in the presentation of a set of "required proofs" to be memorized, a variety of "original proofs" to be constructed, and computational problems in which the theorems are to be applied. Three-dimensional Euclidean space or solid geometry is rarely studied any more (NACOME Report, 7).

The advent of affordable computers capable of rapid calculation and more sophisticated interaction, and in particular, computers able to simulate motion and present mathematical ideas visually by means of graphics capabilities, provides an opportunity for mathematicians and mathematics educators to reexamine all school mathematics curricula and suggests new directions as a result of these new capabilities. The new technology has a potentially great effect on what is taught at every level of mathematics, as well as on how the content is presented. This effect could be especially significant in the teaching of geometry because of its visual nature and because changes in this area are clearly needed and have, in fact, been called for (NACOME Report, 15).

Van Hiele posits five levels of development of thought in geometry, a claim which was subsequently substantiated by research conducted by his followers (Freudenthal 1973) and research conducted in this country (Usiskin 1982). He sees the first level as simply the perception of geometric figures in their totality as entities (units) without any awareness of parts of figures or relationships between these parts. In the second level, parts of the figure are discriminated and properties of the figure described. In the third level, a relationship between the properties of the figures themselves is established, and an attempt is made to see if one property could be deduced from another. It is only in the fourth level that the logical structure of proof and of analysis is grasped and that deduction as a means of constructing a geometric theory is understood. Finally, a standard of rigor such as Hilbert's with its corresponding level of abstraction is attained in the fifth level.

Van Hiele also emphasizes that levels may not be omitted in the learning of geometry. As in Piaget's developmental stages, it is necessary to pass through each lower level in order to attain a higher one. Furthermore, Van Hiele's research (Wirszup 1974) indicates that a strategy of structuring experiences according to levels can speed up the attainment of specific levels. He found that in an average school situation, only 10-15% of fifth graders reached the second level (perception, cognition, memory of classes.) The large leap noted in the seventh grade implies that this stage may be reached incidentally, but further statistics show that instruction relying on the incidental learning theory is highly inefficient. With instruction that emphasizes visual, nonmetric aspects of geometry patterned on Van Hiele's levels, pupils are able to reach the level of divergent thinking and evaluation desired in geometry sooner and far more consistently.

An area of interest to Soviet researchers quite foreign to researchers in America is that of the development of spatial abilities. Exercises to this end suggested by Yakimanskaya (1971) and Vladimirskii (1971) are in accord with the Van Hiele thesis. The former emphasizes that the use of visual aids is not merely auxiliary but essential to the development of ability in geometry. She further advocates drill and practice with spatial concepts similar to that used in consolidating the concepts of number and operation. Vladimirskii emphasizes the need for mastery of elementary rules for representing spatial figures and constructs a detailed series of exercises for developing spatial imagination. The progression of these exercises is closely related to Van Hiele's levels. Vladimirskii finds that pupils who could not complete spatial tasks corresponding to Van Hiele's level four "based their judgment of the figures on direct impression from the diagram without using properly the task's condition" (p. 77). This implies that the unsuccessful pupils were operating on a lower level, which adds weight to Van Hiele's belief that there can be no communication between levels.

The purpose of this paper is to suggest directions for a curriculum in geometry given the results of research on the teaching and learning of geometry and the potential influence of the computer. These suggestions strive to make effective use of the computer by making its integration into the teaching of geometry an essential part of the curriculum. We believe that the computer can serve as an excellent visual aid and as a means for giving students experience with exercises at various levels of geometric ability.

Suggestions for Change in the Geometry Curriculum

Towards a More Intuitive Approach to the Teaching of Geometric Concepts: Proceeding from the Concrete to the Abstract

Concepts must be mastered and relationships between structures understood before an individual is able to perform operations or reason with understanding. For example, students must clearly understand the notions of congruence and similarity before they attempt to construct or even understand proofs dealing with these relations.

Let me illustrate the intuitive approach to teaching the congruence relation using a program written for the Apple II called CONGRUENCE. Many modern textbooks introduce the congruence relation in the section on triangles. As a result, geometry students often think of congruence as a relation that can be applied only to pairs of triangles rather than one which is relevant to any pair of geometric entities. In CONGRUENCE, by contrast, the microcomputer constructs a polygon with 3, 4, 5, 6, or 7 sides (exactly how many is determined by random selection), then positions a second, congruent polygon in a random orientation on the screen. By using a pair of paddles, the student is able to position one polygon atop the other or align the two polygons so that corresponding sides are parallel. Such an exercise using the computer as manipulator not only helps the student master the concept of congruence as a one-to-one correspondence between the sides and angles of _any_ polygon, but also gives the student practice in transforming one figure into another using operations that can be dealt with metrically and more formally later on. Practice with CONGRUENCE helps the student form a general notion of congruence and allows the student to become acquainted with the variety of shapes of convex polygons and the fact that irregular polygons seem to take on different appearances when placed in different positions in the plane.

Similarly, the concept of similarity can be introduced using very simple LOGO procedures in which students simply construct figures (not necessarily triangles) having the same angle measures and different side lengths in various positions on the screen, and then compare the shapes of the figures.

Simple LOGO procedures also allow students to draw angle bisectors, medians, altitudes, and perpendicular bisectors of the sides of a triangle and to see which of these lines are concurrent. Such procedures for scalene triangles as well as special cases (right triangles, isosceles triangles, equilateral triangles) not only confirm concurrence in all cases, but also help develop in students an intuition for where points of concurrence may occur. Students at one level of interest or ability might test polygons other than triangles for corresponding properties and proceed to proofs of theorems while students at another level would gain geometric knowledge from observations of regularities.

Such movement from concrete examples to the definition of a mathematical idea gives the student greater assurance of mastery of an idea. Furthermore, the ability to interact with the computer, particularly where there is significant decision-making in the interaction, gives the student a feeling of more active participation in the learning process.

Toward a More Intuitive Presentation of Theorems

In the traditional geometry course, some definitions are given in general terms at the beginning of a chapter and followed by statements of theorems to be proved and the proofs of those theorems. This approach is closer to the teaching of history than the teaching of mathematics. If we are to teach students how to do mathematics, we must give them some feeling about how mathematics is done. CONGRUENCE provides us with a tool by means of which students can experiment by generating data — both graphic and numerical — that lead them to conjecture and draw hypotheses. This is something we have never had before. Conjectures about necessary and sufficient correspondences leading to hypotheses about congruence could then be derived from visual images of various pairs of polygons.

After interaction with the computer, questions such as "What would we need to know to be certain that one triangle is congruent to another?" lead naturally to hypotheses for the congruence theorems. Moreover, other questions could also be asked, for example, "What correspondences would be necessary to make one quadrilateral (or pentagon or hexagon) congruent to another?" "What about 'special' quadrilaterals such as rectangles or parallelograms?" "What about regular polygons?"

Students who are active participants in such a process of making conjectures and in stating theorems gain a much better understanding of the theorems. And the intuition that leads to the statement of a theorem often helps in determining what direction the proof should take.

The computational capability of the computer also provides an opportunity for students to see numerous sample applications of theorems that can give them clues about generalizations that are implied by the data. For example, conjectures about theorems related to similarity can come from direct experience in dealing with equal angles and proportional sides in similar polygons.

The technique of drawing conjectures by means of induction and thus being led to deductive proofs can prove to be a powerful teaching tool. Although the idea of generating data to illustrate a theorem or even suggest a hypothesis for a generalization is not new, its potential for widespread use was never as great as it is now with the availability of high-speed computational tools. But even the quick computational capability of the calculator is not enough when complex multistep calculations are required. Using a computer, moreover, a student can actually select the values to be tried out and then complete the calculations without "getting in the way" of the line of thinking underlying the lesson or his or her own work. That students can themselves decide on the values and then watch as tables are generated rather than simply being presented with prepared tables imparts a degree of spontaneity and gives students the feeling that mathematics is a dynamic subject with much still to be discovered, rather than something they are "learning about." Such activities, not easily implemented before the advent of the computer, give students the sense of the regularity that exists in mathematics and a true experience of "doing mathematics."

Varying the Approaches to Introducing New Content and Theorems: Emphasizing the Spatial-Visual Aspect of Geometry As Well As the Logico-Deductive

Several years ago the transformational approach to geometry briefly reared its head in a few secondary school districts. Similarly, vector and analytic approaches to proof were introduced as optional topics elsewhere. Perhaps these approaches were a bit ahead of their time for the secondary school. Ironically, the essential notions of transformation geometry are very simple ("stretchers and shrinkers" as presented to junior high school students in Beberman's UICSM mathematics program), but once the initial concepts were introduced, work with more complex ideas such as composition of transformations became too time-consuming for very extensive study because of the many drawings involved, or too complex because of the calculations required. By providing the capability to depict transformations graphically — even compositions of transformations of any given geometric figure in the plane — the microcomputer has opened new doors for the introduction of geometric content by means of the transformational approach. Two mathematics educators have recently introduced programs which illustrate how to use the microcomputer to present a transformational approach to geometry. Shigalis (1982) uses a program written in BASIC by means of which students can transform figures using distance-preserving isometries (translations, rotations, ordinary and glide reflections) or similarity dilations or perform compositions of transformations. In a similar but more comprehensive set of programs using LOGO, Thompson (1982a) applies a transformation approach to teaching an entire unit on motion geometry through student manipulation of a small flag on the screen. Again, students can perform translations, rotations, and reflections as well as compositions of these transformations.

At first glance, such prototype programs look like interesting supplementary exercises in which the microcomputer serves to illustrate concepts once they have been taught. A more significant use of the microcomputer, however, is as an essential tool in the initial period of instruction to help students see relations such as congruence and similarity from yet another perspective and provide them with practice in visually estimating the relative positions of structures in the plane under given transformations.

The Shigalis program allows students to enter the coordinates of the vertices of a triangle and then select a particular transformation and the parameters for the transformation. As in CONGRUENCE, students gain experience working with pairs of congruent triangles of diverse shapes in a variety of orientations in the plane. Work with this program not only helps them develop skill in recognizing congruent triangles, but also helps them develop the ability to predict the position of a triangle after given transformations have been applied.

The Thompson programs provide an interesting contrast. Since the same flag is used in all the transformations, the student's attention is drawn to the new position of the flag (particularly the flagpole). Both programs can illustrate how to express congruence dynamically in metric terms as a transformation of one figure into another in the plane, and both programs are excellent prototypes of use of the computer as an integral component of the instructional process.

Programs such as these bridge the gap between the synthetic and analytic, the transformational and the analytic, in the study of geometry. So far, discussions of the use of the computer have been limited to its use as an interactive tool where a student or group of students interacts with it by means of programs or procedures which have already been written by others. Being required to construct their own programs is an excellent way for students to learn how to deal with the mathematical notions underlying what they are learning. It is one thing for a student to select a transformation and parameters for the transformation. It is a far more difficult thing to have to define what is happening in the plane using the Cartesian coordinate system. In fact, the very first translation the student is confronted with is that of moving the origin, since the origin in the plane of the video screen is in the upper left hand corner. The notion of translation of axes, a notion which is very useful in high school mathematics, is often ignored in the secondary school, for there is usually little motivation to apply it. Students who write their own programs for geometric transformations rapidly make this technique their very own, a result having great potential for the later study of mathematics. Even the relationship between the polar and Cartesian coordinate systems takes on new meaning for the student who is forced to write a program to generate some rotation.

The transformational approach to geometry also provides motivation for the introduction of operations with matrices. Students who write their own programs for isometric or similarity transformations will be motivated to learn how to handle matrices. These programs become a tangible application of the utility of matrices.

The transformational approach is extremely graphic. It is an excellent tool for helping the student master concepts intuitively and make conjectures that could lead to hypothetical theorems. But more important, perhaps, the transformation approach makes geometry a dynamic subject which can help develop the student's spatial visualization ability as well as the ability to reason.

The subject of vector analysis is often very difficult for students at the post-secondary level, again because here students are being introduced to new operations (e.g., vector addition and multiplication) and asked to reason abstractly about them. Many college students go through the motions of working with vectors, grinding out proofs and applications but never really understanding what they are doing. Several LOGO procedures, such as the classical procedure known as POLY, which generates a polygon with sides of a given length and angles of a given size, can be interpreted in terms of vectors, and thus used to generate a closed curve for a vector sum of 0, or regular polygons for constant sides and angles (factors of 360°). Addition and subtraction of vectors and multiplication of a vector by a scalar are easily illustrated. A very nice feature of this program is its simplicity. Students can experiment by writing their own modifications of the procedure if they wish to find out what happens to the vectors under different conditions. As they are experimenting, they gain experience with the concept of vectors, and this prepares them for more abstract work with the subject.

Many theorems in geometry can be proven quite nicely by means of vectors. The theorem that the median of a trapezoid is parallel to the bases of the trapezoid and equal to one-half the sum of the bases is an example of such a theorem. The theorem that the line joining the midpoints of two sides of a triangle is parallel to the third side and equal to one-half of this side can be proven in the same way and clearly becomes a limiting case of the trapezoid theorem where the upper base is the zero vector. The proofs or explanations of many theorems in plane geometry could be accomplished with the aid of the computer using vector methods; in this way students would be led to a better understanding of theorems by learning that there is more than one approach to the proof of any given theorem.

The analytic approach to the study of geometry has never really taken hold in the secondary curriculum. Curriculum projects such as the School Mathematics Study Group Project attempted some integration of the analytic approach into the curriculum, but only very few school systems accepted SMSG materials. Most geometry courses are basically synthetic geometry, with analytic topics being covered elsewhere in the secondary curriculum. In a paper reprinted in the Mathematics Teacher in 1967, Oswald Veblen discusses the geometry curriculum and gives interesting historical insights as to why the emphasis has remained so strongly Euclidean. He notes, however, that in the application of geometry to physics, the emphasis is on the analytic. Veblen views both as distinct and useful disciplines in themselves, and suggests that both points of view are important. Perhaps, in a paraphrase of G.K. Chesterton, an integration of methods of geometry has not failed; it has never been tried!

The availability of the computer is giving us another opportunity to attempt this integration. Many theorems in geometry can be treated analytically as well as synthetically so as to emphasize concepts that students are taught in algebra and relate them to what they are learning in geometry. Theorems related to parallelism and perpendicularity could be applied using the analytic approach and computer methods. Again, much of the tedium and possibility of computational error are eliminated by having the computer complete the calculations, allowing the student to focus his or her attention on the mathematical idea.

Placing Greater Emphasis on Special Cases and Limiting Cases

Most students can tell us that "A circle contains 360 degrees," but their realization of the geometric significance of "360 degrees" stops there. There is a myriad of important geometric principles related to this idea, yet many students complete their study of geometry without any insight into the central role played by this notion. Students, even very young children who work with Turtle

Geometry and write their own LOGO procedures, soon begin to see the significance of measures of angles that are factors or multiples of 360°. Striking geometric patterns created by recursive procedures are seen to depend on how many times the procedure has to be repeated before a multiple of 360° is reached. After some trial and error, students learn that in drawing regular polygons the direction at each vertex must change by $(360/n)^o$ to form a regular polygon. Relationships between certain triangles such as the equilateral and 30-60-90 right triangle and the regular hexagon become clear, and the significance of the equilateral triangle as a "special case" begins to make sense.

Because of the ability of the computer to generate data, isoperimetric theorems can be studied in a way not possible before. Some isolated questions (e.g., find a rectangle of greatest area for given perimeter) are generally included as exercises in geometry, but the elegant mathematical idea of the regular n-gon as the polygon of greatest area for given perimeter and given n is seldom treated. The topic of maximizing a given area is such a central one in the calculus as to call for introductory treatment during the study of geometry.

Isoperimetric theorems, such as those involving the isosceles triangle (the triangle of greatest area for a given common base and perimeter and triangle of least perimeter for a given common base and area) are interesting topics in an inductive discovery lesson, and variations of isoperimetric problems make very good "real world" application problems. Using Heron's formula, it is possible to generate a great deal of data for conjectures to be generalized on — and even get an idea of the direction of a proof if the data are generated graphically as well.

Experiences with sets of theorems such as isoperimetric theorems will, furthermore, introduce students to the concept of a limiting case, a concept that can be an important point of departure for the solution of many problems.

Restructuring the Geometry Curriculum to Take Advantage of the Computer: Including Geometric Topics Now Treated Elsewhere in the Mathematics Curriculum, and Placing Broader Emphasis on Certain Topics and De-Emphasizing Others

Some geometric topics are conceptually quite simple but require complex calculations in order to apply the principles underlying the topic. Topics such as computing the area of a triangle given two sides and the included angle and the laws of sines, cosines, and tangents are generally included in a second-year algebra or combined algebra and trigonometry course probably because solving problems based on these theorems requires the use of logarithms to perform complex calculations. Logically, however, the trigonometric laws for the solution of triangles are metric applications of relationships between the sides and angles of a triangle and could be introduced following the treatment of congruence, perhaps together with a discussion of the triangle inequality. The laws themselves and some of the derivations of these laws are relatively simple. Much of the complexity involved occurs in calculations that now can be handled with ease by the computer. Most students are introduced to the trigonometric ratios in elementary algebra, before studying geometry. Once the various trigonometric formulas can be applied without tedium and complicated computations, they can be easily included in the geometry course. Even Heron's formula, which can be forbidding when unwieldy numbers are involved, can be introduced to give students more power in finding the area of a triangle. Today many students enter college courses in mathematics thinking that the only way to find the area of a triangle is to calculate half the product of its base and altitude. It's not that all students should be familiarized with all of these topics, or even that all students who do learn about them should acquire the same amount of knowledge about them. Rather, what is required is a rethinking of where these topics belong in the curriculum so that students derive the maximum benefit from their study of geometry.

Many notions in geometry have received limited treatment because of a lack of good examples of applications with manageable calculations. One case in point is the development and use of the formula for the area of a regular polygon as a function of its perimeter and apothem. Typically, the formula is given, or perhaps derived, and students are then asked to find the areas of equilateral triangles, squares, and regular hexagons, because these polygons have apothems which are easily calculated. Using the ability of the computer to perform complex calculations, a student using a LOGO procedure, for example, can construct an arbitrary regular n-sided polygon, find the length of the apothem, and then calculate the area of the polygon. (A fringe benefit of this procedure would be discovering an "n" for which such a construction is impossible and making conjectures about why this is so.) A "unit" created in this way could include the entire discovery (and proof?) of the measure of the interior angles of a regular polygon, since the construction of the polygons depends on the size of the angles, the sum of the interior angles of any polygon, and the sum of the exterior angles of any polygon.

The student could then be encouraged to let the number of sides of a regular polygon become very large, in fact to let this number approach infinity, a procedure that can be easily implemented by means of LOGO. In this way the student is led to observe that the length of the apothem approaches the radius of the circumscribed circle. The relationship of the circumference of a circle to its radius and area, hence the definition of π, is an easy step from there. This is another example of an intuitive introduction to a topic in a case where a rigorous proof would be meaningless to many students but where the mathematical idea — the fact that the circumference is a function of the radius — is a very important one.

Summary and Conclusions

Geometry is now taught as a deductive science. To many secondary school students, it is unrelated to any other area of the mathematics curriculum. In some sense this is true. The emphasis in geometry is on proof and even the form of proof in geometry is different from the form of the proofs students have seen or done before — and any proofs they will ever see or do later on. Helping a student learn to reason deductively may be one objective of geometry, but it should not be the only objective. The development of spatial intuition, the ability to make reasonable conjectures, and the ability to deal with geometric figures metrically are all worthwhile objectives of geometry — objectives which, if fulfilled effectively, could, in fact, foster the development of the ability to reason logically in more and more students. The majority of students at the early secondary school level may not be ready for the rigor of reasoning in a mathematical system now attempted in the study of geometry.

Geometry is a basic skill. It is, therefore, necessary for all students — and not merely those who are college-bound or those who are interested in mathematics or science-related careers — to have some measure of exposure to geometry. Our suggestions for the geometry curriculum are intended to create an environment in which students at all levels of interest or ability in the subject could learn geometry. Although the computer should play an integral part in any new geometry curriculum, it is not necessary for all students to use this technology in the same way. Some may learn by using programs written by someone else. Others may want to delve more deeply and learn how to make the computer produce desired results, such as those in transformation programs. Even in the latter group, some students will write programs to generate translations and/or reflections, while others will be interested in exploring the relationship between the Cartesian and polar coordinate systems by writing programs that produce rotations and compositions of functions. Similarly, in working with programs that make conjectures for hypotheses, some students may simply look for patterns in data generated by existing programs and state conjectures based on those data, while others will not only write programs to generate data for conjectures, but will go on to prove those conjectures. The particulars of any geometry course for a given group of students are left to those responsible for developing the syllabi, though all students should be able to have some experience with the content of geometry, at least on an intuitive level.

The microcomputer can provide excellent interactive experiences for students at all levels of interest and ability in geometry. Even before the coming of the new technology, there had been calls for changes in the curriculum that would place greater emphasis on the intuitive and spatial-visual and utilize a variety of approaches, such as analytic and vector, as well as the purely synthetic (Osserman 1981; NACOME Report 1975, 15; Dieudonne 1981). The advent of the computer has given new dimensions to the discussions of the need for change in the geometry curriculum. It has been responsible for a renewed interest in the transformational and analytic approaches to the subject and, with the discovery of a computer-aided proof of the four-color problem, has led to wide-ranging discussion on the nature of an acceptable mathematical proof (Appel and Haken 1981; Manin 1981; Renz 1981). The proof of the four-color problem is meaningful not only in and of itself, but also because of its potential effect on future discoveries in mathematics. It is a watershed event that will encourage instruction in a variety of forms of proof, besides providing further impetus for the integration of the computer into the teaching and learning of mathematics.

All of these factors support the need for changes in the geometry curriculum. These changes should not be merely cosmetic, nor is it simply a matter of integrating the computer into the existing curriculum. Essential changes in many of the objectives and approaches to the teaching of the subject are what is needed.

Finally, any curriculum that provides innovation based on interaction with a computer should be viewed as a curriculum in flux. Experiences that will be new to secondary school students today, next year, or even five years from now because of implementation of new developments in the technology are experiences that many elementary school students could be having now. If we use the technology correctly at the elementary school level, the secondary school students of the future will come to us with a better intuition of geometric ideas and, therefore, with the potential to explore even newer and more complex ideas. Certainly, secondary school students will be coming to us with far more programming ability, perhaps in more than just one programming language. Our approach to building a curriculum, therefore, must not be one of forming a static curriculum, but one in which we plan for continual change to meet the potential of the next generation of students, a generation whose education will be so greatly influenced by the ever-changing technology of the computer. The technological revolution of today should lead to a corresponding educational revolution.

References

Ableson and DiSessa. 1981. Turtle Geometry. Cambridge, MA: The MIT Press.

Appel, K. and W. Haken. 1981. "The Nature of Proof: Limits and Opportunities." Two-Year College Mathematics Journal 12: 118-119.

Brown, R.G. 1982. "Making Geometry a Personal and Inventive Experience." Mathematics Teacher 75: 442-446.

Chetverukhin, N.F. 1971. "An Experimental Investigation of Pupils' Spatial Concepts and Spatial Imagination." In Soviet Studies in the Psychology of Learning and Teaching Mathematics V, 5-56, J. Kilpatrick and I. Wirszup, eds. Stanford, CA: School Mathematics Study Group, and Chicago: Survey of Recent East European Mathematical Literature.

Dieudonne, J. 1981. "The Universal Domination of Geometry." Two-Year College Mathematics Journal 12: 227-231.

Freudenthal, H. 1973. Mathematics as an Educational Task. Dordrecht, Holland: D. Reidel Publishing Co.

Galda, K. 1981. "An Informal History of Formal Proofs: From Vigor to Rigor?" Two-Year College Mathematics Journal 12: 126-140.

"Geometry." 1967. Report of the K-13 Geometry Committee of the Development Division. Toronto: The Ontario Institute for Studies in Education.

"Goals for School Mathematics." 1963. Report of the Cambridge Conference on School Mathematics. Boston: Houghton Mifflin Co.

Grunbaum, B. 1981. "Shouldn't We Teach GEOMETRY?" Two-Year College Mathematics Journal 12: 232-238.

Hoffer, A. 1981. "Geometry Is More Than Proof." Mathematics Teacher 74: 11-18.

Manin, Y.I. 1981. "A Digression of Proof." Two-Year College Mathematics Journal 12: 104-107.

National Advisory Committee on Mathematical Education. 1975. Overview and Analysis of School Mathematics Grades K-12. Washington, D.C.: Conference Board of the Mathematical Sciences. (NACOME Report.)

Norris, D. 1981. "Let's Put Computers into the Mathematics Classroom." Mathematics Teacher 74: 24-26.

Osserman, R. 1981. "Structure vs. Substance: The Fall and Rise of Geometry." Two-Year College Mathematics Journal 12: 239-246.

Renz, P. 1981. "Mathematical Proof: What It Is and What It Ought To Be." Two-Year College Mathematics Journal 12: 83–103.

Sherard, W.H. 1981. "Why Is Geometry a Basic Skill?" Mathematics Teacher 74: 19–23.

Shigalis, T.W. 1982. "Geometric Transformations on a Microcomputer." Mathematics Teacher 75: 16–19.

Thompson, P. 1982a. Motion Geometry. A Module for the Teaching of Geometry to Pre-service Elementary School Teachers.

Thompson, P. 1982b. Geometry. A Module for the Teaching of Geometry to Elementary School Teachers.

Tymoczko, T. 1981. "Computer Use to Computer Proof: A Rational Reconstruction." Two-Year College Mathematics Journal 12: 120–125.

Usiskin, Z. 1981. "What Should Not Be in the Algebra and Geometry Curricula of Average College Bound Students." Mathematics Teacher 73: 413–424.

Usiskin, Z. 1982. "Van Hiele Levels and Achievement in Secondary School Geometry." Paper presented at the Annual Meeting of the American Educational Research Association, New York.

Veblen, O. 1967. "The Modern Approach to Elementary Geometry." Mathematics Teacher 60: 98–104.

Vladimirskii, G.A. 1971. "An Experimental Investigation of Pupils' Spatial Concepts and Spatial Imagination." In Soviet Studies in the Psychology of Learning and Teaching Mathematics V, 5–56, J. Kilpatrick and I. Wirszup, eds. Stanford, CA: School Mathematics Study Group.

Vredenduin, P.G.J. 1958. "A Method of Initiation Into Geometry at Secondary Schools." In Report on Methods of Initiation into Geometry, 31–45, H. Freudenthal, ed. Groninghen: J. Walters.

Wirszup, I. 1974. "Some Breakthroughs in the Psychology of Learning and Teaching Geometry." Address given at closing session of annual meeting of National Council of Teachers of Mathematics (April). Atlantic City.

Yakimanskaya, I.S. 1971. "The Development of Spatial Concepts and Their Role in the Mastery of Elementary Geometric Knowledge." In The Development of Spatial Abilities, Soviet Studies in the Psychology of Teaching and Learning Mathematics 5, J. Kilpatrick and I. Wirszup, eds. Stanford: School Mathematics Study Group.

Microcomputer-Based Courses for School Geometry

Max S. Bell

The University of Chicago

Introduction

For part of 1983-85 I directed a small team of teachers, programmers, mathematics educators and data entry people in the development of a comprehensive microcomputer-based high school course in plane geometry (Bell, et al. 1986). My group also worked with a team directed by Warren Crown on a geometry strand in a microcomputer-based program for elementary school mathematics (Crown, et al. 1985). The development work was done at WICAT Systems, Orem, Utah. Besides myself, the team included Kaye Crown, Dale Underwood, and Diane Hill as assistant directors, Charles Knowlton and Hoanganh Thuy Nguyen as programmers, Kevin Westover as assistant programmer, and Donald Danner, Daniel Dick, Alain Mehl, and Chad Van Orden as writers; Warren Crown, Peter Fairweather, Sheila Sconiers, Sharon Senk, and Theodore Wight served as consultants. Year-long microcomputer-based courses for schools are still rare. Hence, it may be useful to describe our results and the process by which they were achieved, and to put them in the context of present-day textbook-based courses in school geometry.

School Geometry in the United States

Since about 1965, the standard geometry course in the United States has been a plane Euclidean course emphasizing synthetic proofs based on "metric" postulates about length and angle measure. That standard course derives from the School Mathematics Study Group Geometry Course (SMSG 1960), which is based on metric postulates as pioneered by G.D. Birkhoff in the 1930's (Birkhoff and Beatley 1940). Most U.S. textbooks for plane geometry courses have the same basic content, with careful definitions and explicitly stated metric postulates and with heavy reliance on two-column proof. Most such books feature a brief introduction to coordinate geometry, and most now have a chapter on transformations. It is pretty much the case that textbooks in the U.S. that depart from this norm have not survived in the marketplace. The course is normally taken in tenth grade.

But school geometry is very much on the curriculum reform agenda again, in part because of the generally acknowledged efforts of the director of this conference, Izaak Wirszup, to make us aware of seminal work done outside the United States. Two foreign scholars who have influenced the work here are Kutuzov (1960), who offers broad, varied, and very clear approaches to standard high school content, and Pyshkalo (1968), who provides an example of a rich primary school experience. Many mathematics educators have been influenced by the work of the Van Hieles (1958), also brought to our attention by Wirszup (1976).

In particular, attention to the Van Hieles' stage models for learning geometry has resulted in theories and research information to support what teachers of proof-oriented high school geometry courses have long suspected: many otherwise capable students come to that course with serious deficiencies in basic geometry knowledge and intuition that greatly inhibit learning how to do proofs. Put briefly, that research has shown that about 40% of U.S. students complete a proof-oriented high school geometry course. Of those, more than half enter the course greatly deficient in geometry knowledge and intuition, and it can be assumed that most of those who do not attempt a proof-oriented high school course are even less well prepared in geometry. Only about 30% of that 40% of U.S. students surviving a proof-oriented course achieve reasonable mastery of the proof-writing objectives of the standard school course (Senk 1983, Senk 1986, Usiskin 1982).

The poor performance of students in standard high school geometry courses is probably beyond the control of most teachers of that course. Elementary school mathematics instruction in the U.S. pretty much ignores geometry, so serious deficits in geometric intuition are hardly surprising. The wide range of preparation and ability of students in most high school geometry courses presents teachers with great difficulties. Most students need considerable feedback as they are learning to construct proofs. To see why they are unlikely to get what they need, consider a teacher with only 60 geometry students who might wish to spend three minutes per student per day outside of class checking, say, one proof each day. That would require three hours each day, which is just not possible, given the workloads of most teachers.

Furthermore, geometry is a highly visual subject, but appropriately rich and precise graphics presentations are difficult with only chalkboard or printed materials. Such limited graphics capabilities also make it difficult for students to "see" the key elements on which they need to focus.

To put it briefly, poor student preparation and poor pedagogic technologies combine to make poor courses, even with capable teaching.

As we began planning for the course described below, we felt that we could provide via microcomputers new instruction technologies that offer sharper, more dynamic graphics under user control; a gentler introduction into the arts of proof; more practice in doing proofs, with correction and feedback; more varied content and approaches to geometry; and mechanisms by which students who need more time can take more time. We felt that interactive microcomputer-based geometry exercises could be provided from kindergarten on. Therefore, we undertook to give teachers the tools needed to overcome the severe difficulties in teaching geometry we have discussed.

Overview Description of the Microcomputer-Based Geometry Course

Details about the rationale for and actual content of the courses we produced are given below. There is a geometry strand (among other strands) in all the Math Concepts courses for grades K-8. The High School Geometry course is in two parts:

Geometry 1, Foundations includes many lessons that build intuition for basic geometry ideas, then a transitional set of lessons using transformations to develop ideas of system building and the meanings of "proof."

Geometry 2, Proofs and Extensions mainly covers the "synthetic proof" content of standard school courses, exploiting an intelligent Proofchecker to give accurate step-by-step correction and assistance to users. This part of the course also includes lessons developing coordinate geometry and extends standard course content in a number of ways through a series of "problems and recreations."

The significant instructional innovations in the geometry course are made possible by two rather fine programs that support essentially all lessons in the course and in addition give users considerable power to explore geometry ideas on their own. Proofchecker verifies logical and syntactic correctness of synthetic proofs entered by users. As far as we know, this has not been accomplished elsewhere for microcomputers. (A fine minicomputer-based proof checker does exist; see Anderson, et al. 1985). GeoDraw offers considerable user control over construction and transformation of geometric figures, supporting most of the non-proof lessons of the course and giving considerable exploratory power to users independent of lessons.

The diagram below indicates the scope and organization of the course.

Geometry 1   Geometry 2
Foundations   Proofs and Extensions

Geometry   Geometric   Synthetic   Additional Topics
Overview   Reasoning   Proof   and Problems

These geometry programs play on IBM PC or compatible microcomputers with at least 128K (kilobytes) of memory and one disk drive. Geometry 1 and Geometry 2 are each stored on two 360K diskettes or comparable storage media, with about a third of the storage space for program code and the remainder for data that permit the many activities of the courses. On the one hand, that is rather remarkable economy achieved by fine programmers. On the other hand, the reply to almost any query a reader might have about why this or that additional function is not available is that we ran out of room. Relatively inexpensive microcomputers without such limitations are increasingly available, and even more powerful programs than these should soon be feasible.

Each of the Geometry 1 and Geometry 2 courses also includes a printed Student Guide and a Teacher Guide. That makes them usable independent of textbooks, but it should also be possible to use them with any of a number of standard or more innovative text materials.

These programs are easier to use than to explain in print, but in the next few sections of this report I will talk about many of their features and the on-screen activities that exploit them.

Overview of the GeoDraw Program

The keys on the top row (the number row) of the microcomputer keyboard and the "function keys" are assigned special functions active in all the GeoDraw subsystems. Figure 1 shows part of a typical keyboard with a cardboard keyboard guide in place that identifies each of these special functions. Users move a cursor with arrow keys and FAST, MEDIUM, or SLOW sets how far the cursor moves on each key press; CLEAR takes all user input off the screen; FILL permits coloring of closed figures with any of the colors chosen by the COLOR key; LABELS allows users to identify points on the screen using letters or coordinates; MEASURE permits finding length or degree measure of segments or angles.

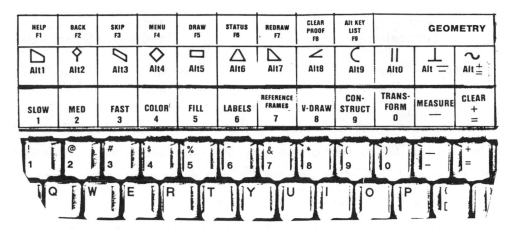

A Typical Keyboard and Keyboard Guide to Special Functions

Figure 1

CONSTRUCT, V-DRAW, and TRANSFORM enable the user to choose how geometric figures are constructed and changed on the screen. With the DEL (delete) key, users can cancel a key press or "undo" the most recent step of whatever they have constructed on the screen. We find that this undo function promotes an experimental approach to constructions — it is easy to correct things that seemed worth trying but didn't work out.

If the key assigned to REFERENCE FRAMES is pressed, a menu appears at the bottom of the screen, as shown in Figure 2. If users then press **S** for **S**quare-lattice, for example, they will be asked to specify how far apart the dots should be, with an interval of 30 pixels displayed to give scale. (A pixel, for "picture

608

element," is the smallest part of a monitor screen under the control of the computer; in our programs, for example, it is 1/320th of the horizontal dimension of the screen.) With that information supplied, a square array of dots is put on the screen, and users have the choice of restricting cursor movement only to those dots or permitting the cursor to move freely to any screen point. Circle lattice, Grid, and Iso-lattice give other on-screen guides for drawing. Activating Axes puts a standard coordinate system on the screen with scale as requested by users. Whenever axes are on the screen all measures will be given in terms of the unit specified for the axes rather than the basic "pixel" unit. Figure 2 shows an example.

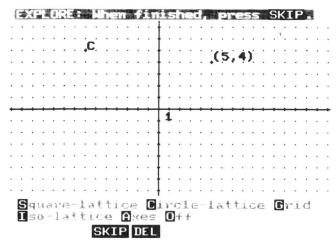

Reference Frames Menu and Example of Axes

Figure 2

Figure 2 also illustrates how users are given access to many graphics capabilities without being overwhelmed with dozens of special key press sequences to learn or remember. The available drawing tools are spelled out at the bottom of the screen and made available by pressing a single letter as highlighted in that display. Classroom trials indicate that this feature makes the programs quite easy to use.

The last line of the screen, as shown in Figure 2, reminds users of available general system functions. When in doubt about what to do next, users learn to look at "the bottom line" to see what can be done to move on.

Details of the GeoDraw Program

GeoDraw consists of three main subsystems: Construct, Transform, and V-Draw, the special functions just described, and the capacity to display on the screen the reference frames listed in Figure 2. GeoDraw was designed to reflect the various ways in which Plane Geometry can be organized:

1. CONSTRUCT models to some extent centuries-old Euclidean geometry; with MEASURE it reflects twentieth-century modernization of Euclidean geometry.
2. V-DRAW models "turtle geometry" construction methods and can also be made to model vector approaches to geometry.
3. TRANSFORM models transformation approaches to geometry.
4. CONSTRUCT with AXES and the point-slope option for drawing lines models coordinate approaches to geometry.

The descriptions below are for the high school program. Two simpler systems are available for children in the geometry strand of Math Concepts, generally with keyboard control of certain constructions and transformations rather than response to parameter prompts.

The Construct Subsystem

Figure 3 shows a typical screen display for the Construct subsystem of GeoDraw. Each of the drawing functions is activated by pressing the letter on the keyboard that is capitalized and highlighted in the menu at the bottom of the screen. Press **P** for **P**oint, and a point is marked wherever the cursor is on the screen and named with the next available letter of the alphabet — 52 points nameable with upper and lower case letters. With points named on the screen, users may draw the other indicated figures by responding to prompts that appear when the appropriate letter key is pressed. For example, the prompt after pressing **Y** for ra**Y** is "Ray from point __ through point __"; the ray is drawn when the blanks are properly filled by the user.

N-gon prompts for the number of sides. When that number is supplied by the user, a regular polygon is drawn with center wherever the cursor is on the screen and a default "radius" of 40 pixels and labeled with the next available letters of the alphabet. Circle prompts for a center and radius, Midpoint prompts for the endpoints of a segment, and anGle prompts for an initial ray and an angle measure. The aRc key simulates a drawing compass by first prompting for a center, radius, and bearing at which to start the arc, then instructs users to press the arrow keys to draw whatever length arc they want. Line has two options: one prompt asks for two points and the other for a point and slope. The point-slope option used with Axes from REFERENCE FRAMES has obvious uses in exploring coordinate geometry. Pressing **A** for p**A**rallel gets a prompt of "Parallel to line __ through point __; **E** for p**E**rpendicular evokes a similar prompt.

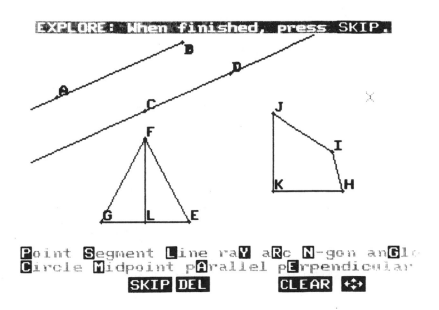

CONSTRUCT Subsystem Menu and Example

Figure 3

The Transform Subsystem

Figure 4 shows a typical screen for transformations. Pressing the high-lighted letters activates prompts for the function named. This subsystem was designed to teach the essential parameters for various transformations by requiring users to respond to them in prompts. For example, the Translate prompt requests the endpoints of a translation vector; the **Scale** prompt requests center of scaling and scale factor; the **Rotate** prompt requests center and amount of rotation; and the r**E**flect prompt requests two points on a mirror line. The p**U**ll prompt tells users the fixed line goes through wherever the cursor is, then asks whether a horizontal or vertical stretch is wanted, then asks for a scale factor. Since the effects of parameters for s**H**ear are rather obscure, the prompt tells users that the fixed line is through the cursor and that the actual (horizontal) shearing is done by pressing arrow keys.

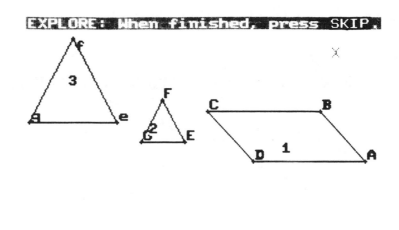

TRANSFORM Subsystem Menu and Examples

Figure 4

If several figures are on the screen, it is essential to specify which is to be transformed. Figures constructed within the Transform mode are assigned numbers. Users can also use the Construct subsystem to build a figure, then switch to the Transform subsystem, **Define** and assign a number to that figure by listing all points on the figure, then transform it.

Pressing **K** for **Keep** lets users decide whether or not to retain pre-images on the screen. For example, if pre-images are kept, it is easy to observe the effects of composition of transformations.

Point, **N**-gon, and Circle respond pretty much as they do in the Construct subsystem. They are included mainly for users to get figures quickly for transformation without going to the Construct subsystem. For example, some of our lessons ask users to start with this or that N-gon, then to use the transformations to make other specified figures; parallelogram ABCD in Figure 4 results from a square (4-gon) that has been transformed with pUll and sHear, with pre-images not retained.

The general system functions are available in all subsystems. For example, users can specify **Axes** from REFERENCE FRAMES, then explore the effects of various transformations on coordinates of points by choosing the Coordinates option in the LABELS function.

The V-Draw Subsystem

V-Draw resembles what is often called a "turtle geometry," named for the "turtle" in LOGO programs. Hence, the effects of the functions shown in Figure 5 will probably be obvious to readers familiar with LOGO implementations. V-Draw has a double origin, first in the V shape of our drawing cursor, which we wanted to be strongly directional, and second in the fact that the basic drawing elements are vectors. These vectors are drawn by setting direction with **Left** or **Right** and distance with **Go**. Every figure drawn in V-Draw is composed of these vectors, or segments. **Point** labels a point at the cursor; p**En** is a toggle switch that determines whether or not a visible screen trace is left by the movement of the cursor; **Send** prompts for a point label, then takes the cursor to that point; **Hide** is a toggle that makes the cursor visible or invisible; **Flip** turns the cursor 180 degrees to point it in the opposite direction. Hence, Go with Flip is equivalent to the "forward" and "back" of most LOGO implementations.

In many LOGO implementations the distance metric is not clearly indicated. Also, it is often difficult to tell in what direction the cursor is pointing, and trial and error may be required to ascertain whether the cursor ("pen") is or is not leaving a trace on the screen. The status information on the right at the bottom of Figure 5 is intended to correct these problems. It shows the cursor bearing counterclockwise from the horizontal, indicates how long 30 pixels is on the monitor screen, shows whether the pen is or is not leaving a trace, and shows the status of Hide.

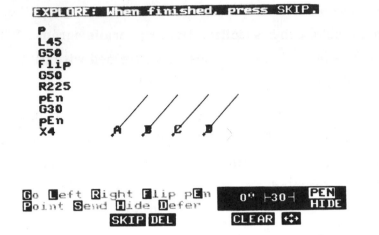

V-DRAW Subsystem Menu and Example of a Defer Sequence

Figure 5

By pressing **D** for **Defer**, users can define a procedure with a list of V-Draw commands, then have that list executed as many times as they like. For example, a list with entries "Go 80, Left 90, eXecute 4 times" gives a square with sides 80 pixels long. Another list and its results are shown in Figure 5. As in other turtle geometry implementations, this feature provides much of the fun and interest in V-Draw and is perhaps as good an introduction as any to using simple programming languages.

Learning Activities Using GeoDraw and Testing Sequences

With GeoDraw designed and implemented, we designed and programmed an activity authoring and playback program to present quite a large number of lessons with specific content objectives. A typical activity begins with "Exposition" frames giving background or stating a concept. It continues with "Guided Practice" frames in which only specified keys respond in a specified order, often as a tutorial to learn a certain program function (such as Scale) as a tool in learning a particular concept (such as "similar"). It may continue with "checked practice" frames where users can do as they like to try to achieve a specified result, with checking of whether the figure they thus produce is exactly what was specified. After such guidance, a typical lesson would continue with "Explore" frames, with the full array of GeoDraw powers available for users to attempt solutions of problems posed or to explore on their own. Much of the intuition-building intentions of these courses are expressed in the invitations to solve specific graphics problems with the GeoDraw tools. Many activities then conclude with "Check up" fill-in or multiple-choice questions, with correct answer feedback and scoring.

A fairly comprehensive geometry course is given by the on-line activities directed at specific concepts. Students also learn a lot by free explorations using GeoDraw. We hope teachers and others will also generate printed activity cards and exercise sheets posing problems or guiding students through the learning of additional concepts.

The Proofchecker Program

Proofchecker provides for intelligent checking of proofs entered by users. There is no internal "key" to correctness of any proof — whatever correct sequence of steps is entered will be counted as correct by Proofchecker. If the diagram given for a proof needs additional "auxiliary lines," then those can be drawn within Proofchecker, and their attributes automatically become additional properties usable in the proof. Proofchecker also provides for intelligent checking of syntax of inputs. For example, if there are several points on each ray of a certain angle so that there are many ways of naming the angle by listing three points, Proofchecker will accept any of the correct names for the angle.

Proofchecker's Three-Column Format: Claim, Reference, Rule

Synthetic proofs in essentially all U.S. high school geometry textbooks ask for proofs in two-column "statement-reason" form. Each step of a proof consists in a statement followed by citation of some axiom, theorem, or definition which serves as logical justification of the statement. At first we tried to mimic that two-column format, but validating the logical links between "statements" and "reasons" by searching through all previous steps sometimes took our Proofchecker too much time. For example, if a user claimed triangles were congruent and justified that by the side-angle-side congruence proposition, Proofchecker needed to find particular prior steps in the proof that had already established equal measures for the appropriate pairs of sides and angles. To simplify that, we decided to require users to identify those essential prior steps, and this led us to the three-column Claim-Reference-Rule format that is used in Proofchecker.

I now believe that this middle step of providing specific references to support the links between "statement" and "reason" (or between the "claim" and "rule" in Proofchecker) should become a regular part of school instruction even when our Proofchecker is not used. This is one of a number of cases where working through the requirements of sound teaching using computers led us to insights about sound pedagogies more generally.

Using the Proofchecker Program

As with GeoDraw, it is easier to use Proofchecker than to explain how to use it. The explanation here involves a series of typical screens that might appear in an actual proof. For example, Figure 6 shows the problem statement that users see after selecting from the course menu the problem of proving the <u>Pons Asinorum</u> theorem.

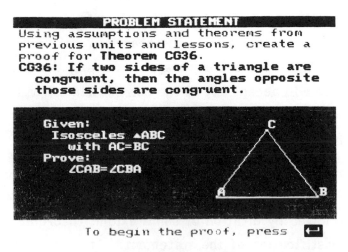

Problem Statement for <u>Pons Asinorum</u> Theorem

Figure 6

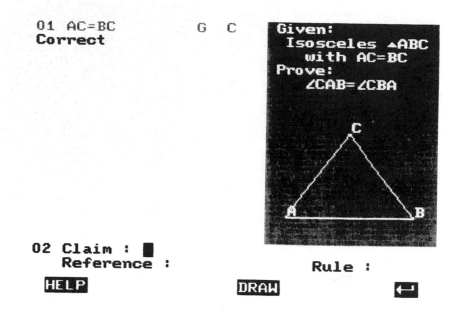

Proof Entry Screen

Figure 7

Figure 7 shows the screen on which users enter their proofs, with one step already entered and checked; the reference "G" is to a given fact and "C" indicates the claim was "copied" directly from the Given, rather than requiring justification by some rule.

The system prompt line at the bottom of the screen indicates the existence of two aids to users. If users press the HELP key, a brief "plan" for one way to do the proof will be provided; in the case at hand, users are advised to draw a median to the base and prove two triangles congruent. If users press the DRAW key, a screen such as shown in Figure 8 is displayed, so that users can take the advice to draw a median, using Midpoint and Segment. On return to the proof, as shown in Figure 9, the diagram used in the proof will include the median, Proofchecker will display AD=DB as a step in the proof, and Proofchecker will add this information to its internal routines for checking validity of steps in the proof. Figure 9 also shows a completed proof of the proposition.

617

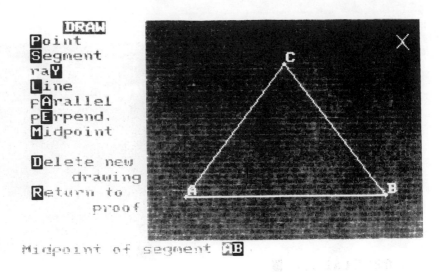

The DRAW Program for Adding to Figures

Figure 8

Each "rule" is entered as a code, since it would obviously be very difficult to enter and check full statements of propositions. Users always have before them a printed "reference card" that gives statements of propositions with appropriate code. In Figure 9, **AL**1 is the code for the **AL**gebra rule that any number (measure) is equal to itself and **LA**11 is the code for the definition of a midpoint in a **L**ines and **A**ngles group of rules.

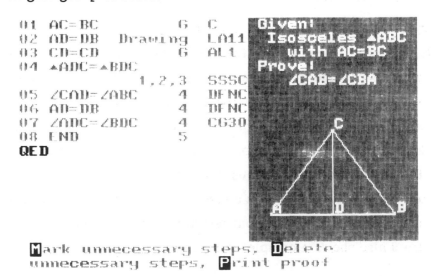

A Completed Proof of <u>Pons Asinorum,</u> with Extra Steps

Figure 9

Figure 10 illustrates one of the most important features of Proofchecker. Any correct step is accepted, and hence any of a variety of proofs can be constructed, as indicated by the two additional proofs of <u>Pons Asinorum</u> embedded in the steps shown in Figure 10.

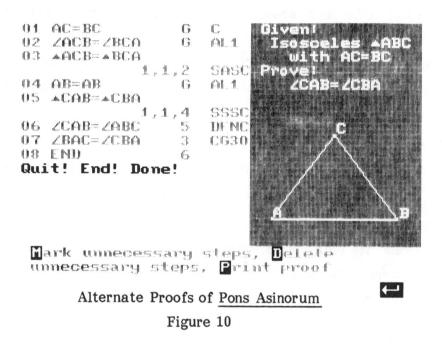

Alternate Proofs of <u>Pons Asinorum</u>

Figure 10

Proofchecker provides feedback on every step as it is entered and does not permit a user to continue until a given step is correct. Proofchecker does not attempt verification that users have completed a proof until they enter END in the Claim space and give a Reference to the step they believe is equivalent to what was to have been proved. If indeed the proof is valid, Proofchecker responds with any one of several variations on QED. In the example shown here, the user could also have used step 7 as reference in the END statement. DFNC (step 5) and CG30 (step 7) are alternate codes for the definition of congruent triangles on the Reference Card.

Figure 10 also indicates the existence of three additional options for users. Users can see their completed proof with superfluous steps marked, can have the computer delete those unnecessary steps, and can have their completed proof printed as written or with unnecessary steps marked or with unnecessary steps deleted. The "print" option is gratifying to users — they have tangible evidence of

619

their accomplishments. It should also make life easier for teachers. Neat printouts of certified correct proofs should considerably ease the task of keeping track of student progress.

The proof units in the Geometry 2 course cover the standard content of a proof-oriented U.S. high school plane geometry course: basic line and angle propositions, parallel and perpendicular lines, congruent triangles, similar triangles, quadrilaterals, and circles. Early proofs are done with fill-in-the-blanks, with on-screen tutorial comments. Most other proofs are done with Proofchecker.

The current version of Proofchecker can deal only with direct proofs, and its algebra checking capacities are still inadequate. Fill-in-the-blanks with checked responses is used for proofs that Proofchecker cannot check.

Unlike GeoDraw, Proofchecker is a moderately difficult program to use. Like GeoDraw, its entry requirements amount to yet another logical system for students to learn along with the plane geometry system. It seems to me that using such programs as these and learning to exploit their full power may teach as much about system building as the geometry subject matter itself. Axiomatic systems are, after all, essentially systems with definitely stated constraints and capabilities, just like these microcomputer-based geometry tools.

Summary

I have outlined here some of the problems of high school geometry courses as we see them in the United States, and especially of the proof oriented course that is standard fare for about half of our students. I have given in some detail an account of microcomputer-based materials created to help users gain excellent geometric intuition and to learn to do formal synthetic proofs. Those materials depend on two basic programs, GeoDraw and Proofchecker, designed by the development team and implemented by fine programmers to work within the constraints of a microcomputer with 128K of random access memory and one diskette drive.

Such programs as these should open up fine new research opportunities. Instructional input can be more uniform, more predictable, and more easily monitored. It will no doubt be possible now to write programs that keep much more detailed records of user responses.

Such programs require the collaboration of subject matter curriculum specialists and excellent programmers, each willing to try to understand the particular demands of the other. The time, human resources, and financial investment needed should not be underestimated as more such programs are attempted.

Geometry 1 and Geometry 2 with Proofchecker and GeoDraw seem to me to provide excellent existence proofs for powerful programs for mathematics instruction playable on existing and relatively inexpensive microcomputers. Other relatively comprehensive instruction programs in school mathematics, language arts, and reading were produced at WICAT at the same time as these programs, and are now available from IBM. Overall they should encourage production of software that goes well beyond the useful but relatively limited special purpose teaching programs that have been available for some time now and should move us to consider again how computers can best be used in teaching.

References

Anderson, J.R., C.F. Boyle, and B.J. Reiser. 1985. "Intelligent Tutoring Systems." Science, 228, 456–462.

Bell, M.S. 1986. Geometry One: Foundations. Geometry Two: Proofs and Extensions. Diskettes, Student Guides, Teacher Guides. Atlanta, GA: IBM Educational Systems.

Birkhoff, G.D. and R. Beatley. 1940. Basic Geometry. Chicago: Scott, Foresman and Company.

Crown, W. 1985. Math Concepts, Levels P-IV. Diskettes, Teacher Guides. Atlanta, GA: IBM Educational Systems.

Kutuzov, B.V. 1960. "Geometry." Studies in Mathematics 4. Stanford, CA: The School Mathematics Study Group.

Moise, E., et al. 1960. Geometry. Stanford, CA: The School Mathematics Study Group.

Pyshkalo, A.M. 1968. Geometry in Grades 1-4, 2nd ed. Moscow: Prosveshchenie Publishing House.

Senk, S.L. 1986. "How Well Do Students Write Mathematical Proofs?" In The Mathematics Teacher 79: 448–456.

Senk, S.L. 1983. "Proof Writing Achievement and Van Hiele Levels Among Secondary School Geometry Students." Doctoral dissertation, University of Chicago.

Usiskin, Z. 1982. "Van Hiele Levels and Achievement in Secondary School Geometry Project." Final report of the Cognitive Development and Achievement in Secondary School Geometry Project. Department of Education, University of Chicago.

Van Hiele, P.M. and D. Van Hiele-Geldof. 1958. "A Method of Initiation into Geometry at Secondary School." In Report on Methods of Initiation into Geometry, H. Freudenthal, ed. Groningen, Netherlands: J.B. Wolters.

Wirszup, I. 1976. "Breakthroughs in the Psychology of Learning and Teaching Geometry." In Space and Geometry. Columbus, OH: ERIC/SMEAC.

THE GEOMETRIC SUPPOSER:
Using Microcomputers to Restore Invention to the Learning of Mathematics*

Judah L. Schwartz

Massachusetts Institute of Technology and

Harvard Graduate School of Education

Michal Yerushalmy

University of Haifa

What Problem Was THE GEOMETRIC SUPPOSER Designed to Solve?

There are two lines of reasoning that led to the development of THE GEOMETRIC SUPPOSER. The first centers on the almost total absence of the making of conjectures by students (or teachers, for that matter) in the teaching of mathematics.

There is something odd about the way we teach mathematics. We teach it as if assuming our students will themselves never have occasion to make new mathematics.

We don't teach language that way. If we did, we would never require students to write an original piece of prose or poetry. We would simply require them to recognize and appreciate great pieces of language of the past, the literary equivalents of the Pythagorean theorem and the Law of Cosines.

The fact remains that the nature of mathematics instruction is such that when a teacher assigns a theorem to prove, the student ordinarily assumes that the theorem is true and that a proof can be found. This constitutes a kind of satire on the nature of mathematical thinking and the way new mathematics is made. The central activity in the making of new mathematics lies in making and testing conjectures. THE GEOMETRIC SUPPOSER is designed to help the student become a potent and nimble conjecture maker.

*An earlier version of this paper was presented at the Second International Conference on Thinking at Harvard in August 1984, the Proceedings of which are published by Lawrence Erlbaum Associates, Inc. 1986.

To make conjectures in geometry, clearly it helps to know how to construct and manipulate geometrical "objects" so as to be able to explore the relationships that do (or do not hold) among these "objects". It is in the exploration of relationships that may or may not hold between geometrical "objects" that we are led to the second line of reasoning that underlies the development of THE GEOMETRIC SUPPOSER.

We do not have a reasonably accessible general notation scheme for geometric constructs in the way that we do for algebraic constructs. The kth term in a sequence represents any term in the sequence. But can one draw a triangle that can represent any triangle without the particularity of the drawn triangle intruding? As a consequence of this difference between geometry and other branches of mathematics, in geometry we are obliged to use representations of particular cases and infer from the particular to the general. That this procedure is not totally satisfactory is amply attested to by the fact that everyone has had the experience of being fooled by a diagram.

Because we do not have a general notation scheme for geometric constructs nor, for that matter, for images in general, exploring conjectures in geometry becomes a matter of exploring a sequence of particular instances in the hope of educing generality. Normally, this would be a dismal prospect. With THE GEOMETRIC SUPPOSER, however, the consequences of one's constructions may be explored across an ensemble of equivalent diagrams, thereby reducing (although not eliminating) the dependence on the particular. For example, the reader may remember the remarkable fact that the line segments joining the midpoints of adjacent sides of any quadrilateral form a parallelogram. Suppose this is discovered to be true in a particular case, say the isosceles trapezoid with base angles of 45° each. What would be a convincing demonstration that it is true for other isosceles trapezoids, indeed true for any trapezoid and in fact true for any quadrilateral? Simply carrying out a large number of straightedge and compass constructions by hand, assuming one is willing to undertake the task, would prove nothing, for there would always be another case to consider.

In the normal course of events, it is not feasible to ask students (or anyone else) to do a great deal of construction for at least two reasons. First, accurate constructions are difficult to make, and second, constructions are just difficult to do. Nor is it reasonable to expect people to be willing to repeat constructions over and over again to generate a repertoire of cases on which to base a conjecture.

THE GEOMETRIC SUPPOSER allows users to make any construction they wish on a primitive structure (e.g., point and line, triangle, quadrilateral) of their choice. The program records the construction as a procedure which can then be executed with any exemplar of the primitive structure as argument. As a result, it becomes a simple matter to explore the consequences of a given construction on a given structure to see if they depend on some particular and peculiar property of that structure or if a more general result may be obtained. In the examples cited above, the program would remember

label midpoint of AB with the letter E

label midpoint of BC with the letter F

label midpoint of CD with the letter G

label midpoint of DA with the letter H

draw line segment EF

draw line segment FG

draw line segment GH

draw line segment HE

as a procedure that can be carried out on any quadrilateral ABCD (Figure 1). The user of the program may now proceed to execute this procedure on a wide variety of quadrilaterals and in each instance ascertain whether the newly formed quadrilateral EFGH is a parallelogram.

Among the possible constructions within THE GEOMETRIC SUPPOSER are the construction of triangles and quadrilaterals as well as the drawing of segments, medians, altitudes, parallels, perpendiculars, perpendicular bisectors, angle bisectors, and inscribed and circumscribed circles. In addition, the user can measure lengths, angles, areas, and distances as well as arithmetic combinations of these measures (e.g., sum of two angles, product of two lengths, ratio of two areas).

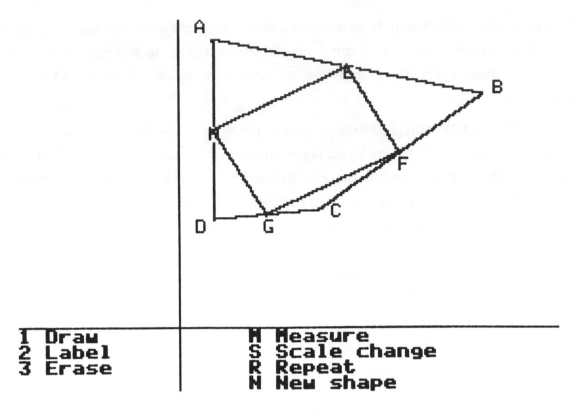

Figure 1

A program that provides extensive construction facilities to a student and/or teacher is clearly useful in the learning of geometry in that it makes it easy to produce accurate constructions without chancing either the intellectual or the pedagogic nature of the enterprise. The power of THE GEOMETRIC SUPPOSER lies in its ability to remember and repeat constructions.

As described above, any construction on a primitive structure such as a quadrilateral made with THE GEOMETRIC SUPPOSER may be repeated on a new quadrilateral of the user's construction, a previously used quadrilateral, or a random parallelogram, trapezoid, rhombus, rectangle, kite, or quadrilateral that may be inscribed in or circumscribe a circle.

Needless to say, neither possibility nor plausibility constitutes proof. Proof remains central to both the creating and the learning of mathematics. But conjecture, in this instance with the aid of THE GEOMETRIC SUPPOSER as intellectual amplifier, can assume its proper role as the key activity in the making of mathematics.

By the same token, just because THE SUPPOSER makes it possible to produce constructions easily and quickly does not mean students should never have to use a straightedge and compass to bisect angles or erect perpendiculars. We believe that actual constructions with straightedge and compass are important for students. Indeed, it is only after some physical manipulations with these tools that the power of THE SUPPOSER becomes apparent.

What Can Students (of All Ages) Discover with THE GEOMETRIC SUPPOSER?

Let us start by seeing how THE GEOMETRIC SUPPOSER helps in introducing conjecture and invention into the teaching and learning of mathematics.

Using this facility, beginning geometry students have discovered that if a median on a triangle is drawn, it wil bisect the area of the triangle, and that this seems to be true if they repeat the construction with triangles of all shapes and sizes. The plausibility of this conjecture having been established, they are in a position to devise a proof with some conviction about just what it is they are trying to prove.

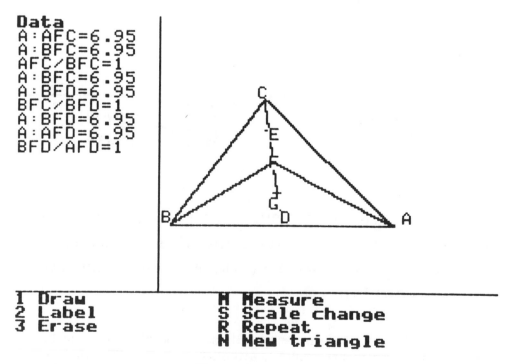

Figure 2a

Students have discovered for themselves that a midsegment in a triangle is parallel to the third side of the triangle, and that the three midsegments of a triangle partition it into four triangles congruent to one another and similar to the original triangle. In addition, they have made discoveries about the ratios of perimeters and areas frequently presented as theorems.

Perhaps most striking of all is that a growing number of students are discovering new theorems. For example, consider the problem of dividing any triangle into n triangles of equal area. A tenth-grade student who hitherto has not had a distinguished mathematical career in school has just produced a novel solution to this problem (Figures 2a and 2b).

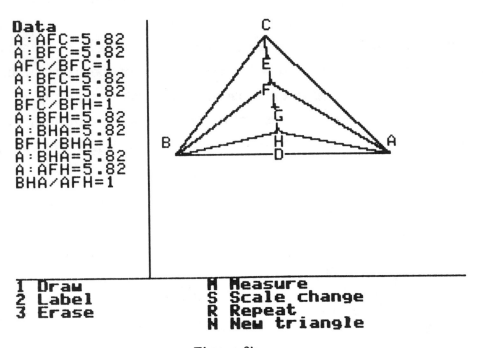

Figure 2b

The exploration that THE SUPPOSER makes possible is not appropriate to high school students alone. What follows is an example of a conjecture the authors have found. It is part of a family of conjectures about quadrilaterals for which we do not yet have a proof.

It is well known that in general it is not possible to inscribe a circle in an arbitrary quadrilateral ABCD, although it is possible to inscribe a circle in any triangle. Consider, then, the triangles ABC, BCD, CDA, DAB that are defined by the vertices of an arbitrary quadrilateral ABCD. These are the triangles formed by the sides of the quadrilateral and its diagonals. Suppose one were to inscribe circles in these triangles and label the centers E, F, G and H respectively. The following figures (Figures 3a and 3b) suggest the quadrilateral EFGH might bear an interesting relationship to quadrilateral ABCD.

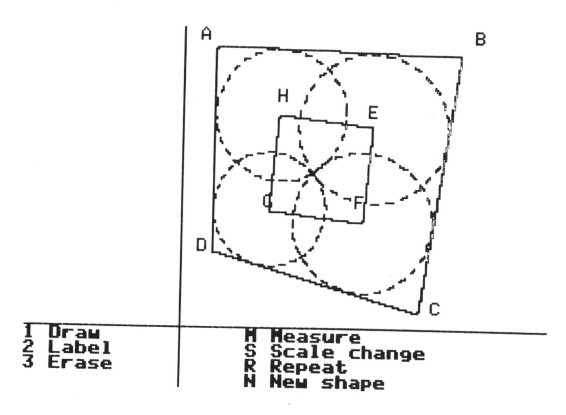

Figure 3a

In devising new mathematics in this way, students of all ages and degrees of intellectual engagement can come to understand that making new mathematics is a continuing and exciting enterprise, and one that with the right tools is accessible to them.

THE GEOMETRIC SUPPOSER and Formal Proof

It is likely that many readers at this point will see THE SUPPOSER as a mixed blessing. Specifically, one may fear that a tool this useful may weaken the already feeble appreciation of formal proof many students have. After all, why bother to prove something if you can try out your assertion as often and as easily as you like?

One of the delightful side effects of THE SUPPOSER is the realization on the part of students that demonstration, even with a very large number of cases, is not tantamount to proof. Another anecdote supports this interpretation of the evolution of student attitudes. During the 1984–85 school year, one of the authors (MY) noticed a distinct shift in the nature of the students' requests when she came to visit the classroom. This shift took place over a four-to-six-week period. At the beginning, students would ask her to "come look at my conjecture." Toward the end they would say "come look at my proof!" It would seem that rather than do away with the need for proof, using THE SUPPOSER seems to sharpen students' perception of the need for proof.

Some Details of the Program

To give the reader a clearer idea of the nature of the program, we describe here the capability of the version of THE SUPPOSER that works with triangles as primitive structure. A similar set of capabilities obtains in the case of the quadrilateral, circle, and preSUPPOSER versions of the program.

To begin with, THE SUPPOSER allows users to construct a triangle of their own or choose a random right, acute, obtuse, isosceles, or equilateral triangle as the shape on which they will make further constructions. If users opt to construct their own triangle, they are asked how they wish to specify their triangle, i.e., by SIDE-SIDE-SIDE, or SIDE-ANGLE-SIDE, or ANGLE-SIDE-ANGLE. They then give the values of these quantities after which THE SUPPOSER constructs their triangle just as they would with straightedge and compass.

THE GEOMETRIC SUPPOSER allows the user to draw upon a repertoire of primitive constructions, each of which is classically possible with straightedge and compass. These include:

- a line segment between any two labeled points on the screen
- a circle, either circumscribed about or inscribed in a triangle of the user's choice, or centered on any labeled point on the screen with radius a multiple of any line segment (drawn or undrawn) on the screen

630

- a median in any triangle, from any vertex
- an altitude in any triangle, from any vertex
- through any labeled point on the screen a parallel to any line segment (drawn or undrawn) with length a multiple of any line segment (drawn or undrawn) on the screen
- a perpendicular through any labeled point on the screen to any line segment (drawn or undrawn) with length a multiple of any line segment (drawn or undrawn) on the screen
- the bisector of any angle on the screen
- the perpendicular bisector of any segment on the screen
- a specified midsegment in any triangle on the screen
- the extension of any segment on the screen from either end, by an amount a multiple of any line segment (drawn or undrawn) on the screen.

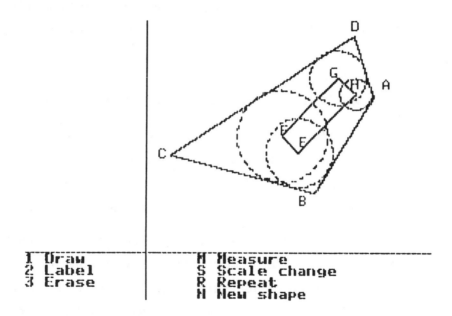

Figure 3b

THE GEOMETRIC SUPPOSER also allows the user to label the intersection of any two line segments, subdivide any line segment into as many as 6 equal segments, reflect a point or line segment through a line, and place a point at random on a line segment, either inside or outside a specified triangle.

The measurement functions of THE SUPPOSER include the ability to measure lengths, perimeters, areas, angles, distances from points to lines, and distances between lines. In addition, the sums, differences, products, or ratios of any two similar quantities can be measured directly.

In THE SUPPOSER, the user has the capability of moving a cursor while monitoring the changing values of up to three angles, lengths, or distances. Because of this facility, it is possible to do such things as locate the center of circumscribing or inscribed circles, erect equilateral triangles on line segments, and so on.

Finally, THE SUPPOSER repeats the constructions that are on the screen on another exemplar of the primitive shape (here, the triangle). It is by virtue of this feature that THE SUPPOSER provokes both students and teachers into making conjectures. It is very tempting to explore whether some newly discovered property of one's construction is true in general or is only an artifact of the specific case.

How Has THE GEOMETRIC SUPPOSER Been Used?

It is hard to say with whom THE GEOMETRIC SUPPOSER is best used. At the time of this writing, several middle school classes as well as a dozen or so high school classes, both public and private, have used various versions of the program, along with at least one group of students in a vocational-technical high school.

The middle school classes are using the preSUPPOSER version to explore a set of geometric ideas that would otherwise not be accessible to them. In one of these middle school classes, the teacher posed the problem of discovering exactly what a right triangle is. Shortly afterwards two eager children were pleased to report that, "A right triangle has three different angles!" This conclusion is understandable; even though there are an infinite number of distinct right triangles, the statement is true for all but one type of right triangle, the isosceles right triangle. The teacher showed the children an acute triangle that had three different angles and asked them whether their definition still held. There then ensued a discussion as to how many properties of an entity have to be part of its definition for the definition to be adequate.

The high school classes have used the triangles version, and to a lesser extent, the quadrilateral version of THE SUPPOSER to approach the traditional content of geometry courses along a variety of paths. Some of these constitute rather modest departures from ordinary instruction in this area, while others are truly revolutionary in the way they seem to draw mathematics out of the students. Clearly a class in which students are sufficiently engaged in the enterprise to come to class with conjectures they are willing to sign and post for all to see is a class in which mathematics is being cast in an unusual and exciting role.

In one class (otherwise of average ability) in which THE SUPPOSER was used, the students were given the following quiz without any prior acquaintance with midsegments:

> A midsegment in a triangle is the line segment joining the midpoints of two sides of the triangle. Make a list of all the things you believe to be true about midsegments and triangles.

By the end of the hour the class had concluded that, in a triangle:

- The midsegments are parallel to the respective opposite sides of the triangle
- the midsegments partition the original triangle into four congruent triangles
- each of these congruent triangles is similar to the original triangle
- the area of one of these triangles is equal to one-fourth the area of the original triangle
- the perimeter of one of these triangles is equal to one-half the perimeter of the original triangle

and so on.

Another example of student-discovered geometry is the conjecture that in a triangle ABC, the length of the median AD drawn from A is less than the average of the lengths of the other two sides AB and AC. It is indeed rare for high school geometry students to engage the subtlety of geometric inequality.

One final example of student-discovered geometry: Consider the square ABCD and the points that trisect the sides of the square. The line segments joining appropriately corresponding "trisective" points form a new square whose area is 5/9 the area of the original square (Figure 4). The students put forward the conjecture, proved it, generalized it to arbitrary subdivision of the sides of the square, and were last seen trying to generalize the theorem to the case of a parallelogram whose sides have been so subdivided.

THE SUPPOSER has also been used in settings not ordinarily those in which traditional geometry courses are taught. In one vocational and technical high school, students have used THE GEOMETRIC SUPPOSER to augment their studies in drafting and design and to help develop a stronger set of spatial reasoning techniques.

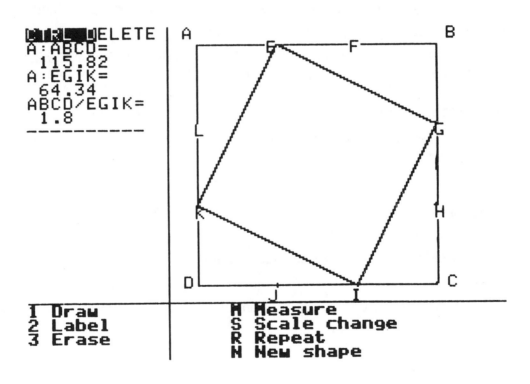

Figure 4

Some of the settings in which THE GEOMETRIC SUPPOSER has been used have been laboratory settings with a collection of microcomputers or terminals available for class use as well as classrooms with only a single microcomputer. Needless to say, different teachers have evolved different styles of using THE GEOMETRIC SUPPOSER in their teaching because of constraints of hardware availability. It does not seem that there is any single most desirable arrangement, nor should there be.

Afterword

The first version of THE GEOMETRIC SUPPOSER was developed during the 1981-82 school year. The early trials were exciting, challenging, and frustrating. Although we believed deeply that we were engaged in an important endeavor in mathematics education, we were impatient with our own inability to clarify issues and define things as crisply as we would have liked. At first, it was hard to explain to teachers what we were up to, and why we thought it important.

In the intervening period we have gained much experience working with a wide variety of teachers and classroom settings. We are now more persuaded than ever that THE GEOMETRIC SUPPOSER offers the mathematics teacher a new path to the teaching of geometry. More generally, we believe that the larger idea that underlies this program, i.e., that students can make their own mathematics and that microcomputers can help them to do so, can change the way mathematics is taught and learned at all levels.

Moreover, we believe that the kind of use of the microcomputer that THE GEOMETRIC SUPPOSER exemplifies can be extended beyond Euclidean plane geometry. Why not make such programs that work on constant-curvature non-Euclidean manifolds? Why restrict the dimensionality of the spaces in which such programs operate?

It is likely that this approach to the use of microcomputers in mathematics education can be extended beyond the study of geometry. The making and exploring of conjecture is clearly desirable throughout mathematics. For that matter, it is quite likely that the approach we have taken with THE SUPPOSER can be extended to other areas of intellectual inquiry in which it would be helpful to be able to explore the consequences of one's conjectures and investigate the domain of their application. Our hope is that, ultimately, THE GEOMETRIC SUPPOSER will help engage the intellectual attention of a new generation of students.

References

Schwartz, J.L. & M. Yerushalmy. In press. <u>The College Mathematics Journal.</u>

Technology-Supported Learning

Alan Hoffer

University of Oregon and University of Chicago

Introduction

The two major topics of this conference are the development of an applications-oriented mathematics curriculum, and the exploration of innovative instructional strategies. In these days of rapid technological achievements, it is essential for curriculum innovation that we consider ways in which we can best utilize new and increasingly available technology. This is especially true as we develop a curriculum that emphasizes applications. With calculators and computers as tools our students are freed from the shackles of tedious computation as well as the many hours of exercises that we deemed necessary to master the skills to apply to even the simplest problem: the so-called "textbook problem" that usually has as its solution a small positive integer. The purpose of this paper is to touch on one aspect of using technology in an applications-oriented curriculum: namely, to describe a phenomenon that, while not new, is entering a new era with the increased use of calculators and computers in mathematics education. There are descriptions here of (1) some of the essential features of technology-supported learning environments; (2) crucial aspects of the documentation that enables such environments to function; and (3) a particular computer environment for algebra.

TSL Environments

I use the word "technology" in a broad sense to include not only mechanical and electrical equipment, but also nonphysical methods, tools and utilities. A technology is simply a procedure for accomplishing a family of tasks. An activity is supported by a technology if the technology aids users by performing tasks that require a low level of thought or that might be time-consuming, and thereby enables users to extend or increase their level of thought and to apply their energies to more important tasks.

By technology-supported learning (TSL) I mean specific and general learning environments that are supported by technology. Examples include:

- Numeration. Numeration systems offer us procedures to encode, record, and manipulate numerical information. A symbol such as 578 frees the mind for tasks other than collecting hatch marks of bundles of tens and hundreds. We manipulate these symbols to learn algorithms and perform calculations.

- Grids. These are extremely useful for teaching and learning about several topics such as fractions, areas, similar shapes, and slopes.

- Abacus. An abacus is a useful technology not only for shopkeepers but also for children who are learning about numeration and understanding how algorithms work.

- Slide rule. Some years ago the slide rule was a popular technology used not only by engineers but also by students of elementary functions who were learning about logarithms.

- Heuristics. Champions of so-called problem-solving activities advocate teaching heuristics such as those that were developed by George Polya (1957), thereby providing users with procedures for attacking problems.

As noted above, an essential feature of a TSL is the capability to aid users, to free them from menial tasks. The more useful TSLs also provide users with a structure through which they can accomplish other actions related to the family of tasks that gave birth to the TSL.

Calculators and computers now offer a multitude of TSLs that show promise for exciting possibilities, especially in mathematics education. I do not distinguish among the various capabilities of calculators, such as solar versus battery power, memory types, scientific functions, programmability, and so forth. Nor do I take account of the multitude of differences between computers, although each additional capability of course enlarges the families of tasks that the TSL can accomplish. Instead, I use calculators and computers generally as a vehicle to focus on some important aspects of TSLs per se.

LOGO is a prototype TSL for both children and adults. A specific LOGO TSL system consists of a computer program, which usually translates a surface language to LISP, an early Artificial Intelligence language that has list and graphics capabilities. In addition to the program, the system contains an instruction procedure, usually a written manual, that enables the user to command the computer to perform various graphics and list processing tasks (e.g., Fuerzeig 1969, Ross 1983). I use LOGO as a generic term to include all the various dialects of the language. What makes LOGO a prototype of TSLs is that the computer program is designed so that the users are able to give simple instructions or commands for functions that the program has the computer execute.

Even the finest LOGO program has little value for those who do not know how to use it, just as ingenious languages such as Modula 2, Prolog, and Smalltalk are of little value to users who do not know their capabilities. The transition from lack of knowledge of the capabilities and method of interaction of a TSL is provided by the documentation.

Documentation Schemes for TSL

A documentation scheme is simply a set of instructions in how to begin to use a TSL. Popular documentations that are now available include:

- Textbooks. These provide information about the technology and (supposedly) explain how to use the procedures. For example, elementary school mathematics texts explain how to use numeration schemes to code and encode numbers in base ten, and possibly other bases.

- Owners' manuals. These explain how to set up the technology, turn it on, and use it to perform certain tasks. For example, there are manuals on how to use computer programs such as word processor and database systems, as well as on how to operate a power saw or a video cassette recorder.

It is unlikely that a documentation scheme provides users with information about all the capabilities of the TSL; but a documentation scheme should provide users with considerably more information than one finds today in most manuals for computer utilities.

The features of a documentation scheme for calculators and computers are outlined below (see also Hoffer WPa, WPb). These are aspects of documentation schemes that are essential for some degree of reliability, so that users will indeed be able to engage the TSL and use it for much more than just the initial stages or levels of activity in the family of activities for which the TSL was constructed. The following is recommended as a desirable sequence for documentation schemes and is in many ways one that is suggestive for instructional use more generally (see also van Hiele 1959, Wirszup 1976, Hoffer 1983).

1. Acknowledge. Users are shown some problems, questions or topics that might occur in the subject under discussion. This serves as an introduction to an overview of the subject and acknowledges that there are indeed reasons why the subject should be studied. Users are introduced to some of the vocabulary, symbols and concepts of the subject as they become alert to its rudiments.

2. (optional) Diagnose. Users are asked to solve some problems and to answer questions as appropriate. If the users are able to answer the question, they might be advised to skip to certain other places in the documentation; otherwise, the results of the diagnosis prescribe a course of study for the user.

3. Direct. Users are shown specific actions that the documentation directs them to repeat on items very similar to the ones that are demonstrated for them. These are essentially separated into two types, depending on whether the specific actions require one or more than one step:

   a. Imitate mode. The documentation shows a one-step example. This is followed immediately by a very similar exercise that users are asked to complete.

   b. Guide mode. The documentation shows a multi-step example. This is followed immediately by a very similar exercise that users are prompted to complete one step at a time.

There should be a sufficient number of examples with accompanying imitations to enable users to enact the procedures comfortably. Experiments are likely needed to guide the authors of documentations as to how many examples are sufficient. At this stage in the documentation, and as the users learn about the subject, it is desirable to give the imitate mode preference over the guide mode (for most

640

users), but there are topics for which multi–step actions are necessary and cannot be avoided in a natural way. More complete documentations provide several examples at this stage along with the companion directed tasks for users. These may be all imitated, all guided, or a combination. The essential aspect of this stage is that users are freed from making involved decisions. They need only imitate what is shown to them. They are concerned with performing actions, rather than making decisions.

4. Explain. Users now receive a verbal description of the skills, concepts or structures that were treated in the previous directed stage of the documentation. This is a formalization step, during which users come to know the structure of the subject at a particular stage in its development. In an exchange between a teacher and a class, this explicatory phase (van Hiele 1959, Hoffer 1983) can be developed so that the students come to recognize and verbalize the structures and algorithms themselves, with only gentle guidance from the teacher. At present this can be accomplished with computers only in a limited way, but as artificial intelligence capability increases we will see such exchanges with machines become increasingly possible (Winston 1984).

5. Practice. Users now have opportunities to practice the skills, vocabulary, simple applications, and the like that are similar to — or at most only slight extensions of — those that occurred in the previous stages. It is essential for the documentation to allow users to develop instinctive reflexes concerning the skills, vocabulary, and so on that are found at this point in the subject. Think of this as a mastery stage: users can perform tasks and operate with vocabulary and symbols automatically.

6. Apply. Users are now able to decide for themselves when and how to use the capabilities of the TSL that they have experienced up to this point in the development. The documentation provides them with examples of questions and problems, along with worked examples as appropriate, that they can complete by applying what has already been studied. As in the directed stage, there are two modes that are

available to users, depending on whether they want to perform a task as a sequence of smaller tasks or do it instantaneously as a single task.

a. Control mode. Users decide which sequence of tasks to perform, and they command the TSL to perform the tasks one at a time. After each step users evaluate the result of the action and then either command the TSL to perform the next task in the sequence or modify the sequence according to the result of the previous action.

b. Miracle mode. Users decide on an action that completes a task for them and then command the TSL to perform the action. Usually this is an action composed of several smaller tasks that users could command the TSL to perform, but the users wish to avoid witnessing the intermediate steps.

It is important at this stage in the documentation that users have the opportunity to make decisions themselves about what tasks to command the TSL to perform for them. If it is in the nature of the subject, examples that require the control mode as well as examples that allow the miracle mode should be offered to users.

7. Extend. Users are here given the opportunity, often by way of suggested examples or recommendations, to investigate their own questions, problems and patterns by using the capabilities of the TSL. The documentation might recall a task that users performed earlier and then suggest that they investigate what would happen if certain aspects of the task were changed. In this way users not only make decisions about what steps or sequences of steps to perform in order to accomplish certain tasks; they also make decisions for themselves as to what kind of investigations they wish to pursue, and make discoveries on their own as they integrate topics and internalize the subject at hand.

A TSL for Elementary Algebra

What follows is a summary description of a TSL for algebra (Hoffer WPa). The program runs on a microcomputer and could be available for individuals at home or at work as well as in a school setting. The TSL consists of two parts: a computer program that we refer to as a utility (an algebra utility) and the documentation as outlined above that serves as a doorway for users to learn how to begin working with the utility and how to bring it under their control.

The algebra utility actually does algebra; that is, it is capable of performing all of the operations and transformations that arise in an elementary algebra setting. Users learn in the documentation what functions the utility can perform, and they learn how to command the utility to perform these tasks.

We take as an example the task of simplifying polynomials in several variables. After first explaining why it is useful to be able to perform such tasks, the documentation moves into the direct instruction phase, which includes an imitation sequence like the following.

- Notice how the like terms are combined to simplify the expression.
  $$2xy + 3y - 4x + 6xy - 5x - 8y + 2w^3 = 2w^3 + 8xy - 9x - 5y$$
- Your turn. Combine all the like terms to simplify this expression.
  $$4xy + 2x - 6y + 9xy - 3y + 4x + 3w^4 =$$

Such an imitation sequence is preceded by preliminary examples as appropriate and is followed by more involved topics. Instead of the imitate mode that is exemplified here, the documentation could have offered the exercise in the guide mode where like terms are combined step by step.

After the direct instruction phase, the documentation provides a verbal description of the main ideas of the topic in this case, what it means to simplify the polynomial expressions, and the rules for doing so. This is then followed by a set of practice exercises.

Now the documentation moves to the application phase, where users decide for themselves what commands to give to the utility. There may be examples that lead toward a generalization, suggestions for topics to explore, or simply statements of problems for the users to investigate.

Continuing the sample task of simplifying polynomials, we offer two examples of the application phase.

- Find the zeros of this polynomial.
  $$(2x - 495)^2 - (x^2 - 3,226 + 520,450)$$

Users now instruct the utility to perform transformations on the polynomial. For example, in the control mode the following instructions can be given (shown on the left below), with the corresponding output (shown on the right). The instructions can be entered into the computer by special key functions, by selection from a menu, or by typing in words or abbreviations.

Command	Output
	$(2x - 495)^2 - (x^2 - 3226x + 520450)$
EXPAND $\rightarrow$	$4x^2 - 1980x + 245025 - x^2 + 3226x - 520450$
COMBINE $\rightarrow$	$3x^2 + 1246x - 275425$
FACTOR $\rightarrow$	$(3x - 479)(x + 575)$
$3x - 479$, SOLVE $\rightarrow$	$x = 479/3$
$x + 575$, SOLVE $\rightarrow$	$x = -575$.

Users could, of course, opt for the miracle mode, where they could obtain the solution of the original polynomial with the input SOLVE, and hence bypass all the intermediate steps. Such an option would be especially appropriate in an applications setting where the main thrust is to obtain information about an applied problem, rather than to look for a pattern about solution sets of polynomials.

The next example shows how the utility can be used to obtain information from which the users can observe patterns and generalizations.

- Use the program to find the powers of these binomials. Make up some more of your own. Look for a pattern and some surprising relationships.

  $(x + y)^2 =$

  $(x + y)^3 =$

  $(x + y)^4 =$

  ....

- Predict what result you would obtain if you expanded $(x + y)^9$. Check your prediction with the computer. Investigate powers of other polynomials such as $(x + 2y)^n$ and $(x + y + z)^n$. Make up some of your own.

The last example hints at how the extension phase of the documentation might look. The documentation could contain samples for the users to work out and analyze and from which generalizations might result, or there could be some suggested investigations for the users. At this stage in the development the users have available to them all of the functions in the utility that they have studied, and can experiment with them at will. Hoffer (WPa) gives a discussion about input, output, and screen display for such an algebra TSL. It is best to have the algebraic expressions appear on the screen in "pretty print" as they do in a book, i.e., with superscripts and subscripts in correct positions, exponents without up arrows, the vinculum for rational expressions instead of the slash, and so forth.

Summary

The increased availability of computers for educational purposes enables us to take a new look at mathematics pedagogy in the framework of past curriculum development efforts and research observations. This is especially true as we consider how to infuse innovative instructional strategies into an applications-oriented curriculum. Calculators and computers now provide powerful utilities that, among other things, perform menial or relatively low level processes and free people to focus on deep concepts and to apply their knowledge to genuinely interesting problems. A technology-supported learning environment or technology-supported system consists of a utility, such as a calculator or a powerful computer program, together with the documentation that first enables users to begin working with the utility; and then provides them with operations and transformations that they can command the technology to perform for them. There are some essential features of documentation schemes for TSLs that are suggestive for instructional purposes, as exemplified in an algebra TSL. Given the availability of such technology-supported systems, there arise deep educational questions concerning what skills we should continue to teach children. For example, what aspects of long division should we teach, now that calculators are available? What aspects of factoring polynomials should we teach, now that computer utilities are available to perform the task quickly and with large numbers? These questions must now be at the forefront of discussions in mathematics pedagogy.

References

Fuerzeig, W. 1969. <u>The LOGO Language</u>. Cambridge, Mass.: Bolt Beranek and Newman.

Hoffer, A. 1983. "Van Hiele-Based Research." In <u>Acquisition of Mathematics Concepts and Processes,</u> R. Lesh and M. Landau, eds. New York: Academic Press.

Hoffer, A. WPa. <u>A TSL for Algebra</u>. UCSMP Working Paper.

Hoffer, A. WPb. <u>A TSL for Geometry</u>. UCSMP Working Paper.

Hoffer, A. WPc. <u>Calculators in a Technology-Supported Learning Environment.</u> UCSMP Working Paper.

Hoffer, A. WPd. <u>Documentation Schemes for Technology-Supported Learning.</u> UCSMP Working Paper.

Hoffer, A. WPe. <u>Technology-Supported Learning Environments for Probability and Statistics.</u> UCSMP Working Paper.

Polya, G. 1957. <u>How to Solve It</u>. New York: Wiley.

Ross, P. 1983. <u>Introducing LOGO</u>. Reading, Mass.: Addison-Wesley.

van Hiele, P.M. 1959. "La pensée de l'enfant et la géometrie." <u>Bulletin de l'Association des Professeurs Mathématiques de l'Enseignement Public,</u> 198-205.

Winston, P. 1984. <u>Artificial Intelligence</u>. Reading, Mass.: Addison-Wesley.

Wirszup, I. 1976. "Breakthroughs in the Psychology of Learning and Teaching Geometry." In <u>Space and Geometry: Papers from a Research Workshop,</u> J. Martin, ed. Columbus, Ohio: ERIC/SMEAC.

Dienes Revisited: Multiple Embodiments in Computer Environments

Richard Lesh

WICAT and Northwestern University

Thomas Post

University of Minnesota

Merlyn Behr

Northern Illinois University

In the early 1960s, the names Cuisenaire, Dienes, Gottegno, Montessori, and Piaget had become fashionable in mathematics education theory development. Each was associated with a "mathematics laboratory" approach to instruction, based on activities for children using concrete materials. However, these theoretical perspectives had little impact on actual classroom practice.

This paper will describe significant ways in which computer-based instruction can encourage teachers and students to make greater use of activities with concrete materials, while at the same time providing a useful context in which to implement some of the best instructional strategies associated with mathematics laboratories — including some strategies which have never worked well using concrete materials.

A Closer Look at Dienes' Instructional Principles

Among the "mathematics laboratory" theorists, Dienes (1960) was the most specific in his recommendations for mathematics teachers. Dienes' perspective was based on the following four principles:

1. The constructive principle: Dienes claimed that mathematical ideas must be constructed. He treated mathematical ideas as abstract structural metaphors; that is, he believed the structure of a given mathematical idea cannot be abstracted from concrete objects, but instead must be abstracted from relational/operational/organizational systems that humans impose on sets of objects. For example, Dienes claimed that when his famous arithmetic blocks (see Figure 1) are used to teach the "regrouping structure" of our base-ten numeration system, children must first organize the blocks using an appropriate system of relations

and operations. Only after these organizational systems have been <u>constructed</u> can children use the materials as a model that embodies the underlying structure.

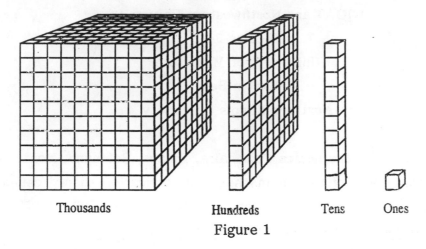

Thousands   Hundreds   Tens   Ones

Figure 1

Part of what Dienes meant when he claimed that mathematical ideas must be constructed is that when arithmetic blocks, for example, are used to teach the structure of our numeration system, the relevant systems must first be "read into" the set of materials in the form of concrete activities. Only after this "reading in" has taken place can the structure be "read out" as abstract relational/operational networks.

Let's consider an example other than arithmetic blocks. Figure 2 shows a diagram that Dienes used to describe a network of relations that his students constructed when working with material designed to help them learn about an algebraic finite group. Notice in Figure 2 that the mathematical structure is pictured not simply as a network of relations and operations; it is also a network which has a <u>pattern</u> (or structure) in which the whole is more than the sum of its parts. According to Dienes, it is the

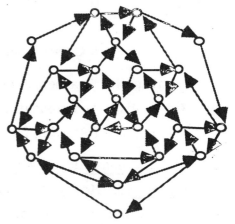

Figure 2

<u>pattern</u> itself (and properties of the pattern) that must be abstracted. So, in activities with blocks, it is not so much the blocks that the child must organize as it is his or her own activities on the blocks; and it is the <u>pattern</u> of the activities

648

that forms the basis for the abstraction. When individual activities cease to be treated as isolated actions and start to be treated as part of a systematic pattern of activities, the student begins to shift from playing with blocks to playing with mathematical structures.

Piaget furnished some of the best evidence to show that the first relational/operational/organizational systems that children use are often based on overt actions (e.g., ordering, combining, separating) performed on concrete objects. Yet, even at an early age, children's organizing and ordering activities are often carried out "in their heads" (often accompanied by spoken language) rather than through overt actions (Vygotsky, 1962) — and often are applied to pictures or diagrams rather than to concrete objects. In fact, when youngsters become too involved in the details of object manipulation, they often "lose sight of the forest because attention is focused on the trees" and fail to notice the patterns or structures that underly their own behaviors.

Using concrete materials does not guarantee the use of "concrete activities." In fact, using concrete materials may actually hinder conceptual development at the stages of learning where attention must shift from isolated actions toward systems of actions. Yet, when concrete materials have been used in instruction, more concern is often given to the "concreteness" of the materials than to the "activeness" of the activity — as though the abstraction were from the materials rather than from the structure that must be imposed on the materials.

The fact that whole mathematical structures have properties that go beyond those of constituent elements means that each individual child must construct these systems on a lower plane (e.g., using concrete objects) before they can be abstracted on a higher plane. Nonetheless, the role of the materials is simply to serve as a support for the student's mental activities, not to serve as the direct basis for abstraction.

Dienes' constructive principle implies two distinct corollaries: (1) mathematical relations and operations are considered to be abstracted from activities rather than from the materials, and (2) systems of activities must be constructed before they attain the status of mathematical structures.

2. The multiple embodiment principle: According to Dienes, mathematical

ideas cannot be abstracted from single isolated patterns (or models which "embody" these patterns) any more than simpler abstractions can occur from single instances. Instead, mathematical abstractions occur when students recognize structural similarities shared by several related models. For example, when base-ten blocks are used to teach arithmetic regrouping operations, Dienes claimed that it is not enough for students to work with a single model; they must also investigate "mappings" to other models, such as bundling sticks or an abacus. So, in laboratory forms of instruction, a primary goal is to help students recognize how patterns of relationships in one model correspond to patterns of relationships in another model, as illustrated in Figure 3.

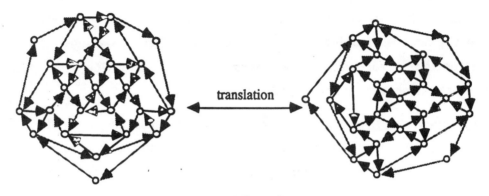

Figure 3

3. The dynamic principle: According to Dienes, the systems that must be abstracted from structurally related models are not simply "static patterns"; they are dynamic, that is, the most important features to be recognized have to do with the way transformations within one model correspond to related transformations in a second model. To reflect this dynamic aspect of mathematical structures, Figure 3 can be modified to include not only translations between models but also transformations within models, as shown in Figure 4.

In Figure 4, a transformation followed by a translation should produce the same result as a translation followed by a transformation; that is, the mathematical property of homomorphism should be true:

$$T(A*B) = T(A)*T(B)$$

4. The perceptual variability principle: Regardless of whether the material used to embody a given model is a set of concrete objects, graphics, written symbols, spoken language, or some other representational system, models always have some characteristics that the modeled system does not have — or conversely, they fail to have some characteristics that the modeled system does have. Consequently, to select a small number of especially appropriate models to embody a given system, the following characteristics should be used:

- Irrelevant characteristics should vary from one model to another so that these characteristics will be "washed out" of the resulting abstraction.
- Collectively, models should illustrate all of the most important structural characteristics of the modeled system.

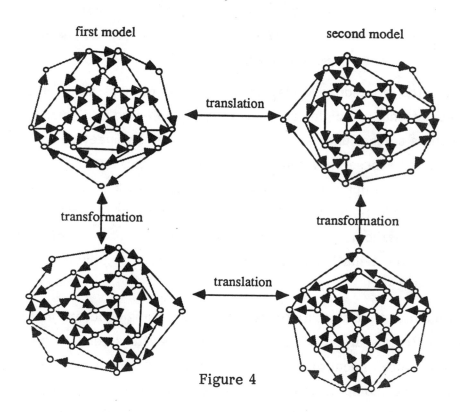

first model   second model

translation

transformation   transformation

translation

Figure 4

The Rare Use of Concrete Mathematical Laboratories Principles

Why have Dienes' mathematical laboratory principles made so little impact on actual classroom construction? Even when concrete materials like arithmetic blocks are used in mathematics instruction,

- They tend to be used only in classroom demonstrations in which students are passive observers rather than constructing participants (a violation of the <u>constructive</u> principle).

- They tend to be the <u>only</u> concrete model used, with attention being focused on the objects themselves rather than "mappings" (or translations) from one model to another (a violation of the <u>multiple embodiment principle</u>).
- They can be cumbersome. To satisfy Dienes' <u>dynamic principle</u>, students must not only investigate transformations within a single model, they must also investigate corresponding transformations within other models. Yet with concrete materials, such correspondences are time-consuming to act out and difficult to keep track of, even in well-rehearsed teacher-demonstrations. It is virtually impossible for students to create these demonstrations on their own and to benefit from them without guidance.

Often more concern is given to the "concreteness" of the materials than to the "activeness" of the activity — as though the abstraction were from the materials rather than the structure that must be imposed on the materials. The purpose of the materials is seldom clear.

Beyond the preceding kinds of difficulties, perhaps the main reason why Dienes' instructional principles are so rarely implemented in classroom instruction is that teachers seldom understand or accept Dienes' perspective about the nature of mathematics because they tend to view mathematics as simply a collection of isolated rules for manipulating symbols. Therefore, work with concrete objects is often viewed as "too baby-ish," when in fact the depth of understanding needed to model dynamic mathematical systems using concrete materials often far exceeds that needed to treat mathematics simply as a set of computational skills (Bell, Fuson, and Lesh 1978).

Another fact that discourages teachers from using concrete materials is that when such materials are used, students' misconceptions tend to become clearly visible. Therefore, because many such misconceptions were formerly hidden behind superficial computational proficiency, the misconceptions that emerge actually seem to be <u>caused by</u> the use of concrete materials (rather than simply being revealed by them). Finally, the misconceptions that emerge when concrete activities are used tend to be unusually difficult to correct because they are so basic.

It does teachers little good to identify students' misunderstandings if they are unable to correct them. So, if a class of errors is easy to ignore by focusing exclusively on low-level skills, and if higher-order errors are viewed as impossible to correct, then a seemingly reasonable teaching strategy is to devote valuable classroom time to objectives that can indeed be attained. Using concrete materials to correct misconceptions or teach conceptions presupposes

- A firm belief in the ultimate payoff that will result.
- A clear understanding about the higher-level misunderstandings that may be avoided.
- Confidence that misunderstandings that <u>do in fact emerge</u> can be corrected.

This paper will briefly describe some ways computers can be used to address the preceding issues. Comments will be based on results from three recent or current NSF-funded projects on Rational Numbers Concepts (RN), Proportional Reasoning (PR), and Applied Mathematical Problem Solving (AMPS), as well as on recent research conducted at the World Institute for Computer Assisted Teaching (WICAT).[1]

Dienes' Multiple Embodiment Principle:

This section will describe some important factors to take into account in computer-based implementation of Dienes' multiple embodiment principle.

Past RN, PR, and AMPS publications (e.g., Lesh 1981; Lesh, Landau and Hamilton 1980; and Behr, Lesh, Post, and Silver 1983) have identified five distinct types of representation systems that occur in mathematics learning and problem solving (see Figure 5). These are (a) "scripts" in which knowledge is organized around "real world" events that serve as models for interpreting and solving other kinds of problem situations; (b) manipulative models (such as Cuisenaire rods, arithmetic blocks, fraction bars, number lines, etc.) in which the "elements" in the system have little meaning per se, but the "built-in" relationships and operations fit many everyday situations; (c) pictures which, like manipulative models, can be internalized as "images"; (d) spoken languages, including specialized sub-languages (e.g., logic); and (e) written symbols which, like spoken languages, can involve specialized sentences and phrases, such as ($x + 3 = 7$, $A' \cup B' = (A \cap B)'$), as well as normal English sentences and phrases.

To illustrate why translations between models in different representational systems are important, consider the following examples.

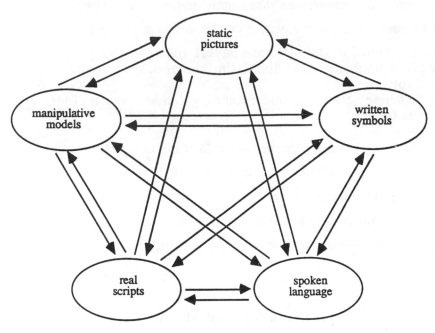

Figure 5

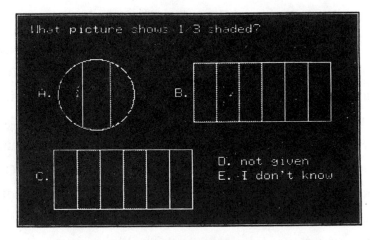

Figure 6

Figure 6 shows a simple type of question that space considerations seldom permit in textbooks and pencil-and-paper tests, but which is quite easy to generate in computer-based instructional materials. The question, taken from a written test on "rational number relations and proportions" used in our RN and PR projects, illustrates a "written symbol to picture" translation. The aim is to require students to establish a relationship from one representational system to another while preserving certain structural characteristics and meaning.

Figure 7 shows another item from the same "relations and proportions" test as Figure 6, but it was adapted from a recent "National Assessment" examination (Carpenter, et al. 1981). To answer the question in Figure 7 correctly, the student's primary task is to perform a computational transformation.

The ratio of boys to girls in a class is 3 to 8. How many girls are in the class if there are 9 boys?

A. 17 B. 14 C. 24 D. not given E. I don't know

Figure 7

Educators familiar with results from recent "National Assessments" (Carpenter, et al. 1981) may not be surprised that our students' success rates for Figure 7 were only 11% for fourth graders, 13% for fifth graders, 30% for sixth graders, 29% for seventh graders, and 51% for eighth graders. Such performances by American students led to "Nation at Risk" reports from a number of federal agencies and professional organizations. However, success rates on the seemingly simpler question in Figure 6 were even lower: 4% for fourth graders, 8% for fifth graders, 19% for sixth graders, 21% for seventh graders, and 24% for eighth graders. On the translation in Figure 6, only one in four students answered correctly! Forty-three percent selected answer choice **a**; 4% selected **b**; 15% selected **c**; 34% selected **d**; 3% selected **e**; and 2% did not give a response.

One major conclusion from our RN and PR research is apparent from the preceding examples: not only do most fourth- through eighth-grade students have seriously deficient understandings with "word problems" and "pencil and paper computations," but many have equally deficient understandings about the models and languages needed to represent and manipulate these ideas. Furthermore, we have found that the ability to do these translations is a significant factor influencing both mathematical learning and problem-solving performance (Behr, Lesh, Post, and Wachsmuth 1985; Post 1986). For example, when we say that

655

students "understand" an idea such as "1/3," we mean that they can (a) recognize the idea embedded in a variety of qualitatively different representational systems, (b) flexibly manipulate the idea within given representational systems, and (c) accurately translate the idea from one system to another. Translation processes are implicit in a variety of common techniques used to investigate whether a student "understands" a given textbook word problem (e.g., "Restate it in your own words." "Draw a diagram to illustrate what it's about." "Act it out with real objects." "Describe a similar problem in a familiar situation."). Similarly, techniques for improving performance on word problems include (1) using several concrete materials to "act out" a given problem situation, (2) describing several everyday problem situations that are similar to a given concrete model, or (3) writing equations to describe a series of word problems, delaying the actual solutions until the student becomes proficient at this descriptive phase.

To diagnose a student's learning difficulties or to identify instructional opportunities, teachers and computers can generate a variety of useful questions by presenting an idea in one representational mode and asking the student to render the same idea in another mode. Then, if diagnostic questions indicate unusual difficulties with one of the processes in Figure 5, other processes in the diagram can be used to strengthen or bypass it. For example, a student who has difficulty translating from situations to written symbols may find it helpful to begin by translating real situations to spoken words, and then translate spoken words to written symbols; or it may be useful to practice the inverse of the troublesome translation (i.e., identifying familiar situations that fit given equations).

Not only are the translation processes in Figure 5 important components of understanding a given idea, they also correspond to some of the most important "modeling" processes needed to use this idea in everyday situations. Essential features of modeling include (1) simplifying the original situation by ignoring irrelevant characteristics in order to focus on more relevant factors, (2) establishing a mapping between the original situation and the "model," (3) investigating the properties of the model in order to generate predictions about the original situation, (4) translating (or mapping) the predictions back into the original situation, and (5) checking to see whether the translated prediction is useful.

Here is an example where the preceding steps are used to solve a standard algebra word problem.

> Al has an after-school job. He earns $6 per hour if he works 15 hours per week. If he works more than 15 hours, he gets paid "time and a half" for overtime. How many hours must Al work to earn $135 during one week?

To solve this problem, students may begin by paraphrasing the given "English sentence" into their own words, perhaps accompanied by a diagram or picture of the situation. Next, the description of the problem may be translated into an "algebraic sentence":

$$(6 \times 15) + 9(x - 15) = 135$$

Then, a series of algebraic transformations may be used to convert this algebraic model into an arithmetic sentence that is sufficient to find the answer. The final transformed description is:

$$x = \frac{135 - (6 \times 15)}{9} + 15$$

Finally, by using a series of arithmetic simplifications, this arithmetic sentence can be reduced to:

$$x = 20$$

So, beyond paraphrasing and diagramming, the entire solution process involves three significant translations: (1) from an English sentence to an algebraic sentence, (2) from an algebraic sentence to an arithmetic sentence, and (3) from an arithmetic sentence back into the original problem situation.

Notice that the algebraic sentence that most naturally describes the preceding problem situation does not immediately fit an arithmetic computation procedure. This possibility of "first describing and then calculating" is one of the key features that makes algebra different from arithmetic.

As the preceding problem illustrates, problem solving often occurs by (1) translating from the "given situation" to a mathematical model, (2) transforming the model so that desired results are apparent, and (3) translating the model-based result back into the original problem situation to see if it is useful. However, the modeling process usually is not this simple. Instead, modeling students frequently use several representation systems (or models) in series or in parallel, with each depicting only a portion of the given problem situation (see Figure 8).

In RN and PR research involving realistic textbook word problems, we found that students seldom work through solutions in a single representational mode (Lesh, Landau, and Hamilton 1983). In fact, many realistic problem-solving situations are inherently multi-modal from the outset. The following two pizza problems illustrate this point.

> Show a sixth grader one-fourth of a real pizza, and then ask, "If I eat this much pizza, and then one-third of another pizza, how much will I have eaten all together?"

> Show a sixth grader one-third of a real pizza, and then ask, "If I already ate one-fourth of a pizza, and now eat this much, how much will I have eaten all together?"

Neither of these pizza problems is a "symbol-symbol" or "word-word" problem. Instead, the "givens" in both problems include (1) a real object (a piece of pizza) and (2) a spoken word (to represent a past or future situation). Like many realistic problems in which mathematics is used, the situation in these two pizza problems is inherently multi-modal. Each of the problems is a "pizza-word" problem in which one of the students' difficulties is translating the two givens into a homogeneous representational mode so that combining is possible.

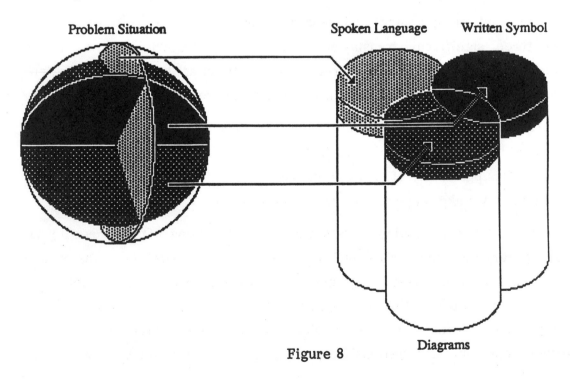

Figure 8

Figure 8 suggests that solutions are often characterized by several mappings from parts of the given situation to parts of several (often partly incompatible) representational systems rather than by one mapping from the whole given situation to only one representational system.

Not only may problems like the above occur in a multi-modal form, but solution paths also may weave back and forth among several representational systems, each of which is typically well-suited for representing some parts of the situation, but ill-suited for representing others. For example, in the above two problems, a student may think about the static quantities (e.g., the two pieces of pizza) in a concrete way (perhaps using pictures), but may switch to spoken language (or to written symbols) to carry out the dynamic "combining" actions (Lesh, Landau, and Hamilton 1983); that is, the student may begin a solution by translating to one representational system and may then map from this system to yet another, as illustrated in Figure 9.

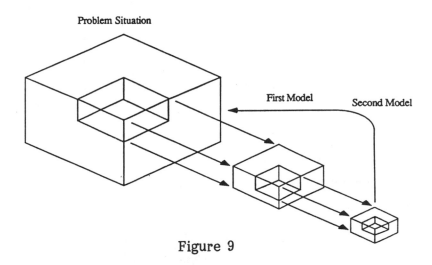

Figure 9

We found that for realistic textbook word problems, the actual solutions students use often combine features depicted in both Figures 7 and 8, and good problem solvers are flexible in their use of various relevant representational systems — they instinctively switch to the most efficient representation at any given point in the solution process.

The main points of this section are:
- Not only is it important for youngsters to work with concrete models that

illustrate mathematical concepts, it is also important to focus on translations from one representational system to another.

- We have extensive evidence from the RN, PR, and AMPS projects and from field tests of computer-based materials developed by WICAT that focusing on translation abilities can significantly influence learning and problem-solving capabilities.

Dienes' Constructive Principle and Perceptual Variability

Concerning Dienes' constructive principle and perceptual variability principle, this section will begin with a series of examples from our own RN and PR research showing that students do not necessarily "see" relationships that seem to be "built into" certain fraction diagrams. Instead, relevant systems of relationships must be organized (constructed) and imposed on the models before their attributes will be noticed. We will then give a series of examples of computer-based lessons designed to help youngsters construct the kind of relational/operational systems needed to correctly interpret fraction and ratio diagrams.

A major goal of our RN and PR projects has been to describe the relational/ organizational systems that youngsters use to make judgments involving rational number concepts or proportional reasoning (Behr, Lesh, Post, and Silver 1983; Lesh, Landau, and Hamilton 1983). The examples that follow will show how such judgments often inherently involve networks of part-whole and part-part comparisons as well as organizational schemes in which composite units are made up of "lower-level" units.

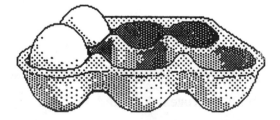

Figure 10

Figure 10 is a fraction situation which involves composite units. The unit of measure (i.e., the "whole") is "a half dozen eggs"; one-third is "two eggs"; and two-thirds is "four eggs."

Figure 11 illustrates how distinct but related types of relationships and unit comparisons often lie at the heart of children's confusion between "fraction judgments" and "ratio judgments." Answer choices A and B involve units made up of discrete pieces, whereas answer choice C involves a continuous quantity. Answer choices B and C both involve a part-part ratio of three-to-four, whereas only answer choice A shows three-fourths shaded. The success rate for eighth graders was only 61.5%. The percentage of students selecting each answer choice is shown in Figure 11.

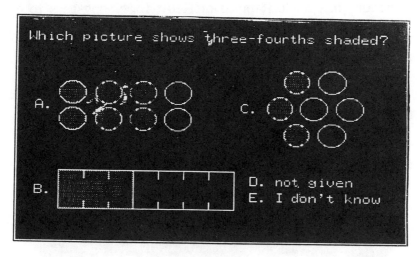

% Correct by grade: 4) 4.6%, 5) 11%, 6) 36.9%, 7) 40%, 8) 61.5%
Answer Option Selected: A. 30.7% B. 7.25% C. 6.87% D. 50.3% E. 3.05%
na. 1.72%

Figure 11

Figure 12 presents a question that was answered incorrectly by more than 50% of the eighth graders. If the two pictures in Figure 12 are compared by the total amount shaded, rather than by the relationships between parts and wholes, then answer choice b would be picked. Or, if the pictures were compared by part-part comparisons, then the ratio of shaded-to-unshaded parts in picture A would be only one-to-two, whereas the ratio for picture B would be two-to-three — again making choice b the answer chosen.

If a picture or diagram draws a student's attention to "perceptual distractors" that increase the possibility of basing rational number judgments on inappropriate cues, then the difficulty of the preceding kinds of questions can be increased significantly (Behr, Lesh, Post, and Silver 1983; Lesh, Landau, and Hamilton 1983; Behr, Post, Lesh, and Wachsmuth 1983). For example, a particularly impressive "perceptual distractor" occurred in the clinical interview phase of our RN and PR testing programs. Students were given a Hershey chocolate bar and were asked to give the interviewer one-third of the chocolate bar. The results showed that this question was significantly more difficult when a plain Hershey bar was used instead of a Hershey bar with nuts. Why? The apparent answer was that the ten subdivisions in the plain bar created a perceptual distractor that made it more difficult for youngsters to divide the bar into thirds. On the other hand, the bar "with nuts" did not have these distracting subdivisions, and so it was easier to divide into thirds.

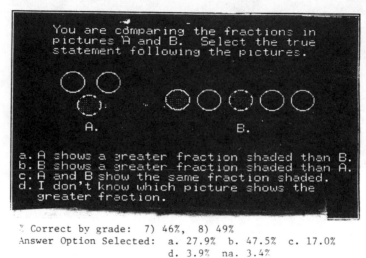

% Correct by grade: 7) 46%, 8) 49%
Answer Option Selected: a. 27.9% b. 47.5% c. 17.0%
d. 3.9% na. 3.4%

Figure 12

The influence of perceptual distractors makes it clear why Dienes' perceptual variability principle, involving the use of more than a single concrete model, is such an important feature of his instructional approach. When students have not yet constructed the relevant system of relationships (i.e., part-whole, part-part, etc.) needed to make rational number judgments, they are more likely to become victims of perceptually compelling (but misleading) cues.

Next, we will give examples of computer-based activities that have helped youngsters gradually use appropriate systems of rational number relations in organized and flexible ways. In Figure 13, the child is asked to fill in a specified

662

fraction of a whole using the computer's arrow keys. If help is requested, then the whole is divided into an appropriate number of parts, and the child's task is simply to fill in the correct number of parts.

Figure 14 is similar to Figure 13 except that it deals with discrete quantities rather than with continuous quantities. The particular question illustrated in Figure 14 is relatively difficult because the "whole" consists of three rows with eighth squares in each row, rather than simply four squares. (Note: For any of the examples given in this section, the difficulty of the items can vary considerably depending on the particular numbers and pictures used; computer programs can adjust difficulty levels to fit the capabilities of individual students.)

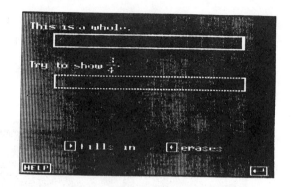

Figure 13

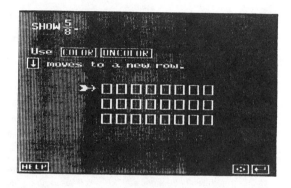

Figure 14

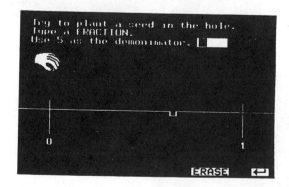

Figure 15

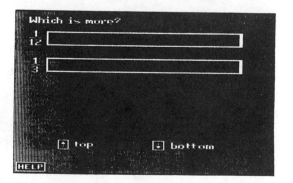

Figure 16

Figure 15 is similar to Figures 13 and 14 except that rational numbers are portrayed as single "things" (as points along a number line) more than as comparisons between pairs of "things" (parts and wholes). Number lines also emphasize position and order relationship (e.g., 3/5 is "between" 2/5 and 4/5, "after" 2/5, and "before" 4/5) more than quantitative relationships (e.g., "more than" and "less than").

Figure 16 involves comparisons of <u>pairs</u> of fractions (or fraction pictures). If help is requested, two pictures are partitioned into an appropriate number of subsections so that the student can more easily compare the relevant attributes. In answer feedbacks, the appropriate number of parts is colored in each picture so that perceptual comparisons are easier to make.

In Figure 17, rather than giving the picture of a "whole" and asking the student to shade in a specified part, the process is reversed; the part is shown and the student is asked to use arrow keys (as in Figure 13) to make the whole. If help is requested, the part is partitioned as shown in Figure 18. A helpful answer feedback is shown in Figure 19.

Figure 17

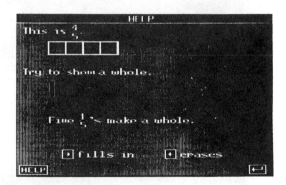

Figure 18

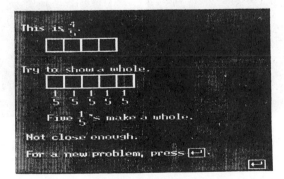

Figure 19

664

Figure 20 is similar to Figure 17 except that discrete quantities are emphasized rather than continuous quantities. A help is shown in Figure 21.

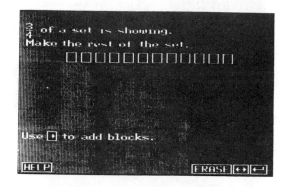

Figure 20

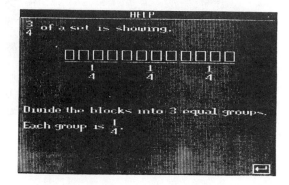

Figure 21

Figure 22 goes beyond comparing pairs of fractions; the student is asked to generate fractions equivalent to a given fraction. Figure 23 shows the additional information given if help is requested

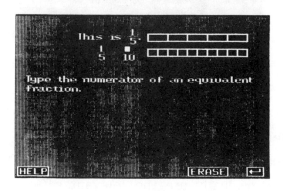

Figure 22

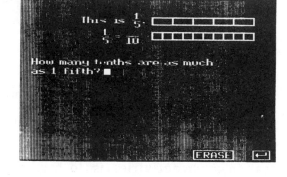

Figure 23

Figure 24 focuses on whole sets of equivalent fractions, and Figure 25 shows one of a sequence of hints that can be given. Figures 26 and 27 show how the types of diagrams in Figures 13 through 25 can be used to go beyond relationships among fractions to deal with operations with pairs of fractions, or with more complex relationships among fractions. Similar diagrams and procedures can also be used to introduce the concepts of decimals, ratios, rates, percent, or

665

proportions.

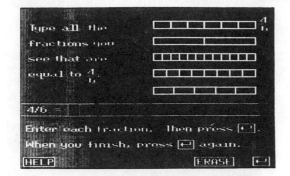

Figure 24

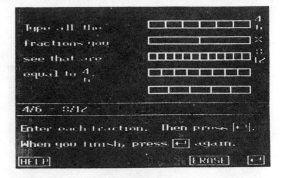

Figure 25

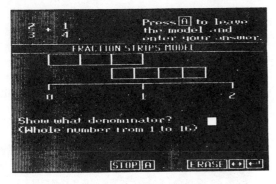

Figure 26

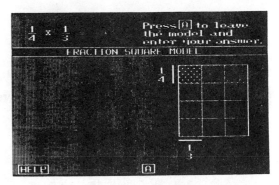

Figure 27

The examples in Figures 13 through 27 show how youngsters can be led to conceptually "take apart" and "re-assemble" rational number diagrams by systematically focusing attention on different relationships and attributes. The goal in such activities is to help students gradually use the relevant relational systems in an organized way.

Helping students construct a system of mathematical relationships is similar to helping students coordinate systems of overt activities like those involved in playing tennis or riding bicycles; that is, the student begins in situations in which the complexity of the system and the degree of coordination are minimal (e.g., all of the balls come waist high on the forehand side just within arm's reach) and gradually progresses to situations that require more complex and well-coordinated systems (e.g., where "getting in position" is important). In general, building more complex systems involves:

1. <u>integration</u> — i.e., simple systems are linked together to build more complex systems, as when a tennis serve is built up by gradually linking together the toss, the hit, the follow-through, etc.
2. <u>differentiation</u> — i.e., a single system is differentiated to produce two or more distinct variations, as when a forehand volley is varied slightly to produce topspin or backspin.

Poorly integrated mathematical systems are also similar to poorly coordinated behavioral systems because:

1. The student will not "read out" all of the available information — e.g., when first learning to ride a bicycle or hit tennis balls, a great deal of relevant information is not noticed.
2. The student "reads in" interpretations that are not objectively given — e.g., when first learning to ride a bicycle or hit tennis balls the student's description of an activity is often distorted and biased.

Both of these factors also appear when, for example, an "eyewitness" to an accident interprets given information in a way that is biased (because only selected pieces of information are noticed) and distorted (because what "made sense" and what was "expected" influenced the interpretation of what actually happened). Similar biased and distorted interpretations also influence students' mathematical judgments in graphics-related problems like the examples in this section.

Next, a few examples will be given to show how the basic approach of "taking apart" and "re-assembling" mathematical ideas can be extended to basic algebraic concepts. We will focus on "unpacking" the systems of operations, relations, and transformations that underlie the basic concepts of linear equations and simple polynomials.

The activities that follow are based on a symbol-manipulator/function-plotter called SAM that WICAT developed to enable students to write, graph, transform and solve algebraic expressions and equations. In lessons, SAM helps students learn some of the most important basic ideas in algebra or calculus, and the algebra ideas can make SAM more useful for problem-solving situations that students want to address. However, SAM is more than a calculator; it has the following characteristics:

1. SAM can serve as an expression checker. We don't have to wait until students give final answers to know whether they are proceeding along correct solution paths. We can, for example, assess whether they "set up"

the equations correctly.

2. SAM is LISP-based, so it not only generates answers, it can also produce solution path "traces" that create many instructional capabilities. For example, it allows us to: a) generate hints by gradually revealing solution steps one at a time, (b) monitor individual steps in students' solution paths, (c) let students examine processes as well as products of solution attempts, and (d) give students the capability to build/edit/store equation-solving routines (like the quadratic formula) in a LOGO-like fashion.

3. SAM's symbol manipulation capabilities interact with its function plotter to produce graphic interpretations of transformations leading to solutions. This gives students ways to visualize symbol transformations, and (in yet another way) to focus on processes as well as "answers" during solution attempts.

4. SAM can reduce answer-giving phases of problem solving so that attention can be focused on "nonanswer-giving" phases (e.g., problem formulation, trial solution evaluation, the quantification of qualitative information, the examination of alternative possibilities, etc.) where "second order" (i.e., thinking about thinking) monitoring and assessing functions often are especially important. So, SAM is not simply an answer-giver; it can help students to go beyond thinking to think about thinking.

For polynomials, it is easy for students to use SAM to carry out the following kinds of investigations:

1. Pick a value for n, between -10 and 10, and investigate the changes that this value produces in the graph of the linear expression nx.

2. Plot the graph of the squared term x^2; then plot the graph of the linear term nx (as in step #1 above); and finally, plot the graph of the polynomial x^2 + nx. Notice that the polynomial crosses the x-axis at the points zero and -n.

For example, Figure 28 shows the graphs of x^2 and 4x. Figure 29 also shows the graph of the polynomial x^2 + 4x.

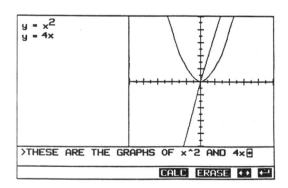

Figure 28

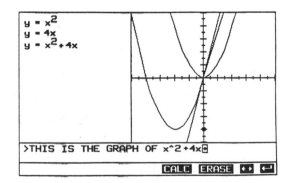

Figure 29

After repeating step 2 for a series of different values for n, it is easy for students to notice that the effect of adding x^2 and nx is to "slide the graph of x^2 downhill along the line nx." Furthermore, it is easy for students to notice that the amount of the slide is just enough to make the polynomial's graph pass through the points zero and –n.

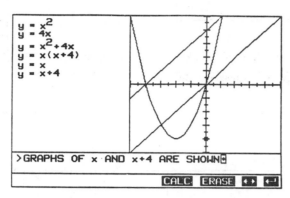

Figure 30

3. Polynomials from step 2 can be factored into the form x(x + b), and each of the linear factors can be graphed as shown in Figure 30. Then notice that the two lines pass through the points zero and –n.

Step 3 shows why polynomials can be solved by: factoring, setting each of the linear factors equal to zero, and then solving these linear equations. The linear terms are equal to zero at exactly the same places as the original polynomial.

Using a symbol-manipulator/function-plotter like SAM, sequences of activities like those in steps 1-3 above can quickly and easily be extended to include more terms and more complex polynomials. For example:

1. First, graph three monomials, like $(-3/8)x^2$ and $(1/4)x$ and 15 (see Figure 31).
2. Next, graph the polynomial formed by taking the sum of the monomials in step 1 (see Figure 32).
3. Then, change the coefficients of the monomials in step 1, and investigate what changes these induce in the polynomial in step 2.
4. Finally, factor the polynomials in steps 2 and 3 and graph each of the linear factors (see Figure 33).

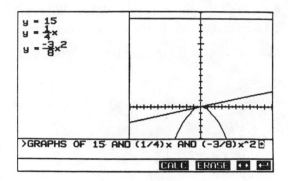

Figure 31

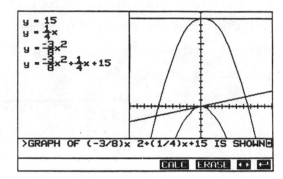

Figure 32

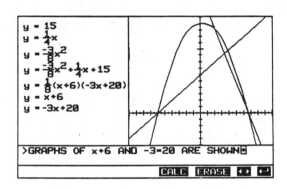

Figure 33

Or reverse the preceding four steps:

1. Graph a pair of linear equations like those shown in Figure 33.
2. Graph the polynomial formed by the product of the linear expressions in step 1 (see Figure 32).
3. Change the linear expressions in step 1, and investigate induced changes in the polynomial at step 2.
4. Graph each of the monomial terms for the polynomials in step 3 (see Figure 31).

In the preceding examples, the two models involved were: (a) written symbols which (although they are on a computer screen) are like those that mathematics teachers write on blackboards, and (b) computer graphics, consisting of graphs of equations in a rectangular coordinate system. Nonetheless, the computer-based activities that use these representations can be based on direct applications of Dienes' instructional principles. For example:

- The <u>constructive principle</u> is involved when we "take apart and then re-assemble" complex mathematical systems related to polynomials.
- The <u>multiple embodiment principle</u> is involved when we focus on mappings between two given models (i.e., written symbols and graphs of equations).
- The <u>dynamic principle</u> can be used to show how transformations performed on algebraic equations are reflected in changes in the graphs of the equations at each step. For example, in the next section, we will show how a slight variation on the preceding sequence of activities can be used to show why the "completing the square" process works in the derivation of the quadratic equation.

The preceding kinds of activities can be used in much the same way as rational number pictures and diagrams to help students build "concrete feelings and imagery" for abstract mathematical ideas. So even though the "materials" used in these examples are computer-based graphics rather than "concrete materials" in the usual sense of this word, the activities can indeed involve overt actions that students can apply to "objects" that they can see and manipulate; and for the first time Dienes' instructional principles can be applied to content areas like "polynomials" which did not seem to lend themselves to a "mathematics laboratory" form of instruction.

Dienes' Dynamic Principle

Earlier in this paper we noted that in classroom activities involving concrete materials students seldom have opportunities to investigate transformations within even a single model. Almost never do they investigate transformations within a second model, or correspondences between transformations in the two models.

The primary point of this section is that computers possess the powerful instructional capabilities to allow students to investigate manipulations within one mathematical model and immediately see the corresponding manipulation in a second or third model.

To begin, it is useful to give an example in which algebraic equations model a typical textbook word problem and then coordinated graphs model the equations. The problem deals with simultaneous linear equations.

Traveling upstream on a river, a boat takes two hours to reach its destination eight miles away. The return trip downstream takes one hour and twenty minutes. What is the speed of the river current?

To translate this problem into an algebraic description, a student might let x represent the speed of the boat and y represent the speed of the current. Then the given relationships can be modeled (or described) using the following two algebraic sentences:

$$2(x - y) = 8 \qquad \text{and} \qquad (1\ 1/3)(x + y) = 8$$

Such equations are models because they are useful simplifications of reality. Once the problem is translated into algebraic sentences, algebraic transformations can be used to solve for x and y, or the algebraic model can be converted to a geometric model by graphing each of the two equations. Then the (x, y) values that satisfy both equations will be given by the coordinates of the point located on both graphs. The solution (5, 1) can be read directly from the coordinates on the graph as shown in Figure 34.

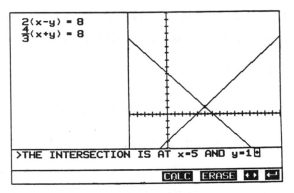

Figure 34

In this case, the graph shows that the algebraic solutions are x = 5 and y = 1, which means that the speed of the current was one mile per hour, and the speed of the boat was five miles per hour. The preceding solution involves three significant translations: (1) describing the problem situation with algebraic sentences, (2) translating the equations into graphic form and reading the solution from the graph, and (3) interpreting the graphic solution in the context of the original problem situation (see Figure 34 above).

Models like coordinate graphs, or systems of linear equations, can be considered to be "conceptual amplifiers" because they help students use their ideas more effectively. They are not simple inert systems that have no meaning; once students learn to meaningfully embed mathematical systems (ideas and principles) or problem situations within them, students are able to "read out" additional meaning. For example, in the preceding graphing-based solution, once students had learned to use equations to describe the problem situation, and to graph these equations, they could solve many problems immediately by simply reading the graphs.

Through most of this paper we have emphasized how ideas can evolve within particular concrete situations when youngsters are helped to construct relevant structural models. However, basic ideas do not always come into being through gradual development within an environment that corresponds to a particular concrete model. Instead, students often learn new ideas in new contexts by mapping to an old model and by relying on meanings that have already developed in the context of the old model to give meaning to the new idea in the new context.

The next example illustrates how the preceding process might work, using a computer-based symbol-manipulator/function-plotter utility to teach the new idea of solving pairs of linear equations with two unknowns. Two old ideas are involved:

1. the ability to solve simultaneous linear equations and single linear equations, and
2. the ability to find the coordinate graphs of linear expressions.

Suppose that students have learned to graph linear equations of the form $y = mx + b$, and that a computer will henceforth graph such equations for them automatically, whenever they want. Furthermore, suppose that the students are learning to perform algebraic transformations by giving commands to a computer such as "add 3 to both sides of the equation," or "substitute $x - 4$ for y in the current expression." Then the algebraic solution to the "boat" problem might look like this:

(1) Divide by 2 to simplify $\quad 2(x-y)=8 \quad$ to $\quad x-y=4.$

Multiply by 3/4 to simplify $\quad 4/3(x+y)=8 \quad$ to $\quad x+y=6.$

(2) Add y and subtract 4 to convert $\quad x-y=4 \quad$ to $\quad x-4=y.$

Subtract x to convert $\quad x+y=6 \quad$ to $\quad y=-x+6.$

(3) Graph the pair of equations $\quad x-4=y \quad$ and $\quad y=-x+6.$

(4) Substitute x – 4 for y from the first equation into the second equation to get

$$x \cdot 4 = -x + 6.$$

Then, notice that the two line graphs from step 3 can be thought of as graphs of the two sides of this equation (see Figure 35).

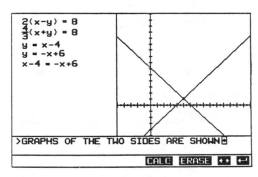

Figure 35

(5) Add 4 to both sides of the previous equation to get

$$x = -x + 10$$

and plot the graphs of the two sides of the equation (see Figure 36)

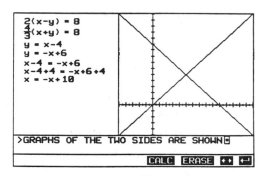

Figure 36

(6) Then add x to both sides of the previous equation to get

$$2x = 10$$

and plot the graphs of the two sides of the equation (see Figure 37).

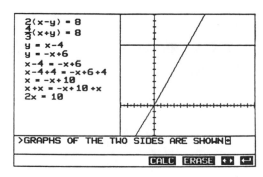

Figure 37

674

(7)   Divide by 2 on both sides to get

$$x = 5$$

and again plot the graph sof the two
sides of the equation (see Figure 38).

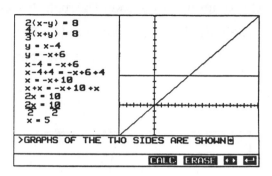

Figure 38

After working on activities of the type described above, many students have
noticed that for any given problem, intersection points for the pairs of lines at
each solution step always lie on a single vertical line: the solution for x.   Some
students have even gone on to think about and describe why this invariant feature
occurs.   So this dynamic representation system, once constructed, actually helps
students generate significant new questions and sophisticated solutions related to
two of the most fundamental ideas in algebra; that is, our students have used
informal language to describe rather deep principles related to: (1) invariance
under mapping among isomorphic systems, and (2) invariance under transformations within a given system.

The examples in this section illustrate how computer environments are well-
suited to Dienes' dynamic principle.   Whether we are dealing with linear equations
and graphs, fraction diagrams and simple proportional reasoning questions, or
polynomials, computers make it easy for the student to manipulate one model and
immediately see corresponding transformations in one or more other models.

Let's end this section with one more series of examples which will be left as
a sort of exercise for the interested reader to interpret.   The example has to do
with the process of "completing the square," which can be used (prior to using the
quadratic formula) to find the roots or factors of quadratic equations like
$x^2 + 2x - 3 = 0$.   Figure 39 shows the graph of $x^2 + 2x - 3 = y$ and $y = 0$.   Figure 40
shows the graph of $x^2 + 2x = y$ and $y = 3$.   Then, Figure 41 shows the graph of
$x^2 + 2x + 1 = y$ and $y = 4$.   Notice that the tip of the parabola just touches the
x-axis.   (Is this significant? Would it happen for other quadratic equations? Which
kinds?)   Figure 42 shows the graphs of $x + 1 = y$, $y = +2$, and $y = -2$.   Notice that

675

the diagonal line goes through the x-axis at the same point where the parabola had touched. (Is this significant?) Figure 43 moves the graphs in Figure 35 so that the diagonal line goes through the origin of the graph. (Is this significant?)

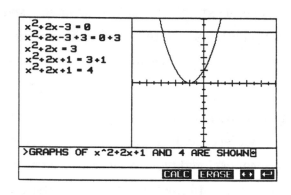

Figure 39

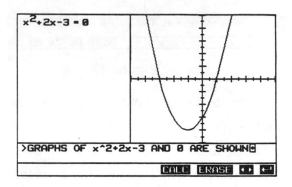

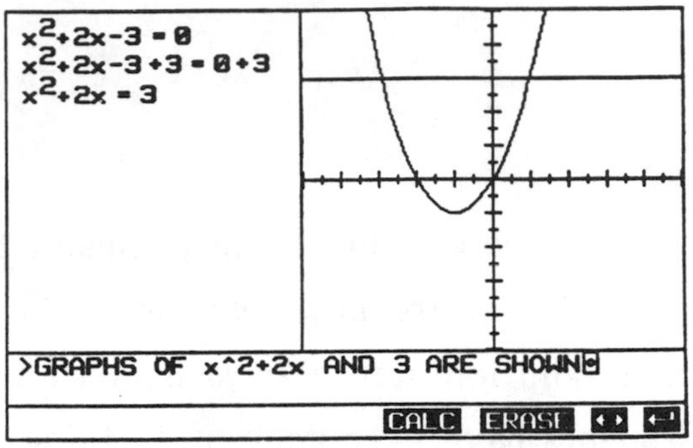

Figure 40

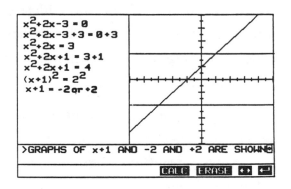

Figure 41

Figure 42

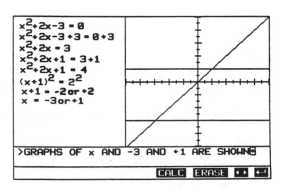

Figure 43

676

Conclusions

In general, we are in sympathy with those LEGO BEFORE LOGO proponents who believe that children's mathematical abstractions should be built on a firm foundation of experiences with real manipulable models and realistic problem-solving situations. However, we also know that even real concrete objects often are used only in very abstract ways and that very few teachers successfully use concrete activities as a significant instructional tool. On the other hand, we have seen that when students use the kind of computer-based activities described in this paper (many of which are electronic versions of the kinds of concrete models that we really hope students will have the opportunity to explore), their teachers actually become more likely to use "mathematical laboratory" activities with real concrete materials. This increased use of real concrete activities seems to occur because computer-based simulations of mathematics laboratories tend to minimize the obstacles to teachers' use of concrete mathematics laboratory principles.

Throughout this article we have illustrated how computer-based activities help teachers implement mathematical laboratory principles such as those described by Dienes. In fact, principles like Dienes' dynamic principle are perfectly suited to computer environments and have never seldom successfully implemented in any other context.

Some factors that favor the use of computer-based mathematics laboratory activities include the following:

Not only can computer utilities serve as powerful tools to help students acquire mathematical ideas and increase their meaningfulness, they can also amplify the power of the acquired concepts. Problem solving in the presence of computer-driven "conceptual amplifiers" (such as the symbol-manipulator/function-plotter referred to in this paper, or even familiar tools like VisiCalc) are becoming indispensable in science, mathematics, and engineering. One can no longer assume that the problem solver is working alone with only a pencil and paper for tools. New "realistic" problem types, such as those involving non-deterministic answers and "stochastic refinement" solution processes, will also be increasingly important (Lesh 1985).

677

Due to the availability of powerful new computer utilities (like WICAT's symbol-manipulator/function-plotter), realistic applications can now be used to introduce a wide variety of mathematical topics. Rather than first attempting to "teach" a given idea and then introducing applied problems involving the idea and associated procedure, computer utilities allow us to "give" the procedural capabilities to the student at the beginning of instruction. Realistic applications can then be used to gradually guide students to build their own comceptualizations of the underlying idea (Fey 1985).

By minimizing tedious answer-giving procedures, computer utilities can focus students' attention on nonanswer-giving phases of problem solving, where higher-order thinking activities related to data gathering, information filtering, problem formulation, and trial solution evaluation are involved. By focusing attention on the underlying conceptualizations of problem situations and on the sensibility of products of thought, the subtle meanings of relevant ideas become apparent. Also, by reducing the conceptual energies devoted to "first-order thinking," higher-order "thinking about thinking" becomes possible. Otherwise, students frequently become so embroiled in "doing" a problem that they are unable to think about what they are doing and why.

Results of a recent Applied Mathematics Problem Solving project (Lesh and Akerstrom 1983; Lesh, Landau, and Hamilton 1983) support the view that the underlying meanings of mathematical ideas tend to be emphasized during nonanswer-giving phases of problem solving; and using computer utilities, nonanswer phases can be addressed even in conceptual areas characterized by complex number-crunching routines or sophisticated conceptualizations. Consequently, computer utilities enable us to use realistic applications to introduce whole new categories of sophisticated mathematical concepts.

Notes

1. The research of the RN, PR, and AMPS projects was supported in part by the National Science Foundation under grants SED 79-20591, SED 80-17771, and SED 82-20591. Any opinions, findings, and conclusions expressed in this report are those of the authors and do not necessarily reflect the views of the National Science Foundation.

2. Perhaps, we should state one of our prejudices about the nature of mathematics for readers who hold different biases. In our opinion, although most mathematical ideas, from early number concepts to algebraic topology, often have computation-type procedures associated with them, "doing the procedure" frequently has little to do with "doing mathematics," nor is it necessarily a good indicator of a student's understanding about the underlying ideas.

   Consider the mega-ideas underlying differentiation or integration in calculus. The procedures needed to compute a given derivative or integral bear virtually no resemblance to the network of relations that define the underlying ideas and constitute the meanings. Unfortunately, the "network of relationships" that characterize (psychologically define) most mathematical ideas are quite difficult to specify. For certain elementary mathematical ideas, considerable progress has been made in this regard in recent years. However, with the exception of certain early number concepts, rational number concepts, and simple spatial/geometric concepts, psychologists and mathematics educators have only begun to map out the specific "conceptual models" that underlie most fundamental mathematical ideas. Certainly, very few details can be given about what it means to "understand" a topic like simultaneous linear equations. Consequently, by default, understanding almost exclusively tends to be assessed in terms of students' abilities to carry out procedures, even though "taking a derivative," say, or solving "one rule under your nose" style word problems, is quite easy to "master" with only the foggiest notion of the underlying ideas.

References

Behr, M., R. Lesh, T. Post, and E. Silver. 1983. "Rational-Number Concepts." In Acquisition of Mathematical Concepts and Processes: 91-126, R. Lesh and M. Landau, eds. New York: Academic Press.

Behr, M., R. Lesh, T. Post, and I. Wachsmuth. 1984 (November). "Order and equivalence of Rational Numbers: A Teaching Experiment." In Journal for Research in Mathematics Education.

Bell, M., K. Fuson, R. Lesh. 1978. Algebraic and Arithmetic Structures: A Concrete Approach. New York: MacMillan Co., Free Press.

Carpenter, T., M. Corbitt, H. Kepner, N. Lindquist, and R. Reys. 1981. Results from the Second Mathematical Assessment of the National Association of Educational Progress. Reston, VA: National Council of Teachers of Mathematics.

Dienes, Z. 1960. Building Up Mathematics (4th edition). London: Hutchinson Educational Ltd.

Fey, J. (Ed.) 1984. Computing and Mathematics: Impact on Secondary School Curricula. Reston, VA: National Council of Teachers of Mathematics.

Fey, J. 1985. Final Report from the National Science Foundation Award (Report No. SED 80-24425). Washington, D.C.: The National Science Foundation.

Lesh, R. 1981 (May). "Applied Mathematical Problem Solving." Educational Studies in Mathematics 12(2): 235-264.

Lesh, R. 1985 (August). "Processes, Skills, and Abilities Needed to Use Mathematics in Everyday Situtations." Education and Urban Society 17(4): 439-446.

Lesh, R., and M. Akerstrom. 1983. "Applied Problem Solving: Priorities for Mathematics Education Research." In Mathematical Problem Solving: Issues in Research: 117-129, F. Lester and J. Garofalo, eds. Philadelphia: Franklin Institute Press.

Lesh, R., M. Landau, and E. Hamilton. 1980. "Rational Number Ideas and the Role of Representational Systems." In Proceedings of the Fourth International Conference for the Psychology of Mathematics Education: 50-59, R. Karplus, ed. Berkeley, CA: Lawrence Hall of Science.

Lesh, R., M. Landau, and E. Hamilton. 1983. "Conceptual Models in Applied Mathematical Problem Solving Research." In Acquisition of Mathematical Concepts and Processes: 263-343, R. Lesh and M. Landau, eds. New York: Academic Press.

Post, T. (In press.) Teaching Mathematics in Grades K-8: Research-based Methods. Boston: Allyn and Bacon.

Post, T., R. Lesh, M. Behr, and I. Wachsmuth. 1985. "Selected Results from the Rational Number Project." In Proceedings of the Ninth International Conference for Psychology of Mathematics Education: Individual Contributions 1: 342-351, L. Streefland, ed. State University of Utrecht, The Netherlands: International Group for the Psychology of Mathematics Education.

Vygotsky, L. 1962. Thought and Language. Cambridge, MA: M.I.T. Press.

Computers and Applications in Secondary School Mathematics[1]

James Fey and Richard A. Good

University of Maryland at College Park

The call for secondary school mathematics curricula that better represent the applications of mathematics is not a new theme in U.S. education. Mathematics teachers are regularly criticized when students are unable to use their skills in science classes. Mathematics teachers themselves frequently complain about the lack of convincing applications of the basic mathematical concepts and skills in teaching materials. Over the past ten to fifteen years there have been a number of thoughtful, imaginative development projects designed to produce materials that meet both needs. However, honest appraisal of the situation in textbooks, tests, and day-to-day instruction will lead most observers to the conclusion that little progress has been made. In all but a few exceptional texts and classes, realistic applications are presented to students only in the form of routine questions about fully mathematized situations. Those applications that require student engagement in the modeling process are commonly drawn from a family of well-known "word problems" involving coins, mixtures, digit puzzles, or rates of work. Teachers find ways to routinize each of these problem types, so even periodic updating of the content does little to motivate students or give them a realistic picture of applied mathematics.

Having made this critical assessment of the situation in U.S. mathematics classes, let me quickly admit that there are plausible justifications. First, one can argue that it does students little good to mathematize a realistic problem situation if they lack the mathematical skill needed to produce answers to the questions in the model. Second, the open discussion and exploration required for realistic problem-solving is not easy to manage within the common institutional constraints of large classes that meet for only 40-50 minutes at a time. Third, high school algebra teachers hear a consistent message that their students must develop proficiency in manipulative skill if they are to succeed in subsequent courses in trigonometry, analytic geometry, and, most of all, calculus. Fourth, many teachers at all levels believe that students cannot acquire the conceptual understanding required for applied problem-solving until they have developed a broad base of factual and skill knowledge.

Each of these justifications for the status quo is challenged by the emergence of calculators and computers as routine problem-solving tools in mathematics. If computers can do algebraic manipulations like factoring, solving equations (including those with literal coefficients), and matrix operations, the importance of student skill in those areas becomes questionable. If programs exist to do differentiation, integration, Taylor series expansion, or limit evaluations in calculus, traditional curricula seem to collapse like a house of cards.

Impressive mathematical tool programs already exist. For example, the muMath package of programs provides symbol manipulation power in an easy-to-use format:

$$\text{SOLVE } (X^2 + 5*X + 6 == 0, X)$$

yields solutions to a quadratic equation, and

$$\text{DIF } (X^2 + 5*X + 6, X)$$

gives the derivative of a function. Commands of equal simplicity yield the other operations of algebra and calculus and readily digest more complex functions. This software is now available for use on moderately-priced microcomputers, and very soon we can expect the same power in handheld hardware.

A natural first reaction to such prospects is to try to diminish the time we spend building students' manipulative skills. At the very least, we should be able to identify a level of performance at which skills are more efficiently done by humans and consign skills requiring a higher level of performance to machine execution. The simple plausibility of this reasoning disguises a quagmire of difficult curricular choices. One faces the almost impossible questions of "How much personal skill is enough?" and "How does proficient performance in algorithmic skills influence understanding of basic concepts?"

After wrestling with these very difficult questions for some time, we formulated an alternative proposal that we feel has tremendous promise. Questioning the future of manipulative skills in algebra, trigonometry, and calculus led us to an analysis of problem situations in which these subjects are commonly applied. We found that a small number of familiar and powerful

682

mathematical ideas are at the heart of most common applications and, further, that a student assisted by the available software need not endure a long skill-building apprenticeship to become an effective problem-solver — if key organizing concepts are well understood.

New Content and Organization for Algebra

The easiest way to describe the spirit and substance of our proposed new look at secondary school mathematics is to begin with some specific illustrative examples in algebra. Consider a familiar and fundamental topic, the quadratic equation $ax^2 + bx + c = 0$. There can be no question that this quantitative model is enormously useful for describing relations among variables in a vast array of situations. Unfortunately, most secondary school students see few of those applications. If they are lucky, they will meet the classic projectile motion problem in a form like

$$-4.9t^2 + 14.7t + 19.6 = 0,$$

complete with parameters carefully chosen to give a solution to the equation by factoring and the answer to the single question: When will a projectile with initial height 19.6 meters and initial velocity 14.7 meters per second return to earth?

In reality, there are many interesting questions that might be asked about the projectile in this situation:

- When is an elevation of m meters reached?
- What will the maximum elevation be and when will it occur?
- How fast will the projectile be falling at any time t?
- What is the average speed?

If a student approaches these application questions with the computer tools now readily available, a whole new world opens up. Simply by graphing or producing a table of values for the function $H(t) = -4.9t^2 + 14.7t + 19.6$, students can determine very good approximations to the desired quantities. Major issues in elementary calculus can be studied by students who have done none of the laborious skill-building that is the ordinary preparation for that subject.

What algebra students _do_ need as preparation for these questions about projectile motion is a fundamental change in their view of the nature and uses of algebraic expressions. Traditional high school algebra deals almost exclusively with questions of the form "Find x that makes a well-behaved expression equal to zero." This "Find x" mentality misses a crucial point about the sources of mathematical questions — the study of relations among two or more quantitative variables, frequently functional relations. The expression $-4.9t^2 + 14.7t + 19.6$ is of interest not only for the single positive value of t that makes it zero. It provides useful information for any value of t within a reasonable domain. It expresses a dynamic relation between quantities that vary, not a static condition disguising a fixed but unknown number.

The questions that one would ask about the quadratic function have structural analogues in the other familiar classes of elementary functions. For instance, the tides in an ocean harbor are described roughly by functions like

$$W(t) = 3 \sin(.5t) + 20.$$

In studying this function, we naturally ask for the value of W at some specified time t, the value of t that yields some specified water depth W, the time and value of maximum or minimum depth, the rate of change in water depth at a specified time t, and the average water depth. Again, a computer-generated graph or table of values for the function provides an effective tool for answering a range of interesting questions.

The patterns underlying these two illustrations as well as many others involving elementary functions are as follows:

For a given function f(x), find

1. f(x) for x = a

2. x so that f(x) = a

3. x so that maximum or minimum values of f(x) occur

4. the rate of change in f near x = a

5. the average value of f over the interval (a,b).

For nearly every interesting function, computer utilities make all five questions accessible in some intellectually honest and mathematically powerful form to students who have not followed the conventional regimen of skill development.

They open a fast track to polynomial, trigonometric, exponential, and algebraic functions that model interesting phenomena in the physical, biological, economic, and social worlds. Computing offers an opportunity to turn the secondary school mathematics curriculum on its head. Instead of meeting applications as a reward for years of preparation, students can begin with the most natural and motivating aspect of mathematics — its applications.

The new curriculum structure that this environment makes possible replaces the traditional manipulative skill algebra program with a study of the major families of elementary functions. For each type of function, we envision a developmental sequence like the following:

Stage 1. Recognition and mathematization of relations.

Through exploration of interesting situations in which quantitative relations are modeled well by rules of the type in focus (linear, quadratic, trigonometric, etc.), students are given an opportunity to develop the ability to identify functional relations and to express these relations in suitable mathematical form. Their mathematization skills should include translation of rules or physical principles into function expressions ($d = rt$, $V = IR$, etc.) but also the realistic and important strategy of fitting functions to experimental data (using a curve-fitting computer utility).

Stage 2. Answering questions about functions.

Strategies for studying function behavior can progress from informal mental approximation to the use of sophisticated computer utilities: guess and test by hand, computer-generated tables of values and graphs, then exact solution by computer programs like muMath or TK!Solver.

Stage 3. Formal organization and verification.

A structured summary of principles and methods discovered in exploratory work can highlight powerful generalizations that will help students retain and expand their mathematical understanding. The important properties of each family of functions and of all functions taken together (number and location of zeroes, extrema, operations of combination, etc.) can be formally stated and verified through deductive mathematics. This phase of activity will also raise new questions and stimulate further investigations in the exploratory-to-confirmatory cycle.

While a curriculum that follows the suggested pattern represents a radical departure from current school mathematics programs, it offers a number of powerful advantages. It begins with real-world situations and promises the best motivation — learning something obviously useful. It sidesteps the almost unsolvable question of "How much technique is enough?" by allowing students to progress as far as their needs and interests permit. It stresses informal, easy to remember, and powerful successive approximation and graphic methods that should do a lot to build the intuition our students so often miss in our haste to teach manipulative rules for the many algorithms required by formal methods. Finally, it places the function concept at the heart of the curriculum, preparing students for the dynamic, global quantitative thinking that typifies most models they will encounter in future mathematics and in applications of mathematics.

Our research at the University of Maryland indicates that such an approach can be effective — notably with students who have failed at the traditional skill-oriented program. In Utah Dick Lesh is exploring similar approaches with encouraging results. Our students generally feel they are getting a realistic picture of how mathematics is useful, and that our expectations of them are far more appropriate than those of the traditional program.

The array of computer utilities available for mathematics also extends an invitation to include in the algebra curriculum many new topics which, because of their computational complexity, have usually been judged beyond all but the best high school and early college students. One of the most striking new possibilities is the fitting of function rules to scatter plots of realistic data by a least squares procedure. For instance, in many business problems a relation of crucial importance is the demand curve relating price (p) and projected sales (s) of a product. The typical class procedure is to present such a rule with the familiar introduction "Suppose that s = -1000p + 15,000." Such equations routinely drop out of the mathematical sky on skeptical students. As a result, what was intended to be an illustration of mathematics at work becomes simply another exercise in symbol manipulation, since students need not confront the meaning of the variables or the complexities of discovering a good predictive demand curve. Genuine demand curves undoubtedly come from market research surveys that produce approximately linear scatter plots of (price, sales) pairs. Students can readily understand the principle of least squares fit to such data, and, with computer assistance in the calculations, they can use the methods to get a far more realistic and vivid applications experience.

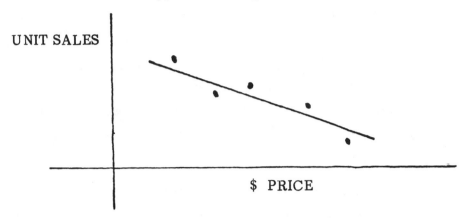

The use of computer mathematics tools relieves the tedium of complex calculations, but it also presents new challenges that demand a higher level of analytical thinking. For instance, the family of optimal assignment problems presents situations of the following sort:

"Given a complex task with m jobs and a pool of n potential workers, what assignment of workers to jobs is 'best'?"

Commonly, each possible pairing of a worker and a job is given a rating (effectiveness or cost), leading to an m x n array from which choices must be made. If each worker can be assigned at most one job, the problem is to select one entry in each job row and at most one in each worker column with the criterion of maximizing total effectiveness (or minimizing cost). The naive approach to this problem is to evaluate the effectiveness or cost of each possible assignment. This strategy might be suitable for very small values of the parameters m and n. For slightly larger m and n, a computer will help in comparison of the possibilities. However, for realistic parameter values the number of combinations to be checked grows so large that such exhaustive searching is impractical, even with a computer. Instead, thoughtful analysis is required to reduce the problem to manageable size. Algebraic concepts, principles, and methods play a crucial role in this process. Appreciation of the need for such analytic simplification and understanding of its effects can be enhanced by exploration on sample data at computer speed.

The optimal assignment problems are only one example of multiple variable linear problems that are important throughout mathematics and in its applications. For most secondary school students today, the only experience that heads in this direction is the solution of two equations in two unknowns. Because students are commonly expected to master all techniques for identifying roots, however, the curriculum scarcely gets past the case of a unique root. Converting the time-consuming algorithms to computer programs removes the shackles that confine problems to simple arithmetic and "nice-looking" answers. A far broader view is gained when students are able to analyze systems with many roots and discover the richness of interrelationships that occur for $n > 2$.

Opportunities in other directions are opened with computers sharing the computational load. For example, many mathematical models are faithful to reality only if integer values for the variables are admitted. In current elementary algebra courses, such restricted answers are achieved by application of "regular" methods to problems with equation parameters that have been carefully planned to yield only integer solutions. With computer assistance, such artificial and misleading restrictions need not dictate the choice of problem material. Similarly, the study of systems of inequalities can be approached in efficient and realistic problem settings, because the complex algebra is no longer a barrier.

We believe that each of the indicated themes should become more prominent in the algebra experience of secondary school students. Initial encounters may involve informal problem recognition and mathematization tasks, with the computer tools helping to find specific answers and suggesting patterns worthy of formalization into theory at a later stage. As each family of concepts and methods is explored and crystallized into an organized body of mathematical results, it will provide raw material for a cycle of deeper investigations at some appropriate later time.

Where to Begin?

The curriculum changes outlined above cannot take place quickly. The new content selection and organization must be translated into curriculum materials of demonstrable effectiveness. Even with that task accomplished, selling the new type of program to teachers, to the public supporters of mathematics education, and to the publishers of influential textbooks and tests will be a major job. Work on prototypes of radical new curricula of the future is beginning. However, for those who are convinced by the logic of arguments for this type of curriculum change, there are ways to move in the right direction immediately.

In algebra, eliminating some particular skills from the usual litany will not be easy. A clean break and movement to something substantially different might be easier than evolution. Nonetheless, certain modifications are possible. First, it should be easy to introduce the function concept early in algebra and use it as a unifying theme throughout the course. The value of successive approximation can be demonstrated at many stages with simple computer programs and graphic methods. The concepts and methods required to deal with situations of linear complexity can be motivated and illustrated by computer-based explorations extending current "small n" studies.

Finally, and perhaps most important, it should be possible for every teacher to place greater stress on situations in which mathematics is used to model the structure of real-life situations. There is a growing supply of resource material for this purpose, and its use will direct attention to important themes in the curriculum of the future.

Note

1. Work on this paper was supported in part by the National Science Foundation under NSF award no. SED80-24425 (DISE). Any opinions, findings, conclusions, or recommendations expressed herein are those of the authors and do not necessarily reflect the views of the National Science Foundation.

References

Conference Board of the Mathematical Sciences. 1983. The Mathematical Sciences Curriculum K-12: What Is Still Fundamental and What Is Not. Washington, DC: CBMS.

Coxford, A. 1985. "School Algebra: What Is Still Fundamental and What Is Not?" In The Secondary School Mathematics Curriculum: 53-64, C. Hirsch, ed. Reston, VA: National Council of Teachers of Mathematics.

Fey, James T., ed. 1984. Computing and Mathematics: The Impact on Secondary School Curricula. Reston, VA: National Council of Teachers of Mathematics.

Fey, J. and R.A. Good. 1985. "Rethinking the Sequence and Priorities of High School Mathematics." In The Secondary School Mathematics Curriculum: 43-52, C. Hirsch, ed. Reston, VA: National Council of Teachers of Mathematics.

Fey, J.T. and M. Kathleen Heid. 1984. "Imperatives and Possibilities for New Curricula in Secondary School Mathematics." In Computers in Mathematics Education: 20-29, V. Hansen, ed. Reston, VA: National Council of Teachers of Mathematics.

Heid, M. Kathleen. 1983. "Calculus with muMath." Computing Teacher 11: 46-49.

Ralston, A. 1985. "The Really New College Mathematics and Its Impact on the High School Curriculum." In The Secondary School Mathematics Curriculum: 29-42, C. Hirsch, ed. Reston, VA: National Council of Teachers of Mathematics.

Ralston, Anthony and Gail S. Young, eds. 1983. The Future of College Mathematics. New York: Springer-Verlag.

Usiskin, Zalman. 1980. "What Should Not Be in the Algebra and Geometry Curricula of Average College-Bound Students?" Mathematics Teacher 73: 413-424.

_____. 1985. "We Need Another Revolution in Secondary School Mathematics." In The Secondary School Mathematics Curriculum: 1-21, C. Hirsch, ed. Reston, VA: National Council of Teachers of Mathematics.

690

ABOUT THE AUTHORS

Merlyn Behr

Dr. Behr is Professor of Mathematical Sciences and Education at Northern Illinois University. His experience in teacher education includes work with undergraduate preservice elementary and secondary mathematics teachers and work at the graduate level.

Professor Behr has been involved extensively in research concerned with how children learn mathematics. His early interests were with children's learning of computational algorithms. For the past ten years his research has been directed at investigations of how children learn fraction and rational number concepts, with more recent attention given to the development of proportional reasoning ability in children. Through his research work Professor Behr has had extensive experience teaching children mathematics concepts.

He is currently involved in the following research projects:
1. The development of computer software to define a Part-Whole Microworld. This microworld will be designed to support instruction to extend children's part-whole schema to include multiplicative relationships among components of a part-whole structure.
2. Continuing work on investigating children's understanding of fraction concepts and proportional reasoning skills.
3. Co-director with Dr. James Hiebert of the NCTM research agenda working group on middle school number concepts.

Max S. Bell

Max S. Bell is Professor of Education at the University of Chicago, where he specializes in mathematics education research and development. His academic degrees in mathematics and mathematics education are from U.C.L.A., the University of Chicago, and the University of Michigan. In general, his teaching, research, development work, and publications focus on making mathematics accessible to most people, in part by emphasizing the uses of mathematics in instruction at all levels.

In early work with the School Mathematics Study Group he was an author or advisor in the writing of <u>Mathematical Uses and Models in Our Everyday World</u>, The New Mathematical Library, and SMSG's "second round" integrated curriculum for grades 7-10. Other work includes studies of the uses of calculators in schooling, books and articles on everyday uses of mathematics, and film or video teaching materials. His recent research concerns the capabilities of 5-9-year-old children, while his development work focuses on pedagogic invention and teaching materials for these children.

Professor Bell has served as Director of the Geometry Project with WICAT Systems and continues as an advisor for IBM Education Systems. He is currently Director of the Elementary Teacher and Classroom Materials subcomponent of the University of Chicago School Mathematics Project.

Hans Brolin

Hans Brolin, Professor of Mathematics at the Institute of Education of the University of Uppsala, Sweden, has long been interested in how the use of pocket calculators in mathematics teaching influences methodology and course content. He has also investigated how the use of calculators in the lower grades affects students' non-algorithmic basic skills and problem solving ability. In addition to writing secondary school textbooks, Dr. Brolin has co-authored <u>THE ARK PROJECT</u> (1984), an analysis of the effects of using calculators, and <u>Computers and the Teaching of Mathematics</u> (1986).

Dr. Brolin is currently head of The ADM Project (Analysis of the Role of the Computer in Mathematics Teaching). This research project is concerned with the following questions:

- How can computers be used in the teaching of mathematics?
- To what extent and how should students learn to use the computer as a tool in different mathematical activities?

- If the computer is to take over more and more routine mathematical tasks, what should the goals and methods of mathematics teaching be?
- How can students at an early stage be made aware of the limitations of the computer as a mathematical tool?

Felix E. Browder

Felix E. Browder has recently been appointed Vice President for Research at Rutgers University after having been the Max Mason Distinguished Service Professor of Mathematics at the University of Chicago. He received his B.S. from M.I.T. in 1946 and his Ph.D. in Mathematics from Princeton in 1948. He has served on the faculty at M.I.T., Boston University, Brandeis University, and Yale University before becoming Professor of Mathematics at the University of Chicago in 1963. He is a member of the National Academy of Sciences and a Fellow of the American Academy of Arts and Sciences. He served as Chairman of the Mathematics Department at the University of Chicago during 1972-77 and 1980-85.

Dr. Browder's mathematical research interests lie in the areas of nonlinear functional analysis and nonlinear partial differential equations. He is the author of over 200 research articles and books.

Maxim Bruckheimer

Maxim Bruckheimer was born in Germany, emigrated first to England, where he received his Ph.D. from Southampton University, and then to Israel, where he now resides. Since 1976 he has been head of the Weizmann Institute's Science Teaching Department, which develops curriculum for the Israeli educational system.

In addition to the many books and articles he has authored, co-authored, edited, and translated, Professor Bruckheimer has founded and helped to edit a Hebrew-language journal for teachers, improved and expanded the in-service teacher training program, and fostered international cooperation in research on curriculum development.

His current research interests include using the history of mathematics in teacher training; developing strategies for instilling in students a deeper interest in mathematics; evaluating the effectiveness of curriculum materials; evaluating and enhancing in-service activities; improving junior high school curriculum; and studying the problems of socially disadvantaged junior high school students.

Alphonse Buccino

Alphonse Buccino is the Dean of the College of Education at the University of Georgia and holds a faculty appointment in the Department of Mathematics Education. He also is recognized as an authority on education policy analysis and scientific and engineering personnel development.

Dr. Buccino earned his B.S., M.S., and Ph.D. degrees in mathematics from the University of Chicago. He joined the mathematics faculty of Roosevelt University in 1961 and became Chairman of the Department of Mathematics at DePaul University in 1963, where he remained until joining the National Science Foundation. At DePaul University he organized revision of the bachelors and masters programs in mathematics, including the introduction of new programs in applied mathematics and computer science.

From 1970 to 1984 Dr. Buccino was a staff member at the National Science Foundation. There he served in various capacities in science and engineering education, including the management and operation of programs and the design and maintenance of organizational capabilities in planning, policy analysis, and evaluation. The object of this work was to monitor the condition of the nation's science and engineering education system and its ability to meet national needs for human resources in science and technology. He was a member of the Foundation's Management Council and has served on agency-wide task groups. While at the Foundation, Dr. Buccino conducted seminars in Science and Technology Policy and Societal Impacts of Science and Technology for the Federal Executive Institute in Charlottesville, Virginia.

Hugh Burkhardt

Hugh Burkhardt is Professor of Mathematical Education in the Department of Mathematics and Director of the Shell Centre for Mathematical Education at the University of Nottingham. An applied mathematician and theoretical physicist specializing in the theory of elementary particles, he graduated from Balliol College, Oxford and the University of Birmingham.

In the early 1960's Birmingham was involved in significant developments in school mathematics. Hugh Burkhardt was invited to organize the applied element of courses for teachers, which focused initially on Newtonian mechanics, the major area of application in British high schools. His work with teachers led him to develop mathematical modeling of standard and unfamiliar situations as a curricular component at all school and undergraduate levels.

Since 1976 Professor Burkhardt has led the Shell Centre's programs of linked research and curriculum development in mathematical education, interpreted broadly to cover the uses of mathematics in other school subjects and, particularly, in the outside world. Some of the Centre's work, particularly with microcomputers, is concerned with structured thinking in subject areas as diverse as language arts and music. The Centre directs numerous projects, including work on the diagnostic approach to teaching, the use of a single microcomputer as "teaching assistant" in the classroom, and curriculum development linked to the public examination system and aimed at introducing a broader range of curricular activities, including non-routine pure problem solving and real world problem solving.

Professor Burkhardt is the United Kingdom National Representative on the International Commission on Mathematical Instruction.

Thomas P. Carpenter

Thomas P. Carpenter is Professor of Curriculum and Instruction at the University of Wisconsin-Madison. He graduated from Stanford University with a B.S. in Mathematics and from the University of Wisconsin with a Ph.D. in

Curriculum and Instruction. He has taught at Boston University and San Diego State University. He has recently been appointed editor of the <u>Journal for Research in Mathematics Education.</u>

Professor Carpenter's research has focused on how children acquire quantitative concepts. Most recently he has been studying how addition and subtraction concepts develop in children. He is currently involved in a National Science Foundation project to study how cognitive and instructional science research can be applied to designing instruction in elementary mathematics.

Elmar Cohors-Fresenborg

Elmar Cohors-Fresenborg studied mathematics, physics, and economics at the University of Münster (West Germany) and the University of Fribourg (Switzerland). He received his doctorate in foundations of mathematics/theoretical computer science from the University of Münster. He taught mathematics education in Münster and Flensburg and is now serving as Professor of Mathematics Education at the University of Osnabrück, where he heads the group "Foundations of Mathematics and Mathematic Education."

Professor Cohors-Fresenborg's main research interests are in two areas, curriculum development and experimental research on mathematical thinking processes. In the first area, he is exploring ways to introduce the fundamental ideas of automata and computer programming to primary and secondary school students. For this purpose he has worked on the development of "Dynamic Mazes" and the "Registermachine" model computer. In the second area, Professor Cohors-Fresenborg's research interests are the role of different (verbal and non-verbal) forms of representing mathematical concepts, the structure of mathematical thinking, and the structure of cognitive strategies in algorithmic thinking.

F. Joe Crosswhite

Joe Crosswhite holds a joint appointment in the Department of Mathematics and the Center for Excellence in Education at Northern Arizona University. He was previously Professor of Mathematics Education at The Ohio State University. Dr. Crosswhite is current Chairman of the Conference Board of the Mathematical Sciences, immediate Past President of the National Council of Teachers of Mathematics, and a member of the Executive Committee of the Mathematical Sciences Education Board of the National Research Council, the Steering Committee of the Triangle Coalition for Science and Technology Education, and the Advisory Committee for the Science and Engineering Education Directorate of the National Science Foundation.

Dr. Crosswhite's current research interests are mathematics curriculum and instruction at the secondary school and undergraduate college levels and mathematics teacher education.

Warren D. Crown

Warren Crown holds a bachelor's degree in physics from Carnegie-Mellon University in Pittsburgh and a Ph.D. in Teacher Education from the University of Chicago. He has taught mathematics in public elementary schools and high schools in Philadelphia and at Governors State University in Illinois. Currently he is Associate Professor of Mathematics Education at Rutgers, The State University of New Jersey. From 1983 through 1985 Dr. Crown was on leave from Rutgers serving as the Project Director of the WICAT Elementary Mathematics Curriculum Project at WICAT Systems in Orem, Utah. This project is discussed in his paper in this volume.

In addition to the IBM Math Concepts and Math Practice programs which grew out of the work at WICAT, Professor Crown is an author of the 1985 Scott, Foresman elementary mathematics program, Invitation to Mathematics, as well as editorial director of the NEA Library Series, Parent Plus Child Equals Success. His interests in math education include the use of computers as a means of delivering instruction, curriculum development in general, and diagnosis.

Renée de Graeve

Renée de Graeve has been teaching mathematics at the University of Grenoble for twenty years. Mlle. de Graeve taught the first biology and mathematics cycle at the university, where programmable pocket calculators were introduced to facilitate observation, visualization, conjectures and simulations. The development of microcomputers has given her the opportunity to experiment with new pedagogical methods and look for applications of information science in the classroom.

Mlle. de Graeve and Professor Bernard Cornu introduced a course at the Institut Fourier of Grenoble called "Information Science for Teaching Mathematics." The course is designed for future mathematics teachers, but is attended also by many in-service teachers.

Since 1983 Mlle. de Graeve has also been working at I.R.E.M. (Institute for Research on Mathematical Education). There, the "Élémentaire" group specializes in the use of computers in primary school (their research focuses on LOGO and geometry). This group also participates in the in-service training of secondary teachers. In 1985–86 the topics were LOGO and school difficulties, and also LOGO and HANDICAP.

Mlle. de Graeve is a member of the selection committee of the journal Grand N, a review of primary school mathematics with a great deal of influence on teachers because of its focus on new experiments directly applicable in the classroom.

John A. Dossey

John A. Dossey received his B.S. and M.S. degrees in mathematics at Illinois State University. Following a period of teaching at the junior and senior high school level, he received a Ph.D. in mathematics education from the University of Illinois at Urbana-Champaign. His research work in mathematics education has focused on concept learning and teaching, evaluation, and comparative achievement studies.

Dr. Dossey has co-authored three collegiate textbooks and produced a number of research articles concerning mathematics education. He is currently serving as President of the National Council of Teachers of Mathematics. In addition, he is a member of the National Research Council's Mathematical Sciences Education Board and the Conference Board of the Mathematical Sciences.

Elizabeth Fennema

Elizabeth Fennema is Professor of Curriculum and Instruction at the University of Wisconsin-Madison. She graduated from Kansas State University with a B.A. in Psychology, and from the University of Wisconsin with an M.A. and a Ph.D. in Curriculum and Instruction. She was editor of Mathematics Education Research: Implications for the 1980s.

Professor Fennema's primary research focus has been gender-related differences in mathematics. This research has included documentation of sex-related differences and identification of affective, cognitive, and educational variables related to these differences, as well as development and evaluation of intervention programs designed to eliminate them. She is currently the director of a National Science Foundation project to study how cognitive and instructional science research can be applied to designing instruction in elementary mathematics.

James T. Fey

James T. Fey is Professor in the Departments of Curriculum and Instruction and Mathematics at the University of Maryland in College Park. He joined the Maryland faculty in 1969 after receiving his Ph.D. in Mathematics Education at Teachers College, Columbia University. Beginning with work on the Secondary School Mathematics Curriculum Improvement Study, he has focused special attention on innovative curricula for secondary school mathematics. Most recently this has included two projects that have produced curriculum materials integrating computing and mathematics — first at the beginning college level and currently in elementary algebra.

As a member of the National Research Council Mathematical Sciences Education Board, Dr. Fey is now involved in projects related to professional standards for school mathematics and prospects for future curricula.

Hans Freudenthal

Hans Freudenthal was born in Luckenwald, Germany in 1905. In 1923 he undertook a course of study in Mathematics, with Physics and Philosophy as minor subjects. In 1930 he received his Ph.D. after defending his thesis on topological groups. Between 1930 and 1946 he worked as Assistant to the outstanding Dutch topologist and logician Brouwer at Amsterdam University, except during the war years 1941-1945. Between 1946 and 1976 he served as Professor and then Director of Mathematics at Utrecht. Since 1970 he has been Director and Advisor of IOWO in Utrecht (Institute for the Development of Mathematics Education). He has served as President of the International Commission on Mathematical Information and was one of the principal originators of the International Congresses on Mathematics Education, starting with ICME 1 at Lyons, France in 1969.

Professor Freudenthal has published voluminously on topology, Lie groups, abstract analysis, probability, geometry, history, philosophy, and mathematics education. His chief interest and main task at present is advising researchers on mathematics education.

Professor Hans Freudenthal is generally regarded as the world's foremost mathematics educator of the past 25 years.

Hiroshi Fujita

Hiroshi Fujita is Professor of Mathematics at the University of Tokyo. After receiving his degree of Doctor of Science from the University of Tokyo in 1961, he held teaching positions in the Faculty of Science and Faculty of Engineering there and presently holds an additional professorship at the Research

700

Institute of Mathematical Sciences of Kyoto University. He has taught at some U.S. universities, including Stanford University, the University of Wisconsin and New York University. He served for two years (1981-1983) as the President of the Mathematical Society of Japan and took the role of Chief Coordinator for the U.S.-Japan seminar on applied analysis. His original specialty was applied analysis; he currently leads a study group on mathematical education in Japan.

Professor Fujita is a member of IPC for ICME 6, to be held in 1988. In addition to a number of research papers and textbooks in mathematics, he has published several papers on mathematical education, one of which is a joint paper with Professor F. Terada in the <u>Proceedings</u> of the ICMI-JSME Regional Conference on Mathematical Education, Tokyo, 1983.

Karen C. Fuson

Karen Fuson is an Associate Professor in the School of Education at Northwestern University. She received her Ph.D. degree in Mathematics Education from the University of Chicago. Her research has focused on the development of mathematical concepts and the development of self-regulating aspects of speech in young children. Her professional work has included in-service and pre-service training of teachers in mathematics.

Her recent mathematics education research has been concentrated in two areas. Several papers and a forthcoming book trace the development of concepts of number in children from age 2 to age 8. A second body of papers (including the chapter here) reports on the effectiveness of teaching packages developed to accelerate and improve the teaching of addition and subtraction of whole numbers in elementary school.

Claude Gaulin

Claude Gaulin is Professor in Mathematics Education at Laval University in Quebec City. For many years he has been the director of PPMM-Laval, an important in-service long-distance teacher education program in mathematics for elementary teachers in the Greater Quebec area.

His main research interests are trends in mathematics education at the international level; psychological and didactic aspects of spatial abilities; socio-cultural factors influencing mathematics education at various levels; effects of hand-held calculators; and problem solving and problem posing in mathematics.

Professor Gaulin is currently Vice-President of the Canadian Mathematics Education Study Group (CMESG). He is also Vice-President of the Comite Interamericano de Educacion Matematica. He has served as president of the Commission international pour l'etude et l'amerlioration de l'enseignement des mathematiques in Europe and a member of the Program Committee of ICME 4 (Berkeley, 1980).

Professor Gaulin is a member of the editorial board of Recherches en didactique des mathematiques (France); For the Learning of Mathematics (Canada); L'Insegnamento della matematica e delle scienze integrate (Italy); and Educatio Mathematica (Poland). He has been involved for many years in programs of cooperation with Morocco, Venezuela, Brazil and Spain.

Richard A. Good

Richard A. Good has served in the Department of Mathematics at the University of Maryland-College Park since 1945. Undergraduate training at Ashland College was followed by doctoral study at the University of Wisconsin-Madison with a specialization in abstract algebra. Good pioneered in the television medium, first with a calculus course at Walter Reed Army Medical Center and later with an on-campus course for freshmen. Under the sponsorship of the National Science Foundation several Summer Institutes and many In-Service Institutes involved him both as director and as an instructor. In curriculum development he has participated in the University of Maryland Mathematics Project, written for the School Mathematics Study Group, and consulted for the Secondary School Mathematics Curriculum Improvement Study. In recent years, besides struggling with the opportunities for mixing computers and elementary algebra, he has tried to enhance several collegiate mathematics courses with the computer using APL.

Heini Halberstam

Since 1980 Professor Heini Halberstam has been Professor and Head of the Mathematics Department of the University of Illinois at Urbana-Champaign. Previously he held similar positions at various universities in England and Ireland, most recently in Nottingham. His Ph.D. is from the University of London. Professor Halberstam has served on the Council of the London Mathematics Society, the Schools Council of the British Mathematics Committee, the British National Committee for Mathematics, and the Royal Society - IMA Committee on Mathematical Education. He is also former Chairman of the Joint Mathematical Council of the United Kingdom.

Professor Halberstam's research lies in the theory of numbers, but he has a long-standing interest in mathematics education; he was a founding member of the Shell Centre for Mathematical Education, Nottingham University, and acting director of this Centre in 1968.

Rina Hershkowitz

Rina Hershkowitz has been involved in science education in Israel for many years. Her experience in teaching mathematics in junior high and high school led her to become involved in curriculum development. She wrote scripts for mathematics programs for educational television, and in 1967 she joined the newly-formed Mathematics Group of the Science Teaching Department of the Weizmann Institute of Science.

In its early years the Mathematics Group focused mainly on creating and implementing a curriculum for all students. More recently, however, the emphasis has been shifting to creating materials and strategies that meet the needs of particular subgroups of the student population, such as the talented and the culturally disadvantaged. The Group's materials are now being used in about 200 junior high schools and 450 elementary schools. Mrs. Hershkowitz has written mathematics textbooks and teacher's guides, conducted inservice teacher training, coordinated research, and developed and implemented remedial materials.

Peter J. Hilton

Peter Hilton was born in London, England. He was educated at Oxford University and spent the years 1942–45 of the Second World War as a cryptanalyst at Bletchley Park, where he became a close friend of J.H.C. (Henry) Whitehead. On demobilization he returned to Oxford to obtain a Ph.D. in algebraic topology under Whitehead's supervision. After an academic career in England, culminating in his holding the Mason Chair of Pure Mathematics at the University of Birmingham, he came to the U.S. in 1962 to become Professor of Mathematics at Cornell University. He is now Distinguished Professor of Mathematics at SUNY-Binghamton.

Hilton's research interests are, as they have always been, in algebraic topology and homological algebra, but he has taken an increasing interest in mathematics education, at all levels, since coming to the U.S. He has been Chairman of the U.S. Commission on Mathematical Instruction, Chairman of the National Research Council Committee on Applied Mathematics Training, Secretary of the International Commission, and First Vice-President of the Mathematical Association of America.

Alan Hoffer

Alan Hoffer has taught mathematics to students from the primary school level through graduate school. He has worked with computers intermittently for nearly thirty years. He is now a Professor of Mathematics and Computer Science Education at Boston University. Prior to his present appointment, Professor Hoffer was a member of the Department of Mathematics at the University of Oregon where he developed a preservice and inservice education program for mathematics teachers; established the Mathematics Resource Center; and directed various projects, including the Mathematics Resource Project that was funded by the National Science Foundation. In addition he has authored a secondary geometry text as well as conducted research and acted as advisor to projects that deal with the learning of geometry. He has had visiting faculty appointments at the University of California at Berkeley, Ohio State University, the University of Utrecht in the Netherlands, and the University of Chicago. He has also worked in computer software development as a project director in mathematics for WICAT Systems, Inc.

Mary Grace Kantowski

Mary Grace Kantowski received her doctorate in Mathematics Education from the University of Georgia. She is currently Professor of Mathematics Education at the University of Florida, where she teaches classes on using the microcomputer in mathematics instruction in middle and secondary schools. Her experience as a secondary school teacher was a strong asset to her work on the steering committee of the National Council of Teachers of Mathematics project that resulted in the NCTM statement entitled The Impact of Computing Technology on School Mathematics. She is the author of the chapter on geometry in the NCTM publication Computing in Mathematics. She has given numerous invited papers and workshops on the applications of the microcomputer in mathematics instruction and has been the director of several federally funded projects in this area, including a project on using the microcomputer for problem solving with talented students.

Professor Kantowski's research interests include processes involved in mathematical problem solving among middle and secondary school students, with an emphasis on nonroutine problems in geometry and number theory. Another special area of research interest is the role of the computer in problem solving and the ways in which problem solving processes are influenced by use of the computer.

Jeremy Kilpatrick

Jeremy Kilpatrick has been Professor of Mathematics Education at the University of Georgia since 1975. Before that, he taught at Teachers College,

Columbia University. He holds degrees from the University of California, Berkeley, and Stanford University, where he received his doctorate in 1967 under the direction of E.G. Begle. He is currently serving a second term as editor of the Journal for Research in Mathematics Education. He is a member of the Board of Governors of the Mathematical Association of America and the Mathematical Sciences Education Board. He has served on the Mathematical Sciences Advisory Committee of the College Board since 1978 and is currently the chairman of that committee.

Professor Kilpatrick is editor with Izaak Wirszup of the Soviet Studies in the Psychology of Learning and Teaching Mathematics and The Psychology of Mathematical Abilities in Schoolchildren by V.A. Krutetskii. He is author, with George Polya, of The Stanford Mathematics Problem Book and, with Geoffrey Howson and Christine Keitel, of Curriculum Development in Mathematics. He has reviewed trends in evaluation around the world as well as recent research in the United States. Recently, he was the principal writer and consultant for the College Board's publication Academic Preparation in Mathematics. His research interests deal with instruction in problem solving, the structure of mathematical abilities, and the evaluation of learning in mathematics.

Richard Lesh

Dr. Richard Lesh is Professor of Mathematics and Psychology at Northwestern University, currently on leave as the Mathematics and Science Education Director at the World Institute for Computer Assisted Teaching (WICAT) in Orem, Utah. WICAT has produced complete computer-based courses for IBM and for CDC (PLATO) in Primary School Math, Middle School Math, Algebra I and II, Geometry, Calculus, Chemistry, SAT preparation, Adult Basic Skills, and a number of other areas.

Dr. Lesh's research projects have been in the areas of rational number concept development, proportional reasoning processes, problem solving, and mathematics teacher training. He has written textbooks for both children and adults, in addition to computer-based software, and is currently focusing on problem solving, teacher training, and the role of computer utilities and computer representations in mathematics learning and problem solving.

John Ling

John Ling studied mathematics at Cambridge University and taught in secondary schools from 1966 until 1977. As a teacher he became very familiar with the School Mathematics Project books, which started appearing in the early 1960s.

In 1974 he participated in a project based at Nottingham University which was undertaking a critical review of the mathematics curriculum. He led a group working on the relationship between mathematics and other subjects and wrote a book for the project: Mathematics Across the Curriculum (Blackie 1977).

In 1977 Dr. Ling was appointed by the SMP to lead the work on a new course for secondary schools, which has become known as SMP 11-16, and this work has occupied him up to the present. He has been a member of various working groups of the British Mathematical Association and has contributed to ICME 4 and ICME 5.

Tatsuro Miwa

Tatsuro Miwa, currently Professor of Mathematics Education at the University of Tsukuba, received his B.S. and M.S. degrees from Tokyo University. He taught mathematics at the Senior High School connected with Tokyo University before serving as Professor at Osaka Kyoiku University. He has served as Director of the Japan Society of Mathematical Education.

Professor Miwa's recent research has been described in his articles on cooperation between teachers of mathematics and science, the mathematization of situations outside mathematics, mathematical modeling in the school, and mathematics education in the United States in the 1980's.

Anne-Marie Pastel

Anne-Marie Pastel studied at the Scientific, Medical, and Technical University of Grenoble. Her main research interests are in the areas of boolean algebra, particularly the geometrical characterization of some logical functions, and graph theory. She has worked for three years in the C.I.A.P. (Centre pour l'informatique et ses applications pédagogiques), where teachers receive a year of significant training in information science and pedagogical uses of computers. She is now involved in the Grenoble I.R.E.M. (Institut de recherche sur l'enseignement des mathématiques), in the group "Mathématiques et Informatique au college."

Boyan Penkov

Boyan Penkov received his M.A. (1950) and Ph.D. (1958) from Sofia University. In the period 1958-1965 he was a Fellow of the Mathematics Institute of the Bulgarian Academy of Sciences. In 1965 he was appointed Associate Professor of Stochastics in the Mathematics Department of Sofia University, where he served two terms as Head of Stochastics. Boyan Penkov has also been a Visiting Professor in the Department of Statistics, University of California at Berkeley, and at Moscow's Steklov Mathematical Institute. His main research interests are approximation theory and mathematical statistics.

Professor Penkov has served as Vice-President of the Union of Bulgarian Mathematicians and Secretary of the Technical Committee on Computer Education of the International Federation for Information Processing. Since 1984 he has been Scientific Secretary of the Research Group on Education, Sofia. He is vice-chairman of the editorial board of the Physics and Mathematics Journal and of the journal Serdica.

Anthony Peressini

Anthony Peressini received his B.S. at the College of Great Falls, Montana and his M.A. and Ph.D. at Washington State University. In 1961 he joined the mathematics faculty at the University of Illinois.

708

Professor Peressini's research interests are in Banach lattices and operator theory. He is editor of the Illinois Universities Math Bulletin, as well as author of a monograph and college textbooks. At the University of Illinois he has received the Campus Award for Excellence in Undergraduate Teaching.

Penelope L. Peterson

Penelope L. Peterson is Sears-Bascom Professor of Educational Psychology at the University of Wisconsin-Madison. She received a B.S. degree from Iowa State University in Psychology and Philosophy and M.A. and Ph.D. degrees from Stanford University in Educational Psychology. She is editor of Review of Educational Research, and also edited the volume Research on Teaching: Concepts, Findings, and Implications. She received the 1986 American Educational Research Association Raymond B. Cattell Early Career Award for her outstanding programmatic research in education.

Professor Peterson's primary research focus has been on the study of teaching. In particular, she has studied teachers' thought processes and decision making and teaching effectiveness. She is currently working on a National Science Foundation project to study how cognitive and instructional science research can be applied to designing instruction and elementary mathematics.

Thomas R. Post

Thomas R. Post is Professor of Mathematics Education at the University of Minnesota at Minneapolis. He has conducted research dealing with mathematical learning and concept development. He is interested in the implications which psychologically related findings have for the development of instructional activities, particularly those which utilize manipulative materials. Dr. Post has co-authored two texts; one is related to the mathematics laboratory, the other is concerned with interdisciplinary approaches to curriculum. His latest effort is an edited book dealing with research-based methods for teachers of elementary and junior high school mathematics.

Professor Post has served as national co-chairperson of the Special Interest Group — Research in Mathematics Education of the American Education Research Association and as chairperson of the North American Chapter of the International Group for the Psychology of Mathematics Education (NA-PME). He currently serves on the editorial board of the Journal for Research in Mathematics Education.

He has been co-principal investigator (1979-83) (with M. Behr and R. Lesh) of the Rational Number Project, an NSF supported effort, designed to investigate the nature of the cognitive structures employed in children's learning of rational number concepts. In July of 1984 he received another NSF grant (also with Behr and Lesh) to investigate the role of rational number concepts in the development of proportional reasoning skills.

Ewa Puchalska

Ewa Puchalska is currently Assistant Professor at the Faculty of Education of the Université de Montréal, where she has been teaching and conducting research since 1982.

Dr. Puchalska got her Master's Degree in Mathematics at the University of Warsaw and received her Ph.D. from the Faculty of Mathematics, Computer Science and Mechanics of the same university. She is the author of mathematical textbooks for six- and seven-year-olds used throughout Poland and co-author of a four-volume Polish book, Primary Mathematics Teaching, addressed to elementary teachers and educators in Poland.

Since 1981 Dr. Puchalska has been a collaborator in PPMM, a well-known long-distance teaching in-service teacher education program in mathematics operating at Université Laval in Quebec City. She is currently co-director of a research project financed by the Government of Quebec on coded graphical representations of three-dimensional shapes and social interaction. Her other major research interests include geometric thinking in children and non-typical arithmetic word problems.

Anthony Ralston

Anthony Ralston received his B.S. and Ph.D. in Mathematics from MIT. He has held positions at the Bell Telephone Laboratories, Stevens Institute of Technology and the State University of New York at Buffalo, where he is now Professor of Computer Science and Mathematics. He has also held visiting positions at the University of Leeds and at the Institute of Computer Science, University College and Imperial College of the University of London. Professor Ralston is the author or editor of 11 books and author of 50 papers. He has served as President of the Association for Computing Machinery and the American Federation of Information Processing Societies and is currently a member of the Board of Governors of the Mathematical Association of America and the Mathematical Sciences Education Board (MSEB) of the National Research Council.

Professor Ralston's research was in numerical analysis. In the 1970s his research concerns turned to more strictly computer science topics. He became interested in mathematics education, particularly the mathematics which should be taught to undergraduates in computer science. For several years he has actively sought to achieve for discrete mathematics a position coequal to that of calculus in the first two years of undergraduate mathematics. More recently he has become interested in precollege mathematics education and is currently heading an MSEB task force to study how the K-12 mathematics curriculum might be modified as a result of developments in computer science and technology.

Richard L. Scheaffer

Richard Scheaffer is Professor and Chairman of the Department of Statistics at the University of Florida. He is on the editorial board of Communications in Statistics, and has recently served on the Board of Directors of the American Statistical Association and as chairman of the ASA-NCTM Joint Committee on Curriculum in Statistics and Probability.

Professor Scheaffer's work has included research on certain aspects of sampling and estimation, applications in industry and ecology, and the teaching of statistics and probability in the public schools.

Alan H. Schoenfeld

Alan Schoenfeld is Associate Professor of Education and Mathematics at the University of California at Berkeley, where he chairs the Graduate Group in Science and Mathematics Education. Originally trained as a mathematician (his thesis at Stanford was on topology and measure theory), he was attracted to mathematical education by Polya's writings on problem solving. His research is an attempt to understand the nature of productive thinking in mathematics, using the tools of various disciplines (cognitive science, psychology, AI, and anthropology, among others) to explore human cognitive processes.

Professor Schoenfeld has taught at the University of California at Davis, Hamilton College, and the University of Rochester. He is current chairman of the Mathematical Association of America's Committee on the Teaching of Under-graduate Mathematics, and past chairman of the editorial board of the Journal for Research in Mathematics Education. His writings on problem solving include Problem Solving in the Mathematics Curriculum (MAA 1983), Mathematical Problem Solving (Academic Press 1985), and the edited volume Cognitive Science and Mathematics Education (Erlbaum, in press).

Judah L. Schwartz

Judah L. Schwartz is Visiting Professor of Education and Co-Director of the Educational Technology Center in the Harvard Graduate School of Education, and Professor of Engineering Science and Education at M.I.T. His research interests include the use of computers to augment human intuition, the process and substance of undergraduate and continuing education, and cognitive development. He has served as a physicist, honorary research associate, technical staff member, and research scientist at many laboratories, schools, and corporations both in the U.S. and abroad. Recent publications include: "The Lost and Found Holy Grail: Some Reflections on Accountability"; "Science, Mathematics, and Technology: Recognizing the Quantitative Arts"; "People and Computers: Who Teaches Whom"; "The Semantic Calculator: Solving the Word Problem Problem"; and "Personal Problems: Microcomputer Software for

712

Undergraduate Physics." In addition, Professor Schwartz is the author or co-author of several software packages, including The Calculus Toolkit, The Geometric Supposer and M-SS-NG L-NKS, and has collaborated in the production of several series of computer-generated films.

Zbigniew Semadeni

Since 1986 Zbigniew Semadeni has been professor at the University of Warsaw, Poland. He received a master's degree in physics and a Ph.D. in mathematics from the University of Poznan. He taught in that university's Department of Mathematics before becoming a researcher in the Institute of Mathematics of the Polish Academy of Sciences, where he has been a professor since 1971 and deputy director from 1973-1985.

Zbigniew Semadeni has been a visiting professor at several universities, including the University of Washington in Seattle, York University in Toronto, and the University of Sydney. He has served as Vice-President of the International Commission on Mathematical Instruction and delivered numerous talks throughout Europe, Asia, and North America.

Professor Semadeni's research fields are functional analysis (particularly Banach spaces of continuous functions and Saks spaces), theory of categories and functors (mainly applications to functional analysis and general topology), and mathematics education (primary mathematics teaching and teacher training). He is author of two books and about 60 research papers on pure mathematics and several papers on mathematics education. He headed a three-year teaching program for some 60,000 teachers in Poland, using national TV, radio and systematic home assignments. He has also been editor-in-chief of <u>Wiadomości Matematyczne</u>, a journal of the Polish Mathematical Society.

Blagovest Sendov

Blagovest Sendov graduated from Sofia University in with a degree in mathematics. He specialized in numerical methods at Moscow University and in computer science at Imperial College, London. He earned his Ph.D. at Sofia University and a Doctor of Science degree at the Steklov Mathematical Institute in Moscow.

Professor Sendov has been President of the Science Committee of the Bulgarian Council of Ministers since 1986. Since 1980 he has also served as Vice-President and Chief Scientific Secretary of the Bulgarian Academy of Sciences. Earlier, he was Deputy Director of the Mathematics Institute of the Bulgarian Academy of Sciences, where he has recently been serving as Head of the Mathematical Modelling Section.

Professor Sendov's main research activity is in approximation theory. He has published over 150 papers, and over 20 monographs and textbooks. He is Head of the Research Group on Education sponsored by the Bulgarian Academy of Sciences and the Ministry of Education. He is active in international organizations and is Vice-President of the International Federation for Information Processing (IFIP) and Honorary President of the (International Association of Universities.

Roland Stowasser

Roland Stowasser is Professor at the Department of Mathematics of the Technische Universität Berlin (West). He taught for several years at the Carl-Duisberg-Gymnasium. He later served as Director of the Studienseminar in Wuppertal (West Germany) and Associate Professor at the Universität Bielefeld (Institut für Didaktik der Mathematik).

Professor Stowasser's primary research interests are the history of mathematics as related to mathematics education, curriculum development in the light of microcomputer technology, and problem solving and artificial intelligence. He has co-authored secondary school textbooks, published papers in a variety of educational journals, and served as chief editor of the German journal Mathematiklehrer (The Mathematics Teacher). Professor Stowasser was chairman of the ICMI-affiliated "International Study Group on the Relation between History and Pedagogy of Mathematics." He is a member of the "Commission

714

internationale pour l'étude et l'amélioration de l'enseignement des mathématiques."

Fumiyuki Terada

Fumiyuki Terada graduated from the Mathematical Institute of Tohoku University and received his Doctor of Science degree from the same institution in 1955. In 1964 he was appointed Professor at Waseda University.

Professor Terada's chief research interest is principal ideal problems in number theory. He has also been working in mathematics education. In 1980 he began studying computer-assisted instruction and subsequently designed the THE System. He is currently involved in designing Japan's compulsory upper secondary school mathematics curriculum for the Ministry of Education.

Kenneth J. Travers

Kenneth J. Travers, Professor of Mathematics Education at the University of Illinois at Urbana-Champaign, has taught mathematics at the elementary and secondary school levels, as well as at the college level. He now teaches graduate courses in mathematics education at the University of Illinois.

Professor Travers holds bachelor's and master's degrees in mathematics and education from the University of British Columbia and a Ph.D. in mathematics education from the University of Illinois at Urbana-Champaign, where he has been a professor since 1972.

Professor Travers has authored or co-authored textbooks and numerous articles in mathematics education. Since 1976, he has been Chairman of the International Mathematics Committee for the Second IEA Mathematics Study, as well as Director of the U.S. phase of the Study.

Zalman Usiskin

Zalman Usiskin is Professor of Education at the University of Chicago, where he has been a faculty member since 1969. He was born in Chicago, went to

Chicago public schools, and received B.S. degrees in Mathematics and Education from the University of Illinois, an M.A.T. in Mathematics from Harvard University, and a Ph.D. in Curriculum and Instruction in Mathematics from the University of Michigan.

His specialty is the mathematics curriculum in schools, in particular the analysis of issues related to the content and testing of that curriculum, and the development of materials for students. He has directed government-sponsored projects related to applications of first-year algebra, applying arithmetic, and the learning of geometry. Since 1983 he has directed the Secondary Component of the University of Chicago School Mathematics Project.

Before embarking on the present project, he wrote and directed field-testing of textbooks for students for each of the years of high school, including Algebra Through Applications and (with Arthur F. Coxford) Geometry - A Transformation Approach. He is the author of dozens of papers, including lead articles in three yearbooks of the National Council of Teachers of Mathematics. He received the Max Beberman award for his work in curriculum from the Illinois Council of Teachers of Mathematics. In 1983, the Metropolitan Mathematics Club of Chicago gave him its first Distinguished Service Award in its 70-year history.

Tamás Varga

Tamás Varga was born in Hungary. He graduated in mathematics and physics from Budapest University in 1942, and did post-graduate work in the Scuola Normale Superiore in Pisa, Italy. He has professional experience in teaching mathematics at various levels, from elementary schools to universities. He has given courses in mathematics education in virtually every country in Eastern and Western Europe, as well as in Canada, the U.S., and Brazil. He is on the editorial board of Educational Studies in Mathematics, the International Journal of Mathematics Education for Science and Technology, the Journal of Mathematical Behavior, and L'Insegnamento della matematica e delle scienze integrate. He is currently working for the Hungarian National Institute of Education.

Dr. Varga's research interests in mathematics education center mainly on teaching logic and probability. His book Mathematical Logic for Beginners has been published in Hungarian, German, French, Spanish, Italian, and Slovak. He is also co-author of three books, published in several languages, on teaching probability and related subjects. The book Teaching School Mathematics - A UNESCO Source Book (Penguin Press) was edited by Dr. Varga and the late W. Servais. The bulk of Professor Varga's publications are in Hungarian.

James W. Wilson

James W. Wilson is Professor of Mathematics Education at the University of Georgia. He has served as Head of the Department since 1969 and has led the development of various programs within the department. In particular, the department has considerable experience with the preservice and inservice preparation of mathematics teachers in mathematics, mathematics pedagogy, and instructional computing at the secondary, middle school, and elementary levels.

Professor Wilson received his B.S. Ed. degree from Kansas State Teachers College, Emporia, and taught in the public schools in Kansas. He holds masters degrees in mathematics from KSTC, Notre Dame, and Stanford. His doctoral studies were at Stanford with E.G. Begle, and he worked on the staff of the School Mathematics Study Group as Project Director of the Research and Analysis Section. His primary responsibility at SMSG was the analysis and reporting of the National Longitudinal Study of Mathematical Abilities.

Dr. Wilson joined the faculty at the University of Georgia in 1968. During 1974-75 Professor Wilson was on leave to the National Science Foundation as a program manager. He served as editor of the Journal for Research in Mathematics Education from 1976 to 1982, and he was elected to the Board of Directors of the National Council of Teachers of Mathematics in 1978. Professor Wilson is an author for the Scott, Foresman Company elementary mathematics series.

Michal Yerushalmy

Michal Yerushalmy is a Research Associate at the Center for Learning Technology at the Education Development Center. At EDC she has worked closely with Dr. Judah Schwartz on the mathematics software series. She has conducted studies on new approaches to teaching mathematics and implementing specially designed software. Ms. Yerushalmy has over ten years of experience as a high school mathematics and computer science teacher. In addition, she has taught procedural thinking and problem-solving to children in grades four through seven, using a special curriculum that introduces mathematical and logical concepts through graphics and games programming. Ms. Yerushalmy holds B.S. degrees in Mathematics and Education and an M.S. in Education (the Department of Teaching Technology and Sciences) from the Technion in Haifa, Israel. She is currently a doctoral student at the Harvard Graduate School of Education.

Participants in the UCSMP International Conference

Professor Isaac D. Abella, Department of Physics, University of Chicago
Ms. Joan L. Akers, San Diego County Office of Education, CA
Ms. Catharine Anderson, Open Court Publishing Company, IL
Professor Richard D. Anderson, Louisiana State University
Mr. Gregory Anerino, Indian Prairie Community Unit, Naperville, IL
Mr. Bob L. Arganbright, Associate Director, Amoco Foundation, Inc.
Professor Walter L. Baily, Jr., Department of Mathematics, University of
    Chicago
Dr. Betsy Jane Becker, Michigan State University
Professor Jerry P. Becker, Southern Illinois University
* Professor Merlyn J. Behr, Northern Illinois University, DeKalb
* Professor Max S. Bell, University of Chicago & UCSMP; WICAT, Orem, UT
** Dr. I. Edward Block, Managing Director, Society for Industrial and Applied
    Mathematics
** Professor Benjamin S. Bloom, Department of Education, University of
    Chicago
** Professor Jerry L. Bona, Department of Mathematics, University of
    Chicago
** Dr. Robert Bortnick, Community Consolidated School District 59,
    Arlington Heights, IL
Dr. Ronald S. Brandt, Executive Editor, Educational Leadership
Dr. Mervin Brennan, Illinois State Board of Education
* Professor Hans Brolin, University of Uppsala, Sweden
* Professor Felix E. Browder, Department of Mathematics, University of
    Chicago
Professor Richard Brown, Carleton College
* Professor M. Bruckheimer, The Weizmann Institute of Science, Rehovot,
    Israel
Dr. Preston Bryant, District Superintendent, District 10, Chicago, IL
* Professor Alphonse Buccino, University of Georgia, Athens, GA
Professor William F. Burger, Oregon State University
* Professor Hugh Burkhardt, University of Nottingham, Great Britain
Professor Charles A. Cable, Allegheny College
Ms. Albina S. Cannavaciolo, Hamden Hall Country Day School, Hamden, CT
Dr. Joanne Capper, Director, Science Project, Council of Chief State School
    Officers, Hall of States, Washington, D.C.
* Professor Thomas P. Carpenter, University of Wisconsin, Madison, WI
Professor Heather Carter, University of Texas at Austin
Mr. Michael Carter, University of Chicago Lab Schools
Professor Timothy D. Cavanagh, University of Northern Colorado
Dr. Ronald Champagne, President, St. Xavier College, Chicago, IL
* Professor Elmar Cohors-Fresenborg, University of Osnabrück, West
    Germany

*   denotes conference author and/or speaker
**   denotes conference presider

Professor John Conway, Sydney, Australia

Mr. Eric Cooper, The College Board, New York

Professor Jack D. Cowan, Department of Mathematics, University of
Chicago

Professor Arthur F. Coxford, University of Michigan

* Professor John E. Craig, Department of Education, University of Chicago

Mrs. Doreen Crewe, President, School Board, Palos Park, IL

Mr. John Crowl, Wheaton North High School, IL

* Professor F. Joe Crosswhite, Ohio State University, Columbus;
President of NCTM

* Professor Warren D. Crown, Rutgers University & WICAT, Orem, UT

Mr. Jerry Cummins, Proviso West High School, Hillside, IL

Mr. Leroy C. Dalton, Wauwatosa High School, WI

Professor Franklin Demana, Ohio State University

Dr. Walter Denham, State Department of Education, Sacramento, CA

Mr. Bernard Michael Donahoe, Education Director, NASA, Moffett Field,
CA

* Professor John A. Dossey, Illinois State University, Normal

** Professor Jim Douglas, Jr., Department of Mathematics, University of
Chicago

Mr. Floyd Downs, Hillsdale High School, CA

Dr. William Duffie, Board of Education, Chicago, IL

Professor Gloria M. Dugan, Ball State University

Professor Nancy C. Dunigan, University of Southern Mississippi

Dr. Robert W. Dunn, Superintendent, Grant Park Community School
District, Grant Park, IL

Dr. Frank Eastman, Executive Editor, Houghton Mifflin Company

Mr. Elden B. Egbers, Mathematics Supervisor, State of Washington

Dr. Edward Esty, Senior Associate, Department of Education, Washington,
D.C.

Mrs. Joan Fefferman, DePaul University, Chicago, IL

Professor Robert Fefferman, Department of Mathematics, University of
Chicago

* Professor James T. Fey, University of Maryland, College Park

** Ms. Paula Filliman, School District 34, Glenview, IL

Mrs. June Finch, Board of Education, Chicago, IL

** Dr. Naomi Fisher, UCSMP

Professor Janice Flake, Florida State University

Dr. Emmett E. Fleming, Plainfield School District 202, Plainfield, IL

Professor Fred Flener, Northeastern University

* Professor Hans Freudenthal, University of Utrecht, The Netherlands

Professor Michael Fried, University of California at Irvine

* Professor Hiroshi Fujita, University of Tokyo, Japan

* Professor Karen Fuson, Northwestern University & UCSMP, Evanston, IL

Dr. Marjorie Gardner, Lawrence Hall of Science, Director, University of
California at Berkeley

Dr. Sol Garfunkel, COMAP, Lexington, MA

** Dr. James D. Gates, Executive Director, NCTM, Reston, VA
 * Professor Claude Gaulin, Université Laval, Quebec, Canada
   Professor Dorothy Geddes, Brooklyn College - CUNY
   Professor Jacob W. Getzels, Department of Education, University of
     Chicago
   Dr. Walter L. Gillespie, Deputy Assistant Director, National Science
     Foundation
** Professor Gene Golub, President, Society for Industrial and Applied
     Mathematics, Stanford University
   Professor Seymour Goodman, MIS/BPA, University of Arizona
 * Professor Renée de Graeve, Université Grenoble & IREM, France
   Professor Herbert J. Greenberg, University of Denver
   Mr. Jonathan Jay Greenwood, Multnomah Education Service District,
     Portland, OR
   Professor Ken Gross, University of Wyoming
   Professor Douglas A. Grouws, University of Missouri
 * Professor Heini Halberstam, University of Illinois, Champaign
   Ms. Judy Halvorson, President, Minnesota Council of Teachers of
     Mathematics, Bloomington, Minnesota
   Dr. Raymond J. Hannapel, Directorate for Science and Engineering
     Education, National Science Foundation
   Mr. Ronald G. Harley, President, Board of Education, Bremen Community
     High Schools, Midlothian, IL
   Mr. Bill Harms, News and Information, University of Chicago
 * Professor Larry V. Hedges, Department of Education, University of Chicago,
     and UCSMP
** Dr. Gerard J. Heing, Assistant Superintendent, Board of Education, Chicago,
     IL
   Mr. Louis G. Henkel, West Ottawa Middle School, Holland, MI
   Ms. Diane Herrman, Department of Mathematics, University of Chicago
   Ms. Patricia M. Hess, Albuquerque Public Schools
   Professor Donald W. Hight, Pittsburg State University
   Professor Shirley A. Hill, University of Missouri at Kansas, Past NCTM
     President
 * Professor Peter J. Hilton, SUNY at Binghamton, NY
   Dr. Christian Hirsch, Western Michigan University
 * Professor Alan Hoffer, University of Oregon, University of Chicago &
     UCSMP
   Ms. Shirley Hoffer, University of Oregon, University of Chicago & UCSMP
   Ms. Shirley Holbrook, University of Chicago Laboratory Schools
   Dr. Evelyn Blose Holman, Superintendent, Wicomico County School System
   Dr. Grace Hopkins, Governors State University, University Park, IL
   Dr. William Humm, Illinois State Board of Education
   Professor Philip W. Jackson, Department of Education, University of
     Chicago
 * Dr. Edward Jacobsen, Division of Science Education, UNESCO, Paris
   Mrs. Marie Jernigan, Board of Education, Chicago, IL

Mrs. Lynn Johns, Executive Administrator, Cook County Superintendent of
   Schools, Chicago, IL
Dr. Beau Fly Jones, North Central Laboratory, Elmhurst, IL
Professor Robert Kalin, Florida State University, Tallahassee
* Professor Mary Grace Kantowski, University of Florida, Gainesville
Professor James J. Kaput, Southeastern Massachusetts University
Ms. Mary Kay Karl, School District #54, Schaumburg, IL
Professor Harvey Keynes, University of Minnesota
Dr. Mary Kiely, Carnegie Corporation of New York
* Professor Jeremy Kilpatrick, University of Georgia, Athens
Ms. Genevieve M. Knight, Hampton Institute, Hampton, VA
Mrs. Pamela J. Koenig, Amoco Foundation, Inc.
Dr. Mary M. Kohlerman, Directorate for Science and Engineering Education,
   National Science Foundation
Professor Marc W. Konvisser, Wayne State University
Professor Sol H. Krasner, Department of Physics, University of Chicago
Ms. Marie Kraus, Tinley Park, IL
Dr. James Landwehr, AT&T Bell Laboratories, Murray Hill, NJ
Professor Glenda Lappan, Michigan State University
Professor Anneli Lax, New York University
Professor John F. LeBlanc, Indiana University
* Professor Richard Lesh, Northwestern University and WICAT, Orem, UT
Ms. Letitia Lestina, Illinois Science Lecture Association, Chicago
** Dr. William J. LeVeque, Executive Director, American Mathematical
   Society
Professor Betty K. Lichtenberg, University of South Florida
Professor Don R. Lichtenberg, University of South Florida
* Mr. John Ling, The School Mathematics Project, London, Great Britain
Ms. Judy Lipschutz, Director, Area Service Center for the Gifted, IL
Professor Bonnie Litwiller, University of Northern Iowa
* Professor Emilio Lluis, Instituto de Matematicas, Mexico
Professor Saunders Mac Lane, Department of Mathematics, University of
   Chicago
Dr. Richard J. Martwick, Superintendent, Educational Service Region of
   Cook County, Chicago, IL
Professor Walter E. Massey, Department of Physics, University of Chicago,
   and Argonne National Laboratory
Professor Peter J. May, Department of Mathematics, University of Chicago
Dr. Ann McAloon, Senior Examiner, Educational Testing Service, Princeton,
   NJ
** Dr. John McConnell, Glenbrook South High School, President, Illinois
   Council of Teachers of Mathematics, Glenview, IL
* Dr. Keith W. McHenry, Jr., Vice-President for Research and Development,
   Amoco Oil Company
Professor Douglas B. McLeod, San Diego State University
Dr. Raymond McNamee, Chairman, Triton College, River Grove, IL

722

Dr. Wendell Meeks, Illinois State Board of Education, Springfield, IL
Dr. Sheila Megley, Vice-President and Academic Dean, Salve Regina-The
    Newport College, RI
Dr. James Mendenhall, Manager, Educational Innovation and Support, Illinois
    State Board of Education
Professor Michael H. Millar, University of Northern Iowa
Mr. Coley Mills, Science Research Associates, Chicago, IL
* Professor Tatsuro Miwa, University of Tsukuba, Japan
Mr. William Mohrdieck, Prospect High School, Mount Prospect, IL
Ms. Ann Morris, Myna Thompson School, Rantoul, IL
Ms. Mary Lu Muffaletto, State Board of Education, IL
Professor Vedula N. Murty, Pennsylvania State University
Professor Norman Nachtrieb, Department of Chemistry, University of
    Chicago
Professor Raghavan Narasimhan, Department of Mathematics, University of
    Chicago
Ms. Kay Nebel, School District 101, Western Springs, IL
Professor Eugene D. Nichols, Florida State University, Tallahassee
Mr. Clarence Olander, Vice President, Scott, Foresman and Company
Professor Ingram Olkin, Stanford University
Professor David W. Oxtoby, Department of Chemistry, University of
    Chicago
** Professor Warren Page, New York City Technical College, Editor, College
    Mathematics Journal
* Professor Anne-Marie Pastel, Université Grenoble & IREM, France
Professor Joseph N. Payne, University of Michigan
* Professor Boyan Penkov, Bulgarian Academy of Sciences, Sofia, Bulgaria
Professor Faustine Perham, Northeastern Illinois University
Ms. Lydia Polonsky, UCSMP
* Professor Tom Post, University of Minnesota
Dr. Alfred J. Price, Avoca School District 37, Wilmette, IL
* Professor Ewa Puchalska, Université Montreal, Canada and Warsaw, Poland
Professor Alfred Putnam, Department of Mathematics, University of
    Chicago
Mrs. Maryann Putnam, University of Chicago Laboratory Schools
Dr. Edward J. Rachford, Superintendent of Schools, Flossmoor, IL
* Professor Anthony Ralston, SUNY at Buffalo, NY
Ms. Mary Margaret Rappe, Walt Disney Magnet School, Chicago, IL
Professor Haim Reingold, Mundelein College, Chicago, IL
Professor Robert E. Reys, University of Missouri - Columbia
* Professor Stuart A. Rice, Department of Chemistry, University of Chicago
Professor David F. Robitaille, University of British Columbia
Professor Thomas A. Romberg, University of Wisconsin, Madison
Mr. Richard W. Ronvik, Director, Region 1 Gifted Area Service Center,
    Chicago, IL
Ms. Paula Rossino, Walt Disney Magnet School, Chicago, IL
Dr. Thomas Rowan, Montgomery County Schools, Rockville, MD
Dr. Rheta Rubenstein, Renaissance High School, Detroit, MI
Mr. Peter Saecker, Science Research Associates, Chicago, IL

Professor Yoram Sagher, University of Illinois at Chicago
* Professor Paul J. Sally, Jr., Department of Mathematics, University of
    Chicago, and UCSMP
  Dr. Robert Sampson, Illinois State Board of Education
* Professor Richard L. Scheaffer, University of Florida, Gainesville
  Mr. Marvin Schlichting, Triton College, River Grove, IL
  Dr. Harold Schoen, University of Iowa
* Professor Alan H. Schoenfeld, University of California, Berkeley
  Miss Merrie L. Schroeder, Mathematics Program Facilitator, Cedar Rapids,
    IA
* Mr. Donald G. Schroeter, Executive Director, Amoco Foundation, Inc.
  Mrs. Candace Schultz, Wheaton-Warrenville Middle School, Wheaton, IL
* Professor Judah L. Schwartz, M.I.T. and Harvard University, Cambridge
** Mrs. Sheila Sconiers, UCSMP
* Professor Zbigniew Semadeni, Polish Academy of Sciences, Warsaw, Poland
  Professor Sharon Senk, Syracuse University, Syracuse, NY
* Dr. Albert Shanker, President, American Federation of Teachers, New York
  Dr. William L. Sharp, Superintendent, Bremen Community High Schools,
    Midlothian, IL
  Dr. Albert P. Shulte, Director, Oakland County Schools, Pontiac, MI
  Mrs. Marcia Sielaff, The Phoenix Gazette
  Professor Jack Silber, Roosevelt University, Chicago, IL
  Ms. Louise M. Smith, Charleston County Schools
  Professor Judith Threagdill-Sowder, Northern Illinois University
  Professor George Springer, Indiana University
  Dr. Elizabeth Stage, Director, Mathematics, Computer Education Programs,
    University of California
  Professor James W. Stigler, Department of Statistics, University of Chicago
* Professor Roland Stowasser, Technische Universität Berlin, West Berlin
  Mr. Robert Streit, Slavic Department, University of Chicago, and UCSMP
** Dr. Dorothy S. Strong, Director, Bureau of Mathematics, Chicago, IL
  Professor Jane Swafford, Northern Michigan University
  Mr. George Swanson, Principal, Grand Park High School, Grand Park, IL
  Dr. Phyllis Tate, Board of Education, Chicago, IL
* Professor Fumiyuki Terada, Waseda University, Japan
  Dr. John A. Thorpe, Directorate for Science and Engineering Education,
    National Science Foundation
  Professor Rosamond Tischler, Brooklyn College, CUNY
  Professor Paul Trafton, National College of Education, Evanston, IL
* Professor Kenneth J. Travers, University of Illinois, Champaign;
    International Mathematics Study
  Dr. David A. Ucko, Museum of Science and Industry, Chicago, IL
  Mrs. Karen Usiskin, Scott, Foresman and Company
* Professor Zalman Usiskin, Department of Education, University of Chicago,
    and UCSMP
  Dr. Salvatore Vallina, Board of Education, Chicago, IL
  Dr. Theodore Van Dorn, Director of Instruction, Mathematics,
    Homewood-Flossmoor High School, Illinois
* Professor Tamás Varga, National Pedagogical Institute, Budapest, Hungary

Dr. Joyce Van Tassel-Baska, Northwestern University

Mr. Charles R. Vietzen, Principal, Hubbard High School, Chicago, IL

Mr. Steven Viktora, Kenwood Academy, Chicago, IL, and UCSMP

Dr. Philip Viso, Superintendent, Vocational and Technical Education, Chicago, IL

Professor Sigrid Wagner, University of Georgia

Professor Philip Wagreich, University of Illinois at Chicago

Dr. Bert Waits, Ohio State University

Professor Avrum I. Weinzweig, University of Illinois at Chicago

Professor Ronald Wenger, University of Delaware

Ms. Linda Wertsch, Education Editor, Chicago Sun-Times

Professor Grayson Wheatley, Purdue University, West Lafayette, IN

Professor Hassler Whitney, Institute for Advanced Study, Princeton, NJ

Mrs. Beverly Whittington, Educational Testing Service, Princeton, NJ

Dr. David Williams, Mathematics Supervisor, Philadelphia, PA

* Professor James D. Wilson, University of Georgia, Athens

* Professor Izaak Wirszup, Department of Mathematics, University of Chicago, and UCSMP

Dr. Lauren G. Woodby, Columbus, OH

Mrs. Lauren G. Woodby, Columbus, OH

Professor Marilyn J. Zweng, University of Iowa